Complex Vector Functional Equations

Ice Risteski

Valéry Covachev

Bulgarian Academy of Sciences

World Scientific
New Jersey • London • Singapore • Hong Kong

Published by

World Scientific Publishing Co. Pte. Ltd.

P O Box 128, Farrer Road, Singapore 912805

USA office: Suite 1B, 1060 Main Street, River Edge, NJ 07661

UK office: 57 Shelton Street, Covent Garden, London WC2H 9HE

British Library Cataloguing-in-Publication Data
A catalogue record for this book is available from the British Library.

COMPLEX VECTOR FUNCTIONAL EQUATIONS

ISBN 981-02-4683-8

Printed in Singapore by Uto-Print

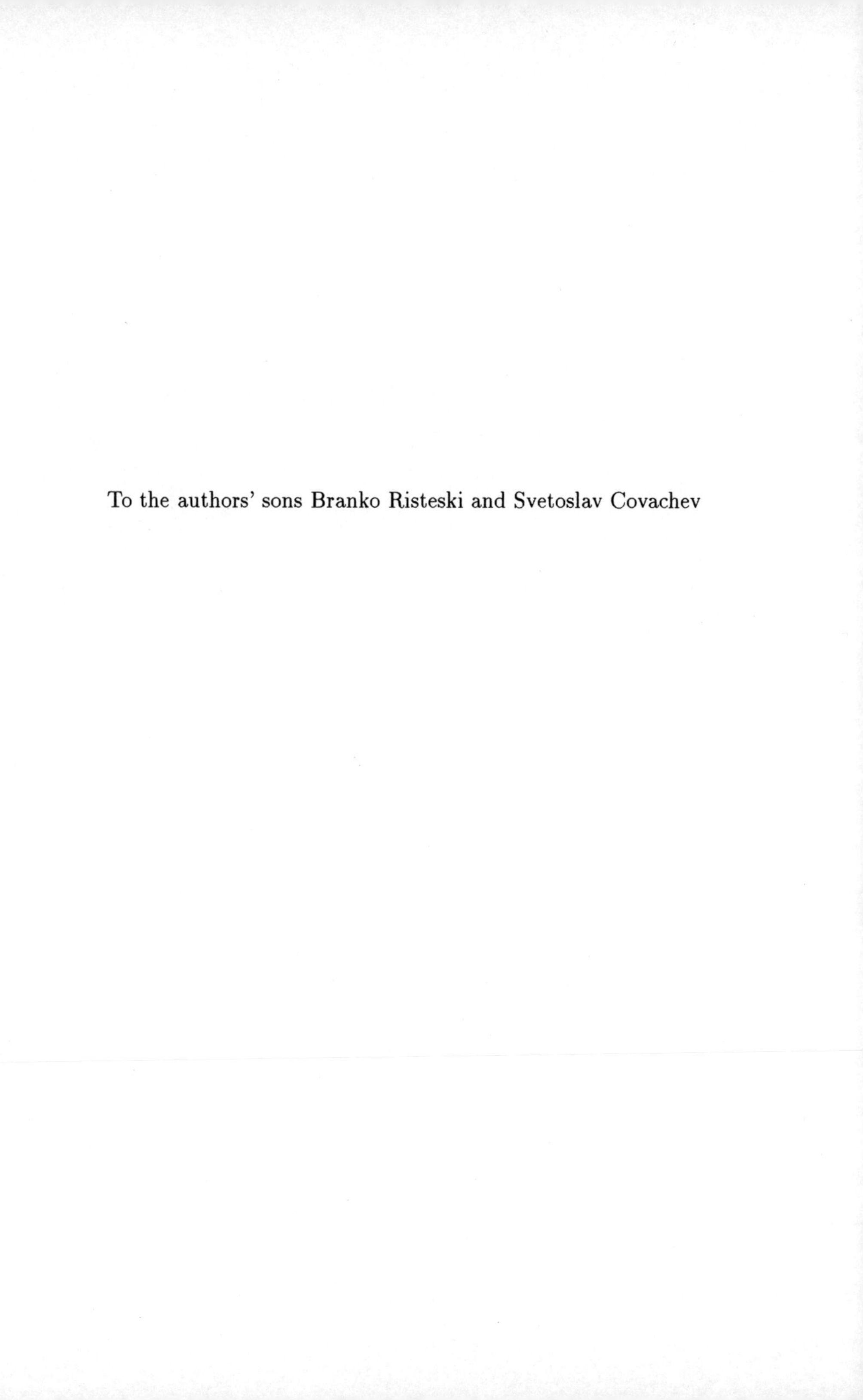

To the authors' sons Branko Risteski and Svetoslav Covachev

Preface

Complex vector functional equations are a new field in the theory of functional equations. Their development up to now has not been particularly dynamical. As every new mathematical discipline, complex vector functional equations call for their more detailed investigation. From this standpoint, in the last few years we dedicated a series of papers to these equations. Our multi-annual experience and the results obtained by the research in this new area show us the necessity to write a strictly scientific systematized monograph.

The object of this monograph is to give an overview to the recently obtained results of the authors from all aspects of utilization, which is obviously not possible with partial results in papers published in distinct mathematical journals.

We have divided our exposition on the field of complex vector functional equations in two parts which are devoted respectively to linear and nonlinear equations.

Part One consists of five chapters.

In the first chapter the general classes of cyclic functional equations in a complex vector form are considered. Among these are: the basic cyclic functional equation, the derived cyclic functional equation, the semicyclic functional equation, the special cyclic functional equation and the condensed cyclic functional equation.

In the second chapter the complex vector functional equations with operations between arguments are considered. First an operator functional equation is solved. Also, in this chapter the generalized equation of operator type is solved. Next, some simple functional equations and functional equations with distinct functions are solved. Of this type of functional equations we can mention Fréchet's functional equations, to which an adequate place in this chapter is given. Finally, some complex vector functional equations with two operations between arguments are also solved.

The functional equations with constant parameters are considered in the third chapter. The equations of this type which are solved here are: the general parametric functional equation, the special parametric functional equation, the expanded parametric functional equation and the general expanded parametric functional equation.

The linear complex vector functional equations with constant coefficients are considered in the fourth chapter by a new matrix approach. In particular, we pay attention both to homogeneous and nonhomogeneous functional equations with constant coefficients. Finally, a paracyclic functional equation with complex constant coefficients is solved.

In the last fifth chapter of Part One two types of systems of functional equations are studied. First some systems in which each equation contains all unknown functions are solved, and then some systems in which not all equations contain all unknown functions.

Part Two is also divided in five chapters which are described below quite briefly.

Chapters 6, 7 and 8 are devoted respectively to quadratic, modified quadratic and expanded quadratic equations. Higher order complex vector functional equations are considered in the ninth chapter. Finally, in Chapter 10 systems of nonlinear (quadratic and higher order) complex vector functional equations are studied.

The present monograph is intended for mathematicians, physicists and engineers who use functional equations in their investigations.

The authors are well aware that all aspects of the complex vector functional equations cannot be included in an adequate way in a newly written book, and they will be grateful for all remarks and suggestions which will be valuable for the correction and completion of a subsequent edition.

The second author acknowledges partial support by Grant MM–706 with the Bulgarian Science Fund. A part of the book was written during the work of the second author at Fatih University, Istanbul, Turkey.

Contents

Preface vii

PART 1   Linear Complex Vector Functional Equations 1

Chapter 1   General Classes of Cyclic Functional Equations   3
1   Basic Cyclic Functional Equation 3
2   Derived Cyclic Functional Equation 6
3   Paracyclic Functional Equation 29
4   Semicyclic Functional Equation 65
5   Special Cyclic Functional Equation 69
6   Condensed Cyclic Functional Equation 72

Chapter 2   Functional Equations with Operations between Arguments   75
7   Operator Functional Equation 75
8   Generalized Functional Equation 81
9   Simple Functional Equations 94
10   Functional Equations with Several Unknown Functions ... 99
11   Fréchet's Functional Equations 106
12   Functional Equations with Two Operations 111

Chapter 3   Functional Equations with Constant Parameters 121
13   General Parametric Functional Equation 121
14   Special Parametric Functional Equation 135
15   Expanded Parametric Functional Equation 142
16   General Expanded Parametric Functional Equation 161

Chapter 4   Functional Equations with Constant Coefficients 173

17   Matrix Equations 173

18   Homogeneous Functional Equations with Constant Coefficients   175

19   Nonhomogeneous Functional Equations with Constant Coefficients 190

20   Paracyclic Functional Equations with Constant Coefficients . .   197

Chapter 5   Systems of Linear Functional Equations   205

21   Systems in Which Each Equation Contains All Unknown
Functions 205

22   Systems in Which Not All Equations Contain All
Unknown Functions 214

PART 2   Nonlinear Complex Vector Functional Equations   217

Chapter 6   Quadratic Functional Equations   219

23   Simple Quadratic Functional Equation 219

24   Special Quadratic Functional Equation 227

Chapter 7   Modified Quadratic Functional Equations   231

25   Basic Quadratic Functional Equation 231

26   Permuted Quadratic Functional Equation 234

27   First Modified Quadratic Functional Equation 238

28   Second Modified Quadratic Functional Equation 241

29   Third Modified Quadratic Functional Equation 244

Chapter 8   Expanded Quadratic Functional Equations   247

30   Expanded Quadratic Functional Equations with
Functional Arguments 247

31   Expanded Quadratic Functional Equations with the Same Signs
between the Functions 251

32   Expanded Quadratic Functional Equation with
Alternating Signs between the Functions 259

33   Generalized Quadratic Functional Equation 262

Chapter 9   Higher Order Functional Equations   265

34   Higher Order Functional Equation without Parameters 265

35   Higher Order Functional Equation with Complex
Parameters 267

36   Nonlinear Operator Functional Equation 273

Chapter 10   Systems of Nonlinear Functional Equations   285

37   Systems of Quadratic Functional Equations 285
38   Systems of Higher Order Functional Equations 311

Bibliography   319

Index   323

PART 1

Linear Complex Vector Functional Equations

Chapter 1

General Classes of Cyclic Functional Equations

In this chapter six general classes of cyclic partial linear complex vector functional equations are solved, namely the basic cyclic functional equation, the derived cyclic functional equation, the paracyclic functional equation, the semicyclic functional equation, the special cyclic functional equation and the condensed cyclic functional equation. The results given here were obtained in [I. B. Risteski (to appear A); I. B. Risteski and V. C. Covachev (2000); I. B. Risteski *et al.* (2000A)].

1 Basic Cyclic Functional Equation

First we will introduce the following notations.

Let $\mathcal{V}$ be a finite dimensional complex vector space and let there exist a mapping $f : \mathcal{V}^n \mapsto \mathcal{V}$, $\mathbf{Z}_i$ $(1 \le i \le n)$ are vectors in $\mathcal{V}$ and C is a constant complex vector in the same space. We assume that $\mathbf{Z}_i = (z_{i1}(t), \cdots, z_{in}(t))^T$, where $z_{ij}(t)$ $(1 \le i \le n)$ are complex functions and $\mathbf{O} = (0, 0, \cdots, 0)^T$ is the zero vector in $\mathcal{V}$.

Now we will give the following results which were obtained in [I. B. Risteski *et al.* (2000A)].

Theorem 1.1 *The general solution of the basic cyclic complex vector functional equation*

$$\sum_{i=1}^{n} f(\mathbf{Z}_i, \mathbf{Z}_{i+1}, \cdots, \mathbf{Z}_{i+n-1}) = \mathbf{O} \qquad (\mathbf{Z}_{n+i} \equiv \mathbf{Z}_i) \qquad (1.1)$$

3

is given by

$$f(\mathbf{Z}_1, \mathbf{Z}_2, \cdots, \mathbf{Z}_n) = F(\mathbf{Z}_1, \mathbf{Z}_2, \cdots, \mathbf{Z}_n) - F(\mathbf{Z}_2, \mathbf{Z}_3, \cdots, \mathbf{Z}_n, \mathbf{Z}_1) \qquad (1.2)$$

where F is an arbitrary complex vector function with values in $\mathcal{V}$.

Proof. From Eq. (1.1) it immediately follows that

$$f(\mathbf{Z}_1, \mathbf{Z}_2, \cdots, \mathbf{Z}_n) = -f(\mathbf{Z}_2, \mathbf{Z}_3, \cdots, \mathbf{Z}_n, \mathbf{Z}_1)$$

$$-f(\mathbf{Z}_3, \mathbf{Z}_4, \cdots, \mathbf{Z}_1, \mathbf{Z}_2) - \cdots - f(\mathbf{Z}_n, \mathbf{Z}_1, \cdots, \mathbf{Z}_{n-1})$$

and

$$nf(\mathbf{Z}_1, \mathbf{Z}_2, \cdots, \mathbf{Z}_n) = f(\mathbf{Z}_1, \mathbf{Z}_2, \cdots, \mathbf{Z}_{n-1}, \mathbf{Z}_n) - f(\mathbf{Z}_2, \mathbf{Z}_3, \cdots, \mathbf{Z}_n, \mathbf{Z}_1)$$

$$+f(\mathbf{Z}_1, \mathbf{Z}_2, \cdots, \mathbf{Z}_{n-1}, \mathbf{Z}_n) - f(\mathbf{Z}_3, \mathbf{Z}_4, \cdots, \mathbf{Z}_1, \mathbf{Z}_2) + \cdots$$

$$+f(\mathbf{Z}_1, \mathbf{Z}_2, \cdots, \mathbf{Z}_{n-1}, \mathbf{Z}_n) - f(\mathbf{Z}_n, \mathbf{Z}_1, \cdots, \mathbf{Z}_{n-2}, \mathbf{Z}_{n-1})$$

$$= f(\mathbf{Z}_1, \mathbf{Z}_2, \cdots, \mathbf{Z}_{n-1}, \mathbf{Z}_n) - f(\mathbf{Z}_2, \mathbf{Z}_3, \cdots, \mathbf{Z}_n, \mathbf{Z}_1)$$

$$+f(\mathbf{Z}_1, \mathbf{Z}_2, \cdots, \mathbf{Z}_{n-1}, \mathbf{Z}_n) - f(\mathbf{Z}_2, \mathbf{Z}_3, \cdots, \mathbf{Z}_n, \mathbf{Z}_1)$$

$$+f(\mathbf{Z}_2, \mathbf{Z}_3, \cdots, \mathbf{Z}_n, \mathbf{Z}_1) - f(\mathbf{Z}_3, \mathbf{Z}_4, \cdots, \mathbf{Z}_1, \mathbf{Z}_2)$$

$$+f(\mathbf{Z}_1, \mathbf{Z}_2, \cdots, \mathbf{Z}_{n-1}, \mathbf{Z}_n) - f(\mathbf{Z}_2, \mathbf{Z}_3, \cdots, \mathbf{Z}_n, \mathbf{Z}_1)$$

$$+f(\mathbf{Z}_2, \mathbf{Z}_3, \cdots, \mathbf{Z}_n, \mathbf{Z}_1) - f(\mathbf{Z}_3, \mathbf{Z}_4, \cdots, \mathbf{Z}_1, \mathbf{Z}_2)$$

$$+f(\mathbf{Z}_3, \mathbf{Z}_4, \cdots, \mathbf{Z}_1, \mathbf{Z}_2) - f(\mathbf{Z}_4, \mathbf{Z}_5, \cdots, \mathbf{Z}_2, \mathbf{Z}_3) + \cdots$$

$$+f(\mathbf{Z}_1, \mathbf{Z}_2, \cdots, \mathbf{Z}_{n-1}, \mathbf{Z}_n) - f(\mathbf{Z}_2, \mathbf{Z}_3, \cdots, \mathbf{Z}_n, \mathbf{Z}_1)$$

$$+f(\mathbf{Z}_2, \mathbf{Z}_3, \cdots, \mathbf{Z}_n, \mathbf{Z}_1) - f(\mathbf{Z}_3, \mathbf{Z}_4, \cdots, \mathbf{Z}_1, \mathbf{Z}_2) + \cdots$$

$$+f(\mathbf{Z}_{n-1}, \mathbf{Z}_n, \cdots, \mathbf{Z}_{n-3}, \mathbf{Z}_{n-2}) - f(\mathbf{Z}_n, \mathbf{Z}_1, \cdots, \mathbf{Z}_{n-2}, \mathbf{Z}_{n-1}).$$

By denoting the sum of members with positive sign by $nF(\mathbf{Z}_1, \mathbf{Z}_2, \cdots, \mathbf{Z}_n)$ we obtain Eq. (1.2).

On the other hand, every function of the form Eq. (1.2) satisfies the complex vector functional equation (1.1). $\qquad\square$

Theorem 1.2 *The general solution of the basic cyclic complex vector functional equation*

$$\sum_{i=1}^{n} f(\mathbf{Z}_i, \mathbf{Z}_{i+1}, \cdots, \mathbf{Z}_{i+p-1}) = \mathbf{O} \qquad (n \geq 2p - 1; \ \mathbf{Z}_{n+i} \equiv \mathbf{Z}_i) \qquad (1.3)$$

is given by

$$f(\mathbf{Z}_1, \mathbf{Z}_2, \cdots, \mathbf{Z}_p) = F(\mathbf{Z}_1, \mathbf{Z}_2, \cdots, \mathbf{Z}_{p-1}) - F(\mathbf{Z}_2, \mathbf{Z}_3, \cdots, \mathbf{Z}_p) \qquad (1.4)$$

where F is an arbitrary complex vector function with values in $\mathcal{V}$.

Proof. A direct calculation shows that every function F of the form Eq. (1.4) satisfies the functional equation (1.3). Now we will prove the converse, i.e. that f has the form Eq. (1.4). For $\mathbf{Z}_i = C$ $(1 \leq i \leq n)$ where C is a constant complex vector from $\mathcal{V}$, the equation (1.3) gives $nf(C, C, \cdots, C) = \mathbf{O}$, i.e. $f(C, C, \cdots, C) = \mathbf{O}$.

By putting $\mathbf{Z}_i = C$ $(p + 1 \leq i \leq n)$ into Eq. (1.3) we obtain

$$f(\mathbf{Z}_1, \mathbf{Z}_2, \cdots, \mathbf{Z}_p) + f(\mathbf{Z}_2, \mathbf{Z}_3, \cdots, \mathbf{Z}_p, C) + \cdots \qquad (1.5)$$

$$+ f(\mathbf{Z}_p, C, C, \cdots, C) + f(C, C, \cdots, C, \mathbf{Z}_1)$$

$$+ f(C, C, \cdots, C, \mathbf{Z}_1, \mathbf{Z}_2) + \cdots + f(C, \mathbf{Z}_1, \mathbf{Z}_2, \cdots, \mathbf{Z}_{p-1}) = \mathbf{O}.$$

If we substitute $\mathbf{Z}_p = C$ into the above equality, we obtain

$$f(\mathbf{Z}_1, \mathbf{Z}_2, \cdots, \mathbf{Z}_{p-1}, C) + f(\mathbf{Z}_2, \mathbf{Z}_3, \cdots, \mathbf{Z}_{p-1}, C, C) + \cdots \qquad (1.6)$$

$$+ f(\mathbf{Z}_{p-1}, C, C, \cdots, C) + f(C, C, \cdots, C, \mathbf{Z}_1)$$

$$+ f(C, C, \cdots, C, \mathbf{Z}_1, \mathbf{Z}_2) + \cdots + f(C, \mathbf{Z}_1, \mathbf{Z}_2, \cdots, \mathbf{Z}_{p-1}) = \mathbf{O}.$$

By subtracting Eq. (1.6) from Eq. (1.5), we get

$$\begin{aligned}
f(\mathbf{Z}_1, \mathbf{Z}_2, \cdots, \mathbf{Z}_p) &= f(\mathbf{Z}_1, \mathbf{Z}_2, \cdots, \mathbf{Z}_{p-1}, C) - f(\mathbf{Z}_2, \mathbf{Z}_3, \cdots, \mathbf{Z}_p, C) \\
+ \ & f(\mathbf{Z}_2, \mathbf{Z}_3, \cdots, \mathbf{Z}_{p-1}, C, C) - f(\mathbf{Z}_3, \mathbf{Z}_4, \cdots, \mathbf{Z}_p, C, C) + \cdots \\
+ \ & f(\mathbf{Z}_{p-1}, C, C, \cdots, C) - f(\mathbf{Z}_p, C, C, \cdots, C).
\end{aligned} \qquad (1.7)$$

By putting

$$F(\mathbf{Z}_1, \mathbf{Z}_2, \cdots, \mathbf{Z}_{p-1}) = f(\mathbf{Z}_1, \mathbf{Z}_2, \cdots, \mathbf{Z}_{p-1}, C)$$
$$+ \quad f(\mathbf{Z}_2, \mathbf{Z}_3, \cdots, \mathbf{Z}_{p-1}, C, C) + \cdots + f(\mathbf{Z}_{p-1}, C, C, \cdots, C),$$

the equality (1.7) takes the form Eq. (1.4). $\qquad\qquad\qquad\qquad$ □

Some particular cases of the functional equation (1.3) are considered in [J. Aczél *et al.* (1960); M. Ghermănescu (1940); M. Hosszú (1961)] under the hypothesis that the functions and the independent variables are real.

2   Derived Cyclic Functional Equation

Let $\mathcal{V}$ be a finite dimensional complex vector space and let there exist mappings $f_i : \mathcal{V}^p \mapsto \mathcal{V}$ $(1 \leq i \leq k)$. Throughout this section $\mathbf{Z}_i$ $(1 \leq i \leq n)$ are vectors in $\mathcal{V}$, and C_i are constant vectors in the same space. We may assume that $\mathbf{Z}_i = (z_{i1}(t), \cdots, z_{in}(t))^T$, where the components $z_{ij}(t)$ $(1 \leq i \leq n; 1 \leq j \leq n)$ are complex functions, and that $\mathbf{O} = (0, 0, \cdots, 0)^T$ is the zero vector in $\mathcal{V}$.

Next we will give the following results obtained in [I. B. Risteski and V. C. Covachev (2000)].

Theorem 2.1   *The general solution of the derived cyclic complex vector functional equation*

$$\sum_{i=1}^{n} f_i(\mathbf{Z}_i, \mathbf{Z}_{i+1}, \cdots, \mathbf{Z}_{i+n-1}) = \mathbf{O} \qquad (\mathbf{Z}_{n+i} \equiv \mathbf{Z}_i, f_{n+i} \equiv f_i) \qquad (2.1)$$

is given by

$$f_i(\mathbf{Z}_1, \mathbf{Z}_2, \cdots, \mathbf{Z}_n) \qquad\qquad\qquad\qquad (2.2)$$
$$= \quad F_i(\mathbf{Z}_1, \mathbf{Z}_2, \cdots, \mathbf{Z}_n) - F_{i+1}(\mathbf{Z}_2, \mathbf{Z}_3, \cdots, \mathbf{Z}_n, \mathbf{Z}_1) \quad (1 \leq i \leq n)$$

where $F_{n+i} \equiv F_i$ $(1 \leq i \leq n)$ are arbitrary complex vector functions with values in $\mathcal{V}$.

Proof.   Using the conventions $f_i \equiv f_{n+i}$ and $\mathbf{Z}_i \equiv \mathbf{Z}_{i+n}$, the equation (2.1) can be written in the following form

$$f_i(Z_i, Z_{i+1}, \cdots, Z_{i+n-1}) + f_{i+1}(Z_{i+1}, Z_{i+2}, \cdots, Z_{i+n-1}, Z_i) + \cdots$$
$$+ \quad f_{i+n-2}(Z_{i+n-2}, Z_{i+n-1}, \cdots, Z_{i+n-3}) \qquad (2.3)$$
$$+ \quad f_{i+n-1}(Z_{i+n-1}, Z_i, \cdots, Z_{i+n-2}) = O.$$

Then

$$f_i(Z_i, Z_{i+1}, \cdots, Z_{i+n-1}) = -f_{i+1}(Z_{i+1}, Z_{i+2}, \cdots, Z_{i+n-1}, Z_i) - \cdots$$

$$-f_{i+n-2}(Z_{i+n-2}, Z_{i+n-1}, \cdots, Z_{i+n-3}) - f_{i+n-1}(Z_{i+n-1}, Z_i, \cdots, Z_{i+n-2})$$

and

$$nf_i(Z_i, Z_{i+1}, \cdots, Z_{i+n-1})$$

$$= f_i(Z_i, Z_{i+1}, \cdots, Z_{i+n-1}) - f_{i+1}(Z_{i+1}, Z_{i+2}, \cdots, Z_{i+n-1}, Z_i)$$

$$+f_i(Z_i, Z_{i+1}, \cdots, Z_{i+n-1}) - f_{i+2}(Z_{i+2}, Z_{i+3}, \cdots, Z_i, Z_{i+1}) + \cdots$$

$$+f_i(Z_i, Z_{i+1}, \cdots, Z_{i+n-1}) - f_{i+n-1}(Z_{i+n-1}, Z_i, \cdots, Z_{i+n-2})$$

$$= f_i(Z_i, Z_{i+1}, \cdots, Z_{i+n-1}) - f_{i+1}(Z_{i+1}, Z_{i+2}, \cdots, Z_{i+n-1}, Z_i)$$

$$+f_i(Z_i, Z_{i+1}, \cdots, Z_{i+n-1}) - f_{i+1}(Z_{i+1}, Z_{i+2}, \cdots, Z_{i+n-1}, Z_i)$$

$$+f_{i+1}(Z_{i+1}, Z_{i+2}, \cdots, Z_{i+n-1}, Z_i) - f_{i+2}(Z_{i+2}, Z_{i+3}, \cdots, Z_i, Z_{i+1})$$

$$+f_i(Z_i, Z_{i+1}, \cdots, Z_{i+n-1}) - f_{i+1}(Z_{i+1}, Z_{i+2}, \cdots, Z_{i+n-1}, Z_i)$$

$$+f_{i+1}(Z_{i+1}, Z_{i+2}, \cdots, Z_{i+n-1}, Z_i) - f_{i+2}(Z_{i+2}, Z_{i+3}, \cdots, Z_i, Z_{i+1})$$

$$+f_{i+2}(Z_{i+2}, Z_{i+3}, \cdots, Z_i, Z_{i+1}) - f_{i+3}(Z_{i+3}, Z_{i+4}, \cdots, Z_{i+1}, Z_{i+2}) + \cdots$$

$$+f_i(Z_i, Z_{i+1}, \cdots, Z_{i+n-1}) - f_{i+1}(Z_{i+1}, Z_{i+2}, \cdots, Z_{i+n-1}, Z_i)$$

$$+f_{i+1}(Z_{i+1}, Z_{i+2}, \cdots, Z_{i+n-1}, Z_i) - f_{i+2}(Z_{i+2}, Z_{i+3}, \cdots, Z_i, Z_{i+1}) + \cdots$$

$$+f_{i+n-2}(\mathbf{Z}_{i+n-2}, \mathbf{Z}_{i+n-1}, \cdots, \mathbf{Z}_{i+n-3}) - f_{i+n-1}(\mathbf{Z}_{i+n-1}, \mathbf{Z}_i, \cdots, \mathbf{Z}_{i+n-2}).$$

By denoting the sum of the members with positive sign by $nF_i(\mathbf{Z}_i, \mathbf{Z}_{i+1}, \cdots, \mathbf{Z}_{i+n-1})$ we obtain Eq. (2.2).

On the other hand, every function of the form Eq. (2.2) satisfies the equation (2.1). $\qquad\qquad\qquad\qquad\qquad\qquad\qquad\qquad\qquad\qquad\quad\square$

Theorem 2.2   *The general solution of the derived cyclic complex vector functional equation*

$$\sum_{i=1}^{n} f_i(\mathbf{Z}_i, \mathbf{Z}_{i+1}, \cdots, \mathbf{Z}_{i+p-1}) = \mathbf{O} \qquad (2.4)$$

$$(n \geq 2p - 1;\ \mathbf{Z}_{n+i} \equiv \mathbf{Z}_i,\ f_{n+i} \equiv f_i)$$

is given by

$$f_i(\mathbf{Z}_1, \mathbf{Z}_2, \cdots, \mathbf{Z}_p) \qquad\qquad\qquad\qquad (2.5)$$

$$= F_i(\mathbf{Z}_1, \mathbf{Z}_2, \cdots, \mathbf{Z}_{p-1}) - F_{i+1}(\mathbf{Z}_2, \mathbf{Z}_3, \cdots, \mathbf{Z}_p) \quad (1 \leq i \leq n)$$

where $F_{n+i} \equiv F_i$ $(1 \leq i \leq n)$ are arbitrary complex vector functions with values in $\mathcal{V}$.

Proof.   If we take into account the conventions $f_i \equiv f_{i+n}$ and $\mathbf{Z}_i \equiv \mathbf{Z}_{i+n}$, then we may rewrite the functional equation (2.4) in the following form

$$f_i(\mathbf{Z}_i, \mathbf{Z}_{i+1}, \cdots, \mathbf{Z}_{i+p-1}) + f_{i+1}(\mathbf{Z}_{i+1}, \mathbf{Z}_{i+2}, \cdots, \mathbf{Z}_{i+p}) + \cdots \qquad (2.6)$$

$$+f_{i+n-p}(\mathbf{Z}_{i+n-p}, \mathbf{Z}_{i+n-p+1}, \cdots, \mathbf{Z}_{i+n-1})$$

$$+f_{i+n-p+1}(\mathbf{Z}_{i+n-p+1}, \mathbf{Z}_{i+n-p+2}, \cdots, \mathbf{Z}_{i+n-1}, \mathbf{Z}_i) + \cdots$$

$$+f_{i+n-1}(\mathbf{Z}_{i+n-1}, \mathbf{Z}_i, \cdots, \mathbf{Z}_{i+p-2}) = \mathbf{O}.$$

By putting $\mathbf{Z}_{i+p} = \mathbf{Z}_{i+p+1} = \cdots = \mathbf{Z}_{i+n-1} = C$ in Eq. (2.6), where C is a fixed vector from $\mathcal{V}$, we obtain

$$f_i(\mathbf{Z}_i, \mathbf{Z}_{i+1}, \cdots, \mathbf{Z}_{i+p-1}) + f_{i+1}(\mathbf{Z}_{i+1}, \mathbf{Z}_{i+2}, \cdots, \mathbf{Z}_{i+p-1}, C) \qquad (2.7)$$

$$+f_{i+2}(\mathbf{Z}_{i+2}, \mathbf{Z}_{i+3}, \cdots, \mathbf{Z}_{i+p-1}, C, C) + \cdots$$

$$+f_{i+p-1}(\mathbf{Z}_{i+p-1}, C, C, \cdots, C) + f_{i+p}(C, C, \cdots, C)$$

$$+f_{i+p+1}(C,C,\cdots,C)+\cdots+f_{i+n-p}(C,C,\cdots,C)$$

$$+f_{i+n-p+1}(C,C,\cdots,C,\mathbf{Z}_i)+f_{i+n-p+2}(C,C,\cdots,C,\mathbf{Z}_i,\mathbf{Z}_{i+1})+\cdots$$

$$+f_{i+n-1}(C,\mathbf{Z}_i,\mathbf{Z}_{i+1},\cdots,\mathbf{Z}_{i+p-2})=\mathbf{O}.$$

The substitution $\mathbf{Z}_{i+p-1}=C$ in Eq. (2.7) yields

$$f_i(\mathbf{Z}_i,\mathbf{Z}_{i+1},\cdots,\mathbf{Z}_{i+p-2},C)+f_{i+1}(\mathbf{Z}_{i+1},\mathbf{Z}_{i+2},\cdots,\mathbf{Z}_{i+p-2},C,C)+\cdots$$

$$+f_{i+p-2}(\mathbf{Z}_{i+p-2},C,C,\cdots,C)+f_{i+p-1}(C,C,\cdots,C) \tag{2.8}$$

$$+f_{i+p}(C,C,\cdots,C)+\cdots+f_{i+n-p}(C,C,\cdots,C)$$

$$+f_{i+n-p+1}(C,C,\cdots,C,\mathbf{Z}_i)+f_{i+n-p+2}(C,C,\cdots,C,\mathbf{Z}_i,\mathbf{Z}_{i+1})+\cdots$$

$$+f_{i+n-1}(C,\mathbf{Z}_i,\mathbf{Z}_{i+1},\cdots,\mathbf{Z}_{i+p-2})=\mathbf{O}.$$

By subtracting Eq. (2.8) from Eq. (2.7), we get the formula

$$f_i(\mathbf{Z}_i,\mathbf{Z}_{i+1},\cdots,\mathbf{Z}_{i+p-1}) \tag{2.9}$$

$$=f_i(\mathbf{Z}_i,\mathbf{Z}_{i+1},\cdots,\mathbf{Z}_{i+p-2},C)-f_{i+1}(\mathbf{Z}_{i+1},\mathbf{Z}_{i+2},\cdots,\mathbf{Z}_{i+p-1},C)$$

$$+f_{i+1}(\mathbf{Z}_{i+1},\mathbf{Z}_{i+2},\cdots,\mathbf{Z}_{i+p-2},C,C)-f_{i+2}(\mathbf{Z}_{i+2},\mathbf{Z}_{i+3},\cdots,\mathbf{Z}_{i+p-1},C,C)+\cdots$$

$$+f_{i+p-2}(\mathbf{Z}_{i+p-2},C,C,\cdots,C)-f_{i+p-1}(\mathbf{Z}_{i+p-1},C,C,\cdots,C)$$

$$+f_{i+p-1}(C,C,\cdots,C) \qquad (1\le i\le n).$$

Let us put now

$$G_i(\mathbf{Z}_1,\mathbf{Z}_2,\cdots,\mathbf{Z}_{p-1}) = f_i(\mathbf{Z}_1,\mathbf{Z}_2,\cdots,\mathbf{Z}_{p-1},C) \tag{2.10}$$

$$+f_{i+1}(\mathbf{Z}_2,\mathbf{Z}_3,\cdots,\mathbf{Z}_{p-1},C,C) + \cdots+f_{i+p-2}(\mathbf{Z}_{p-1},C,C,\cdots,C),$$

$$A_i = f_{i+p-1}(C,C,\cdots,C). \tag{2.11}$$

Since $f_i \equiv f_{i+n}$, we conclude that $G_i \equiv G_{i+n}$. By putting $\mathbf{Z}_i = C$ ($1 \leq i \leq n$) in Eq. (2.4) we obtain

$$\sum_{i=1}^{n} A_i = \mathbf{O}. \tag{2.12}$$

According to Eq. (2.10) and Eq. (2.11), the formula Eq. (2.9) can be written in the following form

$$f_i(\mathbf{Z}_1, \mathbf{Z}_2, \cdots, \mathbf{Z}_p) = G_i(\mathbf{Z}_1, \mathbf{Z}_2, \cdots, \mathbf{Z}_p) - G_{i+1}(\mathbf{Z}_2, \mathbf{Z}_3, \cdots, \mathbf{Z}_p) + A_i, \tag{2.13}$$

$$(1 \leq i \leq n).$$

Finally, with the notations

$$
\begin{aligned}
F_1 &= G_1, \\
F_2 &= G_2 - A_1, \\
F_3 &= G_3 - A_1 - A_2, \\
&\vdots \\
F_n &= G_n - A_1 - A_2 - \cdots - A_{n-1}
\end{aligned}
$$

the expression Eq. (2.13) can be written in the form Eq. (2.5).

Conversely, the direct calculation shows that the function Eq. (2.5) satisfies the derived functional equation Eq. (2.4) for arbitrary F_i. $\qquad\square$

Theorem 2.3 *The general solution of the derived cyclic complex vector functional equation*

$$\sum_{i=1}^{k} f_i(\mathbf{Z}_i, \mathbf{Z}_{i+1}, \ldots, \mathbf{Z}_{i+p-1}) = \mathbf{O} \tag{2.14}$$

$$(p < n < 2p - 1; \ k \leq n; \quad \mathbf{Z}_{n+i} \equiv \mathbf{Z}_i)$$

is given by the formulae

$$
\begin{aligned}
f_r(\mathbf{Z}_1, \mathbf{Z}_2, \ldots, \mathbf{Z}_p) &= \sum_{i=1}^{\min(k-r,n-p)} (-1)^{i-1} F_{ri}(\mathbf{Z}_{i+1}, \mathbf{Z}_{i+2}, \ldots, \mathbf{Z}_p) \\
&+ \sum_{i=n-r+1}^{n-p} (-1)^{i-1} F_{ri}(\mathbf{Z}_{i+1}, \mathbf{Z}_{i+2}, \ldots, \mathbf{Z}_p)
\end{aligned} \tag{2.15}
$$

$$+ \sum_{i=n-p+1}^{\min(k-r,p-1)} (-1)^{i-1} F_{ri}(\mathbf{Z}_{i+1}, \mathbf{Z}_{i+2}, \ldots, \mathbf{Z}_p, \mathbf{Z}_1, \mathbf{Z}_2, \ldots, \mathbf{Z}_{p+i})$$

$$+ \sum_{i=\max(n-r+1,n-p+1)}^{p-1} (-1)^{n-i} F_{i+r,n-i}(\mathbf{Z}_1, \mathbf{Z}_2, \ldots, \mathbf{Z}_{p+i}, \mathbf{Z}_{i+1}, \mathbf{Z}_{i+2}, \ldots, \mathbf{Z}_p)$$

$$+ \sum_{i=p}^{k-r} (-1)^{n-i} F_{i+r,n-i}(\mathbf{Z}_1, \mathbf{Z}_2, \ldots, \mathbf{Z}_{p+i})$$

$$+ \sum_{i=\max(n-r+1,p)}^{n-1} (-1)^{n-i} F_{i+r,n-i}(\mathbf{Z}_1, \mathbf{Z}_2, \ldots, \mathbf{Z}_{p+i}) \quad (1 \leq r \leq k; \ k \leq n),$$

where

$$\sum_a^s = \mathbf{O} \ (a > s); \qquad F_{m+n,i} = F_{mi}; \qquad F'_{i,m+n} = F'_{im}$$

and F_{ij} are arbitrary vector functions from $\mathcal{V}$.

Proof.   The proof of this theorem will be given by induction.

For $k = 2$, Eq. (2.14) becomes

$$f_1(\mathbf{Z}_1, \mathbf{Z}_2, \ldots, \mathbf{Z}_p) + f_2(\mathbf{Z}_2, \mathbf{Z}_3, \ldots, \mathbf{Z}_{p+1}) = \mathbf{O}. \tag{2.16}$$

Putting $\mathbf{Z}_1 = C_1$ into Eq. (2.16), we obtain

$$f_2(\mathbf{Z}_2, \mathbf{Z}_3, \ldots, \mathbf{Z}_{p+1}) = -F_{11}(\mathbf{Z}_2, \mathbf{Z}_3, \ldots, \mathbf{Z}_p), \tag{2.17}$$

where the notation

$$F_{11}(\mathbf{Z}_2, \mathbf{Z}_3, \ldots, \mathbf{Z}_p) = f_1(C_1, \mathbf{Z}_2, \mathbf{Z}_3 \ldots, \mathbf{Z}_p)$$

is introduced. If we put Eq. (2.17) into Eq. (2.16), we have

$$f_1(\mathbf{Z}_1, \mathbf{Z}_2, \ldots, \mathbf{Z}_p) = F_{11}(\mathbf{Z}_2, \mathbf{Z}_3, \ldots, \mathbf{Z}_p). \tag{2.18}$$

Since

$$\begin{aligned}
\min\,(1, n-p) &= 1, & \min\,(1, p-1) &= 1, \\
\max\,(n, n-p+1) &= n, & \max\,(n, p) &= n, \\
\min\,(0, n-p) &= 0, & \min\,(0, p-1) &= 0, \\
\max\,(n-1, n-p+1) &= n-1, & \max\,(n-1, p) &= n-1,
\end{aligned}$$

from Eq. (2.15) we obtain Eqs. (2.18) and (2.17) respectively for $k = 2$, $r = 1$ and for $k = 2$, $r = 2$.

Therefore, the theorem holds for $k = 2$.

Now we will suppose that the theorem holds for some fixed $k < n$, *i.e.* let the general solution of the functional equation

$$\sum_{i=1}^{k} g_i(\mathbf{Z}_i, \mathbf{Z}_{i+1}, \ldots, \mathbf{Z}_{i+p-1}) = \mathbf{O} \quad (k < n) \tag{2.19}$$

be given by Eq. (2.15), where in the place of f_r we have put g_r $(1 \leq r \leq k)$.

Let us consider the following functional equation

$$\sum_{i=1}^{k+1} f_i(\mathbf{Z}_i, \mathbf{Z}_{i+1}, \ldots, \mathbf{Z}_{i+p-1}) = \mathbf{O} \quad (k+1 \leq n). \tag{2.20}$$

We will distinguish the following three cases:

$1°$ Let $2 \leq k < n - p + 1$. If we put $\mathbf{Z}_i = C_i$ for $i \not\equiv k+1, k+2, \ldots, k+p$ (mod n), then Eq. (2.20) becomes

$$f_{k+1}(\mathbf{Z}_{k+1}, \mathbf{Z}_{k+2}, \ldots, \mathbf{Z}_{k+p}) \tag{2.21}$$

$$= \sum_{i=n-k}^{n-1} (-1)^{n-i} F_{i+k+1, n-i}(\mathbf{Z}_{k+1}, \mathbf{Z}_{k+2}, \ldots, \mathbf{Z}_{i+p+k}),$$

where we introduced the notation

$$(-1)^{n-i+1} F_{i+k+1, n-i}(\mathbf{Z}_{k+1}, \mathbf{Z}_{k+2}, \ldots, \mathbf{Z}_{i+p+k})$$

$$= f_{i+k+1-n}(\mathbf{Z}_{i+k+1-n}, \mathbf{Z}_{i+k+2-n}, \ldots, \mathbf{Z}_{i+k+p-n}) \bigg|_{\substack{\mathbf{Z}_i = C_i \text{ for } i \not\equiv k+1, \\ k+2, \ldots, k+p \pmod{n}}} \cdot$$

From Eq. (2.21) it follows that

$$f_{k+1}(\mathbf{Z}_1, \mathbf{Z}_2, \ldots, \mathbf{Z}_p) = \sum_{i=n-k}^{n-1} (-1)^{n-i} F_{i+k+1,n-i}(\mathbf{Z}_1, \mathbf{Z}_2, \ldots, \mathbf{Z}_{i+p}). \quad (2.22)$$

If we substitute Eq. (2.21) into Eq. (2.20) and introduce new notations by the equalities

$$g_i(\mathbf{Z}_i, \mathbf{Z}_{i+1}, \ldots, \mathbf{Z}_{i+p-1}) = f_i(\mathbf{Z}_i, \mathbf{Z}_{i+1}, \ldots, \mathbf{Z}_{i+p-1}) \quad (2.23)$$
$$+ (-1)^{1+k-i} F_{i,1+k-i}(\mathbf{Z}_{k+1}, \mathbf{Z}_{k+2}, \ldots, \mathbf{Z}_{i+p-1}) \quad (i = 1, \ldots, k),$$

we obtain the equation (2.19).

For $2 \le k < n - p$, since

$$\min(k - r, n - p) = \min(k - r, p - 1) = k - r, \qquad n - p + 1 > k - r,$$

on the basis of the expressions Eq. (2.15) the general solution of Eq. (2.19) is

$$g_r(\mathbf{Z}_1, \mathbf{Z}_2, \ldots, \mathbf{Z}_p) = \sum_{i=1}^{k-r} (-1)^{i-1} F_{ri}(\mathbf{Z}_{i+1}, \mathbf{Z}_{i+2}, \ldots, \mathbf{Z}_p) \quad (2.24)$$

$$+ \sum_{i=n-r+1}^{n-p} (-1)^{i-1} F_{ri}(\mathbf{Z}_{i+1}, \mathbf{Z}_{i+2}, \ldots, \mathbf{Z}_p)$$

$$+ \sum_{i=\max(n-r+1,n-p+1)}^{p-1} (-1)^{n-i} F_{i+r,n-i}(\mathbf{Z}_1, \mathbf{Z}_2, \ldots, \mathbf{Z}_{p+i}, \mathbf{Z}_{i+1}, \mathbf{Z}_{i+2}, \ldots, \mathbf{Z}_p)$$

$$+ \sum_{i=\max(n-r+1,p)}^{n-1} (-1)^{n-i} F_{i+r,n-i}(\mathbf{Z}_1, \mathbf{Z}_2, \ldots, \mathbf{Z}_{p+i}) \quad (1 \le r \le k).$$

On the basis of Eqs. (2.22), (2.23) and (2.24), we obtain

$$f_r(\mathbf{Z}_1, \mathbf{Z}_2, \ldots, \mathbf{Z}_p) = \sum_{i=1}^{k+1-r} (-1)^{i-1} F_{ri}(\mathbf{Z}_{i+1}, \mathbf{Z}_{i+2}, \ldots, \mathbf{Z}_p) \quad (2.25)$$

$$+ \sum_{i=n-r+1}^{n-p} (-1)^{i-1} F_{ri}(\mathbf{Z}_{i+1}, \mathbf{Z}_{i+2}, \ldots, \mathbf{Z}_p)$$

$$+ \sum_{i=\max(n-r+1,n-p+1)}^{p-1} (-1)^{n-i} F_{i+r,n-i}(\mathbf{Z}_1, \mathbf{Z}_2, \ldots, \mathbf{Z}_{p+i}, \mathbf{Z}_{i+1}, \mathbf{Z}_{i+2}, \ldots, \mathbf{Z}_p)$$

$$+ \sum_{i=\max(n-r+1,p)}^{n-1} (-1)^{n-i} F_{i+r,n-i}(\mathbf{Z}_1, \mathbf{Z}_2, \ldots, \mathbf{Z}_{p+i}) \qquad (1 \leq r \leq k+1).$$

Therefore, for $2 \leq k < n-p+1$ the theorem holds for $k+1$ if it is true for k. This means that the theorem holds for all such k, and also for $k = n-p+1$.

2°. Let $n-p+1 \leq k < p$. If we put $\mathbf{Z}_i = C_i$ for $i \not\equiv k+1, k+2, \ldots, k+p$ (mod n), then Eq. (2.20) becomes

$$f_{k+1}(\mathbf{Z}_{k+1}, \mathbf{Z}_{k+2}, \ldots, \mathbf{Z}_{k+p}) = \qquad (2.26)$$

$$\sum_{i=n-k}^{p-1} (-1)^{n-i} F_{i+k+1,n-i}(\mathbf{Z}_{k+1}, \mathbf{Z}_{k+2}, \ldots, \mathbf{Z}_{p+k+i}, \mathbf{Z}_{i+k+1}, \mathbf{Z}_{i+k+2}, \ldots, \mathbf{Z}_{p+k})$$

$$+ \sum_{i=p}^{n-1} (-1)^{n-i} F_{i+k+1,n-i}(\mathbf{Z}_{k+1}, \mathbf{Z}_{k+2}, \ldots, \mathbf{Z}_{p+k+i}),$$

where we have introduced the notation

$$f_{i+k+1-n}(\mathbf{Z}_{i+k+1-n}, \mathbf{Z}_{i+k+2-n}, \ldots, \mathbf{Z}_{i+k+p-n}) \Big|_{\substack{\mathbf{Z}_i = C_i \text{ for } i \not\equiv k+1, \\ k+2, \ldots, k+p \ (\text{mod } n)}}$$

$$= \begin{cases} (-1)^{n-i+1} F_{i+k+1,n-i}(\mathbf{Z}_{k+1}, \mathbf{Z}_{k+2}, \ldots, \mathbf{Z}_{p+k+i}, \\ \qquad \mathbf{Z}_{i+k+1}, \mathbf{Z}_{i+k+2}, \ldots, \mathbf{Z}_{p+k}) \quad (i = n-k, n-k+1, \ldots, p-1), \\ (-1)^{n-i+1} F_{i+k+1,n-i}(\mathbf{Z}_{k+1}, \mathbf{Z}_{k+2}, \ldots, \mathbf{Z}_{p+k+i}) \quad (i = p, p+1, \ldots, n-1). \end{cases}$$

Substituting Eq. (2.26) into Eq. (2.20) and introducing new functions by

$$g_i = f_i + (-1)^{1+k-i} F_{i,1+k-i} \quad (i = 1, \ldots, k), \qquad (2.27)$$

we obtain Eq. (2.19).

On the basis of the inductive hypothesis and Eq. (2.27), the general solution of Eq. (2.20) is

$$f_r(\mathbf{Z}_1, \mathbf{Z}_2, \ldots, \mathbf{Z}_p) = \sum_{i=1}^{n-p} (-1)^{i-1} F_{ri}(\mathbf{Z}_{i+1}, \mathbf{Z}_{i+2}, \ldots, \mathbf{Z}_p)$$

$$+ \sum_{i=n-p+1}^{k-r} (-1)^{i-1} F_{ri}(\mathbf{Z}_{i+1}, \mathbf{Z}_{i+2}, \ldots, \mathbf{Z}_p, \mathbf{Z}_1, \mathbf{Z}_2, \ldots, \mathbf{Z}_{p+i})$$

$$+(-1)^{k-r} F_{r,k+1-r}(\mathbf{Z}_{k+2-r}, \mathbf{Z}_{k+3-r}, \ldots, \mathbf{Z}_p, \mathbf{Z}_1, \mathbf{Z}_2, \ldots, \mathbf{Z}_{p+k-r+1})$$

$$+ \sum_{i=n-r+1}^{n-p} (-1)^{i-1} F_{ri}(\mathbf{Z}_{i+1}, \mathbf{Z}_{i+2}, \ldots, \mathbf{Z}_p)$$

$$+ \sum_{i=\max(n-r+1,n-p+1)}^{p-1} (-1)^{n-i} F_{i+r,n-i}(\mathbf{Z}_1, \mathbf{Z}_2, \ldots, \mathbf{Z}_{p+i}, \mathbf{Z}_{i+1}, \mathbf{Z}_{i+2}, \ldots, \mathbf{Z}_p)$$

$$+ \sum_{i=\max(n-r+1,p)}^{n-1} (-1)^{n-i} F_{i+r,n-i}(\mathbf{Z}_1, \mathbf{Z}_2, \ldots, \mathbf{Z}_{p+i}) \quad (r = 1, 2, \ldots, p+k-n);$$

$$(2.28)$$

$$f_r(\mathbf{Z}_1, \mathbf{Z}_2, \ldots, \mathbf{Z}_p) = \sum_{i=1}^{k-r} (-1)^{i-1} F_{ri}(\mathbf{Z}_{i+1}, \mathbf{Z}_{i+2}, \ldots, \mathbf{Z}_p)$$

$$+(-1)^{k-r} F_{r,k+1-r}(\mathbf{Z}_{k+2-r}, \mathbf{Z}_{k+3-r}, \ldots, \mathbf{Z}_p)$$

$$+ \sum_{i=n-r+1}^{n-p} (-1)^{i-1} F_{ri}(\mathbf{Z}_{i+1}, \mathbf{Z}_{i+2}, \ldots, \mathbf{Z}_p)$$

$$+ \sum_{i=\max(n-r+1,n-p+1)}^{p-1} (-1)^{n-i} F_{i+r,n-i}(\mathbf{Z}_1, \mathbf{Z}_2, \ldots, \mathbf{Z}_{p+i}, \mathbf{Z}_{i+1}, \mathbf{Z}_{i+2}, \ldots, \mathbf{Z}_p)$$

$$+ \sum_{i=\max(n-r+1,p)}^{n-1} (-1)^{n-i} F_{i+r,n-i}(\mathbf{Z}_{i+1}, \mathbf{Z}_{i+2}, \ldots, \mathbf{Z}_p) \quad (r = p+k-n+1, \ldots, k).$$

On the basis of Eqs. (2.26) and (2.28), the general solution of Eq. (2.20) is determined by Eq. (2.15), where k must be replaced by $k + 1$.

Therefore, for $n - p + 1 \le k < p$ the theorem holds for $k + 1$ if it holds for k, *i.e.* the theorem is true for all such k, and also for $k = p$.

3° Let $p \le k \le n - 1$. For $\mathbf{Z}_i = C_i$ when $i \not\equiv k + 1, k + 2, \ldots, k + p$ (mod n) Eq. (2.20) becomes

$$f_{k+1}(\mathbf{Z}_{k+1}, \mathbf{Z}_{k+2}, \ldots, \mathbf{Z}_{k+p}) = \sum_{i=n-k}^{n-p}(-1)^{i-1}F_{k+1,i}(\mathbf{Z}_{i+1}, \mathbf{Z}_{i+2}, \ldots, \mathbf{Z}_p) \quad (2.29)$$

$$+ \sum_{i=n-p+1}^{p-1}(-1)^{n-i}F_{i+k+1,n-i}(\mathbf{Z}_{k+1}, \mathbf{Z}_{k+2}, \ldots, \mathbf{Z}_{p+k+i}, \mathbf{Z}_{i+k+1}, \mathbf{Z}_{i+k+2}, \ldots, \mathbf{Z}_{p+k})$$

$$+ \sum_{i=p}^{n-1}(-1)^{n-i}F_{i+k+1,n-i}(\mathbf{Z}_{k+1}, \mathbf{Z}_{k+2}, \ldots, \mathbf{Z}_{p+k+i}),$$

where we have introduced the notation

$$f_{i+k+1-n}(\mathbf{Z}_{i+k+1-n}, \mathbf{Z}_{i+k+2-n}, \ldots, \mathbf{Z}_{i+k+p-n}) \Big|_{\substack{\mathbf{Z}_i = C_i \ \text{for} \ i \not\equiv k+1, \\ k+2, \ldots, k+p \ (\text{mod } n)}}$$

$$= \begin{cases} (-1)^i F_{k+1,i}(\mathbf{Z}_{i+1}, \mathbf{Z}_{i+2}, \ldots, \mathbf{Z}_p) & (i = n - k, \ldots, n - p), \\[2mm] (-1)^{n-i+1}F_{i+k+1,n-i}(\mathbf{Z}_{k+1}, \mathbf{Z}_{k+2}, \ldots, \mathbf{Z}_{p+k+i}, \\ \qquad \mathbf{Z}_{i+k+1}, \mathbf{Z}_{i+k+2}, \ldots, \mathbf{Z}_{p+k}) & (i = n - p + 1, \ldots, p - 1), \\[2mm] (-1)^{n-i+1}F_{i+k+1,n-i}(\mathbf{Z}_{k+1}, \mathbf{Z}_{k+2}, \ldots, \mathbf{Z}_{p+k+i}) & (i = p, p + 1, \ldots, n - 1). \end{cases}$$

Now, we will introduce the following notation

$$g_i = f_i + \begin{cases} (-1)^{i+n-k}F_{k+1,i+n-k-1} & (i = 1, \ldots, k + 1 - p), \\ (-1)^{1+k-i}F_{i,1+k-i} & (i = k + 2 - p, \ldots, k). \end{cases} \quad (2.30)$$

If we substitute the function f_{k+1} determined by Eq. (2.29) into Eq. (2.20), in view of Eq. (2.30) we obtain the functional equation (2.19). On the basis of Eq. (2.30) and the general solution of the functional equation (2.19), we obtain that the general solution of the functional equation (2.20) is determined by

$$f_r(\mathbf{Z}_1, \mathbf{Z}_2, \ldots, \mathbf{Z}_p) = \sum_{i=1}^{n-p}(-1)^{i-1}F_{ri}(\mathbf{Z}_{i+1}, \mathbf{Z}_{i+2}, \ldots, \mathbf{Z}_p)$$

$$+ \sum_{i=n-p+1}^{p-1} (-1)^{i-1} F_{ri}(\mathbf{Z}_{i+1}, \mathbf{Z}_{i+2}, \ldots, \mathbf{Z}_p, \mathbf{Z}_1, \mathbf{Z}_2, \ldots, \mathbf{Z}_{p+i})$$

$$+ \sum_{i=p}^{k-r} (-1)^{n-i} F_{i+r,n-i}(\mathbf{Z}_1, \mathbf{Z}_2, \ldots, \mathbf{Z}_{p+i})$$

$$+ (-1)^{n-k+1-r} F_{k,n-k+1-r}(\mathbf{Z}_1, \mathbf{Z}_2, \ldots, \mathbf{Z}_{p+k-r})$$

$$+ \sum_{i=n-r+1}^{n-p} (-1)^{i-1} F_{ri}(\mathbf{Z}_{i+1}, \mathbf{Z}_{i+2}, \ldots, \mathbf{Z}_p)$$

$$+ \sum_{i=\max(n-r+1,n-p+1)}^{p-1} (-1)^{n-i} F_{i+r,n-i}(\mathbf{Z}_1, \mathbf{Z}_2, \ldots, \mathbf{Z}_{p+i}, \mathbf{Z}_{i+1}, \mathbf{Z}_{i+2}, \ldots, \mathbf{Z}_p)$$

$$+ \sum_{i=\max(n-r+1,p)}^{n-1} (-1)^{n-i} F_{i+r,n-i}(\mathbf{Z}_1, \mathbf{Z}_2, \ldots, \mathbf{Z}_{p+i}) \quad (r = 1, 2, \ldots, n-k-p+1);$$

$$(2.31)$$

$$f_r(\mathbf{Z}_1, \mathbf{Z}_2, \ldots, \mathbf{Z}_p) = \sum_{i=1}^{n-p} (-1)^{i-1} F_{ri}(\mathbf{Z}_{i+1}, \mathbf{Z}_{i+2}, \ldots, \mathbf{Z}_p)$$

$$+ \sum_{i=n-p+1}^{k-r} (-1)^{i-1} F_{ri}(\mathbf{Z}_{i+1}, \mathbf{Z}_{i+2}, \ldots, \mathbf{Z}_p, \mathbf{Z}_1, \mathbf{Z}_2, \ldots, \mathbf{Z}_{p+i})$$

$$+ (-1)^{k-r} F_{r,k-r+1}(\mathbf{Z}_{k-r+2}, \mathbf{Z}_{k-r+3}, \ldots, \mathbf{Z}_p, \mathbf{Z}_1, \mathbf{Z}_2, \ldots, \mathbf{Z}_{p+k-r+1})$$

$$+ \sum_{i=n-r+1}^{n-p} (-1)^{i-1} F_{ri}(\mathbf{Z}_{i+1}, \mathbf{Z}_{i+2}, \ldots, \mathbf{Z}_p)$$

$$+ \sum_{i=\max(n-r+1,n-p+1)}^{p-1} (-1)^{n-i} F_{i+r,n-i}(\mathbf{Z}_1, \mathbf{Z}_2, \ldots, \mathbf{Z}_{p+i}, \mathbf{Z}_{i+1}, \mathbf{Z}_{i+2}, \ldots, \mathbf{Z}_p)$$

$$+ \sum_{i=\max(n-r+1,p)}^{n-1} (-1)^{n-i} F_{i+r,n-i}(\mathbf{Z}_1, \mathbf{Z}_2, \ldots, \mathbf{Z}_{p+i}) \ (r = n-k-p+2, \ldots, p-1);$$

$$f_r(\mathbf{Z}_1, \mathbf{Z}_2, \ldots, \mathbf{Z}_p) = \sum_{i=1}^{k-r} (-1)^{i-1} F_{ri}(\mathbf{Z}_{i+1}, \mathbf{Z}_{i+2}, \ldots, \mathbf{Z}_p)$$

$$+ (-1)^{k-r} F_{r,k+1-r}(\mathbf{Z}_{k+2-r}, \mathbf{Z}_{k+3-r}, \ldots, \mathbf{Z}_p)$$

$$+ \sum_{i=n-r+1}^{n-p} (-1)^{i-1} F_{ri}(\mathbf{Z}_{i+1}, \mathbf{Z}_{i+2}, \ldots, \mathbf{Z}_p)$$

$$+ \sum_{i=\max(n-r+1,n-p+1)}^{p-1} (-1)^{n-i} F_{i+r,n-i}(\mathbf{Z}_1, \mathbf{Z}_2, \ldots, \mathbf{Z}_{p+i}, \mathbf{Z}_{i+1}, \mathbf{Z}_{i+2}, \ldots, \mathbf{Z}_p)$$

$$+ \sum_{i=\max(n-r+1,p)}^{n-1} (-1)^{n-i} F_{i+r,n-i}(\mathbf{Z}_1, \mathbf{Z}_2, \ldots, \mathbf{Z}_{p+i}) \quad (r = p, p+1, \ldots, k).$$

On the basis of Eqs. (2.29) and (2.31) we obtain that the general solution of the functional equation (2.20) in the case $p \leq k \leq n-1$ is determined by Eq. (2.15), where k must be replaced by $k+1$. $\square$

Therefore, the solution of the research problem given in [D. S. Mitrinović and D. Ž. Djoković (1963)] is presented by this theorem. As particular cases see the results given in [M. Hosszú (1961); D. S. Mitrinović (1963B)].

Example 2.1   For $n = 8$, $p = 5$ and $k = 6$ the complex vector functional equation Eq. (2.14) becomes

$$f_1(\mathbf{Z}_1, \mathbf{Z}_2, \mathbf{Z}_3, \mathbf{Z}_4, \mathbf{Z}_5) + f_2(\mathbf{Z}_2, \mathbf{Z}_3, \mathbf{Z}_4, \mathbf{Z}_5, \mathbf{Z}_6) + f_3(\mathbf{Z}_3, \mathbf{Z}_4, \mathbf{Z}_5, \mathbf{Z}_6, \mathbf{Z}_7)$$

$$+ f_4(\mathbf{Z}_4, \mathbf{Z}_5, \mathbf{Z}_6, \mathbf{Z}_7, \mathbf{Z}_8) + f_5(\mathbf{Z}_5, \mathbf{Z}_6, \mathbf{Z}_7, \mathbf{Z}_8, \mathbf{Z}_1) + f_6(\mathbf{Z}_6, \mathbf{Z}_7, \mathbf{Z}_8, \mathbf{Z}_1, \mathbf{Z}_2) = \mathbf{O},$$

whose general solution is given by

$$f_1(\mathbf{Z}_1, \mathbf{Z}_2, \mathbf{Z}_3, \mathbf{Z}_4, \mathbf{Z}_5) = F_{11}(\mathbf{Z}_2, \mathbf{Z}_3, \mathbf{Z}_4, \mathbf{Z}_5) - F_{12}(\mathbf{Z}_3, \mathbf{Z}_4, \mathbf{Z}_5)$$

$$+ F_{13}(\mathbf{Z}_4, \mathbf{Z}_5) - F_{14}(\mathbf{Z}_5, \mathbf{Z}_1) - F_{63}(\mathbf{Z}_1, \mathbf{Z}_2),$$

$$f_2(\mathbf{Z}_1, \mathbf{Z}_2, \mathbf{Z}_3, \mathbf{Z}_4, \mathbf{Z}_5) = F_{21}(\mathbf{Z}_2, \mathbf{Z}_3, \mathbf{Z}_4, \mathbf{Z}_5) - F_{22}(\mathbf{Z}_3, \mathbf{Z}_4, \mathbf{Z}_5)$$

$$+ F_{23}(\mathbf{Z}_4, \mathbf{Z}_5) - F_{24}(\mathbf{Z}_5, \mathbf{Z}_1) - F_{11}(\mathbf{Z}_1, \mathbf{Z}_2, \mathbf{Z}_3, \mathbf{Z}_4),$$

$$f_3(\mathbf{Z}_1, \mathbf{Z}_2, \mathbf{Z}_3, \mathbf{Z}_4, \mathbf{Z}_5) = F_{31}(\mathbf{Z}_2, \mathbf{Z}_3, \mathbf{Z}_4, \mathbf{Z}_5) - F_{32}(\mathbf{Z}_3, \mathbf{Z}_4, \mathbf{Z}_5)$$

$$+ F_{33}(\mathbf{Z}_4, \mathbf{Z}_5) + F_{12}(\mathbf{Z}_1, \mathbf{Z}_2, \mathbf{Z}_3) - F_{21}(\mathbf{Z}_1, \mathbf{Z}_2, \mathbf{Z}_3, \mathbf{Z}_4),$$

$$f_4(\mathbf{Z}_1, \mathbf{Z}_2, \mathbf{Z}_3, \mathbf{Z}_4, \mathbf{Z}_5) = F_{41}(\mathbf{Z}_2, \mathbf{Z}_3, \mathbf{Z}_4, \mathbf{Z}_5) - F_{42}(\mathbf{Z}_3, \mathbf{Z}_4, \mathbf{Z}_5)$$

$$- F_{13}(\mathbf{Z}_1, \mathbf{Z}_2) + F_{22}(\mathbf{Z}_1, \mathbf{Z}_2, \mathbf{Z}_3) - F_{31}(\mathbf{Z}_1, \mathbf{Z}_2, \mathbf{Z}_3, \mathbf{Z}_4),$$

$$f_5(\mathbf{Z}_1, \mathbf{Z}_2, \mathbf{Z}_3, \mathbf{Z}_4, \mathbf{Z}_5) = F_{51}(\mathbf{Z}_2, \mathbf{Z}_3, \mathbf{Z}_4, \mathbf{Z}_5) + F_{14}(\mathbf{Z}_1, \mathbf{Z}_5)$$

$$- F_{23}(\mathbf{Z}_1, \mathbf{Z}_2) + F_{32}(\mathbf{Z}_1, \mathbf{Z}_2, \mathbf{Z}_3) - F_{41}(\mathbf{Z}_1, \mathbf{Z}_2, \mathbf{Z}_3, \mathbf{Z}_4),$$

$$f_6(\mathbf{Z}_1, \mathbf{Z}_2, \mathbf{Z}_3, \mathbf{Z}_4, \mathbf{Z}_5) = F_{63}(\mathbf{Z}_4, \mathbf{Z}_5) + F_{24}(\mathbf{Z}_1, \mathbf{Z}_5)$$

$$- F_{33}(\mathbf{Z}_1, \mathbf{Z}_2) + F_{42}(\mathbf{Z}_1, \mathbf{Z}_2, \mathbf{Z}_3) - F_{51}(\mathbf{Z}_1, \mathbf{Z}_2, \mathbf{Z}_3, \mathbf{Z}_4),$$

where F_{ij} are arbitrary complex vector functions from $\mathcal{V}$.

Now we will give two particular cases of the above theorem.

Theorem 2.4 *The general solution of the derived cyclic complex vector functional equation*

$$\sum_{i=1}^{n} f_i(\mathbf{Z}_i, \mathbf{Z}_{i+1}, \ldots, \mathbf{Z}_{i+p-1}) = \mathbf{O} \quad (p < n < 2p - 1;\ \mathbf{Z}_{n+i} \equiv \mathbf{Z}_i) \quad (2.32)$$

is given by

$$f_r(\mathbf{Z}_1, \mathbf{Z}_2, \ldots, \mathbf{Z}_p) = \sum_{i=1}^{n-p} (-1)^{i-1} F_{ri}(\mathbf{Z}_{i+1}, \mathbf{Z}_{i+2}, \ldots, \mathbf{Z}_p) \qquad (2.33)$$

$$+ \sum_{i=n-p+1}^{\min(n-r,\,p-1)} (-1)^{i-1} F_{ri}(\mathbf{Z}_{i+1}, \mathbf{Z}_{i+2}, \ldots, \mathbf{Z}_p, \mathbf{Z}_1, \mathbf{Z}_2, \ldots, \mathbf{Z}_{p+i})$$

$$+ \sum_{i=\max(n-r+1,\,n-p+1)}^{p-1} (-1)^{n-i} F_{i+r,n-i}(\mathbf{Z}_1, \mathbf{Z}_2, \ldots, \mathbf{Z}_{p+i}, \mathbf{Z}_{i+1}, \mathbf{Z}_{i+2}, \ldots, \mathbf{Z}_p)$$

$$+ \sum_{i=p}^{n-1} (-1)^{n-i} F_{i+r,n-i}(\mathbf{Z}_1, \mathbf{Z}_2, \ldots, \mathbf{Z}_{p+i}) \quad (1 \leq r \leq n),$$

where F_{ij} are arbitrary complex vector functions from $\mathcal{V}$.

Proof. The proof of this theorem immediately follows from the previous theorem for $k = n$. $\square$

This theorem generalizes the results given in [D. Ž. Djoković (1964)].

Theorem 2.5 *The general solution of the basic cyclic complex vector functional equation*

$$\sum_{i=1}^{n} f(\mathbf{Z}_i, \mathbf{Z}_{i+1}, \ldots, \mathbf{Z}_{i+p-1}) = \mathbf{O} \quad (p < n < 2p - 1; \ \mathbf{Z}_{n+i} \equiv \mathbf{Z}_i) \quad (2.34)$$

is given by

$$f(\mathbf{Z}_1, \mathbf{Z}_2, \ldots, \mathbf{Z}_p) = F_0(\mathbf{Z}_1, \mathbf{Z}_2, \ldots, \mathbf{Z}_{p-1}) - F_0(\mathbf{Z}_2, \mathbf{Z}_3, \ldots, \mathbf{Z}_p) \quad (2.35)$$

$$+ \sum_{i=1}^{p-[n/2]} [F_i(\mathbf{Z}_1, \mathbf{Z}_2, \ldots, \mathbf{Z}_i, \mathbf{Z}_{n-p+i+1}, \ldots, \mathbf{Z}_p)$$

$$- F_i(\mathbf{Z}_{p-i+1}, \ldots, \mathbf{Z}_p, \mathbf{Z}_1, \mathbf{Z}_2, \ldots, \mathbf{Z}_{2p-n-i})],$$

where F_i $(0 \leq i \leq p - [n/2])$ are arbitrary complex vector functions from $\mathcal{V}$.

Proof. By summing up the functions f_i $(1 \leq i \leq n)$ determined by Eq. (2.33) and putting $f_1 = f_2 = \cdots = f_n = f$, we obtain Eq. (2.35), where we have introduced the following notations

$$F_0(\mathbf{Z}_1, \mathbf{Z}_2, \ldots, \mathbf{Z}_{p-1}) = \frac{1}{n} \sum_{r=1}^{n-p} \sum_{i=1}^{r} G_r(\mathbf{Z}_i, \mathbf{Z}_{i+1}, \ldots, \mathbf{Z}_{p-r-1+i}),$$

$$G_r(\mathbf{Z}_1, \mathbf{Z}_2, \ldots, \mathbf{Z}_{p-r}) = \sum_{i=1}^{n} (-1)^r F_{ir}(\mathbf{Z}_1, \mathbf{Z}_2, \ldots, \mathbf{Z}_{p-r}),$$

$$F_i(\mathbf{Z}_1, \mathbf{Z}_2, \ldots, \mathbf{Z}_i, \mathbf{Z}_{n-p+i+1}, \ldots, \mathbf{Z}_p)$$

$$= \frac{(-1)^{i+1}}{n} \left[\sum_{r=1}^{n-p+i} F_{r,p-i}(\mathbf{Z}_1, \mathbf{Z}_2, \ldots, \mathbf{Z}_i, \mathbf{Z}_{n-p+i+1}, \ldots, \mathbf{Z}_p) \right.$$

$$\left. - \sum_{r=1}^{p-i} F_{r,n-p+i}(\mathbf{Z}_{n-p+i+1}, \ldots, \mathbf{Z}_p, \mathbf{Z}_1, \mathbf{Z}_2, \ldots, \mathbf{Z}_i) \right]$$

$$(1 \le i \le p - [(n+1)/2]).$$

In particular, if $n = 2m$, we get

$$F_{p-m}(\mathbf{Z}_1, \mathbf{Z}_2, \ldots, \mathbf{Z}_{p-m}, \mathbf{Z}_{m+1}, \ldots, \mathbf{Z}_p)$$

$$= \frac{(-1)^{p-m+1}}{n} \sum_{r=1}^{m} F_{rm}(\mathbf{Z}_1, \mathbf{Z}_2, \ldots, \mathbf{Z}_{p-m}, \mathbf{Z}_{m+1}, \ldots, \mathbf{Z}_p)$$

where

$$F_{r+m,m}(\mathbf{Z}_1, \ldots, \mathbf{Z}_{p-m}, \mathbf{Z}_{m+1}, \ldots, \mathbf{Z}_p)$$

$$= - F_{rm}(\mathbf{Z}_{m+1}, \ldots, \mathbf{Z}_p, \mathbf{Z}_1, \ldots, \mathbf{Z}_{p-m}) \ (1 \le r \le m).$$

$\square$

We have noticed that in [P. M. Vasić and R. Ž. Djordjević (1965)] special generalized cases of Eqs. (2.14), (2.32) and (2.34) are considered. They are solved in a complicated manner using a cyclic operator.

At the end of the present section we give two more general theorems obtained in [I. B. Risteski (to appear A)].

Theorem 2.6 *The general solution of the derived cyclic complex vector functional equation*

$$\sum_{i=1}^{n} f_i(\mathbf{Z}_i, \mathbf{Z}_{i+1}, \ldots, \mathbf{Z}_{i+p-1}) = \mathbf{O} \qquad (\mathbf{Z}_{n+i} \equiv \mathbf{Z}_i) \tag{2.36}$$

is given by

$$f_r(\mathbf{Z}_r, \mathbf{Z}_{r+1}, \ldots, \mathbf{Z}_{r+p-1}) \tag{2.37}$$

$$= \sum_{j=1}^{r-1} (-1)^{r+1} F_{jr}(\{\mathbf{Z}_r, \mathbf{Z}_{r+1}, \ldots, \mathbf{Z}_{r+p-1}\} \cap \{\mathbf{Z}_j, \mathbf{Z}_{j+1}, \ldots, \mathbf{Z}_{j+p-1}\})$$

$$+ \sum_{j=r+1}^{n} (-1)^j F_{rj}(\{\mathbf{Z}_r, \mathbf{Z}_{r+1}, \ldots, \mathbf{Z}_{r+p-1}\} \cap \{\mathbf{Z}_j, \mathbf{Z}_{j+1}, \ldots, \mathbf{Z}_{j+p-1}\})$$

$$(1 \le r \le n),$$

where F_{rj} ($1 \le r \le n-1$, $r+1 \le j \le n$) are arbitrary complex vector functions from $\mathcal{V}$ such that

$$F_{rj}(\{\mathbf{Z}_r, \mathbf{Z}_{r+1}, \ldots, \mathbf{Z}_{r+p-1}\} \cap \{\mathbf{Z}_j, \mathbf{Z}_{j+1}, \ldots, \mathbf{Z}_{j+p-1}\}) = A_{rj}$$

if

$$\{\mathbf{Z}_r, \mathbf{Z}_{r+1}, \ldots, \mathbf{Z}_{r+p-1}\} \cap \{\mathbf{Z}_j, \mathbf{Z}_{j+1}, \ldots, \mathbf{Z}_{j+p-1}\} = \emptyset,$$

where A_{rj} are constant vectors from $\mathcal{V}$ and $\sum\limits_{\sigma}^{\nu} = \mathbf{O}$ for $\sigma > \nu$.

Proof. We will prove the assertion of the theorem by mathematical induction.

For $n = 2$, the equation (2.36) becomes

$$f_1(\mathbf{Z}_1, \mathbf{Z}_2, \cdots, \mathbf{Z}_p) + f_2(\mathbf{Z}_2, \mathbf{Z}_3, \cdots, \mathbf{Z}_{p+1}) = \mathbf{O}. \tag{2.38}$$

Putting $\mathbf{Z}_1 = C_1$ into the equation (2.38), we get

$$f_2(\mathbf{Z}_2, \mathbf{Z}_3, \cdots, \mathbf{Z}_{p+1}) = -F_{12}(\mathbf{Z}_2, \mathbf{Z}_3, \cdots, \mathbf{Z}_p) \tag{2.39}$$

where the following notation is introduced

$$F_{12}(\mathbf{Z}_2, \mathbf{Z}_3, \cdots, \mathbf{Z}_p) = f_1(C_1, \mathbf{Z}_2, \cdots, \mathbf{Z}_p).$$

If we put Eq. (2.39) into Eq. (2.38), we get

$$f_1(\mathbf{Z}_1, \mathbf{Z}_2, \cdots, \mathbf{Z}_p) = F_{12}(\mathbf{Z}_2, \mathbf{Z}_3, \cdots, \mathbf{Z}_p). \tag{2.40}$$

Therefore, for $i = 1, 2$ from Eq. (2.37) we obtain Eqs. (2.40) and (2.39), respectively, which means that the theorem holds for $n = 2$.

For some fixed n let us suppose that the general solution of the functional equation (2.36) is given by Eq. (2.37).

Now, we will consider the functional equation

$$\sum_{i=1}^{n+1} g_i(\mathbf{Z}_i, \mathbf{Z}_{i+1}, \cdots, \mathbf{Z}_{i+p-1}) = \mathbf{O}. \tag{2.41}$$

If we put $\mathbf{Z}_i = C_i$ $(1 \le i \le n)$ into Eq. (2.41), we obtain that the function g_{n+1} has the following form

$$g_{n+1}(\mathbf{Z}_{n+1}, \mathbf{Z}_{n+2}, \cdots, \mathbf{Z}_{n+p}) \tag{2.42}$$
$$= \sum_{j=1}^{n} (-1)^n F_{j,n+1}(\{\mathbf{Z}_{n+1}, \mathbf{Z}_{n+2}, \cdots, \mathbf{Z}_{n+p}\} \cap \{\mathbf{Z}_j, \mathbf{Z}_{j+1}, \cdots, \mathbf{Z}_{j+p-1}\}).$$

Substituting Eq. (2.42) into Eq. (2.41) and introducing the new notations

$$f_i(\mathbf{Z}_i, \mathbf{Z}_{i+1}, \cdots, \mathbf{Z}_{i+p-1}) = g_i(\mathbf{Z}_i, \mathbf{Z}_{i+1}, \cdots, \mathbf{Z}_{i+p-1}) \tag{2.43}$$
$$+ \quad (-1)^n F_{i,n+1}(\{\mathbf{Z}_{n+1}, \mathbf{Z}_{n+2}, \cdots, \mathbf{Z}_{n+p}\} \cap \{\mathbf{Z}_i, \mathbf{Z}_{i+1}, \cdots, \mathbf{Z}_{i+p-1}\})$$
$$(1 \le i \le n),$$

we obtain the equation (2.36). According to the inductive hypothesis, the general solution of this equation is given by the formulae Eq. (2.37). Therefore, from Eqs. (2.37), (2.42) and (2.43) we get

$$g_i(\{\mathbf{Z}_i, \mathbf{Z}_{i+1}, \cdots, \mathbf{Z}_{i+p-1}\})$$

$$= \sum_{j=1}^{i-1} (-1)^{i+1} F_{ji}(\{\mathbf{Z}_i, \mathbf{Z}_{i+1}, \cdots, \mathbf{Z}_{i+p-1}\} \cap \{\mathbf{Z}_j, \mathbf{Z}_{j+1}, \cdots, \mathbf{Z}_{j+p-1}\})$$

$$+ \sum_{j=i+1}^{n} (-1)^j F_{ij}(\{\mathbf{Z}_i, \mathbf{Z}_{i+1}, \cdots, \mathbf{Z}_{i+p-1}\} \cap \{\mathbf{Z}_j, \mathbf{Z}_{j+1}, \cdots, \mathbf{Z}_{j+p-1}\})$$

$$-(-1)^n F_{i,n+1}(\{\mathbf{Z}_{n+1}, \mathbf{Z}_{n+2}, \cdots, \mathbf{Z}_{n+p}\} \cap \{\mathbf{Z}_i, \mathbf{Z}_{i+1}, \cdots, \mathbf{Z}_{i+p-1}\})$$

$$(1 \le i \le n),$$

i.e.,

$$g_i(\mathbf{Z}_i, \mathbf{Z}_{i+1}, \cdots, \mathbf{Z}_{i+p-1})$$

$$= \sum_{j=1}^{i-1} (-1)^{i+1} F_{ji}(\{\mathbf{Z}_i, \mathbf{Z}_{i+1}, \cdots, \mathbf{Z}_{i+p-1}\} \cap \{\mathbf{Z}_j, \mathbf{Z}_{j+1}, \cdots, \mathbf{Z}_{j+p-1}\})$$

$$+ \sum_{j=i+1}^{n+1} (-1)^{j} F_{ij}(\{\mathbf{Z}_i, \mathbf{Z}_{i+1}, \cdots, \mathbf{Z}_{i+p-1}\} \cap \{\mathbf{Z}_j, \mathbf{Z}_{j+1}, \cdots, \mathbf{Z}_{j+p-1}\})$$

$$(1 \le i \le n+1).$$

The theorem holds for $n+1$ if it holds for n. $\square$

This theorem generalizes Theorem 2.4 where it is assumed that $p < n < 2p - 1$. Some other particular cases of the equation (2.36) were considered in [P. M. Vasić (1965); P. M. Vasić (1967)] under the assumption that the functions and the independent variables are real.

Example 2.2   The general solution of the complex vector functional equation

$$f_1(\mathbf{Z}_1, \mathbf{Z}_2) + f_2(\mathbf{Z}_2, \mathbf{Z}_3) + f_3(\mathbf{Z}_3, \mathbf{Z}_4) + f_4(\mathbf{Z}_4, \mathbf{Z}_1) = \mathbf{O},$$

which is a particular case for $n = 4$ and $p = 2$ of the equation (2.36), is given by

$$\begin{aligned}
f_1(\mathbf{Z}_1, \mathbf{Z}_2) &= F_{12}(\mathbf{Z}_2) - A_{13} + F_{14}(\mathbf{Z}_1), \\
f_2(\mathbf{Z}_2, \mathbf{Z}_3) &= -F_{12}(\mathbf{Z}_2) - F_{23}(\mathbf{Z}_3) + A_{24}, \\
f_3(\mathbf{Z}_3, \mathbf{Z}_4) &= A_{13} + F_{23}(\mathbf{Z}_3) + F_{34}(\mathbf{Z}_4), \\
f_4(\mathbf{Z}_4, \mathbf{Z}_1) &= -F_{14}(\mathbf{Z}_1) - A_{24} - F_{34}(\mathbf{Z}_4),
\end{aligned}$$

where F_{ij} $(1 \le i \le 3;\ 2 \le j \le 4)$ are arbitrary complex vector functions from $\mathcal{V}$ and A_{ij} $(i = 1, 2;\ j = 3, 4)$ are arbitrary constant complex vector also from $\mathcal{V}$.

Example 2.3   Now we will consider the functional equation

$$f_1(\mathbf{Z}_1, \mathbf{Z}_2, \mathbf{Z}_3) + f_2(\mathbf{Z}_2, \mathbf{Z}_3, \mathbf{Z}_4) \qquad (2.44)$$
$$+ \quad f_3(\mathbf{Z}_3, \mathbf{Z}_4, \mathbf{Z}_1) + f_4(\mathbf{Z}_4, \mathbf{Z}_1, \mathbf{Z}_2) = \mathbf{O},$$

where $f_i : \mathcal{V}^3 \mapsto \mathcal{V}$ $(1 \leq i \leq 4)$. This equation is a particular case for $p = 3$ and $n = 4$ of the functional equation (2.36).

According to the Theorem 2.6, the general solution of the functional equation (2.44) is

$$
\begin{aligned}
f_1(\mathbf{Z}_1, \mathbf{Z}_2, \mathbf{Z}_3) &= F_{12}(\mathbf{Z}_2, \mathbf{Z}_3) - F_{13}(\mathbf{Z}_1, \mathbf{Z}_3) + F_{14}(\mathbf{Z}_1, \mathbf{Z}_2), \quad (2.45) \\
f_2(\mathbf{Z}_2, \mathbf{Z}_3, \mathbf{Z}_4) &= -F_{12}(\mathbf{Z}_2, \mathbf{Z}_3) - F_{23}(\mathbf{Z}_3, \mathbf{Z}_4) + F_{24}(\mathbf{Z}_2, \mathbf{Z}_4), \quad (2.46) \\
f_3(\mathbf{Z}_3, \mathbf{Z}_4, \mathbf{Z}_1) &= F_{13}(\mathbf{Z}_1, \mathbf{Z}_3) + F_{23}(\mathbf{Z}_3, \mathbf{Z}_4) + F_{34}(\mathbf{Z}_1, \mathbf{Z}_4), \quad (2.47) \\
f_4(\mathbf{Z}_4, \mathbf{Z}_1, \mathbf{Z}_2) &= -F_{14}(\mathbf{Z}_1, \mathbf{Z}_2) - F_{24}(\mathbf{Z}_2, \mathbf{Z}_4) - F_{34}(\mathbf{Z}_1, \mathbf{Z}_4), \quad (2.48)
\end{aligned}
$$

where F_{12}, F_{13}, F_{14}, F_{23} and F_{34} are arbitrary complex vector functions from $\mathcal{V}$.

If we put into Eq. (2.36) $f_i = a_i f$ $(1 \leq i \leq n)$ where a_i are complex constants, then we obtain a special case which will be also solved here in the next theorem.

Theorem 2.7 *The general solution of the basic cyclic complex vector functional equation with complex coefficients*

$$
\sum_{i=1}^{n} a_i f(\mathbf{Z}_i, \mathbf{Z}_{i+1}, \cdots, \mathbf{Z}_{i+p-1}) = \mathbf{O} \qquad (\mathbf{Z}_{n+i} \equiv \mathbf{Z}_i) \tag{2.49}
$$

is given by

$$
f(\mathbf{Z}_1, \mathbf{Z}_2, \cdots, \mathbf{Z}_p) \tag{2.50}
$$

$$
= \sum_{i=1}^{n} \alpha_i \left[\sum_{j=1}^{i-1} (-1)^{i+1} C^{n+1-i} G_{ji}(\{\mathbf{Z}_i, \cdots, \mathbf{Z}_{i+p-1}\} \cap \{\mathbf{Z}_j, \cdots, \mathbf{Z}_{j+p-1}\}) \right.
$$

$$
\left. + \sum_{j=i+1}^{n} (-1)^j C^{n+1-i} G_{ij}(\{\mathbf{Z}_i, \cdots, \mathbf{Z}_{i+p-1}\} \cap \{\mathbf{Z}_j, \cdots, \mathbf{Z}_{j+p-1}\}) \right]
$$

where C is a cyclic operator such that $CG(\mathbf{Z}_1, \cdots, \mathbf{Z}_n) = G(\mathbf{Z}_2, \cdots, \mathbf{Z}_n, \mathbf{Z}_1)$; G_{ij} $(1 \leq i \leq n-1; \ i+1 \leq j \leq n)$ are arbitrary complex vector functions from $\mathcal{V}$ and $\sum_{a}^{s} = \mathbf{O}$ $(a > s)$.

Proof. If we apply the operator C^{n+1-i} to both sides of Eq. (2.37) (with f_i replaced by $a_i f$) and then multiply by still indefinite complex constants

α_i $(1 \le i \le n)$, we have

$$a_i \alpha_i C^{n+1-i} f(\mathbf{Z}_i, \mathbf{Z}_{i+1}, \cdots, \mathbf{Z}_{i+p-1}) \tag{2.51}$$

$$= \alpha_i \left[\sum_{j=1}^{i-1} (-1)^{i+1} C^{n+1-i} F_{ji}(\{\mathbf{Z}_i, \cdots, \mathbf{Z}_{i+p-1}\} \cap \{\mathbf{Z}_j, \cdots, \mathbf{Z}_{j+p-1}\}) \right.$$

$$\left. + \sum_{j=i+1}^{n} (-1)^{j} C^{n+1-i} F_{ij}(\{\mathbf{Z}_i, \cdots, \mathbf{Z}_{i+p-1}\} \cap \{\mathbf{Z}_j, \cdots, \mathbf{Z}_{j+p-1}\}) \right]$$

$$(1 \le i \le n).$$

By summing up the above functions Eq. (2.51), we obtain

$$\left(\sum_{i=1}^{n} a_i \alpha_i \right) f(\mathbf{Z}_1, \mathbf{Z}_2, \cdots, \mathbf{Z}_p) \tag{2.52}$$

$$= \sum_{i=1}^{n} \alpha_i \left[\sum_{j=1}^{i-1} (-1)^{i+1} C^{n+1-i} F_{ji}(\{\mathbf{Z}_i, \cdots, \mathbf{Z}_{i+p-1}\} \cap \{\mathbf{Z}_j, \cdots, \mathbf{Z}_{j+p-1}\}) \right.$$

$$\left. + \sum_{j=i+1}^{n} (-1)^{j} C^{n+1-i} F_{ij}(\{\mathbf{Z}_i, \cdots, \mathbf{Z}_{i+p-1}\} \cap \{\mathbf{Z}_j, \cdots, \mathbf{Z}_{j+p-1}\}) \right].$$

From the equation (2.49) and the equality (2.52) we find

$$\mathbf{O} = \sum_{r=1}^{n} a_r \left(\sum_{i=1}^{n} a_i \alpha_i \right) C^{r-1} f(\mathbf{Z}_1, \mathbf{Z}_2, \cdots, \mathbf{Z}_p)$$

$$= \sum_{r=1}^{n} a_r \left\{ \sum_{i=1}^{n} \alpha_i \left[\sum_{j=1}^{i-1} (-1)^{i+1} C^{n+r-i} F_{ji}(\{\mathbf{Z}_i, \cdots, \mathbf{Z}_{i+p-1}\} \cap \{\mathbf{Z}_j, \cdots, \mathbf{Z}_{j+p-1}\}) \right. \right.$$

$$\left. \left. + \sum_{j=i+1}^{n-1} (-1)^{j} C^{n+r-i} F_{ij}(\{\mathbf{Z}_i, \cdots, \mathbf{Z}_{i+p-1}\} \cap \{\mathbf{Z}_j, \cdots, \mathbf{Z}_{j+p-1}\}) \right] \right\}$$

$$= \sum_{j=1}^{n-1} \sum_{i=j+1}^{n} (-1)^{i+1} \alpha_i \sum_{r=1}^{n} a_r C^{n+r-i} F_{ji}(\{\mathbf{Z}_i, \cdots, \mathbf{Z}_{i+p-1}\} \cap \{\mathbf{Z}_j, \cdots, \mathbf{Z}_{j+p-1}\})$$

$$+ \sum_{i=1}^{n-1} \sum_{j=i+1}^{n} (-1)^j \alpha_i \sum_{r=1}^{n} a_r C^{n+r-i} F_{ij}(\{\mathbf{Z}_i, \cdots, \mathbf{Z}_{i+p-1}\} \cap \{\mathbf{Z}_j, \cdots, \mathbf{Z}_{j+p-1}\})$$

$$= \sum_{i=1}^{n-1} \sum_{j=i+1}^{n} (-1)^j \left\{ \sum_{r=1}^{n} \left[\alpha_i C^{n+r-i} F_{ij}(\{\mathbf{Z}_i, \cdots, \mathbf{Z}_{i+p-1}\} \cap \{\mathbf{Z}_j, \cdots, \mathbf{Z}_{j+p-1}\}) \right. \right.$$

$$\left. \left. -\alpha_j C^{n+r-j} F_{ij}(\{\mathbf{Z}_i, \cdots, \mathbf{Z}_{i+p-1}\} \cap \{\mathbf{Z}_j, \cdots, \mathbf{Z}_{j+p-1}\}) \right] \right\}$$

$$= \sum_{i=1}^{n-1} \sum_{j=i+1}^{n} (-1)^j C^{n+1-i} \times$$

$$\times \left\{ \sum_{r=1}^{n} a_r \left[\alpha_i C^{r-1} F_{ij}(\{\mathbf{Z}_{i,}, \cdots, \mathbf{Z}_{i+p-1}\} \cap \{\mathbf{Z}_j, \cdots, \mathbf{Z}_{j+p-1}\}) \right. \right.$$

$$\left. \left. - \alpha_j C^{r-1} F_{ij}(\{\mathbf{Z}_i, \cdots, \mathbf{Z}_{i+p-1}\} \cap \{\mathbf{Z}_{2i-j}, \mathbf{Z}_{2i-j+1}, \cdots, \mathbf{Z}_{2i-j+p-1}\}) \right] \right\}.$$

Now, we may determine the constants α_i $(1 \le i \le n)$ from the identities

$$\sum_{r=1}^{n} a_r \left[\alpha_i C^{r-1} F_{ij}(\{\mathbf{Z}_i, \cdots, \mathbf{Z}_{i+p-1}\} \cap \{\mathbf{Z}_j, \cdots, \mathbf{Z}_{j+p-1}\}) \right. \tag{2.53}$$

$$\left. -\alpha_j C^{r-1} F_{ij}(\{\mathbf{Z}_i, \cdots, \mathbf{Z}_{i+p-1}\} \cap \{\mathbf{Z}_{2i-j}, \mathbf{Z}_{2i-j+1}, \cdots, \mathbf{Z}_{2i-j+p-1}\}) \right] = O$$

$$(1 \le i \le n-1; \quad i+1 \le j \le n).$$

The equation (2.52) may be rewritten in the following form

$$\left(\sum_{i=1}^{n} a_i \alpha_i \right) T = S, \tag{2.54}$$

and because it has a unique solution T, by introducing the new functions

$$\frac{F_{ij}}{\sum_{i=1}^{n} a_i \alpha_i} = G_{ij} \quad (1 \le i \le n-1; \quad i+1 \le j \le n) \tag{2.55}$$

there follows (2.50). $\qquad\qquad\qquad\qquad\qquad\qquad\qquad$ $\square$

Example 2.4   We consider the functional equation

$$f(\mathbf{Z}_1, \mathbf{Z}_2, \mathbf{Z}_3) - f(\mathbf{Z}_2, \mathbf{Z}_3, \mathbf{Z}_4) + f(\mathbf{Z}_3, \mathbf{Z}_4, \mathbf{Z}_1) - f(\mathbf{Z}_4, \mathbf{Z}_1, \mathbf{Z}_2) = \mathbf{O}. \quad (2.56)$$

According to Eqs. (2.45), (2.46), (2.47) and (2.48) we obtain

$$
\begin{aligned}
f(\mathbf{Z}_1, \mathbf{Z}_2, \mathbf{Z}_3) &= F_{12}(\mathbf{Z}_2, \mathbf{Z}_3) - F_{13}(\mathbf{Z}_1, \mathbf{Z}_3) + F_{14}(\mathbf{Z}_1, \mathbf{Z}_2), \\
-f(\mathbf{Z}_1, \mathbf{Z}_2, \mathbf{Z}_3) &= -F_{12}(\mathbf{Z}_1, \mathbf{Z}_2) - F_{23}(\mathbf{Z}_2, \mathbf{Z}_3) + F_{24}(\mathbf{Z}_1, \mathbf{Z}_3), \\
f(\mathbf{Z}_1, \mathbf{Z}_2, \mathbf{Z}_3) &= F_{13}(\mathbf{Z}_3, \mathbf{Z}_1) + F_{23}(\mathbf{Z}_1, \mathbf{Z}_2) + F_{34}(\mathbf{Z}_3, \mathbf{Z}_2), \\
-f(\mathbf{Z}_1, \mathbf{Z}_2, \mathbf{Z}_3) &= -F_{14}(\mathbf{Z}_2, \mathbf{Z}_3) - F_{24}(\mathbf{Z}_3, \mathbf{Z}_1) - F_{34}(\mathbf{Z}_2, \mathbf{Z}_1).
\end{aligned}
$$

On the basis of the identities Eq. (2.53), we may determine the constants α_i $(i = 1, 2, 3, 4)$ as follows

$$
\begin{aligned}
&\alpha_1 F_{12}(\mathbf{Z}_2, \mathbf{Z}_3) \;-\; \alpha_2 F_{12}(\mathbf{Z}_1, \mathbf{Z}_2) - \alpha_1 F_{12}(\mathbf{Z}_3, \mathbf{Z}_4) + \alpha_2 F_{12}(\mathbf{Z}_2, \mathbf{Z}_3) \\
&+\alpha_1 F_{12}(\mathbf{Z}_4, \mathbf{Z}_1) \;-\; \alpha_2 F_{12}(\mathbf{Z}_3, \mathbf{Z}_4) - \alpha_1 F_{12}(\mathbf{Z}_1, \mathbf{Z}_2) + \alpha_2 F_{12}(\mathbf{Z}_3, \mathbf{Z}_1) = \mathbf{O},
\end{aligned}
$$

$$
\begin{aligned}
&-\alpha_1 F_{13}(\mathbf{Z}_1, \mathbf{Z}_3) \;+\; \alpha_3 F_{13}(\mathbf{Z}_3, \mathbf{Z}_1) + \alpha_1 F_{13}(\mathbf{Z}_2, \mathbf{Z}_4) - \alpha_3 F_{13}(\mathbf{Z}_4, \mathbf{Z}_2) \\
&-\alpha_1 F_{13}(\mathbf{Z}_3, \mathbf{Z}_1) \;+\; \alpha_3 F_{13}(\mathbf{Z}_1, \mathbf{Z}_3) + \alpha_1 F_{13}(\mathbf{Z}_4, \mathbf{Z}_2) - \alpha_3 F_{13}(\mathbf{Z}_2, \mathbf{Z}_4) = \mathbf{O},
\end{aligned}
$$

$$
\begin{aligned}
&\alpha_1 F_{14}(\mathbf{Z}_1, \mathbf{Z}_2) \;-\; \alpha_4 F_{14}(\mathbf{Z}_2, \mathbf{Z}_3) - \alpha_1 F_{14}(\mathbf{Z}_2, \mathbf{Z}_3) + \alpha_4 F_{14}(\mathbf{Z}_3, \mathbf{Z}_4) \\
&+\alpha_1 F_{14}(\mathbf{Z}_3, \mathbf{Z}_4) \;-\; \alpha_4 F_{14}(\mathbf{Z}_4, \mathbf{Z}_1) - \alpha_1 F_{14}(\mathbf{Z}_4, \mathbf{Z}_1) + \alpha_4 F_{14}(\mathbf{Z}_1, \mathbf{Z}_2) = \mathbf{O},
\end{aligned}
$$

$$
\begin{aligned}
&-\alpha_2 F_{23}(\mathbf{Z}_2, \mathbf{Z}_3) \;+\; \alpha_3 F_{23}(\mathbf{Z}_1, \mathbf{Z}_2) + \alpha_2 F_{23}(\mathbf{Z}_3, \mathbf{Z}_4) - \alpha_3 F_{23}(\mathbf{Z}_2, \mathbf{Z}_3) \\
&-\alpha_2 F_{23}(\mathbf{Z}_4, \mathbf{Z}_1) \;+\; \alpha_3 F_{23}(\mathbf{Z}_3, \mathbf{Z}_4) + \alpha_2 F_{23}(\mathbf{Z}_1, \mathbf{Z}_2) - \alpha_3 F_{23}(\mathbf{Z}_4, \mathbf{Z}_1) = \mathbf{O},
\end{aligned}
$$

$$
\begin{aligned}
&\alpha_2 F_{24}(\mathbf{Z}_1, \mathbf{Z}_3) \;-\; \alpha_4 F_{24}(\mathbf{Z}_3, \mathbf{Z}_1) - \alpha_2 F_{24}(\mathbf{Z}_2, \mathbf{Z}_4) + \alpha_4 F_{24}(\mathbf{Z}_4, \mathbf{Z}_2) \\
&+\alpha_2 F_{24}(\mathbf{Z}_3, \mathbf{Z}_1) \;-\; \alpha_4 F_{24}(\mathbf{Z}_1, \mathbf{Z}_3) - \alpha_2 F_{24}(\mathbf{Z}_4, \mathbf{Z}_2) + \alpha_4 F_{24}(\mathbf{Z}_1, \mathbf{Z}_4) = \mathbf{O},
\end{aligned}
$$

$$
\begin{aligned}
&\alpha_3 F_{34}(\mathbf{Z}_3, \mathbf{Z}_2) \;-\; \alpha_4 F_{34}(\mathbf{Z}_2, \mathbf{Z}_1) - \alpha_3 F_{34}(\mathbf{Z}_4, \mathbf{Z}_3) + \alpha_4 F_{34}(\mathbf{Z}_3, \mathbf{Z}_2) \\
&+\alpha_3 F_{34}(\mathbf{Z}_1, \mathbf{Z}_4) \;-\; \alpha_4 F_{34}(\mathbf{Z}_4, \mathbf{Z}_3) - \alpha_3 F_{34}(\mathbf{Z}_2, \mathbf{Z}_1) + \alpha_4 F_{34}(\mathbf{Z}_1, \mathbf{Z}_4) = \mathbf{O},
\end{aligned}
$$

i.e., we obtain

$$\alpha_1 = -\alpha_2, \quad \alpha_1 = \alpha_3, \quad \alpha_1 = -\alpha_4, \quad \alpha_2 = -\alpha_3, \quad \alpha_2 = \alpha_4, \quad \alpha_3 = -\alpha_4,$$

which means that $\alpha_1 = -\alpha_2 = \alpha_3 = -\alpha_4 = 1$.

The general solution of the functional equation (2.56) is

$$f(\mathbf{Z}_1, \mathbf{Z}_2, \mathbf{Z}_3) = F(\mathbf{Z}_1, \mathbf{Z}_2) + F(\mathbf{Z}_2, \mathbf{Z}_3) + G(\mathbf{Z}_1, \mathbf{Z}_3) - G(\mathbf{Z}_3, \mathbf{Z}_1)$$

where

$$
\begin{aligned}
F(\mathbf{Z}_1, \mathbf{Z}_2) &= F_{14}(\mathbf{Z}_1, \mathbf{Z}_2) + F_{12}(\mathbf{Z}_1, \mathbf{Z}_2) + F_{23}(\mathbf{Z}_1, \mathbf{Z}_2) + F_{34}(\mathbf{Z}_2, \mathbf{Z}_1), \\
G(\mathbf{Z}_1, \mathbf{Z}_3) &= -F_{13}(\mathbf{Z}_1, \mathbf{Z}_3) - F_{24}(\mathbf{Z}_1, \mathbf{Z}_3).
\end{aligned}
$$

3 Paracyclic Functional Equation

Let $\mathcal{V}$ be a complex vector space with complex dimension n, and let the complex vectors $\mathbf{X}_i$, $\mathbf{Y}_j \in \mathcal{V}$ ($1 \leq i, j \leq n$) be given as in the previous section. Throughout the section C, D, C_i and D_i are constant vectors in $\mathcal{V}$.

Also, let there exist mappings $f_i : \mathcal{V}^{p+q} \mapsto \mathcal{V}$ ($1 \leq i \leq k$).

Now we will consider the following paracyclic complex vector functional equation of the first kind

$$\sum_{i=1}^{k} f_i(\mathbf{X}_i, \mathbf{X}_{i+1}, \ldots, \mathbf{X}_{i+p-1}, \mathbf{Y}_i, \mathbf{Y}_{i+1}, \ldots, \mathbf{Y}_{i+q-1}) = \mathbf{O} \qquad (3.1)$$

$$(k \leq n; \qquad \mathbf{X}_{n+i} \equiv \mathbf{X}_i, \quad \mathbf{Y}_{n+i} \equiv \mathbf{Y}_i).$$

In order to determine the general solution of the functional equation (3.1), we must distinguish the following six cases:

1^0 $q < 2q - 1 \leq p = n,$ $\qquad\qquad$ 2^0 $q < p = n < 2q - 1,$

3^0 $q < p < n < 2q - 1 < 2p - 1,$

4^0 $q < p < 2q - 1 \leq n \leq p + q - 1 < 2p - 1,$

5^0 $q < p < p + q - 1 < n < 2p - 1,$ $\qquad$ 6^0 $q < p < 2q - 1 < 2p - 1 \leq n.$

Theorem 3.1 *If $q < 2q - 1 \leq p = n$, then the general solution of the* *functional equation (3.1) is given by*

$$f_r(\mathbf{X}_1, \mathbf{X}_2, \ldots, \mathbf{X}_n, \mathbf{Y}_1, \mathbf{Y}_2, \ldots, \mathbf{Y}_q) \tag{3.2}$$

$$= \sum_{i=1}^{\min(k-r,\,q-1)} (-1)^{i-1} F_{ri}(\mathbf{X}_{i+1}, \mathbf{X}_{i+2}, \ldots, \mathbf{X}_i, \mathbf{Y}_{i+1}, \mathbf{Y}_{i+2}, \ldots, \mathbf{Y}_q)$$

$$+ \sum_{i=n-r+1}^{q-1} (-1)^{i-1} F_{ri}(\mathbf{X}_{i+1}, \mathbf{X}_{i+2}, \ldots, \mathbf{X}_i, \mathbf{Y}_{i+1}, \mathbf{Y}_{i+2}, \ldots, \mathbf{Y}_q)$$

$$+ \sum_{i=q}^{\min(k-r,\,n-q)} (-1)^{i-1} F_{ri}(\mathbf{X}_{i+1}, \mathbf{X}_{i+2}, \ldots, \mathbf{X}_i)$$

$$+ \sum_{i=\max(n-r+1,\,q)}^{n-q} (-1)^{n-i} F_{i+r,n-i}(\mathbf{X}_1, \mathbf{X}_2, \ldots, \mathbf{X}_n)$$

$$+ \sum_{i=n-q+1}^{k-r} (-1)^{n-i} F_{i+r,n-i}(\mathbf{X}_1, \mathbf{X}_2, \ldots, \mathbf{X}_n, \mathbf{Y}_1, \mathbf{Y}_2, \ldots, \mathbf{Y}_{q+i})$$

$$+ \sum_{i=\max(n-r+1,\,n-q+1)}^{n-1} (-1)^{n-i} F_{i+r,n-i}(\mathbf{X}_1, \mathbf{X}_2, \ldots, \mathbf{X}_n, \mathbf{Y}_1, \mathbf{Y}_2, \ldots, \mathbf{Y}_{q+i})$$

$$(1 \leq r \leq k),$$

where F_{ij} are arbitrary complex vector functions from $\mathcal{V}$.

Proof.   The proof of this theorem is completely analogous to the proof of Theorem 2.3 given in the previous section. $\square$

Theorem 3.2   *If $q < p = n < 2q - 1$, then the general solution of the functional equation (3.1) will be*

$$f_r(\mathbf{X}_1, \mathbf{X}_2, \ldots, \mathbf{X}_n, \mathbf{Y}_1, \mathbf{Y}_2, \ldots, \mathbf{Y}_q) \tag{3.3}$$

$$= \sum_{i=1}^{\min(k-r,\,n-q)} (-1)^{i-1} F_{ri}(\mathbf{X}_{i+1}, \mathbf{X}_{i+2}, \ldots, \mathbf{X}_i, \mathbf{Y}_{i+1}, \mathbf{Y}_{i+2}, \ldots, \mathbf{Y}_q)$$

$$+ \sum_{i=n-r+1}^{n-q} (-1)^{i-1} F_{ri}(\mathbf{X}_{i+1}, \mathbf{X}_{i+2}, \ldots, \mathbf{X}_i, \mathbf{Y}_{i+1}, \mathbf{Y}_{i+2}, \ldots, \mathbf{Y}_q)$$

$$+ \sum_{i=n-q+1}^{\min(k-r,q-1)} (-1)^{i-1} F_{ri}(\mathbf{X}_{i+1}, \mathbf{X}_{i+2}, \ldots, \mathbf{X}_i,$$

$$\mathbf{Y}_{i+1}, \mathbf{Y}_{i+2}, \ldots, \mathbf{Y}_q, \mathbf{Y}_1, \mathbf{Y}_2, \ldots, \mathbf{Y}_{q+i})$$

$$+ \sum_{i=\max(n-r+1,n-q+1)}^{q-1} (-1)^{n-i} F_{i+r,n-i}(\mathbf{X}_1, \mathbf{X}_2, \ldots, \mathbf{X}_n,$$

$$\mathbf{Y}_1, \mathbf{Y}_2, \ldots, \mathbf{Y}_{q+i}, \mathbf{Y}_{i+1}, \mathbf{Y}_{i+2}, \ldots, \mathbf{Y}_q)$$

$$+ \sum_{i=q}^{k-r} (-1)^{n-i} F_{i+r,n-i}(\mathbf{X}_1, \mathbf{X}_2, \ldots, \mathbf{X}_n, \mathbf{Y}_1, \mathbf{Y}_2, \ldots, \mathbf{Y}_{q+i})$$

$$+ \sum_{i=\max(n-r+1,q)}^{n-1} (-1)^{n-i} F_{i+r,n-i}(\mathbf{X}_1, \mathbf{X}_2, \ldots, \mathbf{X}_n, \mathbf{Y}_1, \mathbf{Y}_2, \ldots, \mathbf{Y}_{q+i})$$

$$(1 \leq r \leq k),$$

where F_{ij} are arbitrary complex vector functions from $\mathcal{V}$.

Proof. The proof of this theorem is analogous to that of the previous Theorem 3.1. $\square$

Theorem 3.3 *If $q < p < n < 2q - 1 < 2p - 1$, then the general solution of the functional equation (3.1) is*

$$f_r(\mathbf{X}_1, \mathbf{X}_2, \ldots, \mathbf{X}_p, \mathbf{Y}_1, \mathbf{Y}_2, \ldots, \mathbf{Y}_q) \qquad (3.4)$$

$$= \sum_{i=1}^{\min(n-p,k-r)} (-1)^{i-1} F_{ri}(\mathbf{X}_{i+1}, \mathbf{X}_{i+2}, \ldots, \mathbf{X}_p, \mathbf{Y}_{i+1}, \mathbf{Y}_{i+2}, \ldots, \mathbf{Y}_q)$$

$$+ \sum_{i=n-r+1}^{n-p} (-1)^{i-1} F_{ri}(\mathbf{X}_{i+1}, \mathbf{X}_{i+2}, \ldots, \mathbf{X}_p, \mathbf{Y}_{i+1}, \mathbf{Y}_{i+2}, \ldots, \mathbf{Y}_q)$$

$$+ \sum_{i=n-p+1}^{\min(k-r,n-q)} (-1)^{i-1} F_{ri}(\mathbf{X}_{i+1}, \mathbf{X}_{i+2}, \ldots, \mathbf{X}_p,$$

$$\mathbf{X}_1, \mathbf{X}_2, \ldots, \mathbf{X}_{i+p}, \mathbf{Y}_{i+1}, \mathbf{Y}_{i+2}, \ldots, \mathbf{Y}_q)$$

$$+ \sum_{i=\max\,(n-p+1,n-r+1)}^{n-q} (-1)^{i-1} F_{ri}(\mathbf{X}_{i+1}, \mathbf{X}_{i+2}, \ldots, \mathbf{X}_p,$$

$$\mathbf{X}_1, \mathbf{X}_2, \ldots, \mathbf{X}_{i+p}, \mathbf{Y}_{i+1}, \mathbf{Y}_{i+2}, \ldots, \mathbf{Y}_q)$$

$$+ \sum_{i=n-q+1}^{\min\,(k-r,\,q-1)} (-1)^{i-1} F_{ri}(\mathbf{X}_{i+1}, \mathbf{X}_{i+2}, \ldots, \mathbf{X}_p,$$

$$\mathbf{X}_1, \mathbf{X}_2, \ldots, \mathbf{X}_{i+p}, \mathbf{Y}_{i+1}, \mathbf{Y}_{i+2}, \ldots, \mathbf{Y}_q, \mathbf{Y}_1, \mathbf{Y}_2, \ldots, \mathbf{Y}_{i+q})$$

$$+ \sum_{i=\max\,(n-q+1,n-r+1)}^{q-1} (-1)^{n-i} F_{i+r,n-i}(\mathbf{X}_1, \mathbf{X}_2, \ldots, \mathbf{X}_{i+p},$$

$$\mathbf{X}_{i+1}, \mathbf{X}_{i+2}, \ldots, \mathbf{X}_p, \mathbf{Y}_1, \mathbf{Y}_2, \ldots, \mathbf{Y}_{i+q}, \mathbf{Y}_{i+1}, \mathbf{Y}_{i+2}, \ldots, \mathbf{Y}_q)$$

$$+ \sum_{i=q}^{\min\,(k-r,p-1)} (-1)^{n-i} F_{i+r,n-i}(\mathbf{X}_1, \mathbf{X}_2, \ldots, \mathbf{X}_{i+p},$$

$$\mathbf{X}_{i+1}, \mathbf{X}_{i+2}, \ldots, \mathbf{X}_p, \mathbf{Y}_1, \mathbf{Y}_2, \ldots, \mathbf{Y}_{i+q})$$

$$+ \sum_{i=\max\,(q,n-r+1)}^{p-1} (-1)^{n-i} F_{i+r,n-i}(\mathbf{X}_1, \mathbf{X}_2, \ldots, \mathbf{X}_{i+p},$$

$$\mathbf{X}_{i+1}, \mathbf{X}_{i+2}, \ldots, \mathbf{X}_p, \mathbf{Y}_1, \mathbf{Y}_2, \ldots, \mathbf{Y}_{i+q})$$

$$+ \sum_{i=p}^{k-r} (-1)^{n-i} F_{i+r,n-i}(\mathbf{X}_1, \mathbf{X}_2, \ldots, \mathbf{X}_{i+p}, \mathbf{Y}_1, \mathbf{Y}_2, \ldots, \mathbf{Y}_{i+q})$$

$$+ \sum_{i=\max\,(p,n-r+1)}^{n-1} (-1)^{n-i} F_{i+r,n-i}(\mathbf{X}_1, \mathbf{X}_2, \ldots, \mathbf{X}_{i+p}, \mathbf{Y}_1, \mathbf{Y}_2, \ldots, \mathbf{Y}_{i+q})$$

$$(1 \leq r \leq k),$$

where F_{ij} are arbitrary complex vector functions from $\mathcal{V}$.

Theorem 3.4 *If $q < p < 2q - 1 \le n \le p + q - 1 < 2p - 1$, then the general solution of the functional equation (3.1) is given by*

$$f_r(\mathbf{X}_1, \mathbf{X}_2, \ldots, \mathbf{X}_p, \mathbf{Y}_1, \mathbf{Y}_2, \ldots, \mathbf{Y}_q) \tag{3.5}$$

$$= \sum_{i=1}^{\min(n-p,k-r)} (-1)^{i-1} F_{ri}(\mathbf{X}_{i+1}, \mathbf{X}_{i+2}, \ldots, \mathbf{X}_p, \mathbf{Y}_{i+1}, \mathbf{Y}_{i+2}, \ldots, \mathbf{Y}_q)$$

$$+ \sum_{i=n-r+1}^{n-p} (-1)^{i-1} F_{ri}(\mathbf{X}_{i+1}, \mathbf{X}_{i+2}, \ldots, \mathbf{X}_p, \mathbf{Y}_{i+1}, \mathbf{Y}_{i+2}, \ldots, \mathbf{Y}_q)$$

$$+ \sum_{i=n-p+1}^{\min(k-r,q-1)} (-1)^{i-1} F_{ri}(\mathbf{X}_{i+1}, \mathbf{X}_{i+2}, \ldots, \mathbf{X}_p,$$

$$\mathbf{X}_1, \mathbf{X}_2, \ldots, \mathbf{X}_{i+p}, \mathbf{Y}_{i+1}, \mathbf{Y}_{i+2}, \ldots, \mathbf{Y}_q)$$

$$+ \sum_{i=\max(n-p+1,n-r+1)}^{q-1} (-1)^{i-1} F_{ri}(\mathbf{X}_{i+1}, \mathbf{X}_{i+2}, \ldots, \mathbf{X}_p,$$

$$\mathbf{X}_1, \mathbf{X}_2, \ldots, \mathbf{X}_{i+p}, \mathbf{Y}_{i+1}, \mathbf{Y}_{i+2}, \ldots, \mathbf{Y}_q)$$

$$+ \sum_{i=q}^{\min(k-r,n-q)} (-1)^{i-1} F_{ri}(\mathbf{X}_{i+1}, \mathbf{X}_{i+2}, \ldots, \mathbf{X}_p, \mathbf{X}_1, \mathbf{X}_2, \ldots, \mathbf{X}_{i+p})$$

$$+ \sum_{i=\max(n-i+1,q)}^{n-q} (-1)^{n-i} F_{i+r,n-i}(\mathbf{X}_1, \mathbf{X}_2, \ldots, \mathbf{X}_{i+p}, \mathbf{X}_{i+1}, \mathbf{X}_{i+2}, \ldots, \mathbf{X}_p)$$

$$+ \sum_{i=n-q+1}^{\min(k-r,p-1)} (-1)^{n-i} F_{i+r,n-i}(\mathbf{X}_1, \mathbf{X}_2, \ldots, \mathbf{X}_{i+p},$$

$$\mathbf{X}_{i+1}, \mathbf{X}_{i+2}, \ldots, \mathbf{X}_p, \mathbf{Y}_1, \mathbf{Y}_2, \ldots, \mathbf{Y}_{i+q})$$

$$+ \sum_{i=\max(n-q+1,n-r+1)}^{p-1} (-1)^{n-i} F_{i+r,n-i}(\mathbf{X}_1,\mathbf{X}_2,\ldots,\mathbf{X}_{i+p},$$

$$\mathbf{X}_{i+1},\mathbf{X}_{i+2},\ldots,\mathbf{X}_p,\mathbf{Y}_1,\mathbf{Y}_2,\ldots,\mathbf{Y}_{i+q})$$

$$+ \sum_{i=p}^{k-r} (-1)^{n-i} F_{i+r,n-i}(\mathbf{X}_1,\mathbf{X}_2,\ldots,\mathbf{X}_{i+p},\mathbf{Y}_1,\mathbf{Y}_2,\ldots,\mathbf{Y}_{i+q})$$

$$+ \sum_{i=\max(p,n-r+1)}^{n-1} (-1)^{n-i} F_{i+r,n-i}(\mathbf{X}_1,\mathbf{X}_2,\ldots,\mathbf{X}_{i+p},\mathbf{Y}_1,\mathbf{Y}_2,\ldots,\mathbf{Y}_{i+q})$$

$$(1 \le r \le k),$$

where F_{ij} are arbitrary complex vector functions from $\mathcal{V}$.

We can prove the previous two theorems in the same way as the following theorem.

Theorem 3.5   *If $q < p < p+q-1 < n < 2p-1$, then the general solution of the functional equation (3.1) is given by the formulae*

$$f_r(\mathbf{X}_1,\mathbf{X}_2,\ldots,\mathbf{X}_p,\mathbf{Y}_1,\mathbf{Y}_2,\ldots,\mathbf{Y}_q) \tag{3.6}$$

$$= \sum_{i=1}^{\min(q-1,k-r)} (-1)^{i-1} F_{ri}(\mathbf{X}_{i+1},\mathbf{X}_{i+2},\ldots,\mathbf{X}_p,\mathbf{Y}_{i+1},\mathbf{Y}_{i+2},\ldots,\mathbf{Y}_q)$$

$$+ \sum_{i=n-r+1}^{q-1} (-1)^{i-1} F_{ri}(\mathbf{X}_{i+1},\mathbf{X}_{i+2},\ldots,\mathbf{X}_p,\mathbf{Y}_{i+1},\mathbf{Y}_{i+2},\ldots,\mathbf{Y}_q)$$

$$+ \sum_{i=q}^{\min(k-r,n-p)} (-1)^{i-1} F_{ri}(\mathbf{X}_{i+1},\mathbf{X}_{i+2},\ldots,\mathbf{X}_p)$$

$$+ \sum_{i=\max(q,n-r+1)}^{n-p} (-1)^{i-1} F_{ri}(\mathbf{X}_{i+1},\mathbf{X}_{i+2},\ldots,\mathbf{X}_p)$$

$$+ \sum_{i=n-p+1}^{\min(k-r,p-1)} (-1)^{i-1} F_{ri}(\mathbf{X}_{i+1},\mathbf{X}_{i+2},\ldots,\mathbf{X}_p,\mathbf{X}_1,\mathbf{X}_2,\ldots,\mathbf{X}_{i+p})$$

$$+ \sum_{i=\max\,(n-p+1,n-r+1)}^{p-1} (-1)^{n-i} F_{i+r,n-i}(\mathbf{X}_1, \mathbf{X}_2, \ldots, \mathbf{X}_{i+p},$$

$$\mathbf{X}_{i+1}, \mathbf{X}_{i+2}, \ldots, \mathbf{X}_p)$$

$$+ \sum_{i=p}^{\min\,(k-r,n-q)} (-1)^{n-i} F_{i+r,n-i}(\mathbf{X}_1, \mathbf{X}_2, \ldots, \mathbf{X}_{i+p})$$

$$+ \sum_{i=\max\,(p,n-r+1)}^{n-q} (-1)^{n-i} F_{i+r,n-i}(\mathbf{X}_1, \mathbf{X}_2, \ldots, \mathbf{X}_{i+p})$$

$$+ \sum_{i=n-q+1}^{k-r} (-1)^{n-i} F_{i+r,n-i}(\mathbf{X}_1, \mathbf{X}_2, \ldots, \mathbf{X}_{i+p}, \mathbf{Y}_1, \mathbf{Y}_2, \ldots, \mathbf{Y}_{i+q})$$

$$+ \sum_{i=\max(n-q+1,n-r+1)}^{n-1} (-1)^{n-i} F_{i+r,n-i}(\mathbf{X}_1, \mathbf{X}_2, \ldots, \mathbf{X}_{i+p}, \mathbf{Y}_1, \mathbf{Y}_2, \ldots, \mathbf{Y}_{i+q})$$

$$(1 \le r \le k),$$

where F_{ij} are arbitrary complex vector functions from $\mathcal{V}$.

Proof. The proof of this theorem is based on mathematical induction.

For $k = 2$ the functional equation (3.1) has the form

$$f_1(\mathbf{X}_1, \mathbf{X}_2, \ldots, \mathbf{X}_p, \mathbf{Y}_1, \mathbf{Y}_2, \ldots, \mathbf{Y}_q) \tag{3.7}$$
$$+ \;\; f_2(\mathbf{X}_2, \mathbf{X}_3, \ldots, \mathbf{X}_{p+1}, \mathbf{Y}_2, \mathbf{Y}_3, \ldots, \mathbf{Y}_{q+1}) = \mathbf{O}.$$

Putting $\mathbf{X}_{p+1} = C$, $\mathbf{Y}_{q+1} = D$ into the equation (3.7), we obtain

$$f_1(\mathbf{X}_1, \mathbf{X}_2, \ldots, \mathbf{X}_p, \mathbf{Y}_1, \mathbf{Y}_2, \ldots, \mathbf{Y}_q) \tag{3.8}$$
$$= \;\; F_{11}(\mathbf{X}_2, \mathbf{X}_3, \ldots, \mathbf{X}_p, \mathbf{Y}_2, \mathbf{Y}_3, \ldots, \mathbf{Y}_q).$$

If we put Eq. (3.8) into Eq. (3.7), we have

$$f_2(\mathbf{X}_2, \mathbf{X}_3, \ldots, \mathbf{X}_{p+1}, \mathbf{Y}_2, \mathbf{Y}_3, \ldots, \mathbf{Y}_{q+1})$$
$$= \;\; - \;\; F_{11}(\mathbf{X}_2, \mathbf{X}_3, \ldots, \mathbf{X}_p, \mathbf{Y}_2, \mathbf{Y}_3, \ldots, \mathbf{Y}_q),$$

i.e.,

$$f_2(\mathbf{X}_1, \mathbf{X}_2, \ldots, \mathbf{X}_p, \mathbf{Y}_1, \mathbf{Y}_2, \ldots, \mathbf{Y}_q) \tag{3.9}$$
$$= - F_{11}(\mathbf{X}_1, \mathbf{X}_2, \ldots, \mathbf{X}_{p-1}, \mathbf{Y}_1, \mathbf{Y}_2, \ldots, \mathbf{Y}_{q-1}).$$

For $k = 2$ and $r = 1$ and $r = 2$, from Eq. (3.6) we deduce Eqs. (3.8) and (3.9), which means that the theorem holds for $k = 2$.

Now we will suppose that the general solution of the functional equation

$$\sum_{i=1}^{k} g_i(\mathbf{X}_i, \mathbf{X}_{i+1}, \ldots, \mathbf{X}_{i+p-1}, \mathbf{Y}_i, \mathbf{Y}_{i+1}, \ldots, \mathbf{Y}_{i+q-1}) = \mathbf{O} \tag{3.10}$$

is given by (3.6) with f_r replaced by g_r.

Let us consider the following functional equation

$$\sum_{i=1}^{k+1} f_i(\mathbf{X}_i, \mathbf{X}_{i+1}, \ldots, \mathbf{X}_{i+p-1}, \mathbf{Y}_i, \mathbf{Y}_{i+1}, \ldots, \mathbf{Y}_{i+q-1}) = \mathbf{O}. \tag{3.11}$$

We will distinguish the following five cases:

1° Let $1 < k < q$. The substitutions

$$\begin{aligned}
\mathbf{X}_i &= C_i \quad \text{for} \quad i \not\equiv k+1, k+2, \ldots, k+p \quad (\text{mod } n), \\
\mathbf{Y}_i &= D_i \quad \text{for} \quad i \not\equiv k+1, k+2, \ldots, k+q \quad (\text{mod } n),
\end{aligned} \tag{3.12}$$

transform the equation (3.11) into

$$f_{k+1}(\mathbf{X}_{k+1}, \mathbf{X}_{k+2}, \ldots, \mathbf{X}_{p+k}, \mathbf{Y}_{k+1}, \mathbf{Y}_{k+2}, \ldots, \mathbf{Y}_{q+k}) \tag{3.13}$$
$$= \sum_{i=n-k}^{n-1} (-1)^{n-i} F_{i+k+1, n-i}(\mathbf{X}_{k+1}, \mathbf{X}_{k+2}, \ldots, \mathbf{X}_{i+p+k},$$
$$\mathbf{Y}_{k+1}, \mathbf{Y}_{k+2}, \ldots, \mathbf{Y}_{i+q+k}).$$

Putting Eq. (3.13) into Eq. (3.11) and introducing new functions by

$$g_i = f_i + (-1)^{k+1-i} F_{i,k+1-i} \qquad (1 \leq i \leq k), \tag{3.14}$$

we obtain the equation (3.10). According to Eqs. (3.6) and (3.14), the general solution of the equation (3.11) is

$$f_r(\mathbf{X}_1, \mathbf{X}_2, \ldots, \mathbf{X}_p, \mathbf{Y}_1, \mathbf{Y}_2, \ldots, \mathbf{Y}_q) \qquad (3.15)$$

$$= \sum_{i=1}^{k-r}(-1)^{i-1}F_{ri}(\mathbf{X}_{i+1}, \mathbf{X}_{i+2}, \ldots, \mathbf{X}_p, \mathbf{Y}_{i+1}, \mathbf{Y}_{i+2}, \ldots, \mathbf{Y}_q)$$

$$+ \sum_{i=n-r+1}^{q-1}(-1)^{i-1}F_{ri}(\mathbf{X}_{i+1}, \mathbf{X}_{i+2}, \ldots, \mathbf{X}_p, \mathbf{Y}_{i+1}, \mathbf{Y}_{i+2}, \ldots, \mathbf{Y}_q)$$

$$+ \sum_{i=\max(q,n-r+1)}^{n-p}(-1)^{i-1}F_{ri}(\mathbf{X}_{i+1}, \mathbf{X}_{i+2}, \ldots, \mathbf{X}_p)$$

$$+ \sum_{i=\max(n-p+1,n-r+1)}^{p-1}(-1)^{n-i}F_{i+r,n-i}(\mathbf{X}_1, \mathbf{X}_2, \ldots, \mathbf{X}_{i+p}, \mathbf{X}_{i+1}, \mathbf{X}_{i+2}, \ldots, \mathbf{X}_p)$$

$$+ \sum_{i=\max(p,n-r+1)}^{n-q}(-1)^{n-i}F_{i+r,n-i}(\mathbf{X}_1, \mathbf{X}_2, \ldots, \mathbf{X}_{i+p})$$

$$+ \sum_{i=\max(n-q+1,n-r+1)}^{n-1}(-1)^{n-i}F_{i+r,n-i}(\mathbf{X}_1, \mathbf{X}_2, \ldots, \mathbf{X}_{i+p}, \mathbf{Y}_1, \mathbf{Y}_2, \ldots, \mathbf{Y}_{i+q})$$

$$+(-1)^{k-r}F_{r,k+1-r}(\mathbf{X}_{k-r+2}, \mathbf{X}_{k-r+3}, \ldots, \mathbf{X}_p, \mathbf{Y}_{k-r+2}, \mathbf{Y}_{k-r+3}, \ldots, \mathbf{Y}_q)$$

$$= \sum_{i=1}^{k+1-r}(-1)^{i-1}F_{ri}(\mathbf{X}_{i+1}, \mathbf{X}_{i+2}, \ldots, \mathbf{X}_p, \mathbf{Y}_{i+1}, \mathbf{Y}_{i+2}, \ldots, \mathbf{Y}_q)$$

$$+ \sum_{i=n-r+1}^{q-1}(-1)^{i-1}F_{ri}(\mathbf{X}_{i+1}, \mathbf{X}_{i+2}, \ldots, \mathbf{X}_p, \mathbf{Y}_{i+1}, \mathbf{Y}_{i+2}, \ldots, \mathbf{Y}_q)$$

$$+ \sum_{i=\max(q,n-r+1)}^{n-p}(-1)^{i-1}F_{ri}(\mathbf{X}_{i+1}, \mathbf{X}_{i+2}, \ldots, \mathbf{X}_p)$$

$$+ \sum_{i=\max(n-p+1,n-r+1)}^{p-1}(-1)^{n-i}F_{i+r,n-i}(\mathbf{X}_1, \mathbf{X}_2, \ldots, \mathbf{X}_{i+p}, \mathbf{X}_{i+1}, \mathbf{X}_{i+2}, \ldots, \mathbf{X}_p)$$

$$+ \sum_{i=\max(p,n-r+1)}^{n-q} (-1)^{n-i} F_{i+r,n-i}(\mathbf{X}_1, \mathbf{X}_2, \ldots, \mathbf{X}_{i+p})$$

$$+ \sum_{i=\max(n-q+1,n-r+1)}^{n-1} (-1)^{n-i} F_{i+r,n-i}(\mathbf{X}_1, \mathbf{X}_2, \ldots, \mathbf{X}_{i+p}, \mathbf{Y}_1, \mathbf{Y}_2, \ldots, \mathbf{Y}_{i+q})$$

$$(1 \leq r \leq k).$$

On the basis of the expressions Eq. (3.13) and (3.15), if $1 < k < q$, the theorem holds for $k + 1$. Thus it holds for all such k, and also for $k = q$.

2° Let $q \leq k < n - p + 1$. For the values Eq. (3.12) the functional equation (3.11) becomes

$$f_{k+1}(\mathbf{X}_{k+1}, \mathbf{X}_{k+2}, \ldots, \mathbf{X}_{p+k}, \mathbf{Y}_{k+1}, \mathbf{Y}_{k+2}, \ldots, \mathbf{Y}_{q+k}) \quad (3.16)$$

$$= \sum_{i=n-k}^{n-q} (-1)^{n-i} F_{i+k+1,n-i}(\mathbf{X}_{k+1}, \mathbf{X}_{k+2}, \ldots, \mathbf{X}_{i+p+k})$$

$$+ \sum_{i=n-q+1}^{n-1} (-1)^{n-i} F_{i+k+1,n-i}(\mathbf{X}_{k+1}, \mathbf{X}_{k+2}, \ldots, \mathbf{X}_{i+p+k},$$

$$\mathbf{Y}_{k+1}, \mathbf{Y}_{k+2}, \ldots, \mathbf{Y}_{i+q+k}).$$

Now we will introduce the notations Eq. (3.14). On the basis of the expressions Eqs. (3.16) and (3.14), the equation (3.11) becomes Eq. (3.10). By using the inductive hypothesis and by virtue of the expression Eq. (3.14), we can conclude that the general solution of the equation (3.11) is given by the following equalities:

$$f_r(\mathbf{X}_1, \mathbf{X}_2, \ldots, \mathbf{X}_p, \mathbf{Y}_1, \mathbf{Y}_2, \ldots, \mathbf{Y}_q)$$

$$= \sum_{i=1}^{q-1} (-1)^{i-1} F_{ri}(\mathbf{X}_{i+1}, \mathbf{X}_{i+2}, \ldots, \mathbf{X}_p, \mathbf{Y}_{i+1}, \mathbf{Y}_{i+2}, \ldots, \mathbf{Y}_q)$$

$$+ \sum_{i=q}^{k-r} (-1)^{i-1} F_{ri}(\mathbf{X}_{i+1}, \mathbf{X}_{i+2}, \ldots, \mathbf{X}_p)$$

$$+ (-1)^{k-r} F_{r,k+1-r}(\mathbf{X}_{k-r+2}, \mathbf{X}_{k-r+3}, \ldots, \mathbf{X}_p)$$

$$+ \sum_{i=n-r+1}^{q-1} (-1)^{i-1} F_{ri}(\mathbf{X}_{i+1}, \mathbf{X}_{i+2}, \ldots, \mathbf{X}_p, \mathbf{Y}_{i+1}, \mathbf{Y}_{i+2}, \ldots, \mathbf{Y}_q)$$

$$+ \sum_{i=\max(q,n-r+1)}^{n-p} (-1)^{i-1} F_{ri}(\mathbf{X}_{i+1}, \mathbf{X}_{i+2}, \ldots, \mathbf{X}_p)$$

$$+ \sum_{i=\max(n-p+1,n-r+1)}^{p-1} (-1)^{n-i} F_{i+r,n-i}(\mathbf{X}_1, \mathbf{X}_2, \ldots, \mathbf{X}_{i+p}, \mathbf{X}_{i+1}, \mathbf{X}_{i+2}, \ldots, \mathbf{X}_p)$$

$$+ \sum_{i=\max(p,n-r+1)}^{n-q} (-1)^{n-i} F_{i+r,n-i}(\mathbf{X}_1, \mathbf{X}_2, \ldots, \mathbf{X}_{i+p})$$

$$+ \sum_{i=\max(n-q+1,n-r+1)}^{n-1} (-1)^{n-i} F_{i+r,n-i}(\mathbf{X}_1, \mathbf{X}_2, \ldots, \mathbf{X}_{i+p}, \mathbf{Y}_1, \mathbf{Y}_2, \ldots, \mathbf{Y}_{i+q})$$

$$(1 \le r \le k - q + 1);$$

$$f_r(\mathbf{X}_1, \mathbf{X}_2, \ldots, \mathbf{X}_p, \mathbf{Y}_1, \mathbf{Y}_2, \ldots, \mathbf{Y}_q)$$

$$= \sum_{i=1}^{k-r} (-1)^{i-1} F_{ri}(\mathbf{X}_{i+1}, \mathbf{X}_{i+2}, \ldots, \mathbf{X}_p, \mathbf{Y}_{i+1}, \mathbf{Y}_{i+2}, \ldots, \mathbf{Y}_q)$$

$$+ \sum_{i=n-r+1}^{q-1} (-1)^{i-1} F_{ri}(\mathbf{X}_{i+1}, \mathbf{X}_{i+2}, \ldots, \mathbf{X}_p, \mathbf{Y}_{i+1}, \mathbf{Y}_{i+2}, \ldots, \mathbf{Y}_q)$$

$$+ \sum_{i=\max(q,n-r+1)}^{n-p} (-1)^{i-1} F_{ri}(\mathbf{X}_{i+1}, \mathbf{X}_{i+2}, \ldots, \mathbf{X}_p)$$

$$+ \sum_{i=\max(n-p+1,n-r+1)}^{p-1} (-1)^{n-i} F_{i+r,n-i}(\mathbf{X}_1, \mathbf{X}_2, \ldots, \mathbf{X}_{i+p}, \mathbf{X}_{i+1}, \mathbf{X}_{i+2}, \ldots, \mathbf{X}_p)$$

$$+ \sum_{i=\max(p,n-r+1)}^{n-q} (-1)^{n-i} F_{i+r,n-i}(\mathbf{X}_1, \mathbf{X}_2, \ldots, \mathbf{X}_{i+p})$$

$$+ \sum_{i=\max(n-q+1,n-r+1)}^{n-1} (-1)^{n-i} F_{i+r,n-i}(\mathbf{X}_1, \mathbf{X}_2, \ldots, \mathbf{X}_{i+p}, \mathbf{Y}_1, \mathbf{Y}_2, \ldots, \mathbf{Y}_{i+q})$$

$$+ (-1)^{k-r} F_{r,k+1-r}(\mathbf{X}_{k-r+2}, \mathbf{X}_{k-r+3}, \ldots, \mathbf{X}_p, \mathbf{Y}_{k-r+2}, \mathbf{Y}_{k-r+3}, \ldots, \mathbf{Y}_q)$$

$$(r = k - q + 2, \ldots, k),$$

or in a general form,

$$f_r(\mathbf{X}_1, \mathbf{X}_2, \ldots, \mathbf{X}_p, \mathbf{Y}_1, \mathbf{Y}_2, \ldots, \mathbf{Y}_q) \tag{3.17}$$

$$= \sum_{i=1}^{\min(q-1,k+1-r)} (-1)^{i-1} F_{ri}(\mathbf{X}_{i+1}, \mathbf{X}_{i+2}, \ldots, \mathbf{X}_p, \mathbf{Y}_{i+1}, \mathbf{Y}_{i+2}, \ldots, \mathbf{Y}_q)$$

$$+ \sum_{i=q}^{k+1-r} (-1)^{i-1} F_{ri}(\mathbf{X}_{i+1}, \mathbf{X}_{i+2}, \ldots, \mathbf{X}_p)$$

$$+ \sum_{i=n-r+1}^{q-1} (-1)^{i-1} F_{ri}(\mathbf{X}_{i+1}, \mathbf{X}_{i+2}, \ldots, \mathbf{X}_p, \mathbf{Y}_{i+1}, \mathbf{Y}_{i+2}, \ldots, \mathbf{Y}_q)$$

$$+ \sum_{i=\max(q,n-r+1)}^{n-p} (-1)^{i-1} F_{ri}(\mathbf{X}_{i+1}, \mathbf{X}_{i+2}, \ldots, \mathbf{X}_p)$$

$$+ \sum_{i=\max(n-p+1,n-r+1)}^{p-1} (-1)^{n-i} F_{i+r,n-i}(\mathbf{X}_1, \mathbf{X}_2, \ldots, \mathbf{X}_{i+p}, \mathbf{X}_{i+1}, \mathbf{X}_{i+2}, \ldots, \mathbf{X}_p)$$

$$+ \sum_{i=\max(p,n-r+1)}^{n-q} (-1)^{n-i} F_{i+r,n-i}(\mathbf{X}_1, \mathbf{X}_2, \ldots, \mathbf{X}_{i+p})$$

$$+ \sum_{i=\max(n-q+1,n-r+1)}^{n-1} (-1)^{n-i} F_{i+r,n-i}(\mathbf{X}_1, \mathbf{X}_2, \ldots, \mathbf{X}_{i+p}, \mathbf{Y}_1, \mathbf{Y}_2, \ldots, \mathbf{Y}_{i+q})$$

$$(1 \le r \le k).$$

Therefore, on the basis of the expressions (3.16) and (3.17), we can conclude that the theorem holds in this case too, and also for $k = n - p + 1$.

3° Let $n - p + 1 \le k < p$. If we substitute Eq. (3.12) into Eq. (3.11), we find

$$f_{k+1}(\mathbf{X}_{k+1}, \mathbf{X}_{k+2}, \ldots, \mathbf{X}_{p+k}, \mathbf{Y}_{k+1}, \mathbf{Y}_{k+2}, \ldots, \mathbf{Y}_{q+k}) \tag{3.18}$$

$$= \sum_{i=n-k}^{p-1} (-1)^{n-i} F_{i+k+1,n-i}(\mathbf{X}_{k+1}, \mathbf{X}_{k+2}, \ldots, \mathbf{X}_{i+p+k}, \mathbf{X}_{k+1+i}, \mathbf{X}_{k+2+i}, \ldots, \mathbf{X}_{k+p})$$

$$+ \sum_{i=p}^{n-q} (-1)^{n-i} F_{i+k+1,n-i}(\mathbf{X}_{k+1}, \mathbf{X}_{k+2}, \ldots, \mathbf{X}_{i+p+k})$$

$$+ \sum_{i=n-q+1}^{n-1} (-1)^{n-i} F_{i+k+1,n-i}(\mathbf{X}_{k+1}, \mathbf{X}_{k+2}, \ldots, \mathbf{X}_{i+p+k},$$

$$\mathbf{Y}_{k+1}, \mathbf{Y}_{k+2}, \ldots, \mathbf{Y}_{i+q+k}).$$

If we substitute Eq. (3.18) into Eq. (3.11) and if we take into account the transformation Eq. (3.14), then we obtain the equation (3.10). On the basis of the expression Eq. (3.14) and the inductive hypothesis, we find that the general solution of the functional equation (3.11) is given by the following formulae

$$f_r(\mathbf{X}_1, \mathbf{X}_2, \ldots, \mathbf{X}_p, \mathbf{Y}_1, \mathbf{Y}_2, \ldots, \mathbf{Y}_q)$$

$$= \sum_{i=1}^{q-1} (-1)^{i-1} F_{ri}(\mathbf{X}_{i+1}, \mathbf{X}_{i+2}, \ldots, \mathbf{X}_p, \mathbf{Y}_{i+1}, \mathbf{Y}_{i+2}, \ldots, \mathbf{Y}_q)$$

$$+ \sum_{i=q}^{n-p} (-1)^{i-1} F_{ri}(\mathbf{X}_{i+1}, \mathbf{X}_{i+2}, \ldots, \mathbf{X}_p)$$

$$+ \sum_{i=n-p+1}^{k-r} (-1)^{i-1} F_{ri}(\mathbf{X}_{i+1}, \mathbf{X}_{i+2}, \ldots, \mathbf{X}_p, \mathbf{X}_1, \mathbf{X}_2, \ldots, \mathbf{X}_{i+p})$$

$$+ (-1)^{k-r} F_{r,k+1-r}(\mathbf{X}_{k+2-r}, \mathbf{X}_{k+3-r}, \ldots, \mathbf{X}_p, \mathbf{X}_1, \mathbf{X}_2, \ldots, \mathbf{X}_{k+1-r+p})$$

$$+ \sum_{i=n-r+1}^{q-1} (-1)^{i-1} F_{ri}(\mathbf{X}_{i+1}, \mathbf{X}_{i+2}, \ldots, \mathbf{X}_p, \mathbf{Y}_{i+1}, \mathbf{Y}_{i+2}, \ldots, \mathbf{Y}_q)$$

$$+ \sum_{i=\max(q,n-r+1)}^{n-p} (-1)^{i-1} F_{ri}(\mathbf{X}_{i+1}, \mathbf{X}_{i+2}, \ldots, \mathbf{X}_p)$$

$$+ \sum_{i=\max(n-p+1,n-r+1)}^{p-1} (-1)^{n-i} F_{i+r,n-i}(\mathbf{X}_1, \mathbf{X}_2, \ldots, \mathbf{X}_{i+p}, \mathbf{X}_{i+1}, \mathbf{X}_{i+2}, \ldots, \mathbf{X}_p)$$

$$+ \sum_{i=\max(p,n-r+1)}^{n-q} (-1)^{n-i} F_{i+r,n-i}(\mathbf{X}_1, \mathbf{X}_2, \ldots, \mathbf{X}_{i+p})$$

$$+ \sum_{i=\max(n-q+1,n-r+1)}^{n-1} (-1)^{n-i} F_{i+r,n-i}(\mathbf{X}_1, \mathbf{X}_2, \ldots, \mathbf{X}_{i+p}, \mathbf{Y}_1, \mathbf{Y}_2, \ldots, \mathbf{Y}_{i+q})$$

$$(1 \le r \le k+p-n);$$

$$f_r(\mathbf{X}_1, \mathbf{X}_2, \ldots, \mathbf{X}_p, \mathbf{Y}_1, \mathbf{Y}_2, \ldots, \mathbf{Y}_q)$$

$$= \sum_{i=1}^{q-1} (-1)^{i-1} F_{ri}(\mathbf{X}_{i+1}, \mathbf{X}_{i+2}, \ldots, \mathbf{X}_p, \mathbf{Y}_{i+1}, \mathbf{Y}_{i+2}, \ldots, \mathbf{Y}_q)$$

$$+ \sum_{i=q}^{k-r} (-1)^{i-1} F_{ri}(\mathbf{X}_{i+1}, \mathbf{X}_{i+2}, \ldots, \mathbf{X}_p)$$

$$+ \sum_{i=n-r+1}^{q-1} (-1)^{i-1} F_{ri}(\mathbf{X}_{i+1}, \mathbf{X}_{i+2}, \ldots, \mathbf{X}_p, \mathbf{Y}_{i+1}, \mathbf{Y}_{i+2}, \ldots, \mathbf{Y}_q)$$

$$+ \sum_{i=\max(q,n-r+1)}^{n-p} (-1)^{i-1} F_{ri}(\mathbf{X}_{i+1}, \mathbf{X}_{i+2}, \ldots, \mathbf{X}_p)$$

$$+ \sum_{i=\max(n-p+1,n-r+1)}^{p-1} (-1)^{n-i} F_{i+r,n-i}(\mathbf{X}_1, \mathbf{X}_2, \ldots, \mathbf{X}_{i+p}, \mathbf{X}_{i+1}, \mathbf{X}_{i+2}, \ldots, \mathbf{X}_p)$$

$$+ \sum_{i=\max(p,n-r+1)}^{n-q} (-1)^{n-i} F_{i+r,n-i}(\mathbf{X}_1, \mathbf{X}_2, \ldots, \mathbf{X}_{i+p})$$

$$+ \sum_{i=\max(n-q+1,n-r+1)}^{n-1} (-1)^{n-i} F_{i+r,n-i}(\mathbf{X}_1, \mathbf{X}_2, \ldots, \mathbf{X}_{i+p}, \mathbf{Y}_1, \mathbf{Y}_2, \ldots, \mathbf{Y}_{i+q})$$

$$+(-1)^{k-r} F_{r,k+1-r}(\mathbf{X}_{k+2-r}, \mathbf{X}_{k+3-r}, \ldots, \mathbf{X}_p)$$

$$(r = k+p-n+1, \ldots, k-q+1);$$

$$f_r(\mathbf{X}_1, \mathbf{X}_2, \ldots, \mathbf{X}_p, \mathbf{Y}_1, \mathbf{Y}_2, \ldots, \mathbf{Y}_q)$$

$$= \sum_{i=1}^{k-r} (-1)^{i-1} F_{ri}(\mathbf{X}_{i+1}, \mathbf{X}_{i+2}, \ldots, \mathbf{X}_p, \mathbf{Y}_{i+1}, \mathbf{Y}_{i+2}, \ldots, \mathbf{Y}_q)$$

$$+ \sum_{i=n-r+1}^{q-1} (-1)^{i-1} F_{ri}(\mathbf{X}_{i+1}, \mathbf{X}_{i+2}, \ldots, \mathbf{X}_p, \mathbf{Y}_{i+1}, \mathbf{Y}_{i+2}, \ldots, \mathbf{Y}_q)$$

$$+ \sum_{i=\max(q,n-r+1)}^{n-p} (-1)^{i-1} F_{ri}(\mathbf{X}_{i+1}, \mathbf{X}_{i+2}, \ldots, \mathbf{X}_p)$$

$$+ \sum_{i=\max(n-p+1,n-r+1)}^{p-1} (-1)^{n-i} F_{i+r,n-i}(\mathbf{X}_1, \mathbf{X}_2, \ldots, \mathbf{X}_{i+p},$$
$$\mathbf{X}_{i+1}, \mathbf{X}_{i+2}, \ldots, \mathbf{X}_p)$$

$$+ \sum_{i=\max(p,n-r+1)}^{n-q} (-1)^{n-i} F_{i+r,n-i}(\mathbf{X}_1, \mathbf{X}_2, \ldots, \mathbf{X}_{i+p})$$

$$+ \sum_{i=\max(n-q+1,n-r+1)}^{n-1} (-1)^{n-i} F_{i+r,n-i}(\mathbf{X}_1, \mathbf{X}_2, \ldots, \mathbf{X}_{i+p}, \mathbf{Y}_1, \mathbf{Y}_2, \ldots, \mathbf{Y}_{i+q})$$

$$+ (-1)^{k-r} F_{r,k+1-r}(\mathbf{X}_{k+2-r}, \mathbf{X}_{k+3-r}, \ldots, \mathbf{X}_p, \mathbf{Y}_{k+2-r}, \mathbf{Y}_{k+3-r}, \ldots, \mathbf{Y}_q)$$

$$(r = k - q + 2, \ldots, k).$$

We can write the above equalities in a general form

$$f_r(\mathbf{X}_1, \mathbf{X}_2, \ldots, \mathbf{X}_p, \mathbf{Y}_1, \mathbf{Y}_2, \ldots, \mathbf{Y}_q) \tag{3.19}$$

$$= \sum_{i=1}^{\min(q-1,k+1-r)} (-1)^{i-1} F_{ri}(\mathbf{X}_{i+1}, \mathbf{X}_{i+2}, \ldots, \mathbf{X}_p, \mathbf{Y}_{i+1}, \mathbf{Y}_{i+2}, \ldots, \mathbf{Y}_q)$$

$$+ \sum_{i=q}^{\min(n-p,k+1-r)} (-1)^{i-1} F_{ri}(\mathbf{X}_{i+1}, \mathbf{X}_{i+2}, \ldots, \mathbf{X}_p)$$

$$+ \sum_{i=n-p+1}^{k+1-r} (-1)^{i-1} F_{ri}(\mathbf{X}_{i+1}, \mathbf{X}_{i+2}, \ldots, \mathbf{X}_p, \mathbf{X}_1, \mathbf{X}_2, \ldots, \mathbf{X}_{i+p})$$

$$+ \sum_{i=n-r+1}^{q-1} (-1)^{i-1} F_{ri}(\mathbf{X}_{i+1}, \mathbf{X}_{i+2}, \ldots, \mathbf{X}_p, \mathbf{Y}_{i+1}, \mathbf{Y}_{i+2}, \ldots, \mathbf{Y}_q)$$

$$+ \sum_{i=\max(q,n-r+1)}^{n-p} (-1)^{i-1} F_{ri}(\mathbf{X}_{i+1}, \mathbf{X}_{i+2}, \ldots, \mathbf{X}_p)$$

$$+ \sum_{i=\max(n-p+1,n-r+1)}^{p-1} (-1)^{n-i} F_{i+r,n-i}(\mathbf{X}_1, \mathbf{X}_2, \ldots, \mathbf{X}_{i+p},$$

$$\mathbf{X}_{i+1}, \mathbf{X}_{i+2}, \ldots, \mathbf{X}_p)$$

$$+ \sum_{i=\max(p,n-r+1)}^{n-q} (-1)^{n-i} F_{i+r,n-i}(\mathbf{X}_1, \mathbf{X}_2, \ldots, \mathbf{X}_{i+p})$$

$$+ \sum_{i=\max(n-q+1,n-r+1)}^{n-1} (-1)^{n-i} F_{i+r,n-i}(\mathbf{X}_1, \mathbf{X}_2, \ldots, \mathbf{X}_{i+p}, \mathbf{Y}_1, \mathbf{Y}_2, \ldots, \mathbf{Y}_{i+q})$$

$$(1 \le r \le k).$$

On the basis of the equalities (3.18) and (3.19), the theorem holds for this case as well, and also for $k = p$.

4° Let $p \le k < n - q + 1$. Putting Eq. (3.12) into Eq. (3.11), we obtain

$$f_{k+1}(\mathbf{X}_{k+1}, \mathbf{X}_{k+2}, \ldots, \mathbf{X}_{p+k}, \mathbf{Y}_{k+1}, \mathbf{Y}_{k+2}, \ldots, \mathbf{Y}_{q+k}) \quad (3.20)$$

$$= \sum_{i=n-k}^{n-p} (-1)^{i-1} F_{k+1,i}(\mathbf{X}_{i+k+1}, \mathbf{X}_{i+k+2}, \ldots, \mathbf{X}_{p+k})$$

$$+ \sum_{i=n-p+1}^{p-1} (-1)^{n-i} F_{i+k+1,n-i}(\mathbf{X}_{k+1}, \mathbf{X}_{k+2}, \ldots, \mathbf{X}_{i+k+p},$$

$$\mathbf{X}_{i+k+1}, \mathbf{X}_{i+k+2}, \ldots, \mathbf{X}_{k+p})$$

$$+ \sum_{i=p}^{n-q} (-1)^{n-i} F_{i+k+1,n-i}(\mathbf{X}_{k+1}, \mathbf{X}_{k+2}, \ldots, \mathbf{X}_{i+p+k})$$

$$+ \sum_{i=n-q+1}^{n-1} (-1)^{n-i} F_{i+k+1,n-i}(\mathbf{X}_{k+1}, \mathbf{X}_{k+2}, \ldots, \mathbf{X}_{i+p+k},$$

$$\mathbf{Y}_{k+1}, \mathbf{Y}_{k+2}, \ldots, \mathbf{Y}_{i+q+k}).$$

If we substitute f_{k+1} given by Eq. (3.20) into Eq. (3.11), by the substitutions

$$g_i = \begin{cases} f_i + (-1)^{n-k+i} F_{k+1,n-k+i-1} & (1 \le i \le k-p+1), \\ f_i + (-1)^{k+1-i} F_{i,k+1-i} & (i = k-p+2, \ldots, k), \end{cases} \qquad (3.21)$$

we obtain the equation (3.10). According to (3.21), the general solution of the functional equation (3.11) is given by

$$f_r(\mathbf{X}_1, \mathbf{X}_2, \ldots, \mathbf{X}_p, \mathbf{Y}_1, \mathbf{Y}_2, \ldots, \mathbf{Y}_q)$$

$$= \sum_{i=1}^{q-1} (-1)^{i-1} F_{ri}(\mathbf{X}_{i+1}, \mathbf{X}_{i+2}, \ldots, \mathbf{X}_p, \mathbf{Y}_{i+1}, \mathbf{Y}_{i+2}, \ldots, \mathbf{Y}_q)$$

$$+ \sum_{i=q}^{n-p} (-1)^{i-1} F_{ri}(\mathbf{X}_{i+1}, \mathbf{X}_{i+2}, \ldots, \mathbf{X}_p)$$

$$+ \sum_{i=n-p+1}^{p-1} (-1)^{i-1} F_{ri}(\mathbf{X}_{i+1}, \mathbf{X}_{i+2}, \ldots, \mathbf{X}_p, \mathbf{X}_1, \mathbf{X}_2, \ldots, \mathbf{X}_{i+p})$$

$$+ \sum_{i=p}^{k-r} (-1)^{n-i} F_{i+r,n-i}(\mathbf{X}_1, \mathbf{X}_2, \ldots, \mathbf{X}_{i+p})$$

$$+ (-1)^{n-k+r-1} F_{k+1,n-k-1+r}(\mathbf{X}_1, \mathbf{X}_2, \ldots, \mathbf{X}_{k+1-r+p})$$

$$+ \sum_{i=n-r+1}^{q-1} (-1)^{i-1} F_{ri}(\mathbf{X}_{i+1}, \mathbf{X}_{i+2}, \ldots, \mathbf{X}_p, \mathbf{Y}_{i+1}, \mathbf{Y}_{i+2}, \ldots, \mathbf{Y}_q)$$

$$+ \sum_{i=\max(q,n-r+1)}^{n-p} (-1)^{i-1} F_{ri}(\mathbf{X}_{i+1}, \mathbf{X}_{i+2}, \ldots, \mathbf{X}_p)$$

$$+ \sum_{i=\max(n-p+1,n-r+1)}^{p-1} (-1)^{n-i} F_{i+r,n-i}(\mathbf{X}_1, \mathbf{X}_2, \ldots, \mathbf{X}_{i+p},$$

$$\mathbf{X}_{i+1}, \mathbf{X}_{i+2}, \ldots, \mathbf{X}_p)$$

$$+ \sum_{i=\max(p,n-r+1)}^{n-q} (-1)^{n-i} F_{i+r,n-i}(\mathbf{X}_1, \mathbf{X}_2, \ldots, \mathbf{X}_{i+p})$$

$$+ \sum_{i=\max(n-q+1,n-r+1)}^{n-1} (-1)^{n-i} F_{i+r,n-i}(\mathbf{X}_1, \mathbf{X}_2, \ldots, \mathbf{X}_{i+p}, \mathbf{Y}_1, \mathbf{Y}_2, \ldots, \mathbf{Y}_{i+q})$$

$$(1 \le r \le k - p + 1);$$

$$f_r(\mathbf{X}_1, \mathbf{X}_2, \ldots, \mathbf{X}_p, \mathbf{Y}_1, \mathbf{Y}_2, \ldots, \mathbf{Y}_q)$$

$$= \sum_{i=1}^{q-1} (-1)^{i-1} F_{ri}(\mathbf{X}_{i+1}, \mathbf{X}_{i+2}, \ldots, \mathbf{X}_p, \mathbf{Y}_{i+1}, \mathbf{Y}_{i+2}, \ldots, \mathbf{Y}_q)$$

$$+ \sum_{i=q}^{n-p} (-1)^{i-1} F_{ri}(\mathbf{X}_{i+1}, \mathbf{X}_{i+2}, \ldots, \mathbf{X}_p)$$

$$+ \sum_{i=n-p+1}^{k-r} (-1)^{i-1} F_{ri}(\mathbf{X}_{i+1}, \mathbf{X}_{i+2}, \ldots, \mathbf{X}_p, \mathbf{X}_1, \mathbf{X}_2, \ldots, \mathbf{X}_{i+p})$$

$$+ (-1)^{k-r} F_{r,k+1-r}(\mathbf{X}_{k+2-r}, \mathbf{X}_{k+3-r}, \ldots, \mathbf{X}_p, \mathbf{X}_1, \mathbf{X}_2, \ldots, \mathbf{X}_{k+1-r+p})$$

$$+ \sum_{i=n-r+1}^{q-1} (-1)^{i-1} F_{ri}(\mathbf{X}_{i+1}, \mathbf{X}_{i+2}, \ldots, \mathbf{X}_p, \mathbf{Y}_{i+1}, \mathbf{Y}_{i+2}, \ldots, \mathbf{Y}_q)$$

$$+ \sum_{i=\max(q,n-r+1)}^{n-p} (-1)^{i-1} F_{ri}(\mathbf{X}_{i+1}, \mathbf{X}_{i+2}, \ldots, \mathbf{X}_p)$$

$$+ \sum_{i=\max(n-p+1,n-r+1)}^{p-1} (-1)^{n-i} F_{i+r,n-i}(\mathbf{X}_1, \mathbf{X}_2, \ldots, \mathbf{X}_{i+p}, \mathbf{X}_{i+1}, \mathbf{X}_{i+2}, \ldots, \mathbf{X}_p)$$

$$+ \sum_{i=\max(p,n-r+1)}^{n-q} (-1)^{n-i} F_{i+r,n-i}(\mathbf{X}_1, \mathbf{X}_2, \ldots, \mathbf{X}_{i+p})$$

$$+ \sum_{i=\max(n-q+1,n-r+1)}^{n-1} (-1)^{n-i} F_{i+r,n-i}(\mathbf{X}_1, \mathbf{X}_2, \ldots, \mathbf{X}_{i+p}, \mathbf{Y}_1, \mathbf{Y}_2, \ldots, \mathbf{Y}_{i+q})$$

$$(r = k - p + 2, \ldots, k + p - n);$$

$$f_r(\mathbf{X}_1, \mathbf{X}_2, \ldots, \mathbf{X}_p, \mathbf{Y}_1, \mathbf{Y}_2, \ldots, \mathbf{Y}_q)$$

$$= \sum_{i=1}^{q-1} (-1)^{i-1} F_{ri}(\mathbf{X}_{i+1}, \mathbf{X}_{i+2}, \ldots, \mathbf{X}_p, \mathbf{Y}_{i+1}, \mathbf{Y}_{i+2}, \ldots, \mathbf{Y}_q)$$

$$+ \sum_{i=q}^{k-r} (-1)^{i-1} F_{ri}(\mathbf{X}_{i+1}, \mathbf{X}_{i+2}, \ldots, \mathbf{X}_p)$$

$$+ (-1)^{k-r} F_{r,k+1-r}(\mathbf{X}_{k+2-r}, \mathbf{X}_{k+3-r}, \ldots, \mathbf{X}_p)$$

$$+ \sum_{i=n-r+1}^{q-1} (-1)^{i-1} F_{ri}(\mathbf{X}_{i+1}, \mathbf{X}_{i+2}, \ldots, \mathbf{X}_p, \mathbf{Y}_{i+1}, \mathbf{Y}_{i+2}, \ldots, \mathbf{Y}_q)$$

$$+ \sum_{i=\max(q,n-r+1)}^{n-p} (-1)^{i-1} F_{ri}(\mathbf{X}_{i+1}, \mathbf{X}_{i+2}, \ldots, \mathbf{X}_p)$$

$$+ \sum_{i=\max(n-p+1,n-r+1)}^{p-1} (-1)^{n-i} F_{i+r,n-i}(\mathbf{X}_1, \mathbf{X}_2, \ldots, \mathbf{X}_{i+p}, \mathbf{X}_{i+1}, \mathbf{X}_{i+2}, \ldots, \mathbf{X}_p)$$

$$+ \sum_{i=\max(p,n-r+1)}^{n-q} (-1)^{n-i} F_{i+r,n-i}(\mathbf{X}_1, \mathbf{X}_2, \ldots, \mathbf{X}_{i+p})$$

$$+ \sum_{i=\max(n-q+1,n-r+1)}^{n-1} (-1)^{n-i} F_{i+r,n-i}(\mathbf{X}_1, \mathbf{X}_2, \ldots, \mathbf{X}_{i+p}, \mathbf{Y}_1, \mathbf{Y}_2, \ldots, \mathbf{Y}_{i+q})$$

$$(r = k + p + 1 - n, \ldots, k - q + 1);$$

$$f_r(\mathbf{X}_1, \mathbf{X}_2, \ldots, \mathbf{X}_p, \mathbf{Y}_1, \mathbf{Y}_2, \ldots, \mathbf{Y}_q)$$

$$= \sum_{i=1}^{k-r} (-1)^{i-1} F_{ri}(\mathbf{X}_{i+1}, \mathbf{X}_{i+2}, \ldots, \mathbf{X}_p, \mathbf{Y}_{i+1}, \mathbf{Y}_{i+2}, \ldots, \mathbf{Y}_q)$$

$$+(-1)^{k-r+1} F_{r,k-r+1}(\mathbf{X}_{k-r+2}, \mathbf{X}_{k-r+3}, \ldots, \mathbf{X}_p, \mathbf{Y}_{k-r+2}, \mathbf{Y}_{k-r+3}, \ldots, \mathbf{Y}_q)$$

$$+ \sum_{i=n-r+1}^{q-1} (-1)^{i-1} F_{ri}(\mathbf{X}_{i+1}, \mathbf{X}_{i+2}, \ldots, \mathbf{X}_p, \mathbf{Y}_{i+1}, \mathbf{Y}_{i+2}, \ldots, \mathbf{Y}_q)$$

$$+ \sum_{i=\max(q,n-r+1)}^{n-p} (-1)^{i-1} F_{ri}(\mathbf{X}_{i+1}, \mathbf{X}_{i+2}, \ldots, \mathbf{X}_p)$$

$$+ \sum_{i=\max(n-p+1,n-r+1)}^{p-1} (-1)^{n-i} F_{i+r,n-i}(\mathbf{X}_1, \mathbf{X}_2, \ldots, \mathbf{X}_{i+p}, \mathbf{X}_{i+1}, \mathbf{X}_{i+2}, \ldots, \mathbf{X}_p)$$

$$+ \sum_{i=\max(p,n-r+1)}^{n-q} (-1)^{n-i} F_{i+r,n-i}(\mathbf{X}_1, \mathbf{X}_2, \ldots, \mathbf{X}_{i+p})$$

$$+ \sum_{i=\max(n-q+1,n-r+1)}^{n-1} (-1)^{n-i} F_{i+r,n-i}(\mathbf{X}_1, \mathbf{X}_2, \ldots, \mathbf{X}_{i+p}, \mathbf{Y}_1, \mathbf{Y}_2, \ldots, \mathbf{Y}_{i+q})$$

$$(r = k - q + 2, \ldots, k).$$

From the above equalities we obtain

$$f_r(\mathbf{X}_1, \mathbf{X}_2, \ldots, \mathbf{X}_p, \mathbf{Y}_1, \mathbf{Y}_2, \ldots, \mathbf{Y}_q) \tag{3.22}$$

$$= \sum_{i=1}^{\min(q-1,k+1-r)} (-1)^{i-1} F_{ri}(\mathbf{X}_{i+1}, \mathbf{X}_{i+2}, \ldots, \mathbf{X}_p, \mathbf{Y}_{i+1}, \mathbf{Y}_{i+2}, \ldots, \mathbf{Y}_q)$$

$$+ \sum_{i=q}^{\min(n-p,k+1-r)} (-1)^{i-1} F_{ri}(\mathbf{X}_{i+1}, \mathbf{X}_{i+2}, \ldots, \mathbf{X}_p)$$

$$+ \sum_{i=n-p+1}^{\min(p-1,k+1-r)} (-1)^{i-1} F_{ri}(\mathbf{X}_{i+1}, \mathbf{X}_{i+2}, \ldots, \mathbf{X}_p, \mathbf{X}_1, \mathbf{X}_2, \ldots, \mathbf{X}_{i+p})$$

$$+ \sum_{i=p}^{k+1-r} (-1)^{n-i} F_{i+r,n-i}(\mathbf{X}_1, \mathbf{X}_2, \ldots, \mathbf{X}_{i+p})$$

$$+ \sum_{i=n-r+1}^{q-1} (-1)^{i-1} F_{ri}(\mathbf{X}_{i+1}, \mathbf{X}_{i+2}, \ldots, \mathbf{X}_p, \mathbf{Y}_{i+1}, \mathbf{Y}_{i+2}, \ldots, \mathbf{Y}_q)$$

$$+ \sum_{i=\max(q,n-r+1)}^{n-p} (-1)^{i-1} F_{ri}(\mathbf{X}_{i+1}, \mathbf{X}_{i+2}, \ldots, \mathbf{X}_p)$$

$$+ \sum_{i=\max(n-p+1,n-r+1)}^{p-1} (-1)^{n-i} F_{i+r,n-i}(\mathbf{X}_1, \mathbf{X}_2, \ldots, \mathbf{X}_{i+p}, \mathbf{X}_{i+1}, \mathbf{X}_{i+2}, \ldots, \mathbf{X}_p)$$

$$+ \sum_{i=\max(p,n-r+1)}^{n-q} (-1)^{n-i} F_{i+r,n-i}(\mathbf{X}_1, \mathbf{X}_2, \ldots, \mathbf{X}_{i+p})$$

$$+ \sum_{i=\max(n-q+1,n-r+1)}^{n-1} (-1)^{n-i} F_{i+r,n-i}(\mathbf{X}_1, \mathbf{X}_2, \ldots, \mathbf{X}_{i+p}, \mathbf{Y}_1, \mathbf{Y}_2, \ldots, \mathbf{Y}_{i+q})$$

$$(1 \le r \le k).$$

Therefore, the theorem is proved for $p \le k \le n - q + 1$.

5° Let $n - q + 1 \le k < n$. If we put Eq. (3.12) into Eq. (3.11), we get

$$f_{k+1}(\mathbf{X}_{k+1}, \mathbf{X}_{k+2}, \ldots, \mathbf{X}_{p+k}, \mathbf{Y}_{k+1}, \mathbf{Y}_{k+2}, \ldots, \mathbf{Y}_{q+k}) \qquad (3.23)$$

$$= \sum_{i=n-k}^{q-1} (-1)^{i-1} F_{k+1,i}(\mathbf{X}_{i+k+1}, \mathbf{X}_{i+k+2}, \ldots, \mathbf{X}_{p+k},$$

$$\mathbf{Y}_{i+k+1}, \mathbf{Y}_{i+k+2}, \ldots, \mathbf{Y}_{q+k})$$

$$+ \sum_{i=q}^{n-p} (-1)^{i-1} F_{k+1,i}(\mathbf{X}_{i+k+1}, \mathbf{X}_{i+k+2}, \ldots, \mathbf{X}_{p+k})$$

$$+ \sum_{i=n-p+1}^{p-1} (-1)^{n-i} F_{i+k+1,n-i}(\mathbf{X}_{k+1}, \mathbf{X}_{k+2}, \ldots, \mathbf{X}_{i+k+p},$$

$$\mathbf{X}_{i+k+1}, \mathbf{X}_{i+k+2}, \ldots, \mathbf{X}_{k+p})$$

$$+ \sum_{i=p}^{n-q} (-1)^{n-i} F_{i+k+1,n-i}(\mathbf{X}_{k+1}, \mathbf{X}_{k+2}, \ldots, \mathbf{X}_{i+p+k})$$

$$+ \sum_{i=n-q+1}^{n-1} (-1)^{n-i} F_{i+k+1,n-i}(\mathbf{X}_{k+1}, \mathbf{X}_{k+2}, \ldots, \mathbf{X}_{i+p+k},$$

$$\mathbf{Y}_{k+1}, \mathbf{Y}_{k+2}, \ldots, \mathbf{Y}_{i+q+k}).$$

Substituting the function f_{k+1} determined by Eq. (3.23) into Eq. (3.11) and using the substitutions Eq. (3.21), we obtain the equation (3.10). Therefore, the general solution of the functional equation (3.11) in the case considered will be

$$f_r(\mathbf{X}_1, \mathbf{X}_2, \ldots, \mathbf{X}_p, \mathbf{Y}_1, \mathbf{Y}_2, \ldots, \mathbf{Y}_q)$$

$$= \sum_{i=1}^{q-1} (-1)^{i-1} F_{ri}(\mathbf{X}_{i+1}, \mathbf{X}_{i+2}, \ldots, \mathbf{X}_p, \mathbf{Y}_{i+1}, \mathbf{Y}_{i+2}, \ldots, \mathbf{Y}_q)$$

$$+ \sum_{i=q}^{n-p} (-1)^{i-1} F_{ri}(\mathbf{X}_{i+1}, \mathbf{X}_{i+2}, \ldots, \mathbf{X}_p)$$

$$+ \sum_{i=n-p+1}^{p-1} (-1)^{i-1} F_{ri}(\mathbf{X}_{i+1}, \mathbf{X}_{i+2}, \ldots, \mathbf{X}_p, \mathbf{X}_1, \mathbf{X}_2, \ldots, \mathbf{X}_{i+p})$$

$$+ \sum_{i=p}^{n-q} (-1)^{n-i} F_{i+r,n-i}(\mathbf{X}_1, \mathbf{X}_2, \ldots, \mathbf{X}_{i+p})$$

$$+ \sum_{i=n-q+1}^{k-r} (-1)^{n-i} F_{i+r,n-i}(\mathbf{X}_1, \mathbf{X}_2, \ldots, \mathbf{X}_{i+p}, \mathbf{Y}_1, \mathbf{Y}_2, \ldots, \mathbf{Y}_{i+q})$$

$$+ \sum_{i=n-r+1}^{q-1} (-1)^{i-1} F_{ri}(\mathbf{X}_{i+1}, \mathbf{X}_{i+2}, \ldots, \mathbf{X}_p, \mathbf{Y}_{i+1}, \mathbf{Y}_{i+2}, \ldots, \mathbf{Y}_q)$$

$$+ \sum_{i=\max(q,n-r+1)}^{n-p} (-1)^{i-1} F_{ri}(\mathbf{X}_{i+1}, \mathbf{X}_{i+2}, \ldots, \mathbf{X}_p)$$

$$+ \sum_{i=\max(n-p+1,n-r+1)}^{p-1} (-1)^{n-i} F_{i+r,n-i}(\mathbf{X}_1, \mathbf{X}_2, \ldots, \mathbf{X}_{i+p}, \mathbf{X}_{i+1}, \mathbf{X}_{i+2}, \ldots, \mathbf{X}_p)$$

$$+ \sum_{i=\max(p,n-r+1)}^{n-q} (-1)^{n-i} F_{i+r,n-i}(\mathbf{X}_1, \mathbf{X}_2, \ldots, \mathbf{X}_{i+p})$$

$$+ \sum_{i=\max(n-q+1,n-r+1)}^{n-1} (-1)^{n-i} F_{i+r,n-i}(\mathbf{X}_1, \mathbf{X}_2, \ldots, \mathbf{X}_{i+p}, \mathbf{Y}_1, \mathbf{Y}_2, \ldots, \mathbf{Y}_{i+q})$$

$$+(-1)^{n-k+r-1} F_{k+1,n-k+r-1}(\mathbf{X}_1, \mathbf{X}_2, \ldots, \mathbf{X}_{k+1-r+p}, \mathbf{Y}_1, \mathbf{Y}_2, \ldots, \mathbf{Y}_{k+1-r+q})$$

$$(1 \leq r \leq q + k - n);$$

$$f_r(\mathbf{X}_1, \mathbf{X}_2, \ldots, \mathbf{X}_p, \mathbf{Y}_1, \mathbf{Y}_2, \ldots, \mathbf{Y}_q)$$

$$= \sum_{i=1}^{q-1} (-1)^{i-1} F_{ri}(\mathbf{X}_{i+1}, \mathbf{X}_{i+2}, \ldots, \mathbf{X}_p, \mathbf{Y}_{i+1}, \mathbf{Y}_{i+2}, \ldots, \mathbf{Y}_q)$$

$$+ \sum_{i=q}^{n-p} (-1)^{i-1} F_{ri}(\mathbf{X}_{i+1}, \mathbf{X}_{i+2}, \ldots, \mathbf{X}_p)$$

$$+ \sum_{i=n-p+1}^{p-1} (-1)^{i-1} F_{ri}(\mathbf{X}_{i+1}, \mathbf{X}_{i+2}, \ldots, \mathbf{X}_p, \mathbf{X}_1, \mathbf{X}_2, \ldots, \mathbf{X}_{i+p})$$

$$+ \sum_{i=p}^{k-r} (-1)^{n-i} F_{i+r,n-i}(\mathbf{X}_1, \mathbf{X}_2, \ldots, \mathbf{X}_{i+p})$$

$$+ (-1)^{n-k+r-1} F_{k+1,n-k+r-1}(\mathbf{X}_1, \mathbf{X}_2, \ldots, \mathbf{X}_{k-r+1+p})$$

$$+ \sum_{i=n-r+1}^{q-1} (-1)^{i-1} F_{ri}(\mathbf{X}_{i+1}, \mathbf{X}_{i+2}, \ldots, \mathbf{X}_p, \mathbf{Y}_{i+1}, \mathbf{Y}_{i+2}, \ldots, \mathbf{Y}_q)$$

$$+ \sum_{i=\max(q,n-r+1)}^{n-p} (-1)^{i-1} F_{ri}(\mathbf{X}_{i+1}, \mathbf{X}_{i+2}, \ldots, \mathbf{X}_p)$$

$$+ \sum_{i=\max(n-p+1,n-r+1)}^{p-1} (-1)^{n-i} F_{i+r,n-i}(\mathbf{X}_1, \mathbf{X}_2, \ldots, \mathbf{X}_{i+p}, \mathbf{X}_{i+1}, \mathbf{X}_{i+2}, \ldots, \mathbf{X}_p)$$

$$+ \sum_{i=\max(p,n-r+1)}^{n-q} (-1)^{n-i} F_{i+r,n-i}(\mathbf{X}_1, \mathbf{X}_2, \ldots, \mathbf{X}_{i+p})$$

$$+ \sum_{i=\max(n-q+1,n-r+1)}^{n-1} (-1)^{n-i} F_{i+r,n-i}(\mathbf{X}_1, \mathbf{X}_2, \ldots, \mathbf{X}_{i+p}, \mathbf{Y}_1, \mathbf{Y}_2, \ldots, \mathbf{Y}_{i+q})$$

$$(r = q + k - n + 1, \ldots, k - p + 1);$$

$$f_r(\mathbf{X}_1, \mathbf{X}_2, \ldots, \mathbf{X}_p, \mathbf{Y}_1, \mathbf{Y}_2, \ldots, \mathbf{Y}_q)$$

$$= \sum_{i=1}^{q-1} (-1)^{i-1} F_{ri}(\mathbf{X}_{i+1}, \mathbf{X}_{i+2}, \ldots, \mathbf{X}_p, \mathbf{Y}_{i+1}, \mathbf{Y}_{i+2}, \ldots, \mathbf{Y}_q)$$

$$+ \sum_{i=q}^{n-p} (-1)^{i-1} F_{ri}(\mathbf{X}_{i+1}, \mathbf{X}_{i+2}, \ldots, \mathbf{X}_p)$$

$$+ \sum_{i=n-p+1}^{k-r} (-1)^{i-1} F_{ri}(\mathbf{X}_{i+1}, \mathbf{X}_{i+2}, \ldots, \mathbf{X}_p, \mathbf{X}_1, \mathbf{X}_2, \ldots, \mathbf{X}_{i+p})$$

$$+ \sum_{i=n-r+1}^{q-1} (-1)^{i-1} F_{ri}(\mathbf{X}_{i+1}, \mathbf{X}_{i+2}, \ldots, \mathbf{X}_p, \mathbf{Y}_{i+1}, \mathbf{Y}_{i+2}, \ldots, \mathbf{Y}_q)$$

$$+ \sum_{i=\max(q,n-r+1)}^{n-p} (-1)^{i-1} F_{ri}(\mathbf{X}_{i+1}, \mathbf{X}_{i+2}, \ldots, \mathbf{X}_p)$$

$$+ \sum_{i=\max(n-p+1,n-r+1)}^{p-1} (-1)^{n-i} F_{i+r,n-i}(\mathbf{X}_1, \mathbf{X}_2, \ldots, \mathbf{X}_{i+p}, \mathbf{X}_{i+1}, \mathbf{X}_{i+2}, \ldots, \mathbf{X}_p)$$

$$+ \sum_{i=\max(p,n-r+1)}^{n-q} (-1)^{n-i} F_{i+r,n-i}(\mathbf{X}_1, \mathbf{X}_2, \ldots, \mathbf{X}_{i+p})$$

$$+ \sum_{i=\max(n-q+1,n-r+1)}^{n-1} (-1)^{n-i} F_{i+r,n-i}(\mathbf{X}_1, \mathbf{X}_2, \ldots, \mathbf{X}_{i+p}, \mathbf{Y}_1, \mathbf{Y}_2, \ldots, \mathbf{Y}_{i+q})$$

$$+(-1)^{k-r} F_{r,k+1-r}(\mathbf{X}_{k+2-r}, \mathbf{X}_{k+3-r}, \ldots, \mathbf{X}_p, \mathbf{X}_1, \mathbf{X}_2, \ldots, \mathbf{X}_{k+1-r+p})$$

$$(r = k - p + 2, \ldots, p - n + k);$$

$$f_r(\mathbf{X}_1, \mathbf{X}_2, \ldots, \mathbf{X}_p, \mathbf{Y}_1, \mathbf{Y}_2, \ldots, \mathbf{Y}_q)$$

$$= \sum_{i=1}^{q-1} (-1)^{i-1} F_{ri}(\mathbf{X}_{i+1}, \mathbf{X}_{i+2}, \ldots, \mathbf{X}_p, \mathbf{Y}_{i+1}, \mathbf{Y}_{i+2}, \ldots, \mathbf{Y}_q)$$

$$+ \sum_{i=q}^{k-r} (-1)^{i-1} F_{ri}(\mathbf{X}_{i+1}, \mathbf{X}_{i+2}, \ldots, \mathbf{X}_p)$$

$$+(-1)^{k-r} F_{r,k+1-r}(\mathbf{X}_{k+2-r}, \mathbf{X}_{k+3-r}, \ldots, \mathbf{X}_p)$$

$$+ \sum_{i=n-r+1}^{q-1} (-1)^{i-1} F_{ri}(\mathbf{X}_{i+1}, \mathbf{X}_{i+2}, \ldots, \mathbf{X}_p, \mathbf{Y}_{i+1}, \mathbf{Y}_{i+2}, \ldots, \mathbf{Y}_q)$$

$$+ \sum_{i=\max(q,n-r+1)}^{n-p} (-1)^{i-1} F_{ri}(\mathbf{X}_{i+1}, \mathbf{X}_{i+2}, \ldots, \mathbf{X}_p)$$

$$+ \sum_{i=\max(n-p+1,n-r+1)}^{p-1} (-1)^{n-i} F_{i+r,n-i}(\mathbf{X}_1, \mathbf{X}_2, \ldots, \mathbf{X}_{i+p},$$

$$\mathbf{X}_{i+1}, \mathbf{X}_{i+2}, \ldots, \mathbf{X}_p)$$

$$+ \sum_{i=\max(p,n-r+1)}^{n-q} (-1)^{n-i} F_{i+r,n-i}(\mathbf{X}_1, \mathbf{X}_2, \ldots, \mathbf{X}_{i+p})$$

$$+ \sum_{i=\max(n-q+1,n-r+1)}^{n-1} (-1)^{n-i} F_{i+r,n-i}(\mathbf{X}_1, \mathbf{X}_2, \ldots, \mathbf{X}_{i+p},$$

$$\mathbf{Y}_1, \mathbf{Y}_2, \ldots, \mathbf{Y}_{i+q})$$

$$(r = p - n + k + 1, \ldots, k - q + 1);$$

$$f_r(\mathbf{X}_1, \mathbf{X}_2, \ldots, \mathbf{X}_p, \mathbf{Y}_1, \mathbf{Y}_2, \ldots, \mathbf{Y}_q)$$

$$= \sum_{i=1}^{k-r} (-1)^{i-1} F_{ri}(\mathbf{X}_{i+1}, \mathbf{X}_{i+2}, \ldots, \mathbf{X}_p, \mathbf{Y}_{i+1}, \mathbf{Y}_{i+2}, \ldots, \mathbf{Y}_q)$$

$$+ (-1)^{k-r} F_{r,k-r+1}(\mathbf{X}_{k+2-r}, \mathbf{X}_{k+3-r}, \ldots, \mathbf{X}_p,$$

$$\mathbf{Y}_{k+2-r}, \mathbf{Y}_{k+3-r}, \ldots, \mathbf{Y}_q)$$

$$+ \sum_{i=n-r+1}^{q-1} (-1)^{i-1} F_{ri}(\mathbf{X}_{i+1}, \mathbf{X}_{i+2}, \ldots, \mathbf{X}_p, \mathbf{Y}_{i+1}, \mathbf{Y}_{i+2}, \ldots, \mathbf{Y}_q)$$

$$+ \sum_{i=\max(q,n-r+1)}^{n-p} (-1)^{i-1} F_{ri}(\mathbf{X}_{i+1}, \mathbf{X}_{i+2}, \ldots, \mathbf{X}_p)$$

$$+ \sum_{i=\max(n-p+1,n-r+1)}^{p-1} (-1)^{n-i} F_{i+r,n-i}(\mathbf{X}_1, \mathbf{X}_2, \ldots, \mathbf{X}_{i+p}, \mathbf{X}_{i+1}, \mathbf{X}_{i+2}, \ldots, \mathbf{X}_p)$$

$$+ \sum_{i=\max(p,n-r+1)}^{n-q} (-1)^{n-i} F_{i+r,n-i}(\mathbf{X}_1, \mathbf{X}_2, \ldots, \mathbf{X}_{i+p})$$

$$+ \sum_{i=\max(n-q+1,n-r+1)}^{n-1} (-1)^{n-i} F_{i+r,n-i}(\mathbf{X}_1, \mathbf{X}_2, \ldots, \mathbf{X}_{i+p},$$

$$\mathbf{Y}_1, \mathbf{Y}_2, \ldots, \mathbf{Y}_{i+q})$$

$$(r = k - q + 2, \ldots, k).$$

We can write the previous equalities in the following general form

$$f_r(\mathbf{X}_1, \mathbf{X}_2, \ldots, \mathbf{X}_p, \mathbf{Y}_1, \mathbf{Y}_2, \ldots, \mathbf{Y}_q) \tag{3.24}$$

$$= \sum_{i=1}^{\min(q-1,k+1-r)} (-1)^{i-1} F_{ri}(\mathbf{X}_{i+1}, \mathbf{X}_{i+2}, \ldots, \mathbf{X}_p, \mathbf{Y}_{i+1}, \mathbf{Y}_{i+2}, \ldots, \mathbf{Y}_q)$$

$$+ \sum_{i=q}^{\min(k+1-r,n-p)} (-1)^{i-1} F_{ri}(\mathbf{X}_{i+1}, \mathbf{X}_{i+2}, \ldots, \mathbf{X}_p)$$

$$+ \sum_{i=n-p+1}^{\min(k+1-r,p-1)} (-1)^{i-1} F_{ri}(\mathbf{X}_{i+1}, \mathbf{X}_{i+2}, \ldots, \mathbf{X}_p, \mathbf{X}_1, \mathbf{X}_2, \ldots, \mathbf{X}_{i+p})$$

$$+ \sum_{i=p}^{\min(k+1-r,n-q)} (-1)^{n-i} F_{i+r,n-i}(\mathbf{X}_1, \mathbf{X}_2, \ldots, \mathbf{X}_{i+p})$$

$$+ \sum_{i=n-q+1}^{k+1-r} (-1)^{n-i} F_{i+r,n-i}(\mathbf{X}_1, \mathbf{X}_2, \ldots, \mathbf{X}_{i+p}, \mathbf{Y}_1, \mathbf{Y}_2, \ldots, \mathbf{Y}_{i+q})$$

$$+ \sum_{i=n-r+1}^{q-1} (-1)^{i-1} F_{ri}(\mathbf{X}_{i+1}, \mathbf{X}_{i+2}, \ldots, \mathbf{X}_p, \mathbf{Y}_{i+1}, \mathbf{Y}_{i+2}, \ldots, \mathbf{Y}_q)$$

$$+ \sum_{i=\max(q,n-r+1)}^{n-p} (-1)^{i-1} F_{ri}(\mathbf{X}_{i+1}, \mathbf{X}_{i+2}, \ldots, \mathbf{X}_p)$$

$$+ \sum_{i=\max(n-p+1,n-r+1)}^{p-1} (-1)^{n-i} F_{i+r,n-i}(\mathbf{X}_1, \mathbf{X}_2, \ldots, \mathbf{X}_{i+p}, \mathbf{X}_{i+1}, \mathbf{X}_{i+2}, \ldots, \mathbf{X}_p)$$

$$+ \sum_{i=\max(p,n-r+1)}^{n-q} (-1)^{n-i} F_{i+r,n-i}(\mathbf{X}_1, \mathbf{X}_2, \ldots, \mathbf{X}_{i+p})$$

$$+ \sum_{i=\max(n-q+1,n-r+1)}^{n-1} (-1)^{n-i} F_{i+r,n-i}(\mathbf{X}_1, \mathbf{X}_2, \ldots, \mathbf{X}_{i+p}, \mathbf{Y}_1, \mathbf{Y}_2, \ldots, \mathbf{Y}_{i+q})$$

$$(1 \leq r \leq k).$$

Therefore, the theorem holds for $n - q + 1 \leq k \leq n$. $\qquad\square$

Now we will solve two particular cases of the equation (3.1).

a) By putting $k = n$ into equation (3.1), we obtain the functional equation

$$\sum_{i=1}^{n} f_i(\mathbf{X}_i, \mathbf{X}_{i+1}, \ldots, \mathbf{X}_{i+p-1}, \mathbf{Y}_i, \mathbf{Y}_{i+1}, \ldots, \mathbf{Y}_{i+q-1}) = \mathbf{O}. \tag{3.25}$$

Therefore, if we put $k = n$ into Eqs. (3.2), (3.3), (3.4), (3.5) and (3.6), then we obtain the general solution of the functional equation (3.25) in the cases considered. For example, if we put $k = n$ into Eq. (3.4), we obtain that the general solution of the functional equation (3.25) for $q < p < n < 2q - 1 < 2p - 1$ is given by the formulae

$$f_r(\mathbf{X}_1, \mathbf{X}_2, \ldots, \mathbf{X}_p, \mathbf{Y}_1, \mathbf{Y}_2, \ldots, \mathbf{Y}_q) \tag{3.26}$$

$$= \sum_{i=1}^{n-p} (-1)^{i-1} F_{ri}(\mathbf{X}_{i+1}, \mathbf{X}_{i+2}, \ldots, \mathbf{X}_p, \mathbf{Y}_{i+1}, \mathbf{Y}_{i+2}, \ldots, \mathbf{Y}_q)$$

$$+ \sum_{i=n-p+1}^{n-q} (-1)^{i-1} F_{ri}(\mathbf{X}_{i+1}, \mathbf{X}_{i+2}, \ldots, \mathbf{X}_p,$$

$$\mathbf{X}_1, \mathbf{X}_2, \ldots, \mathbf{X}_{p+i}, \mathbf{Y}_{i+1}, \mathbf{Y}_{i+2}, \ldots, \mathbf{Y}_q)$$

$$+ \sum_{i=n-q+1}^{\min(n-r,\,q-1)} (-1)^{i-1} F_{ri}(\mathbf{X}_{i+1}, \mathbf{X}_{i+2}, \ldots, \mathbf{X}_p, \mathbf{X}_1 \mathbf{X}_2, \ldots, \mathbf{X}_{p+i},$$

$$\mathbf{Y}_{i+1}, \mathbf{Y}_{i+2}, \ldots, \mathbf{Y}_q, \mathbf{Y}_1, \mathbf{Y}_2, \ldots, \mathbf{Y}_{q+i})$$

$$+ \sum_{i=\max(n-q+1,\,n-r+1)}^{q-1} (-1)^{n-i} F_{i+r,\,n-i}(\mathbf{X}_1, \mathbf{X}_2, \ldots, \mathbf{X}_{p+i}, \mathbf{X}_{i+1}, \mathbf{X}_{i+2}, \ldots, \mathbf{X}_p,$$

$$\mathbf{Y}_1, \mathbf{Y}_2, \ldots, \mathbf{Y}_{q+i}, \mathbf{Y}_{i+1}, \mathbf{Y}_{i+2}, \ldots, \mathbf{Y}_q)$$

$$+ \sum_{i=q}^{p-1} (-1)^{n-i} F_{i+r,\,n-i}(\mathbf{X}_1, \mathbf{X}_2, \ldots, \mathbf{X}_{p+i},$$

$$\mathbf{X}_{i+1}, \mathbf{X}_{i+2}, \ldots, \mathbf{X}_p, \mathbf{Y}_1, \mathbf{Y}_2, \ldots, \mathbf{Y}_{q+i})$$

$$+ \sum_{i=p}^{n-1}(-1)^{n-i}F_{i+r,n-i}(\mathbf{X}_1,\mathbf{X}_2,\ldots,\mathbf{X}_{i+p},\mathbf{Y}_1,\mathbf{Y}_2,\ldots,\mathbf{Y}_{q+i})$$

$$(1 \le r \le n),$$

where F_{ij} are arbitrary complex vector functions from $\mathcal{V}$.

The functional equation (3.1) for $q < p < 2q-1 < 2p-1 \le n$ had not been previously investigated, but for this case we must additionally determine the general solution of the equation (3.25). Now we will give the following result which treats a more general case than the previous one.

Theorem 3.6 *The general solution of the functional equation (3.25) for* $\dfrac{n+1}{2} \ge \max(p,q)$ *is given by the formulae*

$$f_r(\mathbf{X}_1,\mathbf{X}_2,\ldots,\mathbf{X}_p,\mathbf{Y}_1,\mathbf{Y}_2,\ldots,\mathbf{Y}_q) \tag{3.27}$$
$$= F_r(\mathbf{X}_1,\mathbf{X}_2,\ldots,\mathbf{X}_{p-1},\mathbf{Y}_1,\mathbf{Y}_2,\ldots,\mathbf{Y}_{q-1})$$
$$- F_{r+1}(\mathbf{X}_2,\mathbf{X}_3,\ldots,\mathbf{X}_p,\mathbf{Y}_2,\mathbf{Y}_3,\ldots,\mathbf{Y}_q)$$
$$(1 \le r \le n; \quad F_{n+1} \equiv F_1),$$

where F_r are arbitrary complex vector functions from $\mathcal{V}$.

Proof. Using the conventions $f_r \equiv f_{r+n}$, $\mathbf{X}_r \equiv \mathbf{X}_{r+n}$ and $\mathbf{Y}_r \equiv \mathbf{Y}_{r+n}$, the equation (3.25) in an expanded form can be written in the following way

$$f_r(\mathbf{X}_r,\mathbf{X}_{r+1},\ldots,\mathbf{X}_{r+p-1},\mathbf{Y}_r,\mathbf{Y}_{r+1},\ldots,\mathbf{Y}_{r+q-1}) \tag{3.28}$$
$$+\ f_{r+1}(\mathbf{X}_{r+1},\mathbf{X}_{r+2},\ldots,\mathbf{X}_{r+p},\mathbf{Y}_{r+1},\mathbf{Y}_{r+2},\ldots,\mathbf{Y}_{r+q}) + \cdots$$
$$+\ f_{r+n-p}(\mathbf{X}_{r+n-p},\mathbf{X}_{r+n-p+1},\ldots,\mathbf{X}_{r+n-1},$$
$$\mathbf{Y}_{r+n-p},\mathbf{Y}_{r+n-p+1},\ldots,\mathbf{Y}_{r+n-p+q-1})$$
$$+\ f_{r+n-p+1}(\mathbf{X}_{r+n-p+1},\mathbf{X}_{r+n-p+2},\ldots,\mathbf{X}_{r+n-1},\mathbf{X}_r,$$
$$\mathbf{Y}_{r+n-p+1},\mathbf{Y}_{r+n-p+2},\ldots,\mathbf{Y}_{r+n-p+q}) + \cdots$$
$$+\ f_{r+n-1}(\mathbf{X}_{r+n-1},\mathbf{X}_r,\ldots,\mathbf{X}_{r+p-2},\mathbf{Y}_{r+n-1},\mathbf{Y}_r,\ldots,\mathbf{Y}_{r+q-2}) = \mathbf{O}.$$

Assuming that $p > q$ (for $p = q$ there are just slight modifications in the following formulae) and putting $\mathbf{Y}_{r+q} = \mathbf{Y}_{r+q+1} = \cdots = \mathbf{Y}_{r+n-1} = C$

and $\mathbf{X}_{r+p} = \mathbf{X}_{r+p+1} = \cdots = \mathbf{X}_{r+n-1} = C$ into Eq. (3.28), where C is a fixed vector from $\mathcal{V}$, we obtain

$$
\begin{aligned}
&\ f_r(\mathbf{X}_r, \mathbf{X}_{r+1}, \ldots, \mathbf{X}_{r+p-1}, \mathbf{Y}_r, \mathbf{Y}_{r+1}, \ldots, \mathbf{Y}_{r+q-1}) && (3.29)\\
+\ &\ f_{r+1}(\mathbf{X}_{r+1}, \mathbf{X}_{r+2}, \ldots, \mathbf{X}_{r+p-1}, C, \mathbf{Y}_{r+1}, \mathbf{Y}_{r+2}, \ldots, \mathbf{Y}_{r+q-1}, C)\\
+\ &\ f_{r+2}(\mathbf{X}_{r+2}, \mathbf{X}_{r+3}, \ldots, \mathbf{X}_{r+p-1}, C, C, \mathbf{Y}_{r+2}, \mathbf{Y}_{r+3}, \ldots, \mathbf{Y}_{r+q-1}, C, C)\\
+\ &\ \cdots + f_{r+p-1}(\mathbf{X}_{r+p-1}, C, \ldots, C) + f_{r+p}(C, C, \ldots, C)\\
+\ &\ f_{r+p+1}(C, C, \ldots, C) + \cdots + f_{r+n-p}(C, C, C, \ldots, C)\\
+\ &\ f_{r+n-p+1}(C, C, \ldots, C, \mathbf{X}_r, C, C, \ldots, C)\\
+\ &\ f_{r+n-p+2}(C, C, \ldots, C, \mathbf{X}_r, \mathbf{X}_{r+1}, C, C, \ldots, C) + \cdots\\
+\ &\ f_{r+n-1}(C, \mathbf{X}_r, \mathbf{X}_{r+1}, \ldots, \mathbf{X}_{r+p-2}, C, \mathbf{Y}_r, \ldots, \mathbf{Y}_{r+q-2}) = \mathbf{O}.
\end{aligned}
$$

If we further substitute $\mathbf{Y}_{r+q-1} = C$ and $\mathbf{X}_{r+p-1} = C$ in Eq. (3.29), this yields

$$
\begin{aligned}
&\ f_r(\mathbf{X}_r, \mathbf{X}_{r+1}, \ldots, \mathbf{X}_{r+p-2}, C, \mathbf{Y}_r, \mathbf{Y}_{r+1}, \ldots, \mathbf{Y}_{r+q-2}, C) && (3.30)\\
+\ &\ f_{r+1}(\mathbf{X}_{r+1}, \mathbf{X}_{r+2}, \ldots, \mathbf{X}_{r+p-2}, C, C, \mathbf{Y}_{r+1}, \mathbf{Y}_{r+2}, \ldots, \mathbf{Y}_{r+q-2}, C, C)\\
+\ &\ \cdots + f_{r+p-2}(\mathbf{X}_{r+p-2}, C, \ldots, C) + f_{r+p-1}(C, C, \ldots, C)\\
+\ &\ f_{r+p}(C, C, \ldots, C) + \cdots + f_{r+n-p}(C, C, \ldots, C)\\
+\ &\ f_{r+n-p+1}(C, C, \ldots, C, \mathbf{X}_r, C, C, \ldots, C)\\
+\ &\ f_{r+n-p+2}(C, C, \ldots, C, \mathbf{X}_r, \mathbf{X}_{r+1}, C, C, \ldots, C) + \cdots\\
+\ &\ f_{r+n-1}(C, \mathbf{X}_r, \mathbf{X}_{r+1}, \ldots, \mathbf{X}_{r+p-2}, C, \mathbf{Y}_r, \ldots, \mathbf{Y}_{r+q-2}) = \mathbf{O}.
\end{aligned}
$$

Subtracting Eq. (3.30) from Eq. (3.29), we get the formula

$$
\begin{aligned}
&\ f_r(\mathbf{X}_r, \mathbf{X}_{r+1}, \ldots, \mathbf{X}_{r+p-1}, \mathbf{Y}_r, \mathbf{Y}_{r+1}, \ldots, \mathbf{Y}_{r+q-1}) && (3.31)\\
=\ &\ f_r(\mathbf{X}_r, \mathbf{X}_{r+1}, \ldots, \mathbf{X}_{r+p-2}, C, \mathbf{Y}_r, \mathbf{Y}_{r+1}, \ldots, \mathbf{Y}_{r+q-2}, C)\\
-\ &\ f_{r+1}(\mathbf{X}_{r+1}, \mathbf{X}_{r+2}, \ldots, \mathbf{X}_{r+p-1}, C, \mathbf{Y}_{r+1}, \mathbf{Y}_{r+2}, \ldots, \mathbf{Y}_{r+q-1}, C)\\
+\ &\ f_{r+1}(\mathbf{X}_{r+1}, \mathbf{X}_{r+2}, \ldots, \mathbf{X}_{r+p-2}, C, C, \mathbf{Y}_{r+1}, \mathbf{Y}_{r+2}, \ldots, \mathbf{Y}_{r+q-2}, C, C)\\
-\ &\ f_{r+2}(\mathbf{X}_{r+2}, \mathbf{X}_{r+3}, \ldots, \mathbf{X}_{r+p-1}, C, C, \mathbf{Y}_{r+2}, \mathbf{Y}_{r+3}, \ldots, \mathbf{Y}_{r+q-1}, C, C)\\
+\ &\ \cdots + f_{r+p-2}(\mathbf{X}_{r+p-2}, C, \ldots, C)\\
-\ &\ f_{r+p-1}(\mathbf{X}_{r+p-1}, C, \ldots, C) + f_{r+p-1}(C, C, \ldots, C),
\end{aligned}
$$

which holds for every $r = 1, 2, \ldots, n$.

Let us put now

$$
\begin{aligned}
&g_r(\mathbf{X}_1, \mathbf{X}_2, \ldots, \mathbf{X}_{p-1}, \mathbf{Y}_1, \mathbf{Y}_2, \ldots, \mathbf{Y}_{q-1}) \qquad (3.32)\\
&= \ f_r(\mathbf{X}_1, \mathbf{X}_2, \ldots, \mathbf{X}_{p-1}, C, \mathbf{Y}_1, \mathbf{Y}_2, \ldots, \mathbf{Y}_{q-1}, C)\\
&+ \ f_{r+1}(\mathbf{X}_2, \mathbf{X}_3, \ldots, \mathbf{X}_{p-1}, C, C, \mathbf{Y}_2, \mathbf{Y}_3, \ldots, \mathbf{Y}_{q-1}, C, C)\\
&+ \ \cdots \ + \ f_{r+p-2}(\mathbf{X}_{p-1}, C, \ldots, C)
\end{aligned}
$$

and

$$
A_r = f_{r+p-1}(C, C, \ldots, C). \qquad (3.33)
$$

Since $f_r \equiv f_{r+n}$, we conclude that $g_r \equiv g_{r+n}$. Putting in Eq. (3.25) $\mathbf{X}_r = \mathbf{Y}_r = C \ (1 \le r \le n)$, we find

$$
\sum_{r=1}^{n} A_r = \mathbf{O}. \qquad (3.34)
$$

According to Eqs. (3.32) and (3.33), the formula Eq. (3.31) can be written in the form

$$
\begin{aligned}
&f_r(\mathbf{X}_1, \mathbf{X}_2, \ldots, \mathbf{X}_p, \mathbf{Y}_1, \mathbf{Y}_2, \ldots, \mathbf{Y}_q) \qquad (3.35)\\
&= \ g_r(\mathbf{X}_1, \mathbf{X}_2, \ldots, \mathbf{X}_{p-1}, \mathbf{Y}_1, \mathbf{Y}_2, \ldots, \mathbf{Y}_{q-1})\\
&- \ g_{r+1}(\mathbf{X}_2, \mathbf{X}_3, \ldots, \mathbf{X}_p, \mathbf{Y}_2, \mathbf{Y}_3, \ldots, \mathbf{Y}_q) + A_r \quad (1 \le r \le n).
\end{aligned}
$$

Finally, with the notations

$$
\begin{aligned}
F_1 &= g_1,\\
F_2 &= g_2 - A_1,\\
F_3 &= g_3 - A_1 - A_2,\\
&\vdots\\
F_n &= g_n - A_1 - A_2 - \cdots - A_{n-1},
\end{aligned}
$$

the expression Eq. (3.35) can be written in the form Eq. (3.27). Thus we have proved that Eq. (3.27) is a consequence of Eq. (3.25) if $n + 1 \ge 2 \max(p, q)$. Conversely, a straightforward computation shows that the functions Eq. (3.27) satisfy the equation (3.35) for arbitrary functions F_r from $\mathcal{V}$. $\qquad\square$

b) Now we will consider the functional equation

$$\sum_{i=1}^{n} f(\mathbf{X}_i, \mathbf{X}_{i+1}, \ldots, \mathbf{X}_{i+p-1}, \mathbf{Y}_i, \mathbf{Y}_{i+1}, \ldots, \mathbf{Y}_{i+q-1}) = \mathbf{O} \quad (3.36)$$

$$(\mathbf{X}_{n+i} \equiv \mathbf{X}_i, \ \mathbf{Y}_{n+i} \equiv \mathbf{Y}_i)$$

which is a particular case of the functional equation (3.25).

In order to determine the general solution of the functional equation (3.36), we will distinguish the following cases:

$1^0 \ q < 2q - 1 \le p = n,$ $2^0 \ q < p = n < 2q - 1,$

$3^0 \ q < p < n < 2q - 1 < 2p - 1,$

$4^0 \ q < p < 2q - 1 \le n \le p + q - 1 < 2p - 1,$

$5^0 \ q < p < p + q - 1 < n < 2p - 1,$ $6^0 \ q < p < 2q - 1 < 2p - 1 \le n.$

Theorem 3.7 *If $q < 2q - 1 \le p = n$, then the general solution of the functional equation (3.36) is given by*

$$f(\mathbf{X}_1, \mathbf{X}_2, \ldots, \mathbf{X}_n, \mathbf{Y}_1, \mathbf{Y}_2, \ldots, \mathbf{Y}_q) \quad (3.37)$$
$$= \ F_0(\mathbf{X}_1, \mathbf{X}_2, \ldots, \mathbf{X}_n, \mathbf{Y}_1, \mathbf{Y}_2, \ldots, \mathbf{Y}_{q-1})$$
$$- \ F_0(\mathbf{X}_2, \mathbf{X}_3, \ldots, \mathbf{X}_n, \mathbf{X}_1, \mathbf{Y}_2, \mathbf{Y}_3, \ldots, \mathbf{Y}_q),$$

where F_0 is an arbitrary complex vector function from $\mathcal{V}$.

Proof. If we put $k = n$ into Eq. (3.2), then by summing up the functions $f_r \ (1 \le r \le n)$ and putting $f_1 = f_2 = \cdots = f_n = f$, we obtain the formula Eq. (3.37). $\quad\square$

In the same way the following theorems can be proved:

Theorem 3.8 *If $q < p = n < 2q - 1$, then the general solution of the functional equation Eq. (3.36) is given by the formula*

$$f(\mathbf{X}_1, \mathbf{X}_2, \ldots, \mathbf{X}_n, \mathbf{Y}_1, \mathbf{Y}_2, \ldots, \mathbf{Y}_q) \quad (3.38)$$
$$= \ F_0(\mathbf{X}_1, \mathbf{X}_2, \ldots, \mathbf{X}_n, \mathbf{Y}_1, \mathbf{Y}_2, \ldots, \mathbf{Y}_{q-1})$$
$$- \ F_0(\mathbf{X}_2, \mathbf{X}_3, \ldots, \mathbf{X}_n, \mathbf{X}_1, \mathbf{Y}_2, \mathbf{Y}_3, \ldots, \mathbf{Y}_q)$$
$$+ \ \sum_{i=1}^{q-[n/2]} [F_i(\mathbf{X}_{n-q+i+1}, \mathbf{X}_{n-q+i+2}, \ldots, \mathbf{X}_{i+1},$$
$$\mathbf{Y}_1, \mathbf{Y}_2, \ldots, \mathbf{Y}_i, \mathbf{Y}_{n-q+i+1}, \mathbf{Y}_{n-q+i+2}, \ldots, \mathbf{Y}_q)$$

$$- F_i(\mathbf{X}_1, \mathbf{X}_2, \ldots, \mathbf{X}_n, \mathbf{Y}_{q-i+1}, \mathbf{Y}_{q-i+2}, \ldots, \mathbf{Y}_q, \mathbf{Y}_1, \mathbf{Y}_2, \ldots, \mathbf{Y}_{2q-i+n})],$$

where F_i $(i = 0, 1, \ldots, q - [n/2])$ are arbitrary complex vector functions from $\mathcal{V}$.

Theorem 3.9 *If $q < p < n < 2q - 1 < 2p - 1$, then the general solution of the equation (3.36) is*

$$f(\mathbf{X}_1, \mathbf{X}_2, \ldots, \mathbf{X}_p, \mathbf{Y}_1, \mathbf{Y}_2, \ldots, \mathbf{Y}_q) \tag{3.39}$$

$$= F_0(\mathbf{X}_1, \mathbf{X}_2, \ldots, \mathbf{X}_{p-1}, \mathbf{Y}_1, \mathbf{Y}_2, \ldots, \mathbf{Y}_{q-1})$$

$$- F_0(\mathbf{X}_2, \mathbf{X}_3, \ldots, \mathbf{X}_p, \mathbf{Y}_2, \mathbf{Y}_3, \ldots, \mathbf{Y}_q)$$

$$+ \sum_{i=1}^{p-q} [F_i(\mathbf{X}_1, \mathbf{X}_2, \ldots, \mathbf{X}_i, \mathbf{X}_{n-p+i+1}, \mathbf{X}_{n-p+i+2}, \ldots, \mathbf{X}_p,$$

$$\mathbf{Y}_{n-p+i+1}, \mathbf{Y}_{n-p+i+2}, \ldots, \mathbf{Y}_q)$$

$$- F_i(\mathbf{X}_{p-i+1}, \mathbf{X}_{p-i+2}, \ldots, \mathbf{X}_p, \mathbf{X}_1, \mathbf{X}_2, \ldots, \mathbf{X}_{2p-n-i},$$

$$\mathbf{Y}_1, \mathbf{Y}_2, \ldots, \mathbf{Y}_{p+q-n-i})]$$

$$+ \sum_{i=p-q+1}^{p-[n/2]} [F_i(\mathbf{X}_1, \mathbf{X}_2, \ldots, \mathbf{X}_i, \mathbf{X}_{n-p+i+1}, \mathbf{X}_{n-p+i+2}, \ldots, \mathbf{X}_p,$$

$$\mathbf{Y}_1, \mathbf{Y}_2, \ldots, \mathbf{Y}_{i+q-p}, \mathbf{Y}_{n-p+i+1}, \mathbf{Y}_{n-p+i+2}, \ldots, \mathbf{Y}_q)$$

$$- F_i(\mathbf{X}_{p-i+1}, \mathbf{X}_{p-i+2}, \ldots, \mathbf{X}_p, \mathbf{X}_1, \mathbf{X}_2, \ldots, \mathbf{X}_{2p-n-i},$$

$$\mathbf{Y}_{p-i+1}, \mathbf{Y}_{p-i+2}, \ldots, \mathbf{Y}_q, \mathbf{Y}_1, \mathbf{Y}_2, \ldots, \mathbf{Y}_{p+q-n-i})],$$

where F_i $(i = 0, 1, \ldots, p - [n/2])$ are arbitrary complex vector functions from $\mathcal{V}$.

Theorem 3.10 *The general solution of the functional equation (3.36) for $q < p < 2q - 1 \leq n \leq p + q - 1 < 2p - 1$ is given by the formula*

$$f(\mathbf{X}_1, \mathbf{X}_2, \ldots, \mathbf{X}_p, \mathbf{Y}_1, \mathbf{Y}_2, \ldots, \mathbf{Y}_q) \tag{3.40}$$

$$= F_0(\mathbf{X}_1, \mathbf{X}_2, \ldots, \mathbf{X}_{p-1}, \mathbf{Y}_1, \mathbf{Y}_2, \ldots, \mathbf{Y}_{q-1})$$

62   *General Classes of Cyclic Functional Equations*

$$-F_0(\mathbf{X}_2, \mathbf{X}_3, \ldots, \mathbf{X}_p, \mathbf{Y}_2, \mathbf{Y}_3, \ldots, \mathbf{Y}_q)$$

$$+ \sum_{i=1}^{p+q-n-1} [F_i(\mathbf{X}_1, \mathbf{X}_2, \ldots, \mathbf{X}_i, \mathbf{X}_{n-p+i+1}, \mathbf{X}_{n-p+i+2}, \ldots, \mathbf{X}_p,$$

$$\mathbf{Y}_{n-p+i+1}, \mathbf{Y}_{n-p+i+2}, \ldots, \mathbf{Y}_q)$$

$$- F_i(\mathbf{X}_{p-i+1}, \mathbf{X}_{p-i+2}, \ldots, \mathbf{X}_p, \mathbf{X}_1, \mathbf{X}_2, \ldots, \mathbf{X}_{2p-n-i}, \mathbf{Y}_1, \mathbf{Y}_2, \ldots, \mathbf{Y}_{p+q-n-i})]$$

$$+ \sum_{i=p+q-n}^{p-[n/2]} [F_i(\mathbf{X}_1, \mathbf{X}_2, \ldots, \mathbf{X}_i, \mathbf{X}_{n-p+i+1}, \mathbf{X}_{n-p+i+2}, \ldots, \mathbf{X}_p)$$

$$- F_i(\mathbf{X}_{p-i+1}, \mathbf{X}_{p-i+2}, \ldots, \mathbf{X}_p, \mathbf{X}_1, \mathbf{X}_2, \ldots, \mathbf{X}_{2p-n-i})],$$

where F_i $(i = 0, 1, \ldots, p - [n/2])$ are arbitrary complex vector functions from $\mathcal{V}$.

Next we will prove the following theorem.

Theorem 3.11   *The general solution of the equation* (3.36) *for* $q < p < p + q - 1 < n < 2p - 1$ *is given by*

$$f(\mathbf{X}_1, \mathbf{X}_2, \ldots, \mathbf{X}_p, \mathbf{Y}_1, \mathbf{Y}_2, \ldots, \mathbf{Y}_q) \tag{3.41}$$

$$= F_0(\mathbf{X}_1, \mathbf{X}_2, \ldots, \mathbf{X}_{p-1}, \mathbf{Y}_1, \mathbf{Y}_2, \ldots, \mathbf{Y}_{q-1})$$

$$- F_0(\mathbf{X}_2, \mathbf{X}_3, \ldots, \mathbf{X}_p \mathbf{Y}_2, \mathbf{Y}_3, \ldots, \mathbf{Y}_q)$$

$$+ \sum_{i=1}^{p-[n/2]} [F_i(\mathbf{X}_1, \mathbf{X}_2, \ldots, \mathbf{X}_i, \mathbf{X}_{n-p+i+1}, \mathbf{X}_{n-p+i+2}, \ldots, \mathbf{X}_p)$$

$$- F_i(\mathbf{X}_{p-i+1}, \mathbf{X}_{p-i+2}, \ldots, \mathbf{X}_p, \mathbf{X}_1, \mathbf{X}_2, \ldots, \mathbf{X}_{2p-n-i})],$$

where F_i $(i = 0, 1, \ldots, p - [n/2])$ are arbitrary complex vector functions from $\mathcal{V}$.

Proof.   By summing up the functions f_r $(1 \le r \le n)$ determined by Eq. (3.6) and putting $f_1 = f_2 = \cdots f_n = f$, we obtain Eq. (3.41), where we introduced the notations

$$F_0(\mathbf{X}_1, \mathbf{X}_2, \ldots, \mathbf{X}_{p-1}, \mathbf{Y}_1, \mathbf{Y}_2, \ldots, \mathbf{Y}_{q-1})$$

$$= \frac{1}{n}\left[\sum_{r=1}^{q-1}\sum_{i=1}^{r} G_r(\mathbf{X}_i, \mathbf{X}_{i+1}, \ldots, \mathbf{X}_{p-r+i-1}, \mathbf{Y}_i, \mathbf{Y}_{i+1}, \ldots, \mathbf{Y}_{q-r+i-1}) \right.$$

$$\left. + \sum_{r=q}^{n-p}\sum_{i=1}^{r} G_r(\mathbf{X}_i, \mathbf{X}_{i+1}, \ldots, \mathbf{X}_{p-r+i-1})\right],$$

$$G_r(\mathbf{X}_1, \mathbf{X}_2, \ldots, \mathbf{X}_{p-r}, \mathbf{Y}_1, \mathbf{Y}_2, \ldots, \mathbf{Y}_{q-r})$$

$$= \sum_{i=1}^{n}(-1)^r F_{ir}(\mathbf{X}_1, \mathbf{X}_2, \ldots, \mathbf{X}_{p-r}, \mathbf{Y}_1, \mathbf{Y}_2, \ldots, \mathbf{Y}_{q-r})$$

$$(1 \le r \le q-1),$$

$$G_r(\mathbf{X}_1, \mathbf{X}_2, \ldots, \mathbf{X}_{p-r}) = \sum_{i=1}^{n}(-1)^r F_{ir}(\mathbf{X}_1, \mathbf{X}_2, \ldots, \mathbf{X}_{p-r})$$

$$(r = q, q+1, \ldots, n-p);$$

$$F_k(\mathbf{X}_1, \mathbf{X}_2, \ldots, \mathbf{X}_k, \mathbf{X}_{n-p+k+1}, \mathbf{X}_{n-p+k+2}, \ldots, \mathbf{X}_p)$$

$$= \frac{(-1)^{k+1}}{n}\left[\sum_{r=1}^{n-p+k} F_{r,p-k}(\mathbf{X}_1, \mathbf{X}_2, \ldots, \mathbf{X}_k,\right.$$

$$\mathbf{X}_{n-p+k+1}, \mathbf{X}_{n-p+k+2}, \ldots, \mathbf{X}_p)$$

$$\left. - \sum_{r=1}^{p-k} F_{r,n-p+k}(\mathbf{X}_{n-p+k+1}, \mathbf{X}_{n-p+k+2}, \ldots, \mathbf{X}_p, \mathbf{X}_1, \mathbf{X}_2, \ldots, \mathbf{X}_k)\right]$$

$$(1 \le k \le p - [(n+1)/2]).$$

In particular, if $n = 2m$, we obtain

$$F_{p-m}(\mathbf{X}_1, \mathbf{X}_2, \ldots, \mathbf{X}_{p-m}, \mathbf{X}_{m+1}, \mathbf{X}_{m+2}, \ldots, \mathbf{X}_p)$$

$$= \frac{(-1)^{p-m+1}}{n}\sum_{r=1}^{m} F_{rm}(\mathbf{X}_1, \mathbf{X}_2, \ldots, \mathbf{X}_{p-m}, \mathbf{X}_{m+1}, \mathbf{X}_{m+2}, \ldots, \mathbf{X}_p),$$

where

$$F_{r+m,m}(\mathbf{X}_1, \ldots, \mathbf{X}_{p-m}, \mathbf{X}_{m+1}, \ldots, \mathbf{X}_p)$$

$$= -F_{rm}(\mathbf{X}_{m+1}, \ldots, \mathbf{X}_p, \mathbf{X}_1, \ldots, \mathbf{X}_{p-m}) \qquad (1 \le r \le m).$$

$\square$

We have not considered the case $q < p < 2q - 1 < 2p - 1 \le n$, because we will give instead the following more general result.

Theorem 3.12   *If $\dfrac{n+1}{2} \ge \max(p,q)$, then the general solution of the functional equation (3.36) is given by*

$$
\begin{aligned}
& f(\mathbf{X}_1, \mathbf{X}_2, \ldots, \mathbf{X}_p, \mathbf{Y}_1, \mathbf{Y}_2, \ldots, \mathbf{Y}_q) && (3.42) \\
=\; & F(\mathbf{X}_1, \mathbf{X}_2, \ldots, \mathbf{X}_{p-1}, \mathbf{Y}_1, \mathbf{Y}_2, \ldots, \mathbf{Y}_{q-1}) \\
-\; & F(\mathbf{X}_2, \mathbf{X}_3, \ldots, \mathbf{X}_p, \mathbf{Y}_2, \mathbf{Y}_3, \ldots, \mathbf{Y}_q),
\end{aligned}
$$

where F is an arbitrary complex vector function from $\mathcal{V}$.

Proof.   A straightforward calculation shows that every function f of the form Eq. (3.42) satisfies the functional equation (3.36). We have to prove the converse, *i.e.*, that from Eq. (3.36) it follows that f has the form Eq. (3.42).

Let C be a fixed vector from $\mathcal{V}$. For $\mathbf{X}_i = \mathbf{Y}_i = C$ $(1 \le i \le n)$ the equation (3.36) yields

$$f(C, C, \ldots, C) = \mathbf{O}. \qquad (3.43)$$

Assuming that $p > q$ (for $p = q$ there are just slight modifications in the following formulae) and $n \ge 2p - 1$ and putting $\mathbf{Y}_{q+1} = \mathbf{Y}_{q+2} = \cdots = \mathbf{Y}_n = C$ and $\mathbf{X}_{p+1} = \mathbf{X}_{p+2} = \cdots = \mathbf{X}_n = C$ into Eq. (3.36), by virtue of (3.43) we obtain

$$
\begin{aligned}
& f(\mathbf{X}_1, \mathbf{X}_2, \ldots, \mathbf{X}_p, \mathbf{Y}_1, \mathbf{Y}_2, \ldots, \mathbf{Y}_q) && (3.44) \\
+\; & f(\mathbf{X}_2, \mathbf{X}_3, \ldots, \mathbf{X}_p, C, \mathbf{Y}_2, \mathbf{Y}_3, \ldots, \mathbf{Y}_q, C) + \cdots \\
+\; & f(\mathbf{X}_p, C, \ldots, C) + f(C, C, \ldots, C, \mathbf{X}_1, C, C, \ldots, C) \\
+\; & f(C, C, \ldots, C, \mathbf{X}_1, \mathbf{X}_2, C, C, \ldots, C) + \cdots \\
+\; & f(C, \mathbf{X}_1, \mathbf{X}_2, \ldots, \mathbf{X}_{p-1}, C, \mathbf{Y}_1, \mathbf{Y}_2, \ldots, \mathbf{Y}_{q-1}) = \mathbf{O}.
\end{aligned}
$$

If we further substitute $\mathbf{Y}_q = C$ and $\mathbf{X}_p = C$ in the last equation, we get

$$f(\mathbf{X}_1, \mathbf{X}_2, \ldots, \mathbf{X}_{p-1}, C, \mathbf{Y}_1, \mathbf{Y}_2, \ldots, \mathbf{Y}_{q-1}, C) \tag{3.45}$$
$$+ \quad f(\mathbf{X}_2, \mathbf{X}_3, \ldots, \mathbf{X}_{p-1}, C, C, \mathbf{Y}_2, \mathbf{Y}_3, \ldots, \mathbf{Y}_{q-1}, C, C) + \cdots$$
$$+ \quad f(\mathbf{X}_{p-1}, C, \ldots, C) + f(C, C, \ldots, C, \mathbf{X}_1, C, C, \ldots, C)$$
$$+ \quad f(C, C, \ldots, C, \mathbf{X}_1, \mathbf{X}_2, C, C, \ldots, C) + \cdots$$
$$+ \quad f(C, \mathbf{X}_1, \mathbf{X}_2, \ldots, \mathbf{X}_{p-1}, C, \mathbf{Y}_1, \mathbf{Y}_2, \ldots, \mathbf{Y}_{q-1}) = \mathbf{O}.$$

Subtracting Eq. (3.45) from Eq. (3.44), we find

$$f(\mathbf{X}_1, \mathbf{X}_2, \ldots, \mathbf{X}_p, \mathbf{Y}_1, \mathbf{Y}_2, \ldots, \mathbf{Y}_q) \tag{3.46}$$
$$= \quad f(\mathbf{X}_1, \mathbf{X}_2, \ldots, \mathbf{X}_{p-1}, C, \mathbf{Y}_1, \mathbf{Y}_2, \ldots, \mathbf{Y}_{q-1}, C)$$
$$- \quad f(\mathbf{X}_2, \mathbf{X}_3, \ldots, \mathbf{X}_p, C, \mathbf{Y}_2, \mathbf{Y}_3, \ldots, \mathbf{Y}_q, C)$$
$$+ \quad f(\mathbf{X}_2, \mathbf{X}_3, \ldots, \mathbf{X}_{p-1}, C, C, \mathbf{Y}_2, \mathbf{Y}_3, \ldots, \mathbf{Y}_{q-1}, C, C)$$
$$- \quad f(\mathbf{X}_3, \mathbf{X}_4, \ldots, \mathbf{X}_p, C, C, \mathbf{Y}_3, \mathbf{Y}_4, \ldots, \mathbf{Y}_q, C, C)$$
$$+ \quad \cdots + f(\mathbf{X}_{p-1}, C, \ldots, C) - f(\mathbf{X}_p, C, \ldots, C).$$

Putting

$$F(\mathbf{X}_1, \mathbf{X}_2, \ldots, \mathbf{X}_{p-1}, \mathbf{Y}_1, \mathbf{Y}_2, \ldots, \mathbf{Y}_{q-1})$$
$$= \quad f(\mathbf{X}_1, \mathbf{X}_2, \ldots, \mathbf{X}_{p-1}, C, \mathbf{Y}_1, \mathbf{Y}_2, \ldots, \mathbf{Y}_{q-1}, C)$$
$$+ \quad f(\mathbf{X}_2, \mathbf{X}_3, \ldots, \mathbf{X}_{p-1}, C, C, \mathbf{Y}_2, \mathbf{Y}_3, \ldots, \mathbf{Y}_{q-1}, C, C)$$
$$+ \quad \cdots + f(\mathbf{X}_{p-1}, C, \ldots, C),$$

the equality (3.46) takes on the form Eq. (3.42). $\qquad\qquad\square$

For particular cases see the results obtained in [D. S. Mitrinović (1963C); P. M. Vasić *et al.* (1965)].

4 Semicyclic Functional Equation

Let $\mathcal{V}$ be a complex vector space, and let $\mathbf{Z}_i$ $(1 \leq i \leq n)$ be complex vectors as in Sec. 1.

Also, let S_k^n $(0 < k < n)$ be the set of all strictly increasing mappings of the set $\{1, 2, \ldots, k\}$ into $\{1, 2, \ldots, n\}$. Let f_r $(r \in S_k^n)$ be mappings $\mathcal{V}^k \mapsto \mathcal{V}$.

We will solve the following semicyclic complex vector functional equation

$$\sum_{r \in S_k^n} f_r(\mathbf{Z}_{r(1)}, \mathbf{Z}_{r(2)}, \ldots, \mathbf{Z}_{r(k)}) = \mathbf{O}. \tag{4.1}$$

Theorem 4.1 *The general solution of the functional equation (4.1) is given by*

$$f_r(\mathbf{Z}_{r(1)}, \mathbf{Z}_{r(2)}, \ldots, \mathbf{Z}_{r(k)}) \tag{4.2}$$

$$= \sum_{p \in S_{k-1}^k} F_{rp}(\mathbf{Z}_{rp(1)}, \mathbf{Z}_{rp(2)}, \ldots, \mathbf{Z}_{rp(k-1)}) \qquad (r \in S_k^n),$$

where $rp(i) \equiv r(p(i))$, and F_{rp} are arbitrary complex vector functions from the vector space $\mathcal{V}$ such that

$$\sum_{rp=t} F_{rp}(\mathbf{Z}_{t(1)}, \mathbf{Z}_{t(2)}, \ldots, \mathbf{Z}_{t(k-1)}) = \mathbf{O} \qquad (t \in S_{k-1}^n), \tag{4.3}$$

where the sum is extended over all $r \in S_k^n$ and $p \in S_{k-1}^k$ such that $rp = t$.

Proof. Let f_r be defined by Eq. (4.2) and let Eq. (4.3) hold. Then we have

$$\sum_{r \in S_k^n} f_r(\mathbf{Z}_{r(1)}, \mathbf{Z}_{r(2)}, \ldots, \mathbf{Z}_{r(k)})$$

$$= \sum_{r \in S_k^n} \sum_{p \in S_{k-1}^k} F_{rp}(\mathbf{Z}_{rp(1)}, \mathbf{Z}_{rp(2)}, \ldots, \mathbf{Z}_{rp(k-1)})$$

$$= \sum_{t \in S_{k-1}^n} \sum_{rp=t} F_{rp}(\mathbf{Z}_{t(1)}, \mathbf{Z}_{t(2)}, \ldots, \mathbf{Z}_{t(k-1)}) = \mathbf{O}.$$

Hence, such functions satisfy the functional equation (4.1).

Conversely, if f_r $(r \in S_k^n)$ is any solution of Eq. (4.1), we have to prove that the functions f_r admit the representation Eq. (4.2) with the conditions Eq. (4.3). For fixed r let us put $\mathbf{Z}_i = C$ into Eq. (4.1) for $i \neq r(j)$ $(1 \leq j \leq k)$. Then Eq. (4.1) yields the following (not unique) representation

$$f_r(\mathbf{Z}_{r(1)}, \mathbf{Z}_{r(2)}, \ldots, \mathbf{Z}_{r(k)}) = \sum_{p \in S_{k-1}^k} G_{rp}(\mathbf{Z}_{rp(1)}, \mathbf{Z}_{rp(2)}, \ldots, \mathbf{Z}_{rp(k-1)}). \quad (4.4)$$

For an arbitrary $t \in S_{k-1}^n$ let

$$H_t(\mathbf{Z}_{t(1)}, \mathbf{Z}_{t(2)}, \ldots, \mathbf{Z}_{t(k-1)}) = \sum_{rp=t} G_{rp}(\mathbf{Z}_{t(1)}, \mathbf{Z}_{t(2)}, \ldots, \mathbf{Z}_{t(k-1)}). \quad (4.5)$$

The equation (4.1) can be written in the form

$$\sum_{t \in S_{k-1}^n} H_t(\mathbf{Z}_{t(1)}, \mathbf{Z}_{t(2)}, \ldots, \mathbf{Z}_{t(k-1)}) = \mathbf{O}. \quad (4.6)$$

Let P' be the set of all $t \in S_{k-1}^n$ such that $H_t \neq \mathbf{O}$ and $P'' = S_{k-1}^n \setminus P'$. The equation (4.6) is reduced to

$$\sum_{t \in P'} H_t(\mathbf{Z}_{t(1)}, \mathbf{Z}_{t(2)}, \ldots, \mathbf{Z}_{t(k-1)}) = \mathbf{O}. \quad (4.7)$$

We can suppose that among all representations of the form Eq. (4.4) of the functions f_r we have taken that (or one of those) for which the number s of the elements of the set P' is minimal. If $s = 0$, we can take $F_{rp} = G_{rp}$ and the theorem is proved. The case $s = 1$ is impossible in view of Eq. (4.7). Thus we can suppose that $s > 1$. Let t be some fixed element from P'. Putting $\mathbf{Z}_i = C$ into Eq. (4.7) for $i \neq t(j)$ $(1 \leq j \leq k - 1)$, we obtain

$$H_t(\mathbf{Z}_{t(1)}, \mathbf{Z}_{t(2)}, \ldots, \mathbf{Z}_{t(k-1)}) = -\sum J_{rp}(\mathbf{Z}_{t(i_1)}, \mathbf{Z}_{t(i_2)}, \ldots, \mathbf{Z}_{t(i_m)}), \quad (4.8)$$

where $J_{rp}(\mathbf{Z}_{t(i_1)}, \mathbf{Z}_{t(i_2)}, \ldots, \mathbf{Z}_{t(i_m)})$ $(m \leq k - 2)$ is obtained from $G_{rp}(\mathbf{Z}_{rp(1)}, \mathbf{Z}_{rp(2)}, \ldots, \mathbf{Z}_{rp(k-1)})$ by putting $\mathbf{Z}_i = C$ for all i but $t(1), t(2), \ldots, t(k-1)$. The sum on the right-hand side of Eq. (4.8) is extended over certain (not all) pairs of indices r and p.

Consider a certain summand $J_{r_0 p_0}$ on the right-hand side of Eq. (4.8). Let $u_0 \in S_k^n$ and $v_0 \in S_{k-1}^k$ be such that $u_0 v_0 = t$. We can construct a sequence of ordered pairs

$$(u_0, v_0), \ (u_0, w_0), \ (u_1, v_1), \ (u_1, w_1), \ \ldots, \ (u_q, w_q)$$

which satisfy the following conditions

1° $u_i \in S_k^n$, $v_i \in S_{k-1}^k$, $w_i \in S_{k-1}^k$;

2° $(u_q, w_q) = (r_0, p_0)$;

3° $u_{i-1} w_{i-1} = u_i v_i$ $(1 \le i \le q)$;

4° the sequence $u_i w_i(1), \ldots, u_i w_i(k-1)$ $(0 \le i \le q)$ contains the sequence
     $t(i_1), \ldots, t(i_m)$ as a subsequence.

Let us put

$$G_{u_i,v_i}^* = G_{u_i,v_i} + J_{r_0,p_0}, \quad G_{u_i,w_i}^* = G_{u_i,w_i} - J_{r_0,p_0} \quad (0 \le i \le q). \quad (4.9)$$

We note that the representation Eq. (4.4) is still valid with G_{u_i,v_i}^* and G_{u_i,w_i}^* instead of G_{u_i,v_i} and G_{u_i,w_i}, respectively. We have $H_t^* = H_t + J_{r_0 p_0}$. On the other hand, if $t \in P''$, i.e. $H_t \equiv \mathbf{O}$, then also $H_t^* \equiv \mathbf{O}$.

If the same procedure is applied to all summands of the right-hand side of Eq. (4.8), we conclude that the new function H_t is identically the zero vector. This contradicts the minimum property of the number s. Hence, $s = 0$ which proves the theorem. $\qquad\qquad\square$

This theorem generalizes the results given in [D. Ž. Djoković (1965A)].

Example 4.1   If $n = 5$ and $k = 4$, the equation (4.1) is

$$f(\mathbf{Z}_1, \mathbf{Z}_2, \mathbf{Z}_3, \mathbf{Z}_4) + g(\mathbf{Z}_1, \mathbf{Z}_2, \mathbf{Z}_3, \mathbf{Z}_5) + h(\mathbf{Z}_1, \mathbf{Z}_2, \mathbf{Z}_4, \mathbf{Z}_5)$$
$$+ i(\mathbf{Z}_1, \mathbf{Z}_3, \mathbf{Z}_4, \mathbf{Z}_5) + j(\mathbf{Z}_2, \mathbf{Z}_3, \mathbf{Z}_4, \mathbf{Z}_5) = \mathbf{O}.$$

Its general solution is given by

$$
\begin{aligned}
f(\mathbf{Z}_1, \mathbf{Z}_2, \mathbf{Z}_3, \mathbf{Z}_4) &= f_1(\mathbf{Z}_1, \mathbf{Z}_2, \mathbf{Z}_3) + f_2(\mathbf{Z}_1, \mathbf{Z}_2, \mathbf{Z}_4) \\
&+ f_3(\mathbf{Z}_1, \mathbf{Z}_3, \mathbf{Z}_4) + f_4(\mathbf{Z}_2, \mathbf{Z}_3, \mathbf{Z}_4), \\[4pt]
g(\mathbf{Z}_1, \mathbf{Z}_2, \mathbf{Z}_3, \mathbf{Z}_5) &= g_1(\mathbf{Z}_1, \mathbf{Z}_2, \mathbf{Z}_3) + g_2(\mathbf{Z}_1, \mathbf{Z}_2, \mathbf{Z}_5) \\
&+ g_3(\mathbf{Z}_1, \mathbf{Z}_3, \mathbf{Z}_5) + g_4(\mathbf{Z}_2, \mathbf{Z}_3, \mathbf{Z}_5), \\[4pt]
h(\mathbf{Z}_1, \mathbf{Z}_2, \mathbf{Z}_4, \mathbf{Z}_5) &= h_1(\mathbf{Z}_1, \mathbf{Z}_2, \mathbf{Z}_4) + h_2(\mathbf{Z}_1, \mathbf{Z}_2, \mathbf{Z}_5) \\
&+ h_3(\mathbf{Z}_1, \mathbf{Z}_4, \mathbf{Z}_5) + h_4(\mathbf{Z}_2, \mathbf{Z}_4, \mathbf{Z}_5), \\[4pt]
i(\mathbf{Z}_1, \mathbf{Z}_3, \mathbf{Z}_4, \mathbf{Z}_5) &= i_1(\mathbf{Z}_1, \mathbf{Z}_3, \mathbf{Z}_4) + i_2(\mathbf{Z}_1, \mathbf{Z}_3, \mathbf{Z}_5) \\
&+ i_3(\mathbf{Z}_1, \mathbf{Z}_4, \mathbf{Z}_5) + i_4(\mathbf{Z}_3, \mathbf{Z}_4, \mathbf{Z}_5), \\[4pt]
j(\mathbf{Z}_2, \mathbf{Z}_3, \mathbf{Z}_4, \mathbf{Z}_5) &= j_1(\mathbf{Z}_2, \mathbf{Z}_3, \mathbf{Z}_4) + j_2(\mathbf{Z}_2, \mathbf{Z}_3, \mathbf{Z}_5) \\
&+ j_3(\mathbf{Z}_2, \mathbf{Z}_4, \mathbf{Z}_5) + j_4(\mathbf{Z}_3, \mathbf{Z}_4, \mathbf{Z}_5),
\end{aligned}
$$

where

$$f_1(\mathbf{Z}_1, \mathbf{Z}_2, \mathbf{Z}_3) \; + \; g_1(\mathbf{Z}_1, \mathbf{Z}_2, \mathbf{Z}_3) \;\; = \;\; \mathbf{O},$$
$$f_2(\mathbf{Z}_1, \mathbf{Z}_2, \mathbf{Z}_4) \; + \; h_1(\mathbf{Z}_1, \mathbf{Z}_2, \mathbf{Z}_4) \;\; = \;\; \mathbf{O},$$
$$f_3(\mathbf{Z}_1, \mathbf{Z}_3, \mathbf{Z}_4) \; + \; i_1(\mathbf{Z}_1, \mathbf{Z}_3, \mathbf{Z}_4) \;\; = \;\; \mathbf{O},$$
$$f_4(\mathbf{Z}_2, \mathbf{Z}_3, \mathbf{Z}_4) \; + \; j_1(\mathbf{Z}_2, \mathbf{Z}_3, \mathbf{Z}_4) \;\; = \;\; \mathbf{O},$$
$$g_2(\mathbf{Z}_1, \mathbf{Z}_2, \mathbf{Z}_5) \; + \; h_2(\mathbf{Z}_1, \mathbf{Z}_2, \mathbf{Z}_5) \;\; = \;\; \mathbf{O},$$
$$g_3(\mathbf{Z}_1, \mathbf{Z}_3, \mathbf{Z}_5) \; + \; i_2(\mathbf{Z}_1, \mathbf{Z}_3, \mathbf{Z}_5) \;\; = \;\; \mathbf{O},$$
$$g_4(\mathbf{Z}_2, \mathbf{Z}_3, \mathbf{Z}_5) \; + \; j_2(\mathbf{Z}_2, \mathbf{Z}_3, \mathbf{Z}_5) \;\; = \;\; \mathbf{O},$$
$$h_3(\mathbf{Z}_1, \mathbf{Z}_4, \mathbf{Z}_5) \; + \; i_3(\mathbf{Z}_1, \mathbf{Z}_4, \mathbf{Z}_5) \;\; = \;\; \mathbf{O},$$
$$h_4(\mathbf{Z}_2, \mathbf{Z}_4, \mathbf{Z}_5) \; + \; j_3(\mathbf{Z}_2, \mathbf{Z}_4, \mathbf{Z}_5) \;\; = \;\; \mathbf{O},$$
$$i_4(\mathbf{Z}_3, \mathbf{Z}_4, \mathbf{Z}_5) \; + \; j_4(\mathbf{Z}_3, \mathbf{Z}_4, \mathbf{Z}_5) \;\; = \;\; \mathbf{O}.$$

Hence we may take $f_1, f_2, f_3, f_4, g_2, g_3, g_4, h_3, h_4, i_4$ to be arbitrary complex vector functions from the complex vector space $\mathcal{V}$ and

$$\begin{aligned}
g_1 = -f_1, \quad h_1 &= -f_2, \quad i_1 = -f_3, \quad j_1 = -f_4, \\
h_2 &= -g_2, \quad i_2 = -g_3, \quad j_2 = -g_4, \\
& \qquad\qquad\;\; i_3 = -h_3, \quad j_3 = -h_4, \\
& \qquad\qquad\qquad\qquad\qquad\;\; j_4 = -i_4.
\end{aligned}$$

5   Special Cyclic Functional Equation

The notations for the vectors in this section are the same as in Sec. 1.

Let $\mathcal{V}$ be the vector space and let there exist mappings

$$f_i : \mathcal{V}^{i+1} \mapsto \mathcal{V} \quad (1 \le i \le n) \quad \text{and} \quad g_i : \mathcal{V}^{i+2} \mapsto \mathcal{V} \quad (1 \le i \le n-1).$$

Now we will prove the following result.

Theorem 5.1   *The general solution of the special cyclic complex vector functional equation*

$$\sum_{i=1}^{n} f_i(\mathbf{Z}_i, \mathbf{Z}_{i+1}, \cdots, \mathbf{Z}_{2i}) + \sum_{i=1}^{n-1} g_i(\mathbf{Z}_1, \mathbf{Z}_3, \cdots, \mathbf{Z}_{2i+1}, \mathbf{Z}_{2i+2}) = \mathbf{O} \qquad (5.1)$$

is given by

$$f_1(\mathbf{Z}_1, \mathbf{Z}_2) = H_1(\mathbf{Z}_1) - F_1(\mathbf{Z}_2),$$

$$f_i(\mathbf{Z}_i, \mathbf{Z}_{i+1}, \cdots, \mathbf{Z}_{2i}) = (-1)^i F_{i-1}(\mathbf{Z}_i, \mathbf{Z}_{i+1}, \cdots, \mathbf{Z}_{2i-2})$$

$$+(-1)^{i+1} G_{i-1}(\{\mathbf{Z}_i, \mathbf{Z}_{i+1}, \cdots, \mathbf{Z}_{2i}\} \cap \{\mathbf{Z}_1, \mathbf{Z}_3, \cdots, \mathbf{Z}_{2i-1}\}, \mathbf{Z}_{2i})$$

$$+(-1)^i F_i(\mathbf{Z}_{i+1}, \mathbf{Z}_{i+2}, \cdots, \mathbf{Z}_{2i}) \quad (2 \leq i \leq n),$$

$$F_n(\mathbf{Z}_{n+1}, \mathbf{Z}_{n+2}, \cdots, \mathbf{Z}_{2n}) = \mathbf{O}, \tag{5.2}$$

$$g_i(\mathbf{Z}_1, \mathbf{Z}_3, \cdots, \mathbf{Z}_{2i+1}, \mathbf{Z}_{2i+2}) = (-1)^i H_i(\mathbf{Z}_1, \mathbf{Z}_3, \cdots, \mathbf{Z}_{2i-1})$$

$$+(-1)^{i+1} G_i(\{\mathbf{Z}_{i+1}, \mathbf{Z}_{i+2}, \cdots, \mathbf{Z}_{2i+2}\} \cap \{\mathbf{Z}_1, \mathbf{Z}_3, \cdots, \mathbf{Z}_{2i+1}\}, \mathbf{Z}_{2i+2})$$

$$+(-1)^i H_{i+1}(\mathbf{Z}_1, \mathbf{Z}_3, \cdots, \mathbf{Z}_{2i+1}) \quad (1 \leq i \leq n-1),$$

$$H_n(\mathbf{Z}_1, \mathbf{Z}_3, \cdots, \mathbf{Z}_{2n-1}) = \mathbf{O},$$

where F_i, G_i, H_i $(1 \leq i \leq n-1)$ are arbitrary functions with values in $\mathcal{V}$.

Proof. We will derive the proof by mathematical induction on n.

If $n = 2$, then Eq. (5.1) becomes

$$f_1(\mathbf{Z}_1, \mathbf{Z}_2) + f_2(\mathbf{Z}_2, \mathbf{Z}_3, \mathbf{Z}_4) + g_1(\mathbf{Z}_1, \mathbf{Z}_3, \mathbf{Z}_4) = \mathbf{O}, \tag{5.3}$$

whose solution according to [D. S. Mitrinović (1963A)] is

$$\begin{aligned}
f_1(\mathbf{Z}_1, \mathbf{Z}_2) &= H_1(\mathbf{Z}_1) - F_1(\mathbf{Z}_2), \\
f_2(\mathbf{Z}_2, \mathbf{Z}_3, \mathbf{Z}_4) &= F_1(\mathbf{Z}_2) - G_1(\mathbf{Z}_3, \mathbf{Z}_4), \\
g_1(\mathbf{Z}_1, \mathbf{Z}_3, \mathbf{Z}_4) &= -H_1(\mathbf{Z}_1) + G_1(\mathbf{Z}_3, \mathbf{Z}_4).
\end{aligned} \tag{5.4}$$

Thus, the theorem holds for $n = 2$.

Suppose that theorem holds for any fixed n and let us consider the equation

$$\sum_{i=1}^{n+1} f_i(\mathbf{Z}_i, \mathbf{Z}_{i+1}, \cdots, \mathbf{Z}_{2i}) + \sum_{i=1}^{n} g_i(\mathbf{Z}_1, \mathbf{Z}_3, \cdots, \mathbf{Z}_{2i+1}, \mathbf{Z}_{2i+2}) = \mathbf{O}. \tag{5.5}$$

By putting $\mathbf{Z}_i = C_i$ $(1 \leq i \leq n)$, where $C_i = \text{const}$ into Eq. (5.5), we see that f_{n+1} can be represented in the form

$$f_{n+1}(\mathbf{Z}_{n+1}, \mathbf{Z}_{n+2}, \cdots, \mathbf{Z}_{2n+2}) = (-1)^{n+1} F_n(\mathbf{Z}_{n+1}, \mathbf{Z}_{n+2}, \cdots, \mathbf{Z}_{2n}) \tag{5.6}$$

$$+(-1)^n G_n(\{\mathbf{Z}_{n+1}, \mathbf{Z}_{n+2}, \cdots, \mathbf{Z}_{2n+2}\} \cap \{\mathbf{Z}_1, \mathbf{Z}_3, \cdots, \mathbf{Z}_{2n+1}\}, \mathbf{Z}_{2n+2}),$$

where F_n and G_n are arbitrary functions.

By a substitution of Eq. (5.6) into Eq. (5.5), we obtain the equation

$$\sum_{i=1}^{n} A_i(\mathbf{Z}_i, \mathbf{Z}_{i+1}, \cdots, \mathbf{Z}_{2i}) + \sum_{i=1}^{n} B_i(\mathbf{Z}_1, \mathbf{Z}_3, \cdots, \mathbf{Z}_{2i+1}, \mathbf{Z}_{2n+2}) = \mathbf{O}, \quad (5.7)$$

where we introduced the notations

$$A_i = f_i, \quad B_i = g_i \quad (1 \le i \le n-1), \tag{5.8}$$
$$A_n = f_n + (-1)^{n+1} F_n, \quad B_n = g_n + (-1)^n G_n.$$

By putting $\mathbf{Z}_i = C_i$ $(i = 2, 4, \cdots, 2n)$ into Eq. (5.7) we obtain

$$B_n(\mathbf{Z}_1, \mathbf{Z}_3, \cdots, \mathbf{Z}_{2n+1}, \mathbf{Z}_{2n+2}) = (-1)^n H_n(\mathbf{Z}_1, \mathbf{Z}_3, \cdots, \mathbf{Z}_{2n-1}), \tag{5.9}$$

where H_n is an arbitrary function.

On the basis of the expression Eq. (5.9), the equation (5.7) takes on the following form

$$\sum_{i=1}^{n} A_i(\mathbf{Z}_i, \mathbf{Z}_{i+1}, \cdots, \mathbf{Z}_{2i}) + \sum_{i=1}^{n-1} D_i(\mathbf{Z}_1, \mathbf{Z}_3, \cdots, \mathbf{Z}_{2i+1}, \mathbf{Z}_{2i+2}) = \mathbf{O}, \quad (5.10)$$

where we introduced the notations

$$D_i = B_i \quad (1 \le i \le n-2), \qquad D_{n-1} = B_{n-1} + (-1)^n H_n. \tag{5.11}$$

The functional equation (5.10) is an equation of the form Eq. (5.1). According to the inductive hypothesis, the general solution of the equation (5.10) is given by equalities of the form Eq. (5.2) with f_i replaced by A_i and g_i replaced by D_i.

By virtue of Eqs. (5.11), (5.9), (5.8) and (5.6) we deduce that the general solution of the equation (5.5) is given by

$$f_i(\mathbf{Z}_i, \mathbf{Z}_{i+1}, \cdots, \mathbf{Z}_{2i}) = A_i(\mathbf{Z}_i, \mathbf{Z}_{i+1}, \cdots, \mathbf{Z}_{2i}) \qquad (1 \le i \le n-1),$$

$$\begin{aligned} f_n(\mathbf{Z}_n, \mathbf{Z}_{n+1}, \cdots, \mathbf{Z}_{2n}) &= A_n(\mathbf{Z}_n, \mathbf{Z}_{n+1}, \cdots, \mathbf{Z}_{2n}) \\ &+ (-1)^n F_n(\mathbf{Z}_{n+1}, \mathbf{Z}_{n+2}, \cdots, \mathbf{Z}_{2n}), \end{aligned}$$

$$f_{n+1}(\mathbf{Z}_{n+1}, \mathbf{Z}_{n+2}, \cdots, \mathbf{Z}_{2n+2}) = (-1)^{n+1} F_n(\mathbf{Z}_{n+1}, \mathbf{Z}_{n+2}, \cdots, \mathbf{Z}_{2n})$$

$$+(-1)^n G_n(\{\mathbf{Z}_{n+1}, \mathbf{Z}_{n+2}, \cdots, \mathbf{Z}_{2n+2}\} \cap \{\mathbf{Z}_1, \mathbf{Z}_3, \cdots, \mathbf{Z}_{2n+1}\}, \mathbf{Z}_{2n+2}),$$

$$g_i(\mathbf{Z}_1, \mathbf{Z}_3, \cdots, \mathbf{Z}_{2i+1}, \mathbf{Z}_{2i+2}) = D_i(\mathbf{Z}_1, \mathbf{Z}_3, \cdots, \mathbf{Z}_{2i+1}, \mathbf{Z}_{2i+2})$$

$$(1 \le i \le n-2),$$

$$\begin{aligned}
g_{n-1}(\mathbf{Z}_1, \mathbf{Z}_3, \cdots, \mathbf{Z}_{2n-1}, \mathbf{Z}_{2n}) &= D_{n-1}(\mathbf{Z}_1, \mathbf{Z}_3, \cdots, \mathbf{Z}_{2n-1}, \mathbf{Z}_{2n}) \\
&+ (-1)^{n+1} H_n(\mathbf{Z}_1, \mathbf{Z}_3, \cdots, \mathbf{Z}_{2n-1}),
\end{aligned}$$

$$g_n(\mathbf{Z}_1, \mathbf{Z}_3, \cdots, \mathbf{Z}_{2n+1}, \mathbf{Z}_{2n+2}) = (-1)^n H_n(\mathbf{Z}_1, \mathbf{Z}_3, \cdots, \mathbf{Z}_{2n-1})$$

$$+(-1)^{n+1} G_n(\{\mathbf{Z}_{n+1}, \mathbf{Z}_{n+2}, \cdots, \mathbf{Z}_{2n+2}\} \cap \{\mathbf{Z}_1, \mathbf{Z}_3, \cdots, \mathbf{Z}_{2n+1}\}, \mathbf{Z}_{2n+2}). \qquad \square$$

This theorem generalizes the result given in [R. Ž. Djordjević (1965A)].

6   Condensed Cyclic Functional Equation

Consider the condensed cyclic complex vector functional equation

$$\sum_{i=1}^k f_i(\mathbf{Z}_{\bar{i}(1)}, \mathbf{Z}_{\bar{i}(2)}, \ldots, \mathbf{Z}_{\bar{i}(s_i)}) = \mathbf{O}, \tag{6.1}$$

where $1 \le s_i \le n$, $f_i : \mathcal{V}^{s_i} \mapsto \mathcal{V}$ $(1 \le i \le k)$, $\bar{i}(\nu) \in \{1, 2, \ldots, n\}$ for $1 \le \nu \le s_i$.

For two sets of indices $\{\bar{i}(1), \bar{i}(2), \ldots, \bar{i}(s_i)\}$ and $\{\bar{j}(1), \bar{j}(2), \ldots, \bar{j}(s_j)\}$ we denote their intersection by $\{\overline{ij}(1), \overline{ij}(2), \ldots, \overline{ij}(t_{ij})\}$.

Theorem 6.1   *The general solution of the condensed cyclic complex vector functional equation (6.1) is*

$$f_i(\mathbf{Z}_{\bar{i}(1)}, \mathbf{Z}_{\bar{i}(2)}, \ldots, \mathbf{Z}_{\bar{i}(s_i)}) = \sum_{j=1}^{i-1} F_{ji}(\mathbf{Z}_{\overline{ji}(1)}, \ldots, \mathbf{Z}_{\overline{ji}(t_{ji})}) \tag{6.2}$$

$$- \sum_{j=i+1}^k F_{ij}(\mathbf{Z}_{\overline{ij}(1)}, \ldots, \mathbf{Z}_{\overline{ij}(t_{ij})}),$$

where $F_{ij}(Z_{\overline{ij}(1)}, \ldots, Z_{\overline{ij}(t_{ij})})$ are arbitrary functions from $\mathcal{V}$ for $1 \leq i \leq k-1$, $i+1 \leq j \leq k$, and $\sum\limits_{\sigma}^{\nu} = O$ for $\sigma > \nu$.

Proof. We will prove the theorem by induction.

For $k = 2$, equation (6.1) becomes

$$f_1(Z_{\overline{1}(1)}, Z_{\overline{1}(2)}, \ldots, Z_{\overline{1}(s_1)}) + f_2(Z_{\overline{2}(1)}, Z_{\overline{2}(2)}, \ldots, Z_{\overline{2}(s_2)}) = O.$$

It is obvious that the functions f_1 and f_2 depend just on the variables $Z_{\overline{12}(1)}, Z_{\overline{12}(2)}, \ldots, Z_{\overline{12}(t_{12})}$, thus we may write

$$f_1(Z_{\overline{1}(1)}, Z_{\overline{1}(2)}, \ldots, Z_{\overline{1}(s_1)}) = -F_{12}(Z_{\overline{12}(1)}, Z_{\overline{12}(2)}, \ldots, Z_{\overline{12}(t_{12})}),$$
$$f_2(Z_{\overline{2}(1)}, Z_{\overline{2}(2)}, \ldots, Z_{\overline{2}(s_2)}) = F_{12}(Z_{\overline{12}(1)}, Z_{\overline{12}(2)}, \ldots, Z_{\overline{12}(t_{12})})$$

for an arbitrary function $F_{12} : \mathcal{V}^{t_{12}} \mapsto \mathcal{V}$, i.e., for $k = 2$ the general solution of Eq. (6.1) is given by Eq. (6.2).

For some fixed k suppose that the general solution of the functional equation (6.1) is given by the formula Eq. (6.2).

Now, let us consider the equation

$$\sum_{i=1}^{k+1} f_i(Z_{\overline{i}(1)}, Z_{\overline{i}(2)}, \ldots, Z_{\overline{i}(s_i)}) = O. \tag{6.3}$$

If we put $Z_i = C_i$ for $i \notin \{\overline{k+1}(1), \overline{k+1}(2), \ldots, \overline{k+1}(s_{k+1})\}$, we obtain the representation

$$f_{k+1}(Z_{\overline{k+1}(1)}, Z_{\overline{k+1}(2)}, \ldots, Z_{\overline{k+1}(s_{k+1})}) \tag{6.4}$$
$$= \sum_{j=1}^{k} F_{j,k+1}(Z_{\overline{j,k+1}(1)}, Z_{\overline{j,k+1}(2)}, \ldots, Z_{\overline{j,k+1}(t_{j,k+1})})$$

where

$$F_{j,k+1}(Z_{\overline{j,k+1}(1)}, Z_{\overline{j,k+1}(2)}, \ldots, Z_{\overline{j,k+1}(t_{j,k+1})})$$
$$= -\ f_j(Z_{\overline{j}(1)}, Z_{\overline{j}(2)}, \ldots, Z_{\overline{j}(s_j)}) \Bigg|_{\substack{Z_i = C_i \text{ for} \\ i \notin \{\overline{k+1}(1), \overline{k+1}(2), \ldots, \overline{k+1}(s_{k+1})\}}}$$

If we substitute Eq. (6.4) into Eq. (6.3) and denote

$$
\begin{aligned}
g_i(\mathbf{Z}_{\bar{i}(1)}, \mathbf{Z}_{\bar{i}(2)}, \ldots, \mathbf{Z}_{\bar{i}(s_i)}) \;=\; & f_i(\mathbf{Z}_{\bar{i}(1)}, \mathbf{Z}_{\bar{i}(2)}, \ldots, \mathbf{Z}_{\bar{i}(s_i)}) \\
& + \; F_{i,k+1}(\mathbf{Z}_{\overline{i,k+1}(1)}, \mathbf{Z}_{\overline{i,k+1}(2)}, \ldots, \mathbf{Z}_{\overline{i,k+1}(t_{i,k+1})}),
\end{aligned}
$$

then equation (6.3) becomes

$$
\sum_{i=1}^{k} g_i(\mathbf{Z}_{\bar{i}(1)}, \mathbf{Z}_{\bar{i}(2)}, \ldots, \mathbf{Z}_{\bar{i}(s_i)}) = \mathbf{O}.
$$

By assumption its general solution is

$$
\begin{aligned}
g_i(\mathbf{Z}_{\bar{i}(1)}, \mathbf{Z}_{\bar{i}(2)}, \ldots, \mathbf{Z}_{\bar{i}(s_i)}) \;=\; & \sum_{j=1}^{i-1} F_{ji}(\mathbf{Z}_{\overline{ji}(1)}, \ldots, \mathbf{Z}_{\overline{ji}(t_{ji})}) \\
& - \sum_{j=i+1}^{k} F_{ij}(\mathbf{Z}_{\overline{ij}(1)}, \ldots, \mathbf{Z}_{\overline{ij}(t_{ij})}),
\end{aligned}
$$

and hence

$$
\begin{aligned}
f_i(\mathbf{Z}_{\bar{i}(1)}, \mathbf{Z}_{\bar{i}(2)}, \ldots, \mathbf{Z}_{\bar{i}(s_i)}) \;=\; & \sum_{j=1}^{i-1} F_{ji}(\mathbf{Z}_{\overline{ji}(1)}, \ldots, \mathbf{Z}_{\overline{ji}(t_{ji})}) \\
& - \sum_{j=i+1}^{k+1} F_{ij}(\mathbf{Z}_{\overline{ij}(1)}, \ldots, \mathbf{Z}_{\overline{ij}(t_{ij})})
\end{aligned}
$$

for $1 \le i \le k$. Together with Eq. (6.4) this yields Eq. (6.2) with $k + 1$ instead of k. $\qquad\square$

The last theorem which is obtained in [I. B. Risteski and V. C. Covachev (2000)] generalized all previous results given in this chapter.

Chapter 2

Functional Equations with Operations between Arguments

In this chapter some classes of complex vector functional equations are solved, namely such equations in which operations between arguments appear. The results presented in this chapter are given in [I. B. Risteski *et al.* (1999); I. B. Risteski *et al.* (2000B); I. B. Risteski *et al.* (to appear A); K. G. Trenčevski *et al.* (1999)].

7 Operator Functional Equation

Here a new linear operator Ψ is defined such that $\Psi \circ \Psi = \mathbf{O}$. The general analytic solution of the vector functional equation $\Psi f = \mathbf{O}$ is given. The results presented here are obtained in [I. B. Risteski *et al.* (1999)].

Definition 7.1. Let $\mathcal{V}$ and $\mathcal{V}'$ be complex vector spaces. For an arbitrary mapping $f : \mathcal{V}^{n-1} \mapsto \mathcal{V}'$ $(n > 1)$ we define a mapping $\Psi f : \mathcal{V}^n \mapsto \mathcal{V}'$ by

$$
\begin{aligned}
&(\Psi f)(\mathbf{Z}_1, \mathbf{Z}_2, \ldots, \mathbf{Z}_{n-1}, \mathbf{Z}_n) \qquad\qquad (7.1) \\
&= \ (-1)^{n-1} f(\mathbf{Z}_1, \mathbf{Z}_2, \ldots, \mathbf{Z}_{n-1}) - f(\mathbf{Z}_2, \mathbf{Z}_3, \ldots, \mathbf{Z}_n) \\
&+ \ \sum_{i=1}^{n-1} (-1)^{i+1} f(\mathbf{Z}_1, \mathbf{Z}_2, \ldots, \mathbf{Z}_i + \mathbf{Z}_{i+1}, \ldots, \mathbf{Z}_{n-1}, \mathbf{Z}_n).
\end{aligned}
$$

If $n = 1$, we define $\Psi f = \mathbf{O}$.

Remark 7.1. The definition of the operator Ψ is a variation of the formula giving the differential of the bar construction [S. MacLane (1963)].

Theorem 7.1   *For an arbitrary mapping $f : \mathcal{V}^{n-1} \mapsto \mathcal{V}'$ it holds*

$$(\Psi \circ \Psi)f(\mathbf{Z}_1, \mathbf{Z}_2, \ldots, \mathbf{Z}_n, \mathbf{Z}_{n+1}) = \mathbf{O}. \tag{7.2}$$

Proof.   Applying the operator Ψ to the mapping $\Psi f : \mathcal{V}^n \mapsto \mathcal{V}'$, we obtain

$$\begin{aligned}
&(\Psi \circ \Psi)f(\mathbf{Z}_1, \mathbf{Z}_2, \ldots, \mathbf{Z}_n, \mathbf{Z}_{n+1}) \\
=\ & (-1)^n(\Psi f)(\mathbf{Z}_1, \mathbf{Z}_2, \ldots, \mathbf{Z}_{n-1}, \mathbf{Z}_n) - (\Psi f)(\mathbf{Z}_2, \mathbf{Z}_3, \ldots, \mathbf{Z}_n, \mathbf{Z}_{n+1}) \\
+\ & \sum_{i=1}^{n}(-1)^{i+1}(\Psi f)(\mathbf{Z}_1, \mathbf{Z}_2, \ldots, \mathbf{Z}_i + \mathbf{Z}_{i+1}, \ldots, \mathbf{Z}_n, \mathbf{Z}_{n+1}).
\end{aligned}$$

Using the definition of the operator Ψ inside the sum and replacing i by $i + 1$ in some of the terms, we can write down the previous equality as

$$(\Psi \circ \Psi)f(\mathbf{Z}_1, \mathbf{Z}_2, \ldots, \mathbf{Z}_n, \mathbf{Z}_{n+1}) \tag{7.3}$$

$$= (-1)^n(\Psi f)(\mathbf{Z}_1, \mathbf{Z}_2, \ldots, \mathbf{Z}_{n-1}, \mathbf{Z}_n) - (\Psi f)(\mathbf{Z}_2, \mathbf{Z}_3, \ldots, \mathbf{Z}_n, \mathbf{Z}_{n+1})$$

$$+ \left\{ \sum_{i=1}^{n-1}(-1)^{i+1}\Big[(-1)^{n-1}f(\mathbf{Z}_1, \mathbf{Z}_2, \ldots, \mathbf{Z}_i + \mathbf{Z}_{i+1}, \ldots, \mathbf{Z}_{n-1}, \mathbf{Z}_n) \right.$$

$$\left. + f(\mathbf{Z}_2, \mathbf{Z}_3, \ldots, \mathbf{Z}_{i+1} + \mathbf{Z}_{i+2}, \ldots, \mathbf{Z}_n, \mathbf{Z}_{n+1})\Big] \right.$$

$$+ f(\mathbf{Z}_1, \mathbf{Z}_2, \ldots, \mathbf{Z}_{n-1}) - f(\mathbf{Z}_3, \ldots, \mathbf{Z}_n, \mathbf{Z}_{n+1})$$

$$+ \sum_{i=2}^{n-1}\sum_{j=1}^{i-1}(-1)^{i+j}f(\mathbf{Z}_1, \mathbf{Z}_2, \ldots, \mathbf{Z}_j + \mathbf{Z}_{j+1}, \ldots, \mathbf{Z}_{i+1} + \mathbf{Z}_{i+2}, \ldots, \mathbf{Z}_n, \mathbf{Z}_{n+1})$$

$$\left. + \sum_{i=1}^{n-2}\sum_{j=i+2}^{n}(-1)^{i+j}f(\mathbf{Z}_1, \mathbf{Z}_2, \ldots, \mathbf{Z}_i + \mathbf{Z}_{i+1}, \ldots, \mathbf{Z}_j + \mathbf{Z}_{j+1}, \ldots, \mathbf{Z}_n, \mathbf{Z}_{n+1}) \right\}.$$

In the parentheses $\{\cdots\}$ the terms for $j = i$, $i+1$ are omitted because they cancel each other. Further we will consider the double sum

$$\sum_{i=2}^{n-1}\sum_{j=1}^{i-1}(-1)^{i+j}f(Z_1, Z_2, \ldots, Z_j + Z_{j+1}, \ldots, Z_{i+1} + Z_{i+2}, \ldots, Z_n, Z_{n+1})$$

$$= \sum_{i=3}^{n}\sum_{j=1}^{i-2}(-1)^{i+j+1}f(Z_1, Z_2, \ldots, Z_j + Z_{j+1}, \ldots, Z_i + Z_{i+1}, \ldots, Z_n, Z_{n+1})$$

$$= \sum_{j=1}^{n-2}\sum_{i=j+2}^{n}(-1)^{i+j+1}f(Z_1, Z_2, \ldots, Z_j + Z_{j+1}, \ldots, Z_i + Z_{i+1}, \ldots, Z_n, Z_{n+1})$$

$$= -\sum_{i=1}^{n-2}\sum_{j=i+2}^{n}(-1)^{i+j}f(Z_1, Z_2, \ldots, Z_i + Z_{i+1}, \ldots, Z_j + Z_{j+1}, \ldots, Z_n, Z_{n+1}).$$

Since the double sums in Eq. (7.3) cancel each other, we obtain

$$(\Psi \circ \Psi)f(Z_1, Z_2, \ldots, Z_{n-1}, Z_n, Z_{n+1})$$

$$= (-1)^n(\Psi f)(Z_1, Z_2, \ldots, Z_{n-1}, Z_n) - (\Psi f)(Z_2, Z_3, \ldots, Z_n, Z_{n+1})$$

$$+ \sum_{i=1}^{n-1}(-1)^{i+1}\Big[(-1)^{n-1}f(Z_1, Z_2, \ldots, Z_i + Z_{i+1}, \ldots, Z_{n-1}, Z_n)$$

$$+ f(Z_2, Z_3, \ldots, Z_{i+1} + Z_{i+2}, \ldots, Z_n, Z_{n+1})\Big]$$

$$+ f(Z_1, Z_2, \ldots, Z_{n-1}) - f(Z_3, \ldots, Z_n, Z_{n+1})$$

$$= (-1)^n(\Psi f)(Z_1, Z_2, \ldots, Z_{n-1}, Z_n) - (\Psi f)(Z_2, Z_3, \ldots, Z_n, Z_{n+1})$$

$$+ (-1)^{n-1}\Big[(-1)^{n-1}f(Z_1, Z_2, \ldots, Z_{n-1}) - f(Z_2, Z_3, \ldots, Z_n)$$

$$+ \sum_{i=1}^{n-1} (-1)^{i+1} f(\mathbf{Z}_1, \mathbf{Z}_2, \ldots, \mathbf{Z}_i + \mathbf{Z}_{i+1}, \ldots, \mathbf{Z}_{n-1}, \mathbf{Z}_n) \Big]$$

$$+ \Big[(-1)^{n-1} f(\mathbf{Z}_2, \mathbf{Z}_3, \ldots, \mathbf{Z}_n) - f(\mathbf{Z}_3, \ldots, \mathbf{Z}_n, \mathbf{Z}_{n+1})$$

$$+ \sum_{i=1}^{n-1} (-1)^{i+1} f(\mathbf{Z}_2, \mathbf{Z}_3, \ldots, \mathbf{Z}_{i+1} + \mathbf{Z}_{i+2}, \ldots, \mathbf{Z}_n, \mathbf{Z}_{n+1}) \Big]$$

$$= (-1)^n (\Psi f)(\mathbf{Z}_1, \mathbf{Z}_2, \ldots, \mathbf{Z}_{n-1}, \mathbf{Z}_n) - (\Psi f)(\mathbf{Z}_2, \mathbf{Z}_3, \ldots, \mathbf{Z}_n, \mathbf{Z}_{n+1})$$

$$+ (-1)^{n-1} (\Psi f)(\mathbf{Z}_1, \mathbf{Z}_2, \ldots, \mathbf{Z}_{n-1}, \mathbf{Z}_n) + (\Psi f)(\mathbf{Z}_2, \mathbf{Z}_3, \ldots, \mathbf{Z}_n, \mathbf{Z}_{n+1}) = \mathbf{O},$$

which yields the validity of (7.2). $\qquad\qquad\square$

This formula shows that the kernel of the operator Ψ contains all mappings of the form Ψf. The next theorem provides a complete description of this kernel.

Theorem 7.2　*The general solution of the operator equation*

$$\Psi f(\mathbf{Z}_1, \mathbf{Z}_2, \ldots, \mathbf{Z}_n, \mathbf{Z}_{n+1}) = \mathbf{O} \tag{7.4}$$

in the set of analytic functions $f : \mathcal{V}^n \mapsto \mathcal{V}'$ ($n \geq 1$) is given by

$$f(\mathbf{Z}_1, \mathbf{Z}_2, \ldots, \mathbf{Z}_n) = (\Psi F)(\mathbf{Z}_1, \mathbf{Z}_2, \ldots, \mathbf{Z}_{n-1}, \mathbf{Z}_n) + L(\mathbf{Z}_1, \mathbf{Z}_2, \ldots, \mathbf{Z}_n), \tag{7.5}$$

where $F : \mathcal{V}^{n-1} \mapsto \mathcal{V}'$ is an arbitrary analytic function and L is an arbitrary mapping: $\mathcal{V}^n \mapsto \mathcal{V}'$ ($n \geq 1$) linear with respect to each argument.

Proof.　First note that if $n = 1$, the equation

$$(\Psi f)(\mathbf{Z}_1, \mathbf{Z}_2) = \mathbf{O}$$

is the Cauchy functional equation

$$f(\mathbf{Z}_1 + \mathbf{Z}_2) - f(\mathbf{Z}_1) - f(\mathbf{Z}_2) = \mathbf{O}.$$

The general analytic solution of this equation is $f(\mathbf{Z}) = A\mathbf{Z}$, where A is an $(s \times r)$-matrix with arbitrary complex constant entries ($r = \dim \mathcal{V}$ and $s = \dim \mathcal{V}'$). About the solution of the Cauchy matrix functional equation see [O. E. Gheorghiu (1963)] and [A. Kuwagaki (1962)].

Now let $n \geq 2$. The operator equation (7.4) is equivalent to

$$(-1)^n f(\mathbf{Z}_1, \mathbf{Z}_2, \ldots, \mathbf{Z}_n) - f(\mathbf{Z}_2, \mathbf{Z}_3, \ldots, \mathbf{Z}_{n+1}) \qquad (7.6)$$

$$+ \sum_{i=1}^{n} (-1)^{i+1} f(\mathbf{Z}_1, \mathbf{Z}_2, \ldots, \mathbf{Z}_i + \mathbf{Z}_{i+1}, \ldots, \mathbf{Z}_n, \mathbf{Z}_{n+1}) = \mathbf{O}.$$

Note that it is sufficient to prove the theorem if dim $\mathcal{V}' = 1$ and the general case is just a consequence. So let us assume that dim $\mathcal{V}' = 1$. Note also that f given by Eq. (7.5) is a solution of Eq. (7.4), but we want to prove that each solution is included in Eq. (7.5).

Let dim $\mathcal{V} = r$ and let $\mathbf{Z}_i = (z_{i1}, \cdots, z_{ir})^T$ $(1 \leq i \leq n + 1)$. By differentiating the equation (7.6) partially with respect to $z_{n+1,\nu}$ $(1 \leq \nu \leq r)$ at $\mathbf{Z}_{n+1} = \mathbf{O}$, we obtain the following system of r equations

$$\frac{\partial}{\partial z_{n\nu}} f(\mathbf{Z}_1, \mathbf{Z}_2, \ldots, \mathbf{Z}_n) = -p_\nu(\mathbf{Z}_2, \mathbf{Z}_3, \ldots, \mathbf{Z}_n)$$

$$+ \sum_{i=1}^{n-1} (-1)^{i+1} p_\nu(\mathbf{Z}_1, \mathbf{Z}_2, \ldots, \mathbf{Z}_i + \mathbf{Z}_{i+1}, \ldots, \mathbf{Z}_{n-1}, \mathbf{Z}_n) \qquad (1 \leq \nu \leq r),$$

where

$$\left. \frac{\partial}{\partial t_\nu} f(\mathbf{Z}_1, \mathbf{Z}_2, \ldots, \mathbf{Z}_{n-1}, \mathbf{Z}) \right|_{\mathbf{Z}=\mathbf{O}} = (-1)^n p_\nu(\mathbf{Z}_1, \mathbf{Z}_2, \ldots, \mathbf{Z}_{n-1})$$

for $\mathbf{Z} = (t_1, \ldots, t_r)$. After integration of this system we obtain

$$f(\mathbf{Z}_1, \mathbf{Z}_2, \ldots, \mathbf{Z}_n) = R(\mathbf{Z}_1, \mathbf{Z}_2, \ldots, \mathbf{Z}_{n-1}) - P(\mathbf{Z}_2, \mathbf{Z}_3, \ldots, \mathbf{Z}_n) \qquad (7.7)$$

$$+ \sum_{i=1}^{n-1} (-1)^{i+1} P(\mathbf{Z}_1, \mathbf{Z}_2, \ldots, \mathbf{Z}_i + \mathbf{Z}_{i+1}, \ldots, \mathbf{Z}_{n-1}, \mathbf{Z}_n),$$

where

$$\frac{\partial}{\partial z_{n-1,\nu}} P(\mathbf{Z}_1, \mathbf{Z}_2, \ldots, \mathbf{Z}_{n-1}) = p_\nu(\mathbf{Z}_1, \mathbf{Z}_2, \ldots, \mathbf{Z}_{n-1}) \qquad (1 \leq \nu \leq r),$$

and R is an arbitrary analytic function with respect to $\mathbf{Z}_i$ $(1 \leq i \leq n - 1)$. We write

$$R(\mathbf{Z}_1, \mathbf{Z}_2, \ldots, \mathbf{Z}_{n-1}) = (-1)^{n-1} P(\mathbf{Z}_1, \mathbf{Z}_2, \ldots, \mathbf{Z}_{n-1}) + Q(\mathbf{Z}_1, \mathbf{Z}_2, \ldots, \mathbf{Z}_{n-1}),$$

so that Eq. (7.7) becomes

$$
\begin{aligned}
f(\mathbf{Z}_1, \mathbf{Z}_2, \ldots, \mathbf{Z}_n) \;=\;& (\Psi P)(\mathbf{Z}_1, \mathbf{Z}_2, \ldots, \mathbf{Z}_{n-1}, \mathbf{Z}_n) \qquad\qquad (7.8)\\
+\;& Q(\mathbf{Z}_1, \mathbf{Z}_2, \ldots, \mathbf{Z}_{n-1}),
\end{aligned}
$$

with Q analytic function of $\mathbf{Z}_1, \mathbf{Z}_2, \ldots, \mathbf{Z}_{n-1}$.

If $f(\mathbf{Z}_1, \mathbf{Z}_2, \ldots, \mathbf{Z}_{n-1}, \mathbf{Z}_n)$ is a solution of Eq. (7.4), then

$$
(\Psi Q)(\mathbf{Z}_1, \mathbf{Z}_2, \ldots, \mathbf{Z}_{n-1}, \mathbf{Z}_n) = \mathbf{O},
$$

because $(\Psi \circ \Psi)P = \mathbf{O}$. Thus Q satisfies an equation of the form Eq. (7.4) with n replaced by $n - 1$. If $n = 2$, then $Q(\mathbf{Z}) = A\mathbf{Z}$. Otherwise we may assume that Q is given by an equality of the form Eq. (7.8) (n replaced by $n - 1$) and complete the proof by induction. $\qquad\qquad\square$

In other words, the general analytic solution of the functional equation (7.6) is given by

$$
\begin{aligned}
& f(\mathbf{Z}_1, \mathbf{Z}_2, \ldots, \mathbf{Z}_{n-1}, \mathbf{Z}_n) \qquad\qquad\qquad\qquad\qquad (7.9)\\
=\;& (-1)^{n-1} F(\mathbf{Z}_1, \mathbf{Z}_2, \ldots, \mathbf{Z}_{n-1}) - F(\mathbf{Z}_2, \mathbf{Z}_3, \ldots, \mathbf{Z}_n)
\end{aligned}
$$

$$
+ \sum_{i=1}^{n-1} (-1)^{i+1} F(\mathbf{Z}_1, \mathbf{Z}_2, \ldots, \mathbf{Z}_i + \mathbf{Z}_{i+1}, \ldots, \mathbf{Z}_{n-1}, \mathbf{Z}_n) + L(\mathbf{Z}_1, \mathbf{Z}_2, \ldots, \mathbf{Z}_n),
$$

where F is an arbitrary analytic function and L is a linear in each argument mapping.

As particular cases of the operator equation (7.4), we consider the following functional equations given in [S. Kurepa (1956); D. S. Mitrinović (with the collaboration of P. M. Vasić, 1986)] .

Example 7.1 If $n = 2$, then the functional equation (7.6) becomes

$$
f(\mathbf{Z}_1, \mathbf{Z}_2) - f(\mathbf{Z}_2, \mathbf{Z}_3) + f(\mathbf{Z}_1 + \mathbf{Z}_2, \mathbf{Z}_3) - f(\mathbf{Z}_1, \mathbf{Z}_2 + \mathbf{Z}_3) = \mathbf{O}.
$$

According to Eq. (7.9), the general analytic solution of this functional equation is given by

$$
f(\mathbf{Z}_1, \mathbf{Z}_2) = F(\mathbf{Z}_1 + \mathbf{Z}_2) - F(\mathbf{Z}_1) - F(\mathbf{Z}_2) + L(\mathbf{Z}_1, \mathbf{Z}_2).
$$

Example 7.2 If $n = 3$, the functional equation (7.6) is

$$
-f(\mathbf{Z}_1, \mathbf{Z}_2, \mathbf{Z}_3) - f(\mathbf{Z}_2, \mathbf{Z}_3, \mathbf{Z}_4) + f(\mathbf{Z}_1 + \mathbf{Z}_2, \mathbf{Z}_3, \mathbf{Z}_4)
$$

$$-f(Z_1, Z_2 + Z_3, Z_4) + f(Z_1, Z_2, Z_3 + Z_4) = O.$$

The general analytic solution of this equation is given by

$$f(Z_1, Z_2, Z_3) = F(Z_1 + Z_2, Z_3) + F(Z_1, Z_2)$$

$$-F(Z_1, Z_2 + Z_3) - F(Z_2, Z_3) + L(Z_1, Z_2, Z_3).$$

Example 7.3 If $n = 4$, the functional equation (7.6) takes on the form

$$f(Z_1, Z_2, Z_3, Z_4) - f(Z_2, Z_3, Z_4, Z_5) + f(Z_1 + Z_2, Z_3, Z_4, Z_5)$$

$$-f(Z_1, Z_2 + Z_3, Z_4, Z_5) + f(Z_1, Z_2, Z_3 + Z_4, Z_5) - f(Z_1, Z_2, Z_3, Z_4 + Z_5) = O.$$

According to Eq. (7.9), the general analytic solution of this functional equation is given by

$$f(Z_1, Z_2, Z_3, Z_4) = F(Z_1 + Z_2, Z_3, Z_4) + F(Z_1, Z_2, Z_3 + Z_4)$$

$$-F(Z_2, Z_3, Z_4) - F(Z_1, Z_2 + Z_3, Z_4) - F(Z_1, Z_2, Z_3) + L(Z_1, Z_2, Z_3, Z_4),$$

where F is an arbitrary analytic function, and L is an arbitrary linear in each argument mapping.

This method for solving functional equations does not appear in the other references [J. Aczél (1966); M. Ghermănescu (1960); M. Kuczma (1968); G. Valiron (1945)].

8   Generalized Functional Equation

In this section one generalized complex vector functional equation is solved. Further some particular cases are given.

Let $\mathcal{V}$, $\mathcal{V}'$ be arbitrary complex finite dimensional vector spaces. Now we will prove the following result obtained in [I. B. Risteski *et al.* (2000B)].

Theorem 8.1 *The general analytic solution of the functional equation*

$$(-1)^n f_{n+1}(Z_1, Z_2, \cdots, Z_n) - f_{n+2}(Z_2, Z_3, \cdots, Z_{n+1}) \qquad (8.1)$$

$$+ \sum_{i=1}^{n} (-1)^{i+1} f_i(Z_1, Z_2, \cdots, Z_i + Z_{i+1}, \cdots, Z_n, Z_{n+1}) = O$$

is given by

$$f_1(\mathbf{Z}_1, \mathbf{Z}_2, \cdots, \mathbf{Z}_n) = L(\mathbf{Z}_1, \mathbf{Z}_2, \cdots, \mathbf{Z}_n) \tag{8.2}$$

$$+(-1)^{n+1} F_{1n}(\mathbf{Z}_1, \mathbf{Z}_2, \cdots, \mathbf{Z}_{n-1}) - H_1(\mathbf{Z}_2, \mathbf{Z}_3, \cdots, \mathbf{Z}_n)$$

$$+\sum_{i=1}^{n-1}(-1)^{i+1} F_{1i}(\mathbf{Z}_1, \mathbf{Z}_2, \cdots, \mathbf{Z}_i + \mathbf{Z}_{i+1}, \cdots, \mathbf{Z}_{n-1}, \mathbf{Z}_n),$$

$$f_i(\mathbf{Z}_1, \mathbf{Z}_2, \cdots, \mathbf{Z}_n) = L(\mathbf{Z}_1, \mathbf{Z}_2, \cdots, \mathbf{Z}_n) \tag{8.3}$$

$$+(-1)^{n+1} F_{in}(\mathbf{Z}_1, \mathbf{Z}_2, \cdots, \mathbf{Z}_{n-1}) - H_i(\mathbf{Z}_2, \mathbf{Z}_3, \cdots, \mathbf{Z}_n)$$

$$+\sum_{k=1}^{i-1}(-1)^{k+1} F_{k,i-1}(\mathbf{Z}_1, \mathbf{Z}_2, \cdots, \mathbf{Z}_k + \mathbf{Z}_{k+1}, \cdots, \mathbf{Z}_{n-1}, \mathbf{Z}_n)$$

$$+\sum_{k=i}^{n-1}(-1)^{k+1} F_{ik}(\mathbf{Z}_1, \mathbf{Z}_2, \cdots, \mathbf{Z}_k + \mathbf{Z}_{k+1}, \cdots, \mathbf{Z}_{n-1}, \mathbf{Z}_n) \quad (2 \leq i \leq n)$$

$$f_{n+1}(\mathbf{Z}_1, \mathbf{Z}_2, \cdots, \mathbf{Z}_n) = L(\mathbf{Z}_1, \mathbf{Z}_2, \cdots, \mathbf{Z}_n) \tag{8.4}$$

$$+(-1)^{n+1} F_{nn}(\mathbf{Z}_1, \mathbf{Z}_2, \cdots, \mathbf{Z}_{n-1}) - H_{n+1}(\mathbf{Z}_2, \mathbf{Z}_3, \cdots, \mathbf{Z}_n)$$

$$+\sum_{i=1}^{n-1}(-1)^{i+1} F_{in}(\mathbf{Z}_1, \mathbf{Z}_2, \cdots, \mathbf{Z}_i + \mathbf{Z}_{i+1}, \cdots, \mathbf{Z}_{n-1}, \mathbf{Z}_n),$$

$$f_{n+2}(\mathbf{Z}_1, \mathbf{Z}_2, \cdots, \mathbf{Z}_n) = L(\mathbf{Z}_1, \mathbf{Z}_2, \cdots, \mathbf{Z}_n) \tag{8.5}$$

$$+(-1)^{n+1} H_{n+1}(\mathbf{Z}_1, \mathbf{Z}_2, \cdots, \mathbf{Z}_{n-1}) - H_1(\mathbf{Z}_2, \mathbf{Z}_3, \cdots, \mathbf{Z}_n)$$

$$+\sum_{i=1}^{n-1}(-1)^{i+1} H_{i+1}(\mathbf{Z}_1, \mathbf{Z}_2, \cdots, \mathbf{Z}_i + \mathbf{Z}_{i+1}, \cdots, \mathbf{Z}_{n-1}, \mathbf{Z}_n),$$

where the unknown functions $f_1, \cdots, f_{n+2}$ are mappings from $\mathcal{V}^n$ into $\mathcal{V}'$ while F_{ij} $(1 \leq i \leq j \leq n)$ and H_i $(1 \leq i \leq n+1)$ are arbitrary analytic vector functions, and $L(\mathbf{Z}_1, \mathbf{Z}_2, \cdots, \mathbf{Z}_n)$ is a mapping $\mathcal{V}^n \mapsto \mathcal{V}'$ linear with respect to each argument.

Proof. Note that it is sufficient to prove the theorem if dim $\mathcal{V}' = 1$, and then the general case is just a consequence. So let us assume that dim $\mathcal{V}' = 1$, and let dim $\mathcal{V} = m$ and $\mathbf{Z}_i = (z_{i1}, \cdots, z_{im})^T$ $(1 \leq i \leq n+1)$.

We also note that the sum of two solutions of Eq. (8.1) is also a solution of this equation, because of its linearity. Moreover,

$$f_i(\mathbf{Z}_1, \mathbf{Z}_2, \cdots, \mathbf{Z}_n) = L(\mathbf{Z}_1, \mathbf{Z}_2, \cdots, \mathbf{Z}_n) \qquad (1 \leq i \leq n+2)$$

is a solution of this equation. This solution vanishes when we put $\mathbf{Z}_j = \mathbf{O}$ for any $j \in \{1, \cdots, n\}$ what we will often do henceforth, that is why we just add it to the solution obtained below. In order to prove that Eqs. (8.2), (8.3), (8.4) and (8.5) give the general analytic solution of Eq. (8.1), first we differentiate Eq. (8.1) partially with respect to each coordinate $z_{11}, z_{12}, \cdots,$ z_{1m} and then in the result obtained we will put $\mathbf{Z}_1 = \mathbf{O}$. By using the notations

$$\frac{\partial}{\partial t_\nu} f_i(\mathbf{Z}, \mathbf{Z}_2, \cdots, \mathbf{Z}_n)\bigg|_{\mathbf{Z}=\mathbf{O}} = f_{1,i-1,\nu}(\mathbf{Z}_2, \mathbf{Z}_3, \cdots, \mathbf{Z}_n) \quad (2 \leq i \leq n+1)$$

for $\mathbf{Z} = (t_1, \cdots, t_m)$, we obtain

$$\frac{\partial}{\partial z_{2\nu}} f_1(\mathbf{Z}_2, \mathbf{Z}_3, \cdots, \mathbf{Z}_{n+1}) = f_{11\nu}(\mathbf{Z}_2 + \mathbf{Z}_3, \mathbf{Z}_4, \cdots, \mathbf{Z}_{n+1}) \qquad (8.6)$$

$$-f_{12\nu}(\mathbf{Z}_2, \mathbf{Z}_3 + \mathbf{Z}_4, \cdots, \mathbf{Z}_{n+1}) + \cdots$$

$$+(-1)^n f_{1,n-1,\nu}(\mathbf{Z}_2, \mathbf{Z}_3, \cdots, \mathbf{Z}_{n-1}, \mathbf{Z}_n + \mathbf{Z}_{n+1})$$

$$+(-1)^{n+1} f_{1n\nu}(\mathbf{Z}_2, \mathbf{Z}_3, \cdots, \mathbf{Z}_n) \qquad (1 \leq u \leq m).$$

Suppose there exist functions $F_{1i}(\mathbf{Z}_2, \mathbf{Z}_3, \cdots, \mathbf{Z}_{n+1})$, $(1 \leq i \leq n)$ such that

$$\frac{\partial}{\partial z_{2\nu}} F_{1i}(\mathbf{Z}_2, \mathbf{Z}_3, \cdots, \mathbf{Z}_n) = f_{1i\nu}(\mathbf{Z}_2, \mathbf{Z}_3, \cdots, \mathbf{Z}_n) \qquad (1 \leq i \leq n).$$

Then by an integration of Eq. (8.6) we obtain

$$f_1(\mathbf{Z}_2, \mathbf{Z}_3, \cdots, \mathbf{Z}_{n+1}) = F_{11}(\mathbf{Z}_2 + \mathbf{Z}_3, \mathbf{Z}_4, \cdots, \mathbf{Z}_{n+1}) \qquad (8.7)$$

$$-F_{12}(\mathbf{Z}_2, \mathbf{Z}_3 + \mathbf{Z}_4, \cdots, \mathbf{Z}_{n+1}) + \cdots + (-1)^n F_{1,n-1}(\mathbf{Z}_2, \cdots, \mathbf{Z}_{n-1}, \mathbf{Z}_n + \mathbf{Z}_{n+1})$$

$$+(-1)^{n+1}F_{1n}(\mathbf{Z}_2,\mathbf{Z}_3,\cdots,\mathbf{Z}_n)-H_1(\mathbf{Z}_3,\mathbf{Z}_4,\cdots,\mathbf{Z}_{n+1})$$

where $H_1(\mathbf{Z}_3,\mathbf{Z}_4,\cdots,\mathbf{Z}_{n+1})$ is an arbitrary differentiable vector function. By a substitution of Eq. (8.7) into Eq. (8.1), we obtain

$$F_{11}(\mathbf{Z}_1+\mathbf{Z}_2+\mathbf{Z}_3,\mathbf{Z}_4,\cdots,\mathbf{Z}_{n+1})-F_{12}(\mathbf{Z}_1+\mathbf{Z}_2,\mathbf{Z}_3+\mathbf{Z}_4,\mathbf{Z}_5,\cdots,\mathbf{Z}_{n+1}) \quad (8.8)$$

$$+\cdots+(-1)^n F_{1,n-1}(\mathbf{Z}_1+\mathbf{Z}_2,\mathbf{Z}_3,\cdots,\mathbf{Z}_{n-1},\mathbf{Z}_n+\mathbf{Z}_{n+1})$$

$$+(-1)^{n+1}F_{1n}(\mathbf{Z}_1+\mathbf{Z}_2,\mathbf{Z}_3,\cdots,\mathbf{Z}_n)-H_1(\mathbf{Z}_3,\mathbf{Z}_4,\cdots,\mathbf{Z}_n)$$

$$+(-1)^n f_{n+1}(\mathbf{Z}_1,\mathbf{Z}_2,\cdots,\mathbf{Z}_n)-f_{n+2}(\mathbf{Z}_2,\mathbf{Z}_3,\cdots,\mathbf{Z}_{n+1})$$

$$+\sum_{i=2}^{n}(-1)^{i+1}f_i(\mathbf{Z}_1,\mathbf{Z}_2,\cdots,\mathbf{Z}_{i-1},\mathbf{Z}_i+\mathbf{Z}_{i+1},\cdots,\mathbf{Z}_{n+1})=\mathbf{O}.$$

If we put $\mathbf{Z}_2=\mathbf{O}$ in Eq. (8.8) and replace $\mathbf{Z}_{i+1}$ by $\mathbf{Z}_i$ $(2\le i\le n)$, we obtain

$$f_2(\mathbf{Z}_1,\mathbf{Z}_2,\cdots,\mathbf{Z}_n)=F_{11}(\mathbf{Z}_1+\mathbf{Z}_2,\mathbf{Z}_3,\cdots,\mathbf{Z}_n) \quad (8.9)$$

$$-F_{22}(\mathbf{Z}_1,\mathbf{Z}_2+\mathbf{Z}_3,\cdots,\mathbf{Z}_n)+\cdots$$

$$+(-1)^n F_{2,n-1}(\mathbf{Z}_1,\mathbf{Z}_2,\cdots,\mathbf{Z}_{n-2},\mathbf{Z}_{n-1}+\mathbf{Z}_n)$$

$$+(-1)^{n+1}F_{2n}(\mathbf{Z}_1,\mathbf{Z}_2,\cdots,\mathbf{Z}_{n-1})-H_2(\mathbf{Z}_2,\mathbf{Z}_3,\cdots,\mathbf{Z}_n),$$

where

$$F_{2i}(\mathbf{Z}_1,\mathbf{Z}_2,\cdots,\mathbf{Z}_{n-1})=F_{1i}(\mathbf{Z}_1,\mathbf{Z}_2,\cdots,\mathbf{Z}_{n-1})-f_{i+1}(\mathbf{Z}_1,\mathbf{O},\mathbf{Z}_2,\cdots,\mathbf{Z}_{n-1})$$

$$(2\le i\le n)$$

and

$$H_2(\mathbf{Z}_2,\mathbf{Z}_3,\cdots,\mathbf{Z}_n)=H_1(\mathbf{Z}_2,\mathbf{Z}_3,\cdots,\mathbf{Z}_n)+f_{n+2}(\mathbf{O},\mathbf{Z}_2,\mathbf{Z}_3,\cdots,\mathbf{Z}_n).$$

By a substitution of Eq. (8.9) into Eq. (8.8), we obtain

$$-F_{12}(\mathbf{Z}_1+\mathbf{Z}_2,\mathbf{Z}_3+\mathbf{Z}_4,\mathbf{Z}_5,\cdots,\mathbf{Z}_{n+1})+F_{13}(\mathbf{Z}_1+\mathbf{Z}_2,\mathbf{Z}_3,\mathbf{Z}_4+\mathbf{Z}_5,\cdots,\mathbf{Z}_{n+1})$$

$$-\cdots+(-1)^n F_{1,n-1}(\mathbf{Z}_1+\mathbf{Z}_2,\mathbf{Z}_3,\cdots,\mathbf{Z}_{n-1},\mathbf{Z}_n+\mathbf{Z}_{n+1}) \qquad (8.10)$$

$$(-1)^{n+1}F_{1n}(\mathbf{Z}_1+\mathbf{Z}_2,\mathbf{Z}_3,\cdots,\mathbf{Z}_n)+F_{22}(\mathbf{Z}_1,\mathbf{Z}_2+\mathbf{Z}_3+\mathbf{Z}_4,\mathbf{Z}_5,\cdots,\mathbf{Z}_{n+1})$$

$$-F_{23}(\mathbf{Z}_1,\mathbf{Z}_2+\mathbf{Z}_3,\mathbf{Z}_4+\mathbf{Z}_5,\cdots,\mathbf{Z}_{n+1})$$

$$+\cdots+(-1)^{n+1}F_{2,n-1}(\mathbf{Z}_1,\mathbf{Z}_2+\mathbf{Z}_3,\cdots,\mathbf{Z}_{n-1},\mathbf{Z}_n+\mathbf{Z}_{n+1})$$

$$+(-1)^{n+2}F_{2n}(\mathbf{Z}_1,\mathbf{Z}_2+\mathbf{Z}_3,\mathbf{Z}_4,\cdots,\mathbf{Z}_n)$$

$$-H_1(\mathbf{Z}_3,\mathbf{Z}_4,\cdots,\mathbf{Z}_{n+1})+H_2(\mathbf{Z}_2+\mathbf{Z}_3,\mathbf{Z}_4,\cdots,\mathbf{Z}_{n+1})$$

$$+(-1)^n f_{n+1}(\mathbf{Z}_1,\mathbf{Z}_2,\cdots,\mathbf{Z}_n)-f_{n+2}(\mathbf{Z}_2,\mathbf{Z}_3,\cdots,\mathbf{Z}_{n+1})$$

$$+\sum_{i=3}^{n}(-1)^{i+1}f_i(\mathbf{Z}_1,\mathbf{Z}_2,\cdots,\mathbf{Z}_{i-1},\mathbf{Z}_i+\mathbf{Z}_{i+1},\cdots,\mathbf{Z}_n+\mathbf{Z}_{n+1})=\mathbf{O}.$$

Now by induction we will prove that the first k ($2 \leq k \leq n$) unknown functions f_j ($1 \leq j \leq k$) which satisfy the functional equation (8.1) can be represented in the following form Eq. (8.7),

$$f_j(\mathbf{Z}_1,\mathbf{Z}_2,\cdots,\mathbf{Z}_n)=F_{1,j-1}(\mathbf{Z}_1+\mathbf{Z}_2,\mathbf{Z}_3,\cdots,\mathbf{Z}_n) \qquad (8.11)$$

$$-F_{2,j-1}(\mathbf{Z}_1,\mathbf{Z}_2+\mathbf{Z}_3,\cdots,\mathbf{Z}_n)+\cdots$$

$$+(-1)^j F_{j-1,j-1}(\mathbf{Z}_1,\mathbf{Z}_2,\cdots,\mathbf{Z}_{j-1}+\mathbf{Z}_j,\cdots,\mathbf{Z}_n)$$

$$+(-1)^{j+1}F_{jj}(\mathbf{Z}_1,\mathbf{Z}_2,\cdots,\mathbf{Z}_j+\mathbf{Z}_{j+1},\cdots,\mathbf{Z}_n)$$

$$+(-1)^{j+2}F_{j,j+1}(\mathbf{Z}_1,\mathbf{Z}_2,\cdots,\mathbf{Z}_{j+1}+\mathbf{Z}_{j+2},\cdots,\mathbf{Z}_n)$$

$$+\cdots+(-1)^n F_{j,n-1}(\mathbf{Z}_1,\mathbf{Z}_2,\cdots,\mathbf{Z}_{n-2},\mathbf{Z}_{n-1}+\mathbf{Z}_n)$$

$$+(-1)^{n+1}F_{jn}(\mathbf{Z}_1,\mathbf{Z}_2,\cdots,\mathbf{Z}_{n-1})-H_j(\mathbf{Z}_2,\mathbf{Z}_3,\cdots,\mathbf{Z}_n) \quad (2 \leq j \leq k),$$

and the remaining $n - k + 2$ unknown functions will satisfy the following equation

$$(-1)^{k+1} F_{1k}(\mathbf{Z}_1 + \mathbf{Z}_2, \mathbf{Z}_3, \cdots, \mathbf{Z}_{k+1} + \mathbf{Z}_{k+2}, \cdots, \mathbf{Z}_{n+1}) \qquad (8.12)$$

$$+ (-1)^{k+2} F_{1,k+1}(\mathbf{Z}_1 + \mathbf{Z}_2, \mathbf{Z}_3, \cdots, \mathbf{Z}_{k+2} + \mathbf{Z}_{k+3}, \cdots, \mathbf{Z}_{n+1})$$

$$+ \cdots + (-1)^n F_{1,n-1}(\mathbf{Z}_1 + \mathbf{Z}_2, \mathbf{Z}_3, \cdots, \mathbf{Z}_n + \mathbf{Z}_{n+1})$$

$$+ (-1)^{n+1} F_{1n}(\mathbf{Z}_1 + \mathbf{Z}_2, \mathbf{Z}_3, \cdots, \mathbf{Z}_n)$$

$$- [(-1)^{k+1} F_{2k}(\mathbf{Z}_1, \mathbf{Z}_2 + \mathbf{Z}_3, \mathbf{Z}_4, \cdots, \mathbf{Z}_{k+1} + \mathbf{Z}_{k+2}, \cdots, \mathbf{Z}_{n+1})$$

$$+ (-1)^{k+2} F_{2,k+1}(\mathbf{Z}_1, \mathbf{Z}_2 + \mathbf{Z}_3, \mathbf{Z}_4, \cdots, \mathbf{Z}_{k+2} + \mathbf{Z}_{k+3}, \cdots, \mathbf{Z}_{n+1}) + \cdots$$

$$+ (-1)^n F_{2,n-1}(\mathbf{Z}_1, \mathbf{Z}_2 + \mathbf{Z}_3, \mathbf{Z}_4, \cdots, \mathbf{Z}_n + \mathbf{Z}_{n+1})$$

$$+ (-1)^{n+1} F_{2n}(\mathbf{Z}_1, \mathbf{Z}_2 + \mathbf{Z}_3, \mathbf{Z}_4, \cdots, \mathbf{Z}_n)]$$

$$+ \cdots + (-1)^{k-1}[(-1)^{k+1} F_{kk}(\mathbf{Z}_1, \cdots, \mathbf{Z}_k + \mathbf{Z}_{k+1} + \mathbf{Z}_{k+2}, \cdots, \mathbf{Z}_{n+1})$$

$$+ (-1)^{k+2} F_{k,k+1}(\mathbf{Z}_1, \cdots, \mathbf{Z}_k + \mathbf{Z}_{k+1}, \mathbf{Z}_{k+2} + \mathbf{Z}_{k+3}, \cdots, \mathbf{Z}_{n+1})$$

$$+ \cdots + (-1)^n F_{k,n-1}(\mathbf{Z}_1, \cdots, \mathbf{Z}_k + \mathbf{Z}_{k+1}, \mathbf{Z}_{k+2}, \cdots, \mathbf{Z}_n + \mathbf{Z}_{n+1})$$

$$+ (-1)^{n+1} F_{kn}(\mathbf{Z}_1, \cdots, \mathbf{Z}_k + \mathbf{Z}_{k+1}, \cdots, \mathbf{Z}_n)]$$

$$- H_1(\mathbf{Z}_3, \mathbf{Z}_4, \cdots, \mathbf{Z}_{n+1}) + H_2(\mathbf{Z}_2 + \mathbf{Z}_3, \mathbf{Z}_4, \cdots, \mathbf{Z}_{n+1})$$

$$- \cdots + (-1)^k H_k(\mathbf{Z}_2, \cdots, \mathbf{Z}_k + \mathbf{Z}_{k+1}, \cdots, \mathbf{Z}_{n+1})$$

$$+ (-1)^n f_{n+1}(\mathbf{Z}_1, \mathbf{Z}_2, \cdots, \mathbf{Z}_n) - f_{n+2}(\mathbf{Z}_2, \mathbf{Z}_3, \cdots, \mathbf{Z}_{n+1})$$

$$+ \sum_{i=k+1}^{n} (-1)^{i+1} f_i(\mathbf{Z}_1, \cdots, \mathbf{Z}_i + \mathbf{Z}_{i+1}, \cdots, \mathbf{Z}_{n+1}) = \mathbf{O}.$$

Indeed, from Eqs. (8.7), (8.9) and (8.10) it follows that the statement is true for $k = 2$. We will suppose that it holds for k ($2 \leq k \leq n - 1$). If we put $\mathbf{Z}_{k+1} = \mathbf{O}$ and replace $\mathbf{Z}_{i+1}$ by $\mathbf{Z}_i$ $(k + 1 \leq i \leq n)$ in Eq. (8.12), we obtain

$$f_{k+1}(\mathbf{Z}_1, \mathbf{Z}_2, \cdots, \mathbf{Z}_n) = F_{1k}(\mathbf{Z}_1 + \mathbf{Z}_2, \mathbf{Z}_3, \cdots, \mathbf{Z}_n) \qquad (8.13)$$

$$-F_{2k}(\mathbf{Z}_1, \mathbf{Z}_2 + \mathbf{Z}_3, \cdots, \mathbf{Z}_n) + \cdots$$

$$+(-1)^{k+1}F_{kk}(\mathbf{Z}_1, \cdots, \mathbf{Z}_{k-1}, \mathbf{Z}_k + \mathbf{Z}_{k+1}, \cdots, \mathbf{Z}_n)$$

$$+[f_{k+2}(\mathbf{Z}_1, \cdots, \mathbf{Z}_k, \mathbf{O}, \mathbf{Z}_{k+1} + \mathbf{Z}_{k+2}, \cdots, \mathbf{Z}_n)$$

$$-F_{1,k+1}(\mathbf{Z}_1 + \mathbf{Z}_2, \mathbf{Z}_3, \cdots, \mathbf{Z}_k, \mathbf{O}, \mathbf{Z}_{k+1} + \mathbf{Z}_{k+2}, \cdots, \mathbf{Z}_n)$$

$$+F_{2,k+1}(\mathbf{Z}_1, \mathbf{Z}_2 + \mathbf{Z}_3, \mathbf{Z}_4, \cdots, \mathbf{Z}_k, \mathbf{O}, \mathbf{Z}_{k+1} + \mathbf{Z}_{k+2}, \cdots, \mathbf{Z}_n) - \cdots$$

$$+(-1)^{k-1}F_{k-1,k+1}(\mathbf{Z}_1, \cdots, \mathbf{Z}_{k-2}, \mathbf{Z}_{k-1} + \mathbf{Z}_k, \mathbf{O}, \mathbf{Z}_{k+1} + \mathbf{Z}_{k+2}, \mathbf{Z}_{k+3}, \cdots, \mathbf{Z}_n)$$

$$+(-1)^{k}F_{k,k+1}(\mathbf{Z}_1, \cdots, \mathbf{Z}_k, \mathbf{Z}_{k+1} + \mathbf{Z}_{k+2}, \cdots, \mathbf{Z}_n)]$$

$$-[f_{k+3}(\mathbf{Z}_1, \cdots, \mathbf{Z}_k, \mathbf{O}, \mathbf{Z}_{k+1}, \mathbf{Z}_{k+2} + \mathbf{Z}_{k+3}, \cdots, \mathbf{Z}_n)$$

$$-F_{1,k+2}(\mathbf{Z}_1 + \mathbf{Z}_2, \mathbf{Z}_3, \cdots, \mathbf{Z}_k, \mathbf{O}, \mathbf{Z}_{k+1}, \mathbf{Z}_{k+2} + \mathbf{Z}_{k+3}, \cdots, \mathbf{Z}_n)$$

$$+F_{2,k+2}(\mathbf{Z}_1, \mathbf{Z}_2 + \mathbf{Z}_3, \mathbf{Z}_4, \cdots, \mathbf{Z}_k, \mathbf{O}, \mathbf{Z}_{k+1}, \mathbf{Z}_{k+2} + \mathbf{Z}_{k+3}, \cdots, \mathbf{Z}_n) - \cdots$$

$$+(-1)^{k-1}F_{k-1,k+2}(\mathbf{Z}_1, \cdots, \mathbf{Z}_{k-2}, \mathbf{Z}_{k-1} + \mathbf{Z}_k, \mathbf{O}, \mathbf{Z}_{k+1}, \mathbf{Z}_{k+2} + \mathbf{Z}_{k+3}, \cdots, \mathbf{Z}_n)$$

$$+(-1)^{k}F_{k,k+2}(\mathbf{Z}_1, \cdots, \mathbf{Z}_{k+1}, \mathbf{Z}_{k+2} + \mathbf{Z}_{k+3}, \cdots, \mathbf{Z}_n)]$$

$$+ \cdots + (-1)^{n-k-2}[f_n(\mathbf{Z}_1, \cdots, \mathbf{Z}_k, \mathbf{O}, \mathbf{Z}_{k+1}, \cdots, \mathbf{Z}_{n-2}, \mathbf{Z}_{n-1} + \mathbf{Z}_n)$$

$$-F_{1,n-1}(\mathbf{Z}_1 + \mathbf{Z}_2, \mathbf{Z}_3, \cdots, \mathbf{Z}_k, \mathbf{O}, \mathbf{Z}_{k+1}, \cdots, \mathbf{Z}_{n-2}, \mathbf{Z}_{n-1} + \mathbf{Z}_n)$$

$$+F_{2,n-1}(\mathbf{Z}_1, \mathbf{Z}_2 + \mathbf{Z}_3, \mathbf{Z}_4, \cdots, \mathbf{Z}_k, \mathbf{O}, \mathbf{Z}_{k+1}, \cdots, \mathbf{Z}_{n-2}, \mathbf{Z}_{n-1} + \mathbf{Z}_n) - \cdots$$

$$+(-1)^{k-1}F_{k-1,n-1}(\mathbf{Z}_1, \cdots, \mathbf{Z}_{k-2}, \mathbf{Z}_{k-1} + \mathbf{Z}_k, \mathbf{O}, \mathbf{Z}_{k+1}, \cdots, \mathbf{Z}_{n-2}, \mathbf{Z}_{n-1} + \mathbf{Z}_n)$$

$$+(-1)^{k}F_{k,n-1}(\mathbf{Z}_1, \cdots, \mathbf{Z}_{n-2}, \mathbf{Z}_{n-1} + \mathbf{Z}_n)]$$

$$+(-1)^{n-k-1}[f_{n+1}(\mathbf{Z}_1,\cdots,\mathbf{Z}_k,\mathbf{O},\mathbf{Z}_{k+1},\cdots,\mathbf{Z}_{n-1})$$

$$-F_{1n}(\mathbf{Z}_1+\mathbf{Z}_2,\mathbf{Z}_3,\cdots,\mathbf{Z}_k,\mathbf{O},\mathbf{Z}_{k+1},\cdots,\mathbf{Z}_{n-1})$$

$$+F_{2n}(\mathbf{Z}_1,\mathbf{Z}_2+\mathbf{Z}_3,\mathbf{Z}_4,\cdots,\mathbf{Z}_k,\mathbf{O},\mathbf{Z}_{k+1},\cdots,\mathbf{Z}_{n-1})-\cdots$$

$$+(-1)^{k-1}F_{k-1,n}(\mathbf{Z}_1,\cdots,\mathbf{Z}_{k-2},\mathbf{Z}_{k-1}+\mathbf{Z}_k,\mathbf{O},\mathbf{Z}_{k+1},\cdots,\mathbf{Z}_{n-1})$$

$$+(-1)^k F_{kn}(\mathbf{Z}_1,\mathbf{Z}_2,\cdots,\mathbf{Z}_{n-1})]$$

$$+(-1)^k[f_{n+2}(\mathbf{Z}_2,\cdots,\mathbf{Z}_k,\mathbf{O},\mathbf{Z}_{k+1},\cdots,\mathbf{Z}_n)$$

$$+H_1(\mathbf{Z}_3,\cdots,\mathbf{Z}_k,\mathbf{O},\mathbf{Z}_{k+1},\cdots,\mathbf{Z}_n)$$

$$-H_2(\mathbf{Z}_2+\mathbf{Z}_3,\mathbf{Z}_4,\cdots,\mathbf{Z}_k,\mathbf{O},\mathbf{Z}_{k+1},\cdots,\mathbf{Z}_n)$$

$$+\cdots+(-1)^{k-2}H_{k-1}(\mathbf{Z}_1,\cdots,\mathbf{Z}_{k-2},\mathbf{Z}_{k-1}+\mathbf{Z}_k,\mathbf{O},\mathbf{Z}_{k+1},\cdots,\mathbf{Z}_n)$$

$$+(-1)^{k-1}H_k(\mathbf{Z}_2,\mathbf{Z}_3,\cdots,\mathbf{Z}_n)].$$

Now we define

$$(-1)^{k+2}F_{k+1,k+i}(\mathbf{Z}_1,\mathbf{Z}_2,\cdots,\mathbf{Z}_{n-1}) \tag{8.14}$$
$$= \; F_{k+i+1}(\mathbf{Z}_1,\cdots,\mathbf{Z}_k,\mathbf{O},\mathbf{Z}_{k+1},\cdots,\mathbf{Z}_{n-1})$$

$$-F_{1,k+i}(\mathbf{Z}_1+\mathbf{Z}_2,\mathbf{Z}_3,\cdots,\mathbf{Z}_k,\mathbf{O},\mathbf{Z}_{k+1},\cdots,\mathbf{Z}_{n-1})$$

$$+F_{2,k+i}(\mathbf{Z}_1,\mathbf{Z}_2+\mathbf{Z}_3,\mathbf{Z}_4,\cdots,\mathbf{Z}_k,\mathbf{O},\mathbf{Z}_{k+1},\cdots,\mathbf{Z}_{n-1})$$

$$-\cdots+(-1)^{k-1}F_{k-1,k+i}(\mathbf{Z}_1,\cdots,\mathbf{Z}_{k-2},\mathbf{Z}_{k-1}+\mathbf{Z}_k,\mathbf{O},\mathbf{Z}_{k+1},\cdots,\mathbf{Z}_{n-1})$$

$$+(-1)^k F_{k,k+i}(\mathbf{Z}_1,\mathbf{Z}_2,\cdots,\mathbf{Z}_{n-1}) \quad (1\le i\le n-k),$$

$$H_{k+1}(\mathbf{Z}_2,\mathbf{Z}_3,\cdots,\mathbf{Z}_n)=(-1)^{k-1}[f_{n+2}(\mathbf{Z}_2,\cdots,\mathbf{Z}_k,\mathbf{O},\mathbf{Z}_{k+1},\cdots,\mathbf{Z}_n)$$

$$+H_1(\mathbf{Z}_3,\cdots,\mathbf{Z}_k,\mathbf{O},\mathbf{Z}_{k+1},\cdots,\mathbf{Z}_n) \tag{8.15}$$

$$-H_2(\mathbf{Z}_2+\mathbf{Z}_3,\mathbf{Z}_4,\cdots,\mathbf{Z}_k,\mathbf{O},\mathbf{Z}_{k+1},\cdots,\mathbf{Z}_n)$$

$$+ \cdots + (-1)^{k-2} H_{k-1}(\mathbf{Z}_2, \cdots, \mathbf{Z}_{k-2}, \mathbf{Z}_{k-1} + \mathbf{Z}_k, \mathbf{O}, \mathbf{Z}_{k+1}, \cdots, \mathbf{Z}_n)$$

$$+ (-1)^{k-1} H_k(\mathbf{Z}_2, \mathbf{Z}_3, \cdots, \mathbf{Z}_n)].$$

Finally Eq. (8.13) together with Eqs. (8.14) and (8.15) yield

$$f_{k+1}(\mathbf{Z}_1, \mathbf{Z}_2, \cdots, \mathbf{Z}_n) = F_{1k}(\mathbf{Z}_1 + \mathbf{Z}_2, \mathbf{Z}_3, \cdots, \mathbf{Z}_n) \qquad (8.16)$$

$$- F_{2k}(\mathbf{Z}_1, \mathbf{Z}_2 + \mathbf{Z}_3, \mathbf{Z}_4, \cdots, \mathbf{Z}_n) + \cdots$$

$$+ (-1)^{k+1} F_{kk}(\mathbf{Z}_1, \cdots, \mathbf{Z}_{k-1}, \mathbf{Z}_k + \mathbf{Z}_{k+1}, \cdots, \mathbf{Z}_n)$$

$$+ (-1)^{k+2} F_{k+1,k+1}(\mathbf{Z}_1, \cdots, \mathbf{Z}_k, \mathbf{Z}_{k+1} + \mathbf{Z}_{k+2}, \cdots, \mathbf{Z}_n)$$

$$+ (-1)^{k+3} F_{k+1,k+2}(\mathbf{Z}_1, \cdots, \mathbf{Z}_{k+1}, \mathbf{Z}_{k+2} + \mathbf{Z}_{k+3}, \cdots, \mathbf{Z}_n)$$

$$+ \cdots + (-1)^n F_{k+1,n-1}(\mathbf{Z}_1, \cdots, \mathbf{Z}_{n-2}, \mathbf{Z}_{n-1} + \mathbf{Z}_n) +$$

$$+ (-1)^{n+1} F_{k+1,n}(\mathbf{Z}_1, \mathbf{Z}_2, \cdots, \mathbf{Z}_{n-1}) - H_{k+1}(\mathbf{Z}_2, \mathbf{Z}_3, \cdots, \mathbf{Z}_n).$$

By a substitution of Eq. (8.16) into Eq. (8.12), after elimination of $F_{1k}, F_{2k}, \cdots, F_{kk}$ we obtain

$$(-1)^{k+2} F_{1,k+1}(\mathbf{Z}_1 + \mathbf{Z}_2, \mathbf{Z}_3, \cdots, \mathbf{Z}_{k+1}, \mathbf{Z}_{k+2} + \mathbf{Z}_{k+3}, \cdots, \mathbf{Z}_{n+1}) \qquad (8.17)$$

$$+ \cdots + (-1)^n F_{1,n-1}(\mathbf{Z}_1 + \mathbf{Z}_2, \mathbf{Z}_3, \cdots, \mathbf{Z}_n + \mathbf{Z}_{n+1})$$

$$+ (-1)^{n+1} F_{1n}(\mathbf{Z}_1 + \mathbf{Z}_2, \mathbf{Z}_3, \cdots, \mathbf{Z}_n)$$

$$- [(-1)^{k+2} F_{2,k+1}(\mathbf{Z}_1, \mathbf{Z}_2 + \mathbf{Z}_3, \mathbf{Z}_4, \cdots, \mathbf{Z}_{k+2} + \mathbf{Z}_{k+3}, \cdots, \mathbf{Z}_{n+1})$$

$$+ \cdots + (-1)^n F_{2,n-1}(\mathbf{Z}_1, \mathbf{Z}_2 + \mathbf{Z}_3, \mathbf{Z}_4, \cdots, \mathbf{Z}_n + \mathbf{Z}_{n+1})$$

$$+ (-1)^{n+1} F_{2n}(\mathbf{Z}_1, \mathbf{Z}_2 + \mathbf{Z}_3, \mathbf{Z}_4, \cdots, \mathbf{Z}_n)] + \cdots$$

$$+ (-1)^{k+1}[(-1)^{k+2} F_{k,k+1}(\mathbf{Z}_1, \cdots, \mathbf{Z}_{k-1}, \mathbf{Z}_k + \mathbf{Z}_{k+1}, \mathbf{Z}_{k+2} + \mathbf{Z}_{k+3}, \cdots, \mathbf{Z}_{n+1})$$

$$+ \cdots + (-1)^n F_{k,n-1}(\mathbf{Z}_1, \cdots, \mathbf{Z}_{k-1}, \mathbf{Z}_k + \mathbf{Z}_{k+1}, \mathbf{Z}_{k+2}, \cdots, \mathbf{Z}_n + \mathbf{Z}_{n+1})$$

$$+ (-1)^{n+1} F_{kn}(\mathbf{Z}_1, \cdots, \mathbf{Z}_{k-1}, \mathbf{Z}_k + \mathbf{Z}_{k+1}, \cdots, \mathbf{Z}_n)]$$

$$+ (-1)^k [(-1)^{k+2} F_{k+1,k+1}(\mathbf{Z}_1, \cdots, \mathbf{Z}_k, \mathbf{Z}_{k+1} + \mathbf{Z}_{k+2} + \mathbf{Z}_{k+3}, \cdots, \mathbf{Z}_{n+1}) + \cdots$$

$$+ (-1)^n F_{k+1,n-1}(\mathbf{Z}_1, \cdots, \mathbf{Z}_k, \mathbf{Z}_{k+1} + \mathbf{Z}_{k+2}, \mathbf{Z}_{k+3}, \cdots, \mathbf{Z}_n + \mathbf{Z}_{n+1})$$

$$+ (-1)^{n+1} F_{k+1,n}(\mathbf{Z}_1, \cdots, \mathbf{Z}_k, \mathbf{Z}_{k+1} + \mathbf{Z}_{k+2}, \cdots, \mathbf{Z}_n)]$$

$$- H_1(\mathbf{Z}_3, \mathbf{Z}_4, \cdots, \mathbf{Z}_{n+1}) + H_2(\mathbf{Z}_2 + \mathbf{Z}_3, \mathbf{Z}_4, \cdots, \mathbf{Z}_{n+1})$$

$$- \cdots + (-1)^k H_k(\mathbf{Z}_2, \cdots, \mathbf{Z}_{k-1}, \mathbf{Z}_k + \mathbf{Z}_{k+1}, \cdots, \mathbf{Z}_{n+1})$$

$$+ (-1)^{k+1} H_{k+1}(\mathbf{Z}_2, \cdots, \mathbf{Z}_k, \mathbf{Z}_{k+1} + \mathbf{Z}_{k+2}, \cdots, \mathbf{Z}_{n+1})$$

$$+ (-1)^n f_{n+1}(\mathbf{Z}_1, \mathbf{Z}_2, \cdots, \mathbf{Z}_n) - f_{n+2}(\mathbf{Z}_2, \mathbf{Z}_3, \cdots, \mathbf{Z}_{n+1})$$

$$+ \sum_{i=k+2}^{n} (-1)^{i+1} f_i(\mathbf{Z}_1, \cdots, \mathbf{Z}_{i-1}, \mathbf{Z}_i + \mathbf{Z}_{i+1}, \cdots, \mathbf{Z}_{n+1}) = \mathbf{O}.$$

It follows from Eqs. (8.16) and (8.17) that the statement also holds for $k+1$. Hence, Eqs. (8.11) and (8.12) hold for $2 \leq k \leq n$. If we substitute $k = n$ in Eq. (8.12), we obtain

$$(-1)^{n+1} F_{1n}(\mathbf{Z}_1 + \mathbf{Z}_2, \mathbf{Z}_3, \cdots, \mathbf{Z}_n) \tag{8.18}$$

$$+ (-1)^{n+2} F_{2n}(\mathbf{Z}_1, \mathbf{Z}_2 + \mathbf{Z}_3, \mathbf{Z}_4, \cdots, \mathbf{Z}_n) + \cdots$$

$$+ (-1)^{2n-1} F_{n-1,n}(\mathbf{Z}_1, \cdots, \mathbf{Z}_{n-2}, \mathbf{Z}_{n-1} + \mathbf{Z}_n) + (-1)^{2n} F_{nn}(\mathbf{Z}_1, \mathbf{Z}_2, \cdots, \mathbf{Z}_{n-1})$$

$$- H_1(\mathbf{Z}_3, \mathbf{Z}_4, \cdots, \mathbf{Z}_{n+1}) + H_2(\mathbf{Z}_2 + \mathbf{Z}_3, \mathbf{Z}_4, \cdots, \mathbf{Z}_{n+1}) - \cdots$$

$$+ (-1)^n H_n(\mathbf{Z}_2, \cdots, \mathbf{Z}_{n-1}, \mathbf{Z}_n + \mathbf{Z}_{n+1}) + (-1)^n f_{n+1}(\mathbf{Z}_1, \mathbf{Z}_2, \cdots, \mathbf{Z}_n)$$

$$- f_{n+2}(\mathbf{Z}_2, \mathbf{Z}_3, \cdots, \mathbf{Z}_{n+1}) = \mathbf{O}.$$

Now if we substitute $\mathbf{Z}_{n+1} = \mathbf{O}$ in Eq. (8.18), we obtain

$$f_{n+1}(\mathbf{Z}_1, \mathbf{Z}_2, \cdots, \mathbf{Z}_n) = F_{1n}(\mathbf{Z}_1 + \mathbf{Z}_2, \mathbf{Z}_3, \cdots, \mathbf{Z}_n) \tag{8.19}$$

$$-F_{2n}(\mathbf{Z}_1, \mathbf{Z}_2 + \mathbf{Z}_3, \mathbf{Z}_4, \cdots, \mathbf{Z}_n) + \cdots$$

$$+(-1)^n F_{n-1,n}(\mathbf{Z}_1, \cdots, \mathbf{Z}_{n-2}, \mathbf{Z}_{n-1} + \mathbf{Z}_n)$$

$$+(-1)^{n+1} F_{nn}(\mathbf{Z}_1, \mathbf{Z}_2, \cdots, \mathbf{Z}_{n-1}) - H_{n+1}(\mathbf{Z}_2, \mathbf{Z}_3, \cdots, \mathbf{Z}_n)$$

where

$$H_{n+1}(\mathbf{Z}_2, \mathbf{Z}_3, \cdots, \mathbf{Z}_n) = (-1)^{n+1}[H_1(\mathbf{Z}_3, \mathbf{Z}_4, \cdots, \mathbf{Z}_n, \mathbf{O})$$

$$-H_2(\mathbf{Z}_2 + \mathbf{Z}_3, \mathbf{Z}_4, \cdots, \mathbf{Z}_n, \mathbf{O})$$

$$+\cdots+(-1)^n H_{n-1}(\mathbf{Z}_2, \cdots, \mathbf{Z}_{n-2}, \mathbf{Z}_{n-1} + \mathbf{Z}_n, \mathbf{O})$$

$$+(-1)^{n+1} H_n(\mathbf{Z}_2, \mathbf{Z}_3, \cdots, \mathbf{Z}_n) + f_{n+2}(\mathbf{Z}_2, \mathbf{Z}_3, \cdots, \mathbf{Z}_n, \mathbf{O})].$$

By a substitution of Eq. (3.19) into Eq. (8.18) we obtain

$$-H_1(\mathbf{Z}_3, \mathbf{Z}_4, \cdots, \mathbf{Z}_{n+1}) + H_2(\mathbf{Z}_2 + \mathbf{Z}_3, \mathbf{Z}_4, \cdots, \mathbf{Z}_{n+1})$$

$$-\cdots+(-1)^n H_n(\mathbf{Z}_2, \mathbf{Z}_3, \cdots, \mathbf{Z}_{n-1}, \mathbf{Z}_n + \mathbf{Z}_{n+1})$$

$$+(-1)^{n+1} H_{n+1}(\mathbf{Z}_2, \mathbf{Z}_3, \cdots, \mathbf{Z}_n) - f_{n+2}(\mathbf{Z}_2, \mathbf{Z}_3, \cdots, \mathbf{Z}_{n+1}) = \mathbf{O},$$

i.e.,

$$f_{n+2}(\mathbf{Z}_1, \mathbf{Z}_2, \cdots, \mathbf{Z}_n) = H_2(\mathbf{Z}_1 + \mathbf{Z}_2, \mathbf{Z}_3, \cdots, \mathbf{Z}_n) \qquad (8.20)$$

$$-H_3(\mathbf{Z}_1, \mathbf{Z}_2 + \mathbf{Z}_3, \mathbf{Z}_4, \cdots, \mathbf{Z}_n) + \cdots$$

$$+(-1)^n H_n(\mathbf{Z}_1, \cdots, \mathbf{Z}_{n-2}, \mathbf{Z}_{n-1} + \mathbf{Z}_n)$$

$$+(-1)^{n+1} H_{n+1}(\mathbf{Z}_1, \mathbf{Z}_2, \cdots, \mathbf{Z}_{n-1}) - H_1(\mathbf{Z}_2, \mathbf{Z}_3, \cdots, \mathbf{Z}_n).$$

By Eqs. (8.7) and (8.11) for $k = n$, Eq. (8.19) and (8.20) the theorem is proved. $\qquad \square$

This theorem is a generalization of Theorem 7.2. Indeed, in the special case when $f_i = f$ $(1 \le i \le n+2)$ we obtain Theorem 7.2.

As particular cases of the functional equations Eq. (8.1) we consider the following examples.

Example 8.1   If $n = 1$, then the functional equation (8.1) takes the Pexider form

$$-f_2(\mathbf{Z}_1) - f_3(\mathbf{Z}_2) + f_1(\mathbf{Z}_1 + \mathbf{Z}_2) = \mathbf{O}.$$

The general analytic solution of the equation is given by

$$
\begin{aligned}
f_1(\mathbf{Z}) &= C\mathbf{Z} + A + B, \\
f_2(\mathbf{Z}) &= C\mathbf{Z} + A, \\
f_3(\mathbf{Z}) &= C\mathbf{Z} + B,
\end{aligned}
$$

where C is a constant matrix and A and B are constant complex vectors.

Example 8.2   If $n = 2$, then the equation (8.1) becomes

$$f_3(\mathbf{Z}_1, \mathbf{Z}_2) - f_4(\mathbf{Z}_2, \mathbf{Z}_3) + f_1(\mathbf{Z}_1 + \mathbf{Z}_2, \mathbf{Z}_3) - f_2(\mathbf{Z}_1, \mathbf{Z}_2 + \mathbf{Z}_3) = \mathbf{O}.$$

According to the theorem, the general analytic solution of this functional equation is given by

$$f_1(\mathbf{Z}_1, \mathbf{Z}_2) = F_{11}(\mathbf{Z}_1 + \mathbf{Z}_2) - F_{12}(\mathbf{Z}_1) - H_1(\mathbf{Z}_2) + L(\mathbf{Z}_1, \mathbf{Z}_2),$$

$$f_2(\mathbf{Z}_1, \mathbf{Z}_2) = F_{11}(\mathbf{Z}_1 + \mathbf{Z}_2) - F_{22}(\mathbf{Z}_1) - H_2(\mathbf{Z}_2) + L(\mathbf{Z}_1, \mathbf{Z}_2),$$

$$f_3(\mathbf{Z}_1, \mathbf{Z}_2) = F_{12}(\mathbf{Z}_1 + \mathbf{Z}_2) - F_{22}(\mathbf{Z}_1) - H_3(\mathbf{Z}_2) + L(\mathbf{Z}_1, \mathbf{Z}_2),$$

$$f_4(\mathbf{Z}_1, \mathbf{Z}_2) = H_2(\mathbf{Z}_1 + \mathbf{Z}_2) - H_3(\mathbf{Z}_1) - H_1(\mathbf{Z}_2) + L(\mathbf{Z}_1, \mathbf{Z}_2).$$

Example 8.3   If $n = 3$, the general analytic solution of the functional equation (8.1), which takes the form

$$-f_4(\mathbf{Z}_1, \mathbf{Z}_2, \mathbf{Z}_3) - f_5(\mathbf{Z}_2, \mathbf{Z}_3, \mathbf{Z}_4) + f_1(\mathbf{Z}_1 + \mathbf{Z}_2, \mathbf{Z}_3, \mathbf{Z}_4)$$

$$-f_2(\mathbf{Z}_1, \mathbf{Z}_2 + \mathbf{Z}_3, \mathbf{Z}_4) + f_3(\mathbf{Z}_1, \mathbf{Z}_2, \mathbf{Z}_3 + \mathbf{Z}_4) = \mathbf{O},$$

is given by

$$
\begin{aligned}
f_1(\mathbf{Z}_1,\mathbf{Z}_2,\mathbf{Z}_3) &= F_{11}(\mathbf{Z}_1+\mathbf{Z}_2,\mathbf{Z}_3)-F_{12}(\mathbf{Z}_1,\mathbf{Z}_2+\mathbf{Z}_3) \\
&\quad +F_{13}(\mathbf{Z}_1,\mathbf{Z}_2)-H_1(\mathbf{Z}_2,\mathbf{Z}_3)+L(\mathbf{Z}_1,\mathbf{Z}_2,\mathbf{Z}_3),
\end{aligned}
$$

$$
\begin{aligned}
f_2(\mathbf{Z}_1,\mathbf{Z}_2,\mathbf{Z}_3) &= F_{11}(\mathbf{Z}_1+\mathbf{Z}_2,\mathbf{Z}_3)-F_{22}(\mathbf{Z}_1,\mathbf{Z}_2+\mathbf{Z}_3) \\
&\quad +F_{23}(\mathbf{Z}_1,\mathbf{Z}_2)-H_2(\mathbf{Z}_2,\mathbf{Z}_3)+L(\mathbf{Z}_1,\mathbf{Z}_2,\mathbf{Z}_3),
\end{aligned}
$$

$$
\begin{aligned}
f_3(\mathbf{Z}_1,\mathbf{Z}_2,\mathbf{Z}_3) &= F_{12}(\mathbf{Z}_1+\mathbf{Z}_2,\mathbf{Z}_3)-F_{22}(\mathbf{Z}_1,\mathbf{Z}_2+\mathbf{Z}_3) \\
&\quad +F_{33}(\mathbf{Z}_1,\mathbf{Z}_2)-H_3(\mathbf{Z}_2,\mathbf{Z}_3)+L(\mathbf{Z}_1,\mathbf{Z}_2,\mathbf{Z}_3),
\end{aligned}
$$

$$
\begin{aligned}
f_4(\mathbf{Z}_1,\mathbf{Z}_2,\mathbf{Z}_3) &= F_{13}(\mathbf{Z}_1+\mathbf{Z}_2,\mathbf{Z}_3)-F_{23}(\mathbf{Z}_1,\mathbf{Z}_2+\mathbf{Z}_3) \\
&\quad +F_{33}(\mathbf{Z}_1,\mathbf{Z}_2)-H_4(\mathbf{Z}_2,\mathbf{Z}_3)+L(\mathbf{Z}_1,\mathbf{Z}_2,\mathbf{Z}_3),
\end{aligned}
$$

$$
\begin{aligned}
f_5(\mathbf{Z}_1,\mathbf{Z}_2,\mathbf{Z}_3) &= H_2(\mathbf{Z}_1+\mathbf{Z}_2,\mathbf{Z}_3)-H_3(\mathbf{Z}_1,\mathbf{Z}_2+\mathbf{Z}_3) \\
&\quad +H_4(\mathbf{Z}_1,\mathbf{Z}_2)-H_1(\mathbf{Z}_2,\mathbf{Z}_3)+L(\mathbf{Z}_1,\mathbf{Z}_2,\mathbf{Z}_3).
\end{aligned}
$$

Example 8.4 If $n=4$, the given functional equation (8.1) becomes

$$
\begin{aligned}
&f_5(\mathbf{Z}_1,\mathbf{Z}_2,\mathbf{Z}_3,\mathbf{Z}_4)-f_6(\mathbf{Z}_2,\mathbf{Z}_3,\mathbf{Z}_4,\mathbf{Z}_5) \\
+\ &f_1(\mathbf{Z}_1+\mathbf{Z}_2,\mathbf{Z}_3,\mathbf{Z}_4,\mathbf{Z}_5)-f_2(\mathbf{Z}_1,\mathbf{Z}_2+\mathbf{Z}_3,\mathbf{Z}_4,\mathbf{Z}_5) \\
+\ &f_3(\mathbf{Z}_1,\mathbf{Z}_2,\mathbf{Z}_3+\mathbf{Z}_4,\mathbf{Z}_5)-f_4(\mathbf{Z}_1,\mathbf{Z}_2,\mathbf{Z}_3,\mathbf{Z}_4+\mathbf{Z}_5)=\mathbf{O}.
\end{aligned}
$$

The general analytic solution of this equation is

$$
\begin{aligned}
&f_1(\mathbf{Z}_1,\mathbf{Z}_2,\mathbf{Z}_3,\mathbf{Z}_4) \\
=\ &F_{11}(\mathbf{Z}_1+\mathbf{Z}_2,\mathbf{Z}_3,\mathbf{Z}_4)-F_{12}(\mathbf{Z}_1,\mathbf{Z}_2+\mathbf{Z}_3,\mathbf{Z}_4) \\
+\ &F_{13}(\mathbf{Z}_1,\mathbf{Z}_2,\mathbf{Z}_3+\mathbf{Z}_4)-F_{14}(\mathbf{Z}_1,\mathbf{Z}_2,\mathbf{Z}_3) \\
-\ &H_1(\mathbf{Z}_2,\mathbf{Z}_3,\mathbf{Z}_4)+L(\mathbf{Z}_1,\mathbf{Z}_2,\mathbf{Z}_3,\mathbf{Z}_4),
\end{aligned}
$$

$$
\begin{aligned}
&f_2(\mathbf{Z}_1,\mathbf{Z}_2,\mathbf{Z}_3,\mathbf{Z}_4) \\
=\ &F_{11}(\mathbf{Z}_1+\mathbf{Z}_2,\mathbf{Z}_3,\mathbf{Z}_4)-F_{22}(\mathbf{Z}_1,\mathbf{Z}_2+\mathbf{Z}_3,\mathbf{Z}_4) \\
+\ &F_{23}(\mathbf{Z}_1,\mathbf{Z}_2,\mathbf{Z}_3+\mathbf{Z}_4)-F_{24}(\mathbf{Z}_1,\mathbf{Z}_2,\mathbf{Z}_3) \\
-\ &H_2(\mathbf{Z}_2,\mathbf{Z}_3,\mathbf{Z}_4)+L(\mathbf{Z}_1,\mathbf{Z}_2,\mathbf{Z}_3,\mathbf{Z}_4),
\end{aligned}
$$

$$f_3(\mathbf{Z}_1, \mathbf{Z}_2, \mathbf{Z}_3, \mathbf{Z}_4)$$
$$= F_{12}(\mathbf{Z}_1 + \mathbf{Z}_2, \mathbf{Z}_3, \mathbf{Z}_4) - F_{22}(\mathbf{Z}_1, \mathbf{Z}_2 + \mathbf{Z}_3, \mathbf{Z}_4)$$
$$+ F_{33}(\mathbf{Z}_1, \mathbf{Z}_2, \mathbf{Z}_3 + \mathbf{Z}_4) - F_{34}(\mathbf{Z}_1, \mathbf{Z}_2, \mathbf{Z}_3)$$
$$- H_3(\mathbf{Z}_2, \mathbf{Z}_3, \mathbf{Z}_4) + L(\mathbf{Z}_1, \mathbf{Z}_2, \mathbf{Z}_3, \mathbf{Z}_4),$$

$$f_4(\mathbf{Z}_1, \mathbf{Z}_2, \mathbf{Z}_3, \mathbf{Z}_4)$$
$$= F_{13}(\mathbf{Z}_1 + \mathbf{Z}_2, \mathbf{Z}_3, \mathbf{Z}_4) - F_{23}(\mathbf{Z}_1, \mathbf{Z}_2 + \mathbf{Z}_3, \mathbf{Z}_4)$$
$$+ F_{33}(\mathbf{Z}_1, \mathbf{Z}_2, \mathbf{Z}_3 + \mathbf{Z}_4) - F_{44}(\mathbf{Z}_1, \mathbf{Z}_2, \mathbf{Z}_3)$$
$$- H_4(\mathbf{Z}_2, \mathbf{Z}_3, \mathbf{Z}_4) + L(\mathbf{Z}_1, \mathbf{Z}_2, \mathbf{Z}_3, \mathbf{Z}_4),$$

$$f_5(\mathbf{Z}_1, \mathbf{Z}_2, \mathbf{Z}_3, \mathbf{Z}_4)$$
$$= F_{14}(\mathbf{Z}_1 + \mathbf{Z}_2, \mathbf{Z}_3, \mathbf{Z}_4) - F_{24}(\mathbf{Z}_1, \mathbf{Z}_2 + \mathbf{Z}_3, \mathbf{Z}_4)$$
$$+ F_{34}(\mathbf{Z}_1, \mathbf{Z}_2, \mathbf{Z}_3 + \mathbf{Z}_4) - F_{44}(\mathbf{Z}_1, \mathbf{Z}_2, \mathbf{Z}_3)$$
$$- H_5(\mathbf{Z}_2, \mathbf{Z}_3, \mathbf{Z}_4) + L(\mathbf{Z}_1, \mathbf{Z}_2, \mathbf{Z}_3, \mathbf{Z}_4),$$

$$f_6(\mathbf{Z}_1, \mathbf{Z}_2, \mathbf{Z}_3, \mathbf{Z}_4)$$
$$= H_2(\mathbf{Z}_1 + \mathbf{Z}_2, \mathbf{Z}_3, \mathbf{Z}_4) - H_3(\mathbf{Z}_1, \mathbf{Z}_2 + \mathbf{Z}_3, \mathbf{Z}_4)$$
$$+ H_4(\mathbf{Z}_1, \mathbf{Z}_2, \mathbf{Z}_3 + \mathbf{Z}_4) - H_5(\mathbf{Z}_1, \mathbf{Z}_2, \mathbf{Z}_3)$$
$$- H_1(\mathbf{Z}_2, \mathbf{Z}_3, \mathbf{Z}_4) + L(\mathbf{Z}_1, \mathbf{Z}_2, \mathbf{Z}_3, \mathbf{Z}_4).$$

All the particular cases considered here generalize the results given in [S. Kurepa (1956); D. S. Mitrinović and D. Ž. Djoković (1961); D. Ž. Djoković (1962); M. Hosszú (1963)].

9 Simple Functional Equations

Let $\mathcal{V}$ be a finite dimensional complex vector space and let us consider a mapping $f : \mathcal{V}^2 \mapsto \mathcal{V}$. Throughout this section $\mathbf{Z}_i$ $(1 \leq i \leq 4p)$ are vectors in $\mathcal{V}$. We assume that $\mathbf{Z}_i = (z_{i1}(t), \cdots, z_{in}(t))^T$, where $z_{ij}(t)$ $(1 \leq i \leq 4p;\ 1 \leq j \leq n)$ are complex functions and $\mathbf{O} = (0, 0, \cdots, 0)^T$ in the zero vector in $\mathcal{V}$. Further on, we denote by $\mathcal{B}(\mathcal{V})$ the space of continuous mappings $L : \mathcal{V}^2 \mapsto \mathcal{V}$ satisfying

$$L(\mathbf{U}_1 + \mathbf{U}_2, \mathbf{V}) = L(\mathbf{U}_1, \mathbf{V}) + L(\mathbf{U}_2, \mathbf{V}),$$
$$L(\mathbf{U}, \mathbf{V}_1 + \mathbf{V}_2) = L(\mathbf{U}, \mathbf{V}_1) + L(\mathbf{U}, \mathbf{V}_2).$$

Now we will give the results obtained in [I. B. Risteski *et al.* (to appear A)].

Theorem 9.1 *The general continuous solution of the complex vector functional equation*

$$f(\mathbf{Z}_1 + \cdots + \mathbf{Z}_n, \mathbf{Z}_{n+1}) - f(\mathbf{Z}_2 + \cdots + \mathbf{Z}_{n+1}, \mathbf{Z}_1) \quad (9.1)$$
$$+ f(\mathbf{Z}_1 + \cdots + \mathbf{Z}_{n-1}, \mathbf{Z}_n) - f(\mathbf{Z}_2 + \cdots + \mathbf{Z}_n, \mathbf{Z}_{n+1}) + \cdots$$
$$+ f(\mathbf{Z}_1 + \mathbf{Z}_2, \mathbf{Z}_3) - f(\mathbf{Z}_2 + \mathbf{Z}_3, \mathbf{Z}_4) + f(\mathbf{Z}_1, \mathbf{Z}_2) - f(\mathbf{Z}_2, \mathbf{Z}_3) = \mathbf{O}$$

is given by

$$f(\mathbf{U}, \mathbf{V}) = F(\mathbf{U} + \mathbf{V}) - F(\mathbf{U}) - F(\mathbf{V}) + L(\mathbf{U}, \mathbf{V}) + L(\mathbf{V}, \mathbf{U}), \quad (9.2)$$

where F is an arbitrary complex vector function with values in $\mathcal{V}$ and $L \in \mathcal{B}(\mathcal{V})$.

Proof. If we put $\mathbf{Z}_1 = \mathbf{O}$, then Eq. (9.1) reduces to

$$f(\mathbf{O}, \mathbf{Z}_2) = f(\mathbf{Z}_2 + \cdots + \mathbf{Z}_{n+1}, \mathbf{O}),$$

from which we conclude that

$$f(\mathbf{O}, \mathbf{U}) = f(\mathbf{V}, \mathbf{O}) = f(\mathbf{O}, \mathbf{O}) = constant \ complex \ vector. \quad (9.3)$$

By putting $\mathbf{Z}_4 = \mathbf{Z}_5 = \cdots = \mathbf{Z}_{n+1} = \mathbf{O}$, the equation (9.1) according to Eq. (9.3) takes the following form

$$f(\mathbf{Z}_1 + \mathbf{Z}_2, \mathbf{Z}_3) - f(\mathbf{Z}_2 + \mathbf{Z}_3, \mathbf{Z}_1) + f(\mathbf{Z}_1, \mathbf{Z}_2) - f(\mathbf{Z}_2, \mathbf{Z}_3) = \mathbf{O}. \quad (9.4)$$

Rephrasing a result in [I. B. Risteski *et al.* (1999)], we see that Eq. (9.2) is the general continuous solution of the last equation.

By a direct calculation we can verify that the function Eq. (9.2) satisfies the functional equation Eq. (9.1). $\qquad\square$

Theorem 9.2   *The general continuous solution of the complex vector functional equation*

$$
\begin{aligned}
f(\mathbf{Z}_1 + \mathbf{Z}_2, \mathbf{Z}_3 + \mathbf{Z}_4) \;&+\; f(\mathbf{Z}_2 + \mathbf{Z}_3, \mathbf{Z}_4 + \mathbf{Z}_1) \\
+ f(\mathbf{Z}_3 + \mathbf{Z}_4, \mathbf{Z}_1 + \mathbf{Z}_2) \;&+\; f(\mathbf{Z}_4 + \mathbf{Z}_1, \mathbf{Z}_2 + \mathbf{Z}_3) \\
- f(\mathbf{Z}_1, \mathbf{Z}_2 + \mathbf{Z}_3 + \mathbf{Z}_4) \;&-\; f(\mathbf{Z}_2, \mathbf{Z}_3 + \mathbf{Z}_4 + \mathbf{Z}_1) \\
- f(\mathbf{Z}_3, \mathbf{Z}_4 + \mathbf{Z}_1 + \mathbf{Z}_2) \;&-\; f(\mathbf{Z}_4, \mathbf{Z}_1 + \mathbf{Z}_2 + \mathbf{Z}_3)
\end{aligned}
$$

$$
\begin{aligned}
- f(\mathbf{Z}_1 + \mathbf{Z}_2, \mathbf{Z}_3) - f(\mathbf{Z}_2 + \mathbf{Z}_3, \mathbf{Z}_4) \;&-\; f(\mathbf{Z}_3 + \mathbf{Z}_4, \mathbf{Z}_1) - f(\mathbf{Z}_4 + \mathbf{Z}_1, \mathbf{Z}_2) \\
+ f(\mathbf{Z}_1, \mathbf{Z}_4) + f(\mathbf{Z}_2, \mathbf{Z}_1) \;&+\; f(\mathbf{Z}_3, \mathbf{Z}_2) + f(\mathbf{Z}_4, \mathbf{Z}_3) = \mathbf{O}
\end{aligned}
\tag{9.5}
$$

is given by the following formula

$$
f(\mathbf{U}, \mathbf{V}) = F(\mathbf{U} + \mathbf{V}) - F(\mathbf{V}) + G(\mathbf{U}) + L(\mathbf{U}, \mathbf{V}) + L(\mathbf{V}, \mathbf{U}),
\tag{9.6}
$$

where F and G are arbitrary complex vector functions with values in $\mathcal{V}$, and $L \in \mathcal{B}(\mathcal{V})$.

Proof.   If we put $\mathbf{Z}_2 = \mathbf{U}$ and $\mathbf{Z}_1 = \mathbf{Z}_3 = \mathbf{Z}_4 = \mathbf{O}$ into Eq. (9.5), we have

$$
f(\mathbf{O}, \mathbf{U}) = f(\mathbf{O}, \mathbf{O}) = \; constant\ complex\ vector.
\tag{9.7}
$$

By putting $\mathbf{Z}_1 = \mathbf{U}$, $\mathbf{Z}_2 = \mathbf{V}$ and $\mathbf{Z}_3 = \mathbf{Z}_4 = \mathbf{O}$ the equation (9.5) takes the following form

$$
[f(\mathbf{U}, \mathbf{V}) - f(\mathbf{U}, \mathbf{O})] - [f(\mathbf{V}, \mathbf{U}) - f(\mathbf{V}, \mathbf{O})] = \mathbf{O}.
\tag{9.8}
$$

The function

$$
g(\mathbf{U}, \mathbf{V}) = f(\mathbf{U}, \mathbf{V}) - f(\mathbf{U}, \mathbf{O})
\tag{9.9}
$$

satisfies the relations

$$
g(\mathbf{U}, \mathbf{V}) = g(\mathbf{V}, \mathbf{U}) \quad \text{and} \quad g(\mathbf{U}, \mathbf{O}) = g(\mathbf{O}, \mathbf{U}) = \mathbf{O}.
\tag{9.10}
$$

Without difficulty one can verify that the function $g(\mathbf{U}, \mathbf{V})$ also satisfies the equation (9.5)

$$
\begin{aligned}
g(\mathbf{Z}_1 + \mathbf{Z}_2, \mathbf{Z}_3 + \mathbf{Z}_4) \quad &+ \quad g(\mathbf{Z}_2 + \mathbf{Z}_3, \mathbf{Z}_4 + \mathbf{Z}_1) \\
+ g(\mathbf{Z}_3 + \mathbf{Z}_4, \mathbf{Z}_1 + \mathbf{Z}_2) \quad &+ \quad g(\mathbf{Z}_4 + \mathbf{Z}_1, \mathbf{Z}_2 + \mathbf{Z}_3) \\
- g(\mathbf{Z}_1, \mathbf{Z}_2 + \mathbf{Z}_3 + \mathbf{Z}_4) \quad &- \quad g(\mathbf{Z}_2, \mathbf{Z}_3 + \mathbf{Z}_4 + \mathbf{Z}_1) \\
- g(\mathbf{Z}_3, \mathbf{Z}_4 + \mathbf{Z}_1 + \mathbf{Z}_2) \quad &- \quad g(\mathbf{Z}_4, \mathbf{Z}_1 + \mathbf{Z}_2 + \mathbf{Z}_3) \\
- g(\mathbf{Z}_1 + \mathbf{Z}_2, \mathbf{Z}_3) - g(\mathbf{Z}_2 + \mathbf{Z}_3, \mathbf{Z}_4) \quad &- \quad g(\mathbf{Z}_3 + \mathbf{Z}_4, \mathbf{Z}_1) - g(\mathbf{Z}_4 + \mathbf{Z}_1, \mathbf{Z}_2) \\
+ g(\mathbf{Z}_1, \mathbf{Z}_4) + g(\mathbf{Z}_2, \mathbf{Z}_1) \quad &+ \quad g(\mathbf{Z}_3, \mathbf{Z}_2) + g(\mathbf{Z}_4, \mathbf{Z}_3) = \mathbf{O}.
\end{aligned}
$$

If we put $\mathbf{Z}_4 = \mathbf{O}$, we get

$$
g(\mathbf{Z}_2 + \mathbf{Z}_3, \mathbf{Z}_1) - g(\mathbf{Z}_2, \mathbf{Z}_3 + \mathbf{Z}_1) + g(\mathbf{Z}_3, \mathbf{Z}_2) - g(\mathbf{Z}_3, \mathbf{Z}_1) = \mathbf{O}. \qquad (9.11)
$$

Consequently, the function $g(\mathbf{U}, \mathbf{V})$ in the form

$$
g(\mathbf{U}, \mathbf{V}) = F(\mathbf{U} + \mathbf{V}) - F(\mathbf{U}) - F(\mathbf{V}) + L(\mathbf{U}, \mathbf{V}) + L(\mathbf{V}, \mathbf{U}) \qquad (9.12)
$$

satisfies the functional equation (9.11).

From Eqs. (10.9) and (10.12) it follows that

$$
\begin{aligned}
f(\mathbf{U}, \mathbf{V}) \quad &= \quad F(\mathbf{U} + \mathbf{V}) - F(\mathbf{U}) - F(\mathbf{V}) \qquad (9.13) \\
&+ \quad f(\mathbf{U}, \mathbf{O}) + L(\mathbf{U}, \mathbf{V}) + L(\mathbf{V}, \mathbf{U}),
\end{aligned}
$$

which means that $f(\mathbf{U}, \mathbf{V})$ has the form Eq. (9.6).

On the other hand, every function of the form Eq. (9.6) satisfies the complex vector functional equation (9.5). $\qquad \square$

Further let us assume that $f : \mathcal{V}^3 \mapsto \mathcal{V}$.

Theorem 9.3 *The general solution of the functional equation*

$$
\begin{aligned}
f(\mathbf{Z}_1 + \mathbf{Z}_2 + \cdots + \mathbf{Z}_p, \quad & \qquad\qquad (9.14) \\
\mathbf{Z}_{p+1} + \mathbf{Z}_{p+2} + &\cdots + \mathbf{Z}_{2p}, \mathbf{Z}_{2p+1} + \mathbf{Z}_{2p+2} + \cdots + \mathbf{Z}_{4p}) \\
+ \quad f(\mathbf{Z}_2 + \mathbf{Z}_3 + \cdots + \mathbf{Z}_{p+1}, & \\
\mathbf{Z}_{p+2} + \mathbf{Z}_{p+3} + &\cdots + \mathbf{Z}_{2p+1}, \mathbf{Z}_{2p+2} + \mathbf{Z}_{2p+3} + \cdots + \mathbf{Z}_1) + \cdots \\
+ \quad f(\mathbf{Z}_{4p} + \mathbf{Z}_1 + \cdots + \mathbf{Z}_{p-1}, & \\
\mathbf{Z}_p + \mathbf{Z}_{p+1} + &\cdots + \mathbf{Z}_{2p-1}, \mathbf{Z}_{2p} + \mathbf{Z}_{2p+1} + \cdots + \mathbf{Z}_{4p-1}) = \mathbf{O}
\end{aligned}
$$

is given by

$$
\begin{aligned}
f(\mathbf{U}, \mathbf{V}, \mathbf{W}) \quad &= \quad F_1(\mathbf{U}, \mathbf{V} + \mathbf{W}) - F_1(\mathbf{V}, \mathbf{W} + \mathbf{U}) \qquad (9.15) \\
&+ \quad F_2(\mathbf{U} + \mathbf{V}, \mathbf{W}) - F_2(\mathbf{W}, \mathbf{U} + \mathbf{V}),
\end{aligned}
$$

where F_1 and F_2 are arbitrary functions with values in $\mathcal{V}$.

Proof. By a direct calculation we verify that Eq. (9.15) is a solution of the equation (9.14).

Now we assume that f is some solution of the equation (9.14) and we will verify that f has the form Eq. (9.15).

If we put $\mathbf{Z}_1 = \mathbf{U}$, $\mathbf{Z}_{p+1} = \mathbf{V}$, $\mathbf{Z}_{2p+1} = \mathbf{W}$ and $\mathbf{Z}_i = \mathbf{O}$ for $i \neq 1, p+1, 2p+1$ $(1 \leq i \leq 4p)$ into Eq. (9.14), then we obtain

$$p[f(\mathbf{U},\mathbf{V},\mathbf{W}) + f(\mathbf{O},\mathbf{U},\mathbf{V}+\mathbf{W}) + f(\mathbf{W},\mathbf{O},\mathbf{U}+\mathbf{V}) + f(\mathbf{V},\mathbf{W},\mathbf{U})] = \mathbf{O}$$

or in a more convenient form,

$$f(\mathbf{U},\mathbf{V},\mathbf{W})+f(\mathbf{V},\mathbf{W},\mathbf{U}) = -f(\mathbf{W},\mathbf{O},\mathbf{U}+\mathbf{V})-f(\mathbf{O},\mathbf{U},\mathbf{V}+\mathbf{W}). \tag{9.16}$$

By permutations of the vectors $\mathbf{U}$, $\mathbf{V}$ and $\mathbf{W}$ in the above relation, we get

$$-f(\mathbf{V},\mathbf{W},\mathbf{U})-f(\mathbf{W},\mathbf{U},\mathbf{V}) = f(\mathbf{U},\mathbf{O},\mathbf{V}+\mathbf{W})+f(\mathbf{O},\mathbf{V},\mathbf{W}+\mathbf{U}), \tag{9.17}$$

$$f(\mathbf{W},\mathbf{U},\mathbf{V})+f(\mathbf{U},\mathbf{V},\mathbf{W}) = -f(\mathbf{V},\mathbf{O},\mathbf{W}+\mathbf{U})-f(\mathbf{O},\mathbf{W},\mathbf{U}+\mathbf{V}). \tag{9.18}$$

If we add together the last three Eqs. (9.16), (9.17) and (9.18), we obtain

$$\begin{aligned}
2f(\mathbf{U},\mathbf{V},\mathbf{W}) \;=\; & f(\mathbf{U},\mathbf{O},\mathbf{V}+\mathbf{W}) - f(\mathbf{V},\mathbf{O},\mathbf{W}+\mathbf{U}) \\
+ \;& f(\mathbf{O},\mathbf{V},\mathbf{W}+\mathbf{U}) - f(\mathbf{O},\mathbf{U},\mathbf{V}+\mathbf{W}) \\
- \;& f(\mathbf{W},\mathbf{O},\mathbf{U}+\mathbf{V}) - f(\mathbf{O},\mathbf{W},\mathbf{U}+\mathbf{V}).
\end{aligned} \tag{9.19}$$

Formula Eq. (9.19) implies

$$\begin{aligned}
2f(\mathbf{W},\mathbf{O},\mathbf{U}+\mathbf{V}) = & f(\mathbf{W},\mathbf{O},\mathbf{U}+\mathbf{V}) & - & \; f(\mathbf{O},\mathbf{O},\mathbf{U}+\mathbf{V}+\mathbf{W}) \\
& +f(\mathbf{O},\mathbf{O},\mathbf{U}+\mathbf{V}+\mathbf{W}) & - & \; f(\mathbf{O},\mathbf{W},\mathbf{U}+\mathbf{V}) \\
& -f(\mathbf{U}+\mathbf{V},\mathbf{O},\mathbf{W}) & - & \; f(\mathbf{O},\mathbf{U}+\mathbf{V},\mathbf{W})
\end{aligned}$$

or in a more convenient form

$$\begin{aligned}
- \;& f(\mathbf{W},\mathbf{O},\mathbf{U}+\mathbf{V}) - f(\mathbf{O},\mathbf{W},\mathbf{U}+\mathbf{V}) \\
= \;& f(\mathbf{U}+\mathbf{V},\mathbf{O},\mathbf{W}) + f(\mathbf{O},\mathbf{U}+\mathbf{V},\mathbf{W}).
\end{aligned} \tag{9.20}$$

If we use Eqs. (9.19) and (9.20), we find

$$
\begin{aligned}
4f(\mathbf{U}, \mathbf{V}, \mathbf{W}) = {} & 2[f(\mathbf{U}, \mathbf{O}, \mathbf{V} + \mathbf{W}) && - && f(\mathbf{V}, \mathbf{O}, \mathbf{W} + \mathbf{U})] \\
& +2[f(\mathbf{O}, \mathbf{V}, \mathbf{W} + \mathbf{U}) && - && f(\mathbf{O}, \mathbf{U}, \mathbf{V} + \mathbf{W})] \\
& +f(\mathbf{U} + \mathbf{V}, \mathbf{O}, \mathbf{W}) && - && f(\mathbf{W}, \mathbf{O}, \mathbf{U} + \mathbf{V}) \\
& +f(\mathbf{O}, \mathbf{U} + \mathbf{V}, \mathbf{W}) && - && f(\mathbf{O}, \mathbf{W}, \mathbf{U} + \mathbf{V}).
\end{aligned}
$$

If we put

$$
F_1(\mathbf{U}, \mathbf{V}) \;=\; \frac{1}{2}[f(\mathbf{U}, \mathbf{O}, \mathbf{V}) - f(\mathbf{O}, \mathbf{U}, \mathbf{V})],
$$

$$
F_2(\mathbf{U}, \mathbf{V}) \;=\; \frac{1}{4}[f(\mathbf{U}, \mathbf{O}, \mathbf{V}) + f(\mathbf{O}, \mathbf{U}, \mathbf{V})],
$$

we obtain the form of the function $f(\mathbf{U}, \mathbf{V}, \mathbf{W})$. $\qquad\square$

10 Functional Equations with Several Unknown Functions

In this section some functional equations with several unknown functions and one operation of addition between arguments are solved. These results are obtained in [I. B. Risteski *et al.* (to appear A)].

Let $\mathcal{V}$, $\mathcal{V}'$ be finite dimensional complex vector spaces. Denote by $\mathcal{V}^0$ the subspace of all real vectors in $\mathcal{V}$ (thus $\mathcal{V} = \mathcal{V}^0 + i\mathcal{V}^0$) and let $\mathcal{L}(\mathcal{V}^0, \mathcal{V}')$ be the space of all linear mappings $\mathcal{V}^0 \mapsto \mathcal{V}'$.

Lemma 10.1 *Let f be a mapping $\mathcal{V}^2 \mapsto \mathcal{V}'$. Then the general continuous solution of the functional equation*

$$
f(\mathbf{Z}_1, \mathbf{Z}_2 + \mathbf{Z}_3 + \mathbf{Z}_4) - f(\mathbf{Z}_1 + \mathbf{Z}_2, \mathbf{Z}_3 + \mathbf{Z}_4) \qquad (10.1)
$$
$$
- \; f(\mathbf{Z}_1 + \mathbf{Z}_4, \mathbf{Z}_2 + \mathbf{Z}_3) + f(\mathbf{Z}_1 + \mathbf{Z}_2 + \mathbf{Z}_4, \mathbf{Z}_3) = \mathbf{O}
$$

is given by

$$
\begin{aligned}
f(\mathbf{Z}_1, \mathbf{Z}_2) \;=\; & H_1(\mathbf{Z}_1 + \mathbf{Z}_2)\operatorname{Re}\mathbf{Z}_1 + H_2(\mathbf{Z}_1 + \mathbf{Z}_2)\operatorname{Im}\mathbf{Z}_1 \qquad (10.2) \\
& + \; H_3(\mathbf{Z}_1 + \mathbf{Z}_2) + L(\mathbf{Z}_1, \mathbf{Z}_2) - L(\mathbf{Z}_2, \mathbf{Z}_1),
\end{aligned}
$$

where $H_i : \mathcal{V} \mapsto \mathcal{L}(\mathcal{V}^0, \mathcal{V}')$ $(i = 1, 2)$, $H_3 : \mathcal{V} \mapsto \mathcal{V}'$ are arbitrary continuous complex functions, and L is a mapping $\mathcal{V}^2 \mapsto \mathcal{V}'$ linear in each argument.

Proof. First we note that $f(\mathbf{Z}_1, \mathbf{Z}_2) = L(\mathbf{Z}_1, \mathbf{Z}_2) - L(\mathbf{Z}_2, \mathbf{Z}_1)$ is a solution of Eq. (10.1) which vanishes when we put any of its arguments equal to zero. Thus we can just add this solution to the one obtained below.

If we introduce a new unknown function by

$$f(\mathbf{Z}_1, \mathbf{Z}_2) = g(\mathbf{Z}_1, \mathbf{Z}_1 + \mathbf{Z}_2), \tag{10.3}$$

then equation (10.1) becomes

$$\begin{aligned}
g(\mathbf{Z}_1, \mathbf{Z}_1 &+ \mathbf{Z}_2 + \mathbf{Z}_3 + \mathbf{Z}_4) - g(\mathbf{Z}_1 + \mathbf{Z}_2, \mathbf{Z}_1 + \mathbf{Z}_2 + \mathbf{Z}_3 + \mathbf{Z}_4) \\
- \ &g(\mathbf{Z}_1 + \mathbf{Z}_4, \mathbf{Z}_1 + \mathbf{Z}_2 + \mathbf{Z}_3 + \mathbf{Z}_4) \\
+ \ &g(\mathbf{Z}_1 + \mathbf{Z}_2 + \mathbf{Z}_4, \mathbf{Z}_1 + \mathbf{Z}_2 + \mathbf{Z}_3 + \mathbf{Z}_4) = \mathbf{O}.
\end{aligned} \tag{10.4}$$

For $\mathbf{Z}_1 = \mathbf{O}$ and $\mathbf{Z}_2 + \mathbf{Z}_3 + \mathbf{Z}_4 = \mathbf{T}$, Eq. (10.4) takes on the form

$$g(\mathbf{Z}_2 + \mathbf{Z}_4, \mathbf{T}) - g(\mathbf{Z}_2, \mathbf{T}) - g(\mathbf{Z}_4, \mathbf{T}) + g(\mathbf{O}, \mathbf{T}) = \mathbf{O},$$

i.e.,

$$h(\mathbf{Z}_2 + \mathbf{Z}_4, \mathbf{T}) = h(\mathbf{Z}_2, \mathbf{T}) + h(\mathbf{Z}_4, \mathbf{T}), \tag{10.5}$$

where we have denoted

$$h(\mathbf{Z}_1, \mathbf{T}) = g(\mathbf{Z}_1, \mathbf{T}) - g(\mathbf{O}, \mathbf{T}). \tag{10.6}$$

Therefore, the function h is additive in the first argument and the general continuous solution of the functional equation (10.5) is determined by

$$h(\mathbf{Z}_1, \mathbf{Z}_2) = H_1(\mathbf{Z}_2)\mathrm{Re}\, \mathbf{Z}_1 + H_2(\mathbf{Z}_2)\mathrm{Im}\, \mathbf{Z}_1, \tag{10.7}$$

where H_i $(i = 1, 2)$ are arbitrary continuous complex functions with values in $\mathcal{L}(\mathcal{V}^0, \mathcal{V}')$.

On the basis of the expressions Eqs. (10.7) and (10.6), for the function g we obtain

$$g(\mathbf{Z}_1, \mathbf{Z}_2) = H_1(\mathbf{Z}_2)\mathrm{Re}\, \mathbf{Z}_1 + H_2(\mathbf{Z}_2)\mathrm{Im}\, \mathbf{Z}_1 + H_3(\mathbf{Z}_2), \tag{10.8}$$

where we put $g(\mathbf{O}, \mathbf{Z}_2) = H_3(\mathbf{Z}_2)$.

From Eqs. (10.8) and (10.3) there follows Eq. (10.2). $\qquad\square$

In the subsequent Theorems 10.2–10.4 we assume that the mappings

$$f_1 : \mathcal{V}^2 \mapsto \mathcal{V}' \quad \text{and} \quad f_i : \mathcal{V}^3 \mapsto \mathcal{V}' \quad (2 \leq i \leq n - 1).$$

Theorem 10.2 *The functional equation*

$$f_1(Z_1, Z_2+Z_3+Z_4)+f_2(Z_2, Z_3, Z_4+Z_1)+f_3(Z_3, Z_4, Z_1+Z_2) = O \quad (10.9)$$

has a general continuous solution given by

$$
\begin{aligned}
f_2(Z_1, Z_2, Z_3) &= -f_1(Z_3, Z_1+Z_2) + F(Z_2, Z_3+Z_1), & (10.10)\\
f_3(Z_1, Z_2, Z_3) &= f_1(Z_2+Z_3, Z_1) - f_1(Z_3, Z_1+Z_2) - F(Z_1, Z_2+Z_3),
\end{aligned}
$$

where F is an arbitrary continuous function $\mathcal{V}^2 \mapsto \mathcal{V}'$, and the continuous function f_1 satisfies the functional equation (10.1).

Proof.   If we put $Z_4 = O$ into Eq. (10.9), we get

$$f_2(Z_2, Z_3, Z_1) = F(Z_3, Z_1+Z_2) - f_1(Z_1, Z_2+Z_3), \quad (10.11)$$

where $F(Z_1, Z_2) = -f_3(Z_1, O, Z_2)$.

For $Z_2 = O$ the functional equation (10.9) in view of the relation Eq. (10.11) becomes

$$f_3(Z_3, Z_4, Z_1) = f_1(Z_4+Z_1, Z_3) - f_1(Z_1, Z_3+Z_4) - F(Z_3, Z_4+Z_1). \quad (10.12)$$

From Eqs. (10.11) and (10.12) there follows Eq. (10.10).

By putting Eqs. (10.11) and (10.12) into Eq. (10.9), we deduce that the function f_1 must satisfy the equation (10.1). $\square$

Theorem 10.3 *The general continuous solution of the functional equation*

$$
\begin{aligned}
f_1(Z_1, Z_2+Z_3+Z_4+Z_5) &\quad+\quad f_2(Z_2, Z_3, Z_4+Z_5+Z_1) & (10.13)\\
+f_3(Z_3, Z_4, Z_5+Z_1+Z_2) &\quad+\quad f_4(Z_4, Z_5, Z_1+Z_2+Z_3) = O
\end{aligned}
$$

is determined by

$$
\begin{aligned}
f_2(Z_1, Z_2, Z_3) &= -f_1(Z_3, Z_1+Z_2) + F_1(Z_2, Z_3+Z_1), & (10.14)\\
f_3(Z_1, Z_2, Z_3) &= f_1(Z_2+Z_3, Z_1) - f_1(Z_3, Z_1+Z_2)\\
&\quad - F_1(Z_1, Z_2+Z_3) + F_2(Z_2, Z_3+Z_1),\\
f_4(Z_1, Z_2, Z_3) &= f_1(Z_2+Z_3, Z_1) - f_1(Z_3, Z_1+Z_2) - F_2(Z_1, Z_2+Z_3),
\end{aligned}
$$

where F_i $(i = 1, 2)$ are arbitrary continuous functions $\mathcal{V}^2 \mapsto \mathcal{V}'$, and f_1 is the general continuous solution of the equation (10.1).

Proof. If we put $Z_4 = Z_5 = O$ into Eq. (10.13), we obtain

$$f_2(Z_2, Z_3, Z_1) = -f_1(Z_1, Z_2 + Z_3) + F_1(Z_3, Z_1 + Z_2), \qquad (10.15)$$

where

$$F_1(Z_1, Z_2) = -f_3(Z_1, O, Z_2) - f_4(O, O, Z_1 + Z_2).$$

For $Z_5 = Z_2 = O$ the functional equation (10.13) by virtue of the expression Eq. (10.15) becomes

$$\begin{aligned}
f_3(Z_3, Z_4, Z_1) \;=\;& f_1(Z_4 + Z_1, Z_3) - f_1(Z_1, Z_3 + Z_4) \qquad (10.16)\\
-\;& F_1(Z_3, Z_4 + Z_1) + F_2(Z_4, Z_1 + Z_3),
\end{aligned}$$

where we have denoted $F_2(Z_1, Z_2) = -f_4(Z_1, O, Z_2)$.

By putting $Z_2 = Z_3 = O$ into Eq. (10.13), on the basis of Eqs. (10.15) and (10.16) we obtain

$$\begin{aligned}
f_4(Z_4, Z_5, Z_1) \;=\;& f_1(Z_5 + Z_1, Z_4) \qquad\qquad\qquad (10.17)\\
-\;& f_1(Z_1, Z_4 + Z_5) - F_2(Z_4, Z_5 + Z_1).
\end{aligned}$$

From Eqs. (10.15), (10.16) and (10.17) there follows Eq. (10.14).

By setting the functions f_i $(i = 2, 3, 4)$ determined by Eq. (10.14), *i.e.*, Eqs. (10.15), (10.16), (10.17), into Eq. (10.13), we obtain the equation

$$\begin{aligned}
f_1(Z_1, Z_2 + Z_3 + Z_4 + Z_5) \;&-\; f_1(Z_4 + Z_5 + Z_1, Z_2 + Z_3)\\
+ f_1(Z_4 + Z_5 + Z_1 + Z_2, Z_3) \;&-\; f_1(Z_5 + Z_1 + Z_2, Z_3 + Z_4)\\
- f_1(Z_1 + Z_2 + Z_3, Z_4 + Z_5) \;&+\; f_1(Z_5 + Z_1 + Z_2 + Z_3, Z_4) = O.
\end{aligned}$$

For $Z_5 = O$ this equation becomes just Eq. (10.1). $\qquad\square$

Theorem 10.4 *The general continuous solution of the functional equation*

$$f_1\left(Z_1, \sum_{j=2}^{n} Z_j\right) + \sum_{i=2}^{n-1} f_i\left(Z_i, Z_{i+1}, \sum_{j=i+2}^{n+i-1} Z_j\right) = O, \qquad (10.18)$$

where $n \geq 5$ and $\mathbf{Z}_{n+i} \equiv \mathbf{Z}_i$ for $i = 1, \cdots, n-2$, is

$$
\begin{aligned}
f_2(\mathbf{Z}_1, \mathbf{Z}_2, \mathbf{Z}_3) &= -f_1(\mathbf{Z}_3, \mathbf{Z}_1 + \mathbf{Z}_2) + F_1(\mathbf{Z}_2, \mathbf{Z}_3 + \mathbf{Z}_1), \quad (10.19) \\
f_i(\mathbf{Z}_1, \mathbf{Z}_2, \mathbf{Z}_3) &= f_1(\mathbf{Z}_2 + \mathbf{Z}_3, \mathbf{Z}_1) - f_1(\mathbf{Z}_3, \mathbf{Z}_1 + \mathbf{Z}_2) \\
&\quad - F_{i-2}(\mathbf{Z}_1, \mathbf{Z}_2 + \mathbf{Z}_3) + F_{i-1}(\mathbf{Z}_2, \mathbf{Z}_3 + \mathbf{Z}_1) \\
&\quad\quad (3 \leq i \leq n - 2), \\
f_{n-1}(\mathbf{Z}_1, \mathbf{Z}_2, \mathbf{Z}_3) &= f_1(\mathbf{Z}_2 + \mathbf{Z}_3, \mathbf{Z}_1) - f_1(\mathbf{Z}_3, \mathbf{Z}_1 + \mathbf{Z}_2) \\
&\quad - F_{n-3}(\mathbf{Z}_1, \mathbf{Z}_2 + \mathbf{Z}_3),
\end{aligned}
$$

where F_i $(1 \leq i \leq n-3)$ are arbitrary continuous functions $\mathcal{V}^2 \mapsto \mathcal{V}'$, and the continuous function f_1 satisfies the functional equation (10.1).

Proof. For $n = 5$ this theorem holds on the basis of Theorem 10.2. Now, we will assume that this theorem holds for some fixed $n \geq 5$. Then, Eq. (10.18) for $n + 1$ has the form

$$
f_1\left(\mathbf{Z}_1, \sum_{j=2}^{n+1} \mathbf{Z}_j\right) + \sum_{i=2}^{n} f_i\left(\mathbf{Z}_i, \mathbf{Z}_{i+1}, \sum_{j=i+2}^{n+i} \mathbf{Z}_j\right) = \mathbf{O}. \qquad (10.20)
$$

If we put $\mathbf{Z}_{n+1} = \mathbf{O}$ into Eq. (10.20), then we will have

$$
f_1\left(\mathbf{Z}_1, \sum_{j=2}^{n} \mathbf{Z}_j\right) + \sum_{i=2}^{n-2} f_i\left(\mathbf{Z}_i, \mathbf{Z}_{i+1}, \sum_{j=i+2}^{n+i-2} \mathbf{Z}_j\right) \qquad (10.21)
$$

$$
+ \; g_{n-1}\left(\mathbf{Z}_{n-1}, \mathbf{Z}_n, \sum_{j=1}^{n-2} \mathbf{Z}_j\right) = \mathbf{O},
$$

where

$$
g_{n-1}(\mathbf{Z}_1, \mathbf{Z}_2, \mathbf{Z}_3) = f_{n-1}(\mathbf{Z}_1, \mathbf{Z}_2, \mathbf{Z}_3) + f_n(\mathbf{Z}_2, \mathbf{O}, \mathbf{Z}_3 + \mathbf{Z}_1). \qquad (10.22)
$$

On the basis on the inductive hypothesis, the general continuous solution of the functional equation (10.21) is given by Eq. (10.19), where f_{n-1} is replaced by g_{n-1}.

Now introduce the function

$$
F_{n-2}(\mathbf{Z}_1, \mathbf{Z}_2) = -f_n(\mathbf{Z}_1, \mathbf{O}, \mathbf{Z}_2).
$$

By virtue of the expressions for the solution Eq. (10.19) and equality (10.22) we obtain

$$\begin{aligned}
f_{n-1}(\mathbf{Z}_1, \mathbf{Z}_2, \mathbf{Z}_3) \;=\;\; & f_1(\mathbf{Z}_2 + \mathbf{Z}_3, \mathbf{Z}_1) - f_1(\mathbf{Z}_3, \mathbf{Z}_1 + \mathbf{Z}_2) \qquad (10.23)\\
- \;\; & F_{n-3}(\mathbf{Z}_1, \mathbf{Z}_2 + \mathbf{Z}_3) + F_{n-2}(\mathbf{Z}_2, \mathbf{Z}_3 + \mathbf{Z}_1).
\end{aligned}$$

By setting Eq. (10.23) and the functions f_i $(1 \le i \le n-2)$ determined by Eqs. (10.19) into Eq. (10.20), based on the hypothesis that the continuous function f_1 satisfies the equation (10.1), we obtain

$$f_n(\mathbf{Z}_1, \mathbf{Z}_2, \mathbf{Z}_3) = f_1(\mathbf{Z}_2 + \mathbf{Z}_3, \mathbf{Z}_1) - f_1(\mathbf{Z}_3, \mathbf{Z}_1 + \mathbf{Z}_2) - F_{n-2}(\mathbf{Z}_1, \mathbf{Z}_2 + \mathbf{Z}_3).$$
$$\square$$

Theorem 10.5   *Let $f : \mathcal{V}^2 \mapsto \mathcal{V}'$ and $g : \mathcal{V}^3 \mapsto \mathcal{V}'$. Then the general continuous solution of the functional equation*

$$f\left(\mathbf{Z}_1, \sum_{j=2}^{n} \mathbf{Z}_j\right) + \sum_{i=2}^{n-1} g\left(\mathbf{Z}_i, \mathbf{Z}_{i+1}, \sum_{j=i+2}^{n+i-1} \mathbf{Z}_j\right) = \mathbf{O} \qquad (10.24)$$

where $n \ge 4$ and $\mathbf{Z}_{n+i} \equiv \mathbf{Z}_i$ for $i = 1, \cdots, n-2$, is determined by

$$f(\mathbf{Z}_1, \mathbf{Z}_2) = F(\mathbf{Z}_1 + \mathbf{Z}_2), \qquad (10.25)$$
$$g(\mathbf{Z}_1, \mathbf{Z}_2, \mathbf{Z}_3) = -\frac{1}{n-2} F(\mathbf{Z}_1 + \mathbf{Z}_2 + \mathbf{Z}_3),$$

where F is an arbitrary continuous complex vector function $\mathcal{V} \mapsto \mathcal{V}'$.

Proof.   If we put $f_1 = f$, $f_2 = f_3 = \cdots = f_{n-1} = g$ into Eq. (10.18), we obtain Eq. (10.24). Now by the substitution $\begin{pmatrix} 1 & 2 & 3 & \cdots & n-1 & n \\ 1 & 3 & 4 & \cdots & n & 2 \end{pmatrix}$ from Eq. (10.24) we get

$$\begin{aligned}
f\left(\mathbf{Z}_1, \sum_{j=2}^{n} \mathbf{Z}_j\right) \;+\;\; & \sum_{i=3}^{n-1} g\left(\mathbf{Z}_i, \mathbf{Z}_{i+1}, \sum_{j=i+2}^{n+i-1} \mathbf{Z}_j\right) \qquad (10.26)\\
+ \;\; & g\left(\mathbf{Z}_n, \mathbf{Z}_2, \mathbf{Z}_1 + \sum_{j=3}^{n-1} \mathbf{Z}_j\right) = \mathbf{O}.
\end{aligned}$$

From Eqs. (10.25) and (10.26) it follows that the function g satisfies the equation

$$g\left(\mathbf{Z}_2, \mathbf{Z}_3, \mathbf{Z}_1 + \sum_{j=4}^{n} \mathbf{Z}_j\right) - g\left(\mathbf{Z}_n, \mathbf{Z}_2, \mathbf{Z}_1 + \sum_{j=3}^{n-1} \mathbf{Z}_j\right) = \mathbf{O}$$

and, more generally,

$$g\left(\mathbf{Z}_i, \mathbf{Z}_{i+1}, \sum_{j=1}^{i-1} \mathbf{Z}_j + \sum_{j=i+2}^{n} \mathbf{Z}_j\right) = g\left(\mathbf{Z}_n, \mathbf{Z}_2, \mathbf{Z}_1 + \sum_{j=3}^{n-1} \mathbf{Z}_j\right) \qquad (10.27)$$

$$(2 \le i \le n - 1).$$

On the basis of the relation Eq. (10.27), the functional equation Eq. (10.24) becomes

$$f\left(\mathbf{Z}_1, \sum_{j=2}^{n} \mathbf{Z}_j\right) + (n-2)g\left(\mathbf{Z}_n, \mathbf{Z}_3, \mathbf{Z}_1 + \sum_{j=3}^{n-1} \mathbf{Z}_j\right) = \mathbf{O}. \qquad (10.28)$$

If we put $\mathbf{Z}_2 = \mathbf{Z}_n = \mathbf{O}$ and $\sum_{j=3}^{n-1} \mathbf{Z}_j = \mathbf{S}$, we obtain

$$F(\mathbf{Z}_1, \mathbf{S}) + (n-2)g(\mathbf{O}, \mathbf{O}, \mathbf{Z}_1 + \mathbf{S}) = \mathbf{O}.$$

Thus

$$f(\mathbf{Z}_1, \mathbf{S}) = F(\mathbf{Z}_1 + \mathbf{S}), \qquad (10.29)$$

where

$$F(\mathbf{Z}_1) = -(n-2)g(\mathbf{O}, \mathbf{O}, \mathbf{Z}_1).$$

On the basis of Eqs. (10.29) and (10.28) it follows that

$$F\left(\sum_{j=1}^{n} \mathbf{Z}_j\right) + (n-2)g\left(\mathbf{Z}_n, \mathbf{Z}_2, \mathbf{Z}_1 + \sum_{j=3}^{n-1} \mathbf{Z}_j\right) = \mathbf{O},$$

i.e.,

$$g(\mathbf{Z}_1, \mathbf{Z}_2, \mathbf{Z}_3) = -\frac{1}{n-2}F(\mathbf{Z}_1 + \mathbf{Z}_2 + \mathbf{Z}_3).$$

$\square$

11   Fréchet's Functional Equations

First we will introduce the following notations.

Let $\mathcal{V}$ be a finite dimensional complex vector space and let there exist a mapping $f : \mathcal{V}^2 \mapsto \mathcal{V}$. In this section $\mathbf{Z}_i$ $(0 \le i \le n+1)$ will denote vectors in $\mathcal{V}$. We assume that $\mathbf{Z}_i = (z_{i1}(t), \cdots, z_{in}(t))^T$, where the components $z_{ij}(t)$ $(0 \le i \le n+1; 1 \le j \le n)$ are complex functions and $\mathbf{O} = (0, \cdots, 0)^T$ is the zero vector in $\mathcal{V}$. We define multiplication of arbitrary two complex vectors $\mathbf{U} = (u_1(t), \cdots, u_n(t))^T$ and $\mathbf{V} = (v_1(t), \cdots, v_n(t))^T$ in $\mathcal{V}$ as $\mathbf{U}\mathbf{V} = (u_1(t)v_1(t), \cdots, u_n(t)v_n(t))^T$.

Now we will give the result obtained in [I. B. Risteski *et al.* (to appear A); K. G. Trenčevski *et al.* (1999)].

Theorem 11.1   *The general continuous solution of the complex vector functional equation*

$$f(\mathbf{Z}_1 + \mathbf{Z}_2 + \mathbf{Z}_3 + \mathbf{Z}_4, \mathbf{Z}_5) + f(\mathbf{Z}_1 + \mathbf{Z}_2 + \mathbf{Z}_3, \mathbf{Z}_4 + \mathbf{Z}_5) \qquad (11.1)$$

$$-f(\mathbf{Z}_1 + \mathbf{Z}_2, \mathbf{Z}_3 + \mathbf{Z}_4 + \mathbf{Z}_5) - f(\mathbf{Z}_1, \mathbf{Z}_2 + \mathbf{Z}_3 + \mathbf{Z}_4 + \mathbf{Z}_5)$$

$$+f(\mathbf{Z}_2 + \mathbf{Z}_3 + \mathbf{Z}_4 + \mathbf{Z}_5, \mathbf{Z}_1) + f(\mathbf{Z}_2 + \mathbf{Z}_3 + \mathbf{Z}_4, \mathbf{Z}_5 + \mathbf{Z}_1)$$

$$-f(\mathbf{Z}_2 + \mathbf{Z}_3, \mathbf{Z}_4 + \mathbf{Z}_5 + \mathbf{Z}_1) - f(\mathbf{Z}_2, \mathbf{Z}_3 + \mathbf{Z}_4 + \mathbf{Z}_5 + \mathbf{Z}_1)$$

$$+f(\mathbf{Z}_3 + \mathbf{Z}_4 + \mathbf{Z}_5 + \mathbf{Z}_1, \mathbf{Z}_2) + f(\mathbf{Z}_3 + \mathbf{Z}_4 + \mathbf{Z}_5, \mathbf{Z}_1 + \mathbf{Z}_2)$$

$$-f(\mathbf{Z}_3 + \mathbf{Z}_4, \mathbf{Z}_5 + \mathbf{Z}_1 + \mathbf{Z}_2) - f(\mathbf{Z}_3, \mathbf{Z}_4 + \mathbf{Z}_5 + \mathbf{Z}_1 + \mathbf{Z}_2)$$

$$+f(\mathbf{Z}_4 + \mathbf{Z}_5 + \mathbf{Z}_1 + \mathbf{Z}_2, \mathbf{Z}_3) + f(\mathbf{Z}_4 + \mathbf{Z}_5 + \mathbf{Z}_1, \mathbf{Z}_2 + \mathbf{Z}_3)$$

$$-f(\mathbf{Z}_4 + \mathbf{Z}_5, \mathbf{Z}_1 + \mathbf{Z}_2 + \mathbf{Z}_3) - f(\mathbf{Z}_4, \mathbf{Z}_5 + \mathbf{Z}_1 + \mathbf{Z}_2 + \mathbf{Z}_3)$$

$$+f(\mathbf{Z}_5 + \mathbf{Z}_1 + \mathbf{Z}_2 + \mathbf{Z}_3, \mathbf{Z}_4) + f(\mathbf{Z}_5 + \mathbf{Z}_1 + \mathbf{Z}_2, \mathbf{Z}_3 + \mathbf{Z}_4)$$

$$-f(\mathbf{Z}_5 + \mathbf{Z}_1, \mathbf{Z}_2 + \mathbf{Z}_3 + \mathbf{Z}_4) - f(\mathbf{Z}_5, \mathbf{Z}_1 + \mathbf{Z}_2 + \mathbf{Z}_3 + \mathbf{Z}_4)$$

$$+f(\mathbf{Z}_1, \mathbf{Z}_2 + \mathbf{Z}_3 + \mathbf{Z}_4) - f(\mathbf{Z}_1 + \mathbf{Z}_2, \mathbf{Z}_3 + \mathbf{Z}_4) + f(\mathbf{Z}_1 + \mathbf{Z}_2, \mathbf{Z}_3) - f(\mathbf{Z}_1, \mathbf{Z}_2)$$

$$+f(\mathbf{Z}_2, \mathbf{Z}_3 + \mathbf{Z}_4 + \mathbf{Z}_5) - f(\mathbf{Z}_2 + \mathbf{Z}_3, \mathbf{Z}_4 + \mathbf{Z}_5) + f(\mathbf{Z}_2 + \mathbf{Z}_3, \mathbf{Z}_4) - f(\mathbf{Z}_2, \mathbf{Z}_3)$$

$$+f(\mathbf{Z}_3, \mathbf{Z}_4 + \mathbf{Z}_5 + \mathbf{Z}_1) - f(\mathbf{Z}_3 + \mathbf{Z}_4, \mathbf{Z}_5 + \mathbf{Z}_1) + f(\mathbf{Z}_3 + \mathbf{Z}_4, \mathbf{Z}_5) - f(\mathbf{Z}_3, \mathbf{Z}_4)$$

$$+f(\mathbf{Z}_4, \mathbf{Z}_5 + \mathbf{Z}_1 + \mathbf{Z}_2) + f(\mathbf{Z}_4 + \mathbf{Z}_5, \mathbf{Z}_1 + \mathbf{Z}_2) + f(\mathbf{Z}_4 + \mathbf{Z}_5, \mathbf{Z}_1) - f(\mathbf{Z}_4, \mathbf{Z}_5)$$

$$+f(\mathbf{Z}_5, \mathbf{Z}_1 + \mathbf{Z}_2 + \mathbf{Z}_3) - f(\mathbf{Z}_5 + \mathbf{Z}_1, \mathbf{Z}_2 + \mathbf{Z}_3) + f(\mathbf{Z}_5 + \mathbf{Z}_1, \mathbf{Z}_2) - f(\mathbf{Z}_5, \mathbf{Z}_1) = \mathbf{O}$$

is given by

$$
\begin{aligned}
f(\mathbf{U}, \mathbf{V}) \;=\; & F(\mathbf{U} + \mathbf{V}) - F(\mathbf{U}) - F(\mathbf{V}) \\
+ \; & \alpha(3\mathbf{U}^2 + 4\mathbf{V}^2) + \beta(3\mathbf{U} + 4\mathbf{V}) + L(\mathbf{U}, \mathbf{V}),
\end{aligned}
\tag{11.2}
$$

where F is an arbitrary complex vector function with values in $\mathcal{V}$, and α, β are complex constant matrices and $L \in \mathcal{B}(\mathcal{V})$.

Proof. If we put $\mathbf{Z}_1 = \mathbf{U}$ and $\mathbf{Z}_2 = \mathbf{Z}_3 = \mathbf{Z}_4 = \mathbf{Z}_5 = \mathbf{O}$ in Eq. (11.1), then

$$4f(\mathbf{U}, \mathbf{O}) - 3f(\mathbf{O}, \mathbf{U}) = \mathbf{K} = f(\mathbf{O}, \mathbf{O}). \tag{11.3}$$

If we replace $\mathbf{Z}_1 = \mathbf{U}$, $\mathbf{Z}_2 = \mathbf{V}$, $\mathbf{Z}_3 = \mathbf{W}$ and $\mathbf{Z}_4 = \mathbf{Z}_5 = \mathbf{O}$ in (11.1), we find

$$3f(\mathbf{U} + \mathbf{V} + \mathbf{W}, \mathbf{O}) - 2f(\mathbf{O}, \mathbf{U} + \mathbf{V} + \mathbf{W}) - f(\mathbf{O}, \mathbf{O})$$

$$+f(\mathbf{U} + \mathbf{V}, \mathbf{W}) - f(\mathbf{W}, \mathbf{U} + \mathbf{V}) + f(\mathbf{V} + \mathbf{W}, \mathbf{U})$$

$$-f(\mathbf{U}, \mathbf{V} + \mathbf{W}) + f(\mathbf{W} + \mathbf{U}, \mathbf{V}) - f(\mathbf{V}, \mathbf{W} + \mathbf{U}) = \mathbf{O},$$

or, by virtue of Eq. (11.3), the above expression becomes

$$
\begin{aligned}
& f(\mathbf{U} + \mathbf{V} + \mathbf{W}, \mathbf{O}) - f(\mathbf{O}, \mathbf{U} + \mathbf{V} + \mathbf{W}) \\
=\; & f(\mathbf{U} + \mathbf{V}, \mathbf{W}) - f(\mathbf{W}, \mathbf{U} + \mathbf{V}) \\
+\; & f(\mathbf{V} + \mathbf{W}, \mathbf{U}) - f(\mathbf{U}, \mathbf{V} + \mathbf{W}) \\
+\; & f(\mathbf{W} + \mathbf{U}, \mathbf{V}) - f(\mathbf{V}, \mathbf{W} + \mathbf{U}).
\end{aligned}
\tag{11.4}
$$

By putting $\mathbf{Z}_1 = \mathbf{U}$, $\mathbf{Z}_2 = \mathbf{V}$, $\mathbf{Z}_4 = \mathbf{W}$ and $\mathbf{Z}_3 = \mathbf{Z}_5 = \mathbf{O}$ in Eq. (11.1) we obtain

$$2f(\mathbf{U} + \mathbf{V} + \mathbf{W}, \mathbf{O}) - 2f(\mathbf{O}, \mathbf{U} + \mathbf{V} + \mathbf{W}) + f(\mathbf{U} + \mathbf{V}, \mathbf{W}) - 2f(\mathbf{W}, \mathbf{U} + \mathbf{V})$$

$$+2f(\mathbf{V}+\mathbf{W},\mathbf{U}) - f(\mathbf{U},\mathbf{V}+\mathbf{W}) + 2f(\mathbf{W}+\mathbf{U},\mathbf{V}) - 2f(\mathbf{V},\mathbf{W}+\mathbf{U})$$

$$+f(\mathbf{V},\mathbf{W}) - f(\mathbf{U},\mathbf{V}) + f(\mathbf{U}+\mathbf{V},\mathbf{O}) + f(\mathbf{O},\mathbf{U}+\mathbf{V})$$

$$+f(\mathbf{O},\mathbf{W}+\mathbf{U}) - f(\mathbf{V},\mathbf{O}) - f(\mathbf{O},\mathbf{W}) - f(\mathbf{O},\mathbf{U}) = \mathbf{O}.$$

By application of Eq. (11.4), we find

$$4f(\mathbf{U}+\mathbf{V}+\mathbf{W},\mathbf{O}) - 4f(\mathbf{O},\mathbf{U}+\mathbf{V}+\mathbf{W}) + f(\mathbf{U},\mathbf{V}+\mathbf{W}) - f(\mathbf{U}+\mathbf{V},\mathbf{W})$$

$$+f(\mathbf{V},\mathbf{W}) - f(\mathbf{U},\mathbf{V}) + f(\mathbf{U}+\mathbf{V},\mathbf{O}) + f(\mathbf{O},\mathbf{U}+\mathbf{V}) + f(\mathbf{O},\mathbf{W}+\mathbf{U})$$

$$-f(\mathbf{V},\mathbf{O}) - f(\mathbf{O},\mathbf{W}) - f(\mathbf{O},\mathbf{U}) = \mathbf{O},$$

or in view of Eq. (11.3) the above relation becomes

$$
\begin{aligned}
&f(\mathbf{U}+\mathbf{V},\mathbf{W}) - f(\mathbf{U},\mathbf{V}+\mathbf{W}) + f(\mathbf{U},\mathbf{V}) - f(\mathbf{V},\mathbf{W}) \quad\quad (11.5)\\
= {}& \mathbf{K} - f(\mathbf{O},\mathbf{U}+\mathbf{V}+\mathbf{W}) + f(\mathbf{U}+\mathbf{V},\mathbf{O}) + f(\mathbf{O},\mathbf{U}+\mathbf{V})\\
+{}& f(\mathbf{O},\mathbf{U}+\mathbf{W}) - f(\mathbf{V},\mathbf{O}) - f(\mathbf{O},\mathbf{U}) - f(\mathbf{O},\mathbf{W}).
\end{aligned}
$$

By cyclic permutations of the vectors $\mathbf{U}$, $\mathbf{V}$ and $\mathbf{W}$ in the above relation, we obtain

$$
\begin{aligned}
&f(\mathbf{V}+\mathbf{W},\mathbf{U}) - f(\mathbf{V},\mathbf{W}+\mathbf{U}) + f(\mathbf{V},\mathbf{W}) - f(\mathbf{W},\mathbf{U}) \quad\quad (11.6)\\
= {}& \mathbf{K} - f(\mathbf{O},\mathbf{U}+\mathbf{V}+\mathbf{W}) + f(\mathbf{V}+\mathbf{W},\mathbf{O}) + f(\mathbf{O},\mathbf{V}+\mathbf{W})\\
+{}& f(\mathbf{O},\mathbf{U}+\mathbf{V}) - f(\mathbf{W},\mathbf{O}) - f(\mathbf{O},\mathbf{V}) - f(\mathbf{O},\mathbf{U}),
\end{aligned}
$$

$$
\begin{aligned}
&f(\mathbf{W}+\mathbf{U},\mathbf{V}) - f(\mathbf{W},\mathbf{V}+\mathbf{U}) + f(\mathbf{W},\mathbf{U}) - f(\mathbf{U},\mathbf{V}) \quad\quad (11.7)\\
= {}& \mathbf{K} - f(\mathbf{O},\mathbf{U}+\mathbf{V}+\mathbf{W}) + f(\mathbf{W}+\mathbf{U},\mathbf{O}) + f(\mathbf{O},\mathbf{W}+\mathbf{U})\\
+{}& f(\mathbf{O},\mathbf{V}+\mathbf{W}) - f(\mathbf{U},\mathbf{O}) - f(\mathbf{O},\mathbf{W}) - f(\mathbf{O},\mathcal{V}).
\end{aligned}
$$

By adding together Eqs. (11.7), (11.6) and (11.5), and by using Eq. (11.4) we obtain

$$
\begin{aligned}
&f(\mathbf{U}+\mathbf{V}+\mathbf{W},\mathbf{O}) - f(\mathbf{O},\mathbf{U}+\mathbf{V}+\mathbf{W}) \quad\quad\quad\quad (11.8)\\
= {}& 2[f(\mathbf{O},\mathbf{U}+\mathbf{V}) + f(\mathbf{O},\mathbf{V}+\mathbf{W}) + f(\mathbf{O},\mathbf{W}+\mathbf{U})]
\end{aligned}
$$

$$+[f(\mathbf{U}+\mathbf{V},\mathbf{O}) + f(\mathbf{V}+\mathbf{W},\mathbf{O}) + f(\mathbf{W}+\mathbf{U},\mathbf{O})]$$

$$-2[f(\mathbf{O}, \mathbf{U}) + f(\mathbf{O}, \mathbf{V}) + f(\mathbf{O}, \mathbf{W})]$$

$$-[f(\mathbf{U}, \mathbf{O}) + f(\mathbf{V}, \mathbf{O}) + f(\mathbf{W}, \mathbf{O})]$$

$$+3\mathbf{K} - 3f(\mathbf{O}, \mathbf{U} + \mathbf{V} + \mathbf{W}).$$

It immediately follows that the function

$$\varphi(\mathbf{U}) = f(\mathbf{U}, \mathbf{O}) + 2f(\mathbf{O}, \mathbf{U}) - 3\mathbf{K} \tag{11.9}$$

satisfies Fréchet's functional equation [M. Fréchet (1909)]

$$\varphi(\mathbf{U} + \mathbf{V} + \mathbf{W}) - \varphi(\mathbf{U} + \mathbf{V}) - \varphi(\mathbf{V} + \mathbf{W})$$

$$-\varphi(\mathbf{W} + \mathbf{U}) + \varphi(\mathbf{U}) + \varphi(\mathbf{V}) + \varphi(\mathbf{W}) = \mathbf{O}.$$

The general continuous solution of this equation is

$$\varphi(\mathbf{U}) = A\mathbf{U}^2 + B\mathbf{U},$$

where A and B are constant complex matrices.

For determination of the functions $f(\mathbf{U}, \mathbf{O})$ and $f(\mathbf{O}, \mathbf{U})$ we find the equations

$$
\begin{aligned}
4f(\mathbf{U}, \mathbf{O}) - 3f(\mathbf{O}, \mathbf{U}) &= \mathbf{K}, \\
f(\mathbf{U}, \mathbf{O}) + 2f(\mathbf{O}, \mathbf{U}) &= A\mathbf{U}^2 + B\mathbf{U} + 3\mathbf{K}.
\end{aligned}
$$

By solving this system we obtain

$$
\begin{aligned}
f(\mathbf{U}, \mathbf{O}) &= 3\alpha\mathbf{U}^2 + 3\beta\mathbf{U} + \mathbf{K}, \\
f(\mathbf{O}, \mathbf{U}) &= 4\alpha\mathbf{U}^2 + 4\beta\mathbf{U} + \mathbf{K},
\end{aligned}
$$

where $\alpha = A/11$ and $\beta = B/11$. From the equation (11.5) it follows that

$$f(\mathbf{U} + \mathbf{V}, \mathbf{W}) - f(\mathbf{U}, \mathbf{V} + \mathbf{W}) + f(\mathbf{U}, \mathbf{V}) - f(\mathbf{V}, \mathbf{W})$$

$$= 3\alpha\mathbf{U}^2 - 4\alpha\mathbf{W}^2 + 6\alpha\mathbf{U}\mathbf{V} - 8\alpha\mathbf{V}\mathbf{W} + 3\beta\mathbf{U} - 4\beta\mathbf{W}.$$

Using the notation $g(\mathbf{U}, \mathbf{V}) = \alpha(3\mathbf{U}^2 + 4\mathbf{V}^2) + \beta(3\mathbf{U} + 4\mathbf{V})$, the above equations may be written in the form

$$f(\mathbf{U} + \mathbf{V}, \mathbf{W}) - f(\mathbf{U}, \mathbf{V} + \mathbf{W}) + f(\mathbf{U}, \mathbf{V}) - f(\mathbf{V}, \mathbf{W})$$

$$= g(\mathbf{U} + \mathbf{V}, \mathbf{W}) - g(\mathbf{U}, \mathbf{V} + \mathbf{W}) + g(\mathbf{U}, \mathbf{V}) - g(\mathbf{V}, \mathbf{W}).$$

Thus, the function $f(\mathbf{U}, \mathbf{V}) - g(\mathbf{U}, \mathbf{V})$ satisfies a functional equation whose solution is given in [J. Aczél (1966)]:

$$f(\mathbf{U}, \mathbf{V}) - g(\mathbf{U}, \mathbf{V}) = F(\mathbf{U} + \mathbf{V}) - F(\mathbf{U}) - F(\mathbf{V}) + L(\mathbf{U}, \mathbf{V}).$$

This means that Eq. (11.2) is the general continuous solution of Eq. (11.1).

By a direct calculation without difficulty we can verify that the function Eq. (11.2) satisfies the functional equation (11.1). $\qquad\square$

Theorem 11.2 *The general continuous solution of the complex vector functional equation*

$$f(\mathbf{Z}_0 + \mathbf{Z}_1 + \cdots + \mathbf{Z}_n, \mathbf{Z}_{n+1}) - [f(\mathbf{Z}_0 + \mathbf{Z}_1 + \cdots + \mathbf{Z}_{n-1}, \mathbf{Z}_n + \mathbf{Z}_{n+1}) \quad (11.10)$$

$$+ f(\mathbf{Z}_0 + \mathbf{Z}_1 + \cdots + \mathbf{Z}_{n-2} + \mathbf{Z}_n, \mathbf{Z}_{n-1} + \mathbf{Z}_{n+1}) + \cdots$$

$$+ f(\mathbf{Z}_0 + \mathbf{Z}_2 + \mathbf{Z}_3 + \cdots + \mathbf{Z}_n, \mathbf{Z}_1 + \mathbf{Z}_{n+1}) + f(\mathbf{Z}_1 + \mathbf{Z}_2 + \cdots + \mathbf{Z}_n, \mathbf{Z}_0 + \mathbf{Z}_{n+1})]$$

$$+ [f(\mathbf{Z}_0 + \mathbf{Z}_1 + \cdots + \mathbf{Z}_{n-2}, \mathbf{Z}_{n-1} + \mathbf{Z}_n + \mathbf{Z}_{n+1}) + \cdots] - \cdots$$

$$+ (-1)^n [f(\mathbf{Z}_0, \mathbf{Z}_1 + \mathbf{Z}_2 + \cdots + \mathbf{Z}_{n+1}) + f(\mathbf{Z}_1, \mathbf{Z}_2 + \mathbf{Z}_3 + \cdots + \mathbf{Z}_{n+1} + \mathbf{Z}_0)$$

$$+ f(\mathbf{Z}_n, \mathbf{Z}_{n+1} + \mathbf{Z}_0 + \mathbf{Z}_1 + \cdots + \mathbf{Z}_{n-1})] = \mathbf{O}$$

is given by

$$f(\mathbf{U}, \mathbf{V}) = \sum_{i=1}^{n} \mathbf{U}^i F_i(\mathbf{U} + \mathbf{V}), \qquad (11.11)$$

where F_i $(1 \le i \le n)$ are arbitrary continuous complex functions with values in $\mathcal{V}$.

Proof.　Let us introduce the notations

$$f(\mathbf{U}, \mathbf{V}) = F(\mathbf{U}, \mathbf{U} + \mathbf{V}), \quad \sum_{i=0}^{n+1} \mathbf{Z}_i = \mathbf{T}. \qquad (11.12)$$

Using these notations and Eq. (11.10), we obtain Fréchet's functional equation [M. Fréchet (1909)]

$$F(\mathbf{Z}_0 + \mathbf{Z}_1 + \cdots + \mathbf{Z}_n, \mathbf{T}) - [F(\mathbf{Z}_0 + \mathbf{Z}_1 + \cdots + \mathbf{Z}_{n-1}, \mathbf{T}) \qquad (11.13)$$

$$+F(\mathbf{Z}_0 + \mathbf{Z}_1 + \cdots + \mathbf{Z}_{n-2} + \mathbf{Z}_n, \mathbf{T}) + \cdots + F(\mathbf{Z}_0 + \mathbf{Z}_2 + \mathbf{Z}_3 + \cdots + \mathbf{Z}_n, \mathbf{T})$$

$$+F(\mathbf{Z}_1 + \mathbf{Z}_2 + \cdots + \mathbf{Z}_n, \mathbf{T})] + [F(\mathbf{Z}_0 + \mathbf{Z}_1 + \cdots + \mathbf{Z}_{n-2}, \mathbf{T}) + \cdots]$$

$$- \cdots + (-1)^n [F(\mathbf{Z}_0, \mathbf{T}) + F(\mathbf{Z}_1, \mathbf{T}) + F(\mathbf{Z}_2, \mathbf{T}) + \cdots + F(\mathbf{Z}_n, \mathbf{T})] = \mathbf{O},$$

whose general continuous solution according to [J. Aczél (1966)] is given by the formula

$$F(\mathbf{U}, \mathbf{T}) = \sum_{i=1}^{n} \mathbf{U}^i G_i(\mathbf{T}) \tag{11.14}$$

where G_i $(1 \leq i \leq n)$ are arbitrary continuous functions with values in $\mathcal{V}$. We put Eq. (11.14) into Eq. (11.12) and get Eq. (11.11). $\square$

Example 11.1 A particular case of the above theorem for $n = 2$ is the following functional equation

$$f(\mathbf{Z}_0 + \mathbf{Z}_1 + \mathbf{Z}_2, \mathbf{Z}_3) - f(\mathbf{Z}_0 + \mathbf{Z}_1, \mathbf{Z}_2 + \mathbf{Z}_3) - f(\mathbf{Z}_0 + \mathbf{Z}_2, \mathbf{Z}_1 + \mathbf{Z}_3)$$

$$-f(\mathbf{Z}_1 + \mathbf{Z}_2 + \mathbf{Z}_0, \mathbf{Z}_3) + f(\mathbf{Z}_0, \mathbf{Z}_1 + \mathbf{Z}_2 + \mathbf{Z}_3) + f(\mathbf{Z}_1, \mathbf{Z}_2 + \mathbf{Z}_3 + \mathbf{Z}_0)$$

$$+f(\mathbf{Z}_2, \mathbf{Z}_3 + \mathbf{Z}_0 + \mathbf{Z}_1) = \mathbf{O},$$

whose general continuous solution is

$$f(\mathbf{U}, \mathbf{V}) = \mathbf{U}^2 F_2(\mathbf{U} + \mathbf{V}) + \mathbf{U} F_1(\mathbf{U} + \mathbf{V}),$$

where F_1 and F_2 are arbitrary continuous complex functions with values in $\mathcal{V}$.

12 Functional Equations with Two Operations

In this section we will be able to solve some complex vector functional equations supplied by two operations, namely addition and multiplication, between arguments.

For this purpose we use the same notations for the vectors given in the previous section and suppose that $f : \mathcal{V}^2 \mapsto \mathcal{V}$. Now we give the following results.

Theorem 12.1   *The general solution of the functional equation*

$$f(\mathbf{Z}_1 + \mathbf{Z}_2, \mathbf{Z}_3) - f(\mathbf{Z}_1 \cdot \mathbf{Z}_2, \mathbf{Z}_3) - f(\mathbf{Z}_1 + \mathbf{Z}_2, \mathbf{Z}_3 \cdot \mathbf{Z}_4) \tag{12.1}$$
$$+ \quad f(\mathbf{Z}_2 + \mathbf{Z}_3, \mathbf{Z}_4) - f(\mathbf{Z}_2 \cdot \mathbf{Z}_3, \mathbf{Z}_4) - f(\mathbf{Z}_2 + \mathbf{Z}_3, \mathbf{Z}_4 \cdot \mathbf{Z}_1)$$
$$+ \quad f(\mathbf{Z}_3 + \mathbf{Z}_4, \mathbf{Z}_1) - f(\mathbf{Z}_3 \cdot \mathbf{Z}_4, \mathbf{Z}_1) - f(\mathbf{Z}_3 + \mathbf{Z}_4, \mathbf{Z}_1 \cdot \mathbf{Z}_2)$$
$$+ \quad f(\mathbf{Z}_4 + \mathbf{Z}_1, \mathbf{Z}_2) - f(\mathbf{Z}_4 \cdot \mathbf{Z}_1, \mathbf{Z}_2) - f(\mathbf{Z}_4 + \mathbf{Z}_1, \mathbf{Z}_2 \cdot \mathbf{Z}_3) = \mathbf{O}$$

is given by the following formula

$$f(\mathbf{U}, \mathbf{V}) = F(\mathbf{U}) - F(\mathbf{V}), \tag{12.2}$$

where F is an arbitrary function with values in $\mathcal{V}$.

Proof.   If we put $\mathbf{Z}_3 = \mathbf{Z}_4 = \mathbf{O}$ into Eq. (12.1), we obtain

$$
\begin{aligned}
f(\mathbf{Z}_1, \mathbf{Z}_2) \;=\;& f(\mathbf{Z}_1 \cdot \mathbf{Z}_2, \mathbf{O}) + f(\mathbf{O}, \mathbf{Z}_1 \cdot \mathbf{Z}_2) \\
+\;& f(\mathbf{Z}_1, \mathbf{O}) + f(\mathbf{O}, \mathbf{Z}_2) + f(\mathbf{O}, \mathbf{O}).
\end{aligned}
\tag{12.3}
$$

For $\mathbf{Z}_2 = \mathbf{Z}_4 = \mathbf{O}$ from Eq. (12.1) we get

$$
\begin{aligned}
f(\mathbf{Z}_1, \mathbf{Z}_3) + f(\mathbf{Z}_3, \mathbf{Z}_1) \;=\;& f(\mathbf{O}, \mathbf{Z}_1) + f(\mathbf{Z}_1, \mathbf{O}) \\
+\;& f(\mathbf{O}, \mathbf{Z}_3) + f(\mathbf{Z}_3, \mathbf{O}) + 2f(\mathbf{O}, \mathbf{O}).
\end{aligned}
$$

The previous relation also holds if we put $\mathbf{Z}_3 = \mathbf{Z}_2$, *i.e.*,

$$
\begin{aligned}
f(\mathbf{Z}_1, \mathbf{Z}_2) + f(\mathbf{Z}_2, \mathbf{Z}_1) \;=\;& f(\mathbf{O}, \mathbf{Z}_1) + f(\mathbf{Z}_1, \mathbf{O}) \\
+\;& f(\mathbf{O}, \mathbf{Z}_2) + f(\mathbf{Z}_2, \mathbf{O}) + 2f(\mathbf{O}, \mathbf{O}).
\end{aligned}
\tag{12.4}
$$

After a permutation of $\mathbf{Z}_1$ and $\mathbf{Z}_2$ in Eq. (12.3), we obtain

$$
\begin{aligned}
f(\mathbf{Z}_2, \mathbf{Z}_1) \;=\;& f(\mathbf{Z}_2 \cdot \mathbf{Z}_1, \mathbf{O}) + f(\mathbf{O}, \mathbf{Z}_2 \cdot \mathbf{Z}_1) \\
+\;& f(\mathbf{Z}_2, \mathbf{O}) + f(\mathbf{O}, \mathbf{Z}_1) + f(\mathbf{O}, \mathbf{O}).
\end{aligned}
\tag{12.5}
$$

Since the multiplication between the vectors is commutative, we find

$$f(\mathbf{Z}_1, \mathbf{Z}_2) + f(\mathbf{Z}_2, \mathbf{Z}_1) = 2f(\mathbf{Z}_1 \cdot \mathbf{Z}_2, \mathbf{O}) + 2f(\mathbf{O}, \mathbf{Z}_1 \cdot \mathbf{Z}_2) \tag{12.6}$$
$$+ \quad f(\mathbf{Z}_1, \mathbf{O}) + f(\mathbf{O}, \mathbf{Z}_1) + f(\mathbf{Z}_2, \mathbf{O}) + f(\mathbf{O}, \mathbf{Z}_2) + 2f(\mathbf{O}, \mathbf{O}).$$

By comparison of Eqs. (12.4) and (12.6), we have

$$f(\mathbf{Z}_1 \cdot \mathbf{Z}_2, \mathbf{O}) + f(\mathbf{O}, \mathbf{Z}_1 \cdot \mathbf{Z}_2) = \mathbf{O}. \tag{12.7}$$

If we put $\mathbf{U} = \mathbf{Z}_1 \cdot \mathbf{Z}_2$, the equality (12.7) takes on the form

$$f(\mathbf{U}, \mathbf{O}) + f(\mathbf{O}, \mathbf{U}) = \mathbf{O}. \tag{12.8}$$

If we substitute Eq. (12.7) into Eq. (12.3), we get

$$f(\mathbf{Z}_1, \mathbf{Z}_2) = f(\mathbf{Z}_1, \mathbf{O}) + f(\mathbf{O}, \mathbf{Z}_2) + f(\mathbf{O}, \mathbf{O}). \tag{12.9}$$

By virtue of Eq. (12.8), we may rewrite Eq. (12.9) as

$$f(\mathbf{Z}_1, \mathbf{Z}_2) = f(\mathbf{Z}_1, \mathbf{O}) - f(\mathbf{Z}_2, \mathbf{O}) + f(\mathbf{O}, \mathbf{O}). \tag{12.10}$$

If we put $\mathbf{U} = \mathbf{O}$ into Eq. (12.8), we obtain

$$2f(\mathbf{O}, \mathbf{O}) = \mathbf{O}, \quad i.e., \quad f(\mathbf{O}, \mathbf{O}) = \mathbf{O},$$

and thus Eq. (12.10) becomes

$$f(\mathbf{Z}_1, \mathbf{Z}_2) = f(\mathbf{Z}_1, \mathbf{O}) - f(\mathbf{Z}_2, \mathbf{O}). \tag{12.11}$$

Now, Eq. (12.2) follows immediately from Eq. (12.11), *i.e.,* any solution of Eq. (12.1) has the form Eq. (12.2).

Conversely, a direct calculation shows that the function Eq. (12.2) satisfies the equation (12.1) for an arbitrary F with values in $\mathcal{V}$. $\square$

Theorem 12.2   *The general solution of the functional equation*

$$\sum_{i=1}^{n} [f(\mathbf{Z}_i + \mathbf{Z}_{i+1}, \mathbf{Z}_{i+2}) - f(\mathbf{Z}_i \cdot \mathbf{Z}_{i+1}, \mathbf{Z}_{i+2}) \tag{12.12}$$
$$- \quad f(\mathbf{Z}_i + \mathbf{Z}_{i+1}, \mathbf{Z}_{i+2} \cdot \mathbf{Z}_{i+3})] = \mathbf{O} \qquad (\mathbf{Z}_{i+n} \equiv \mathbf{Z}_i;\ n \geq 5)$$

is

$$f(\mathbf{U}, \mathbf{V}) = F(\mathbf{U}) - F(\mathbf{V}) \tag{12.13}$$

where F is an arbitrary complex vector function with values in $\mathcal{V}$.

Proof.   If we put $\mathbf{Z}_i = \mathbf{O}$ $(1 \leq i \leq n)$ into Eq. (12.12), we find

$$nf(\mathbf{O}, \mathbf{O}) = \mathbf{O}, \quad i.e., \quad f(\mathbf{O}, \mathbf{O}) = \mathbf{O}.$$

By putting $\mathbf{Z}_i = \mathbf{O}$ $(3 \leq i \leq n)$ into Eq. (12.12), we obtain

$$f(\mathbf{Z}_1, \mathbf{Z}_2) = f(\mathbf{Z}_1 \cdot \mathbf{Z}_2, \mathbf{O}) + f(\mathbf{O}, \mathbf{Z}_1 \cdot \mathbf{Z}_2) + f(\mathbf{Z}_1, \mathbf{O}) + f(\mathbf{O}, \mathbf{Z}_2). \tag{12.14}$$

Now, if we put $\mathbf{Z}_i = \mathbf{O}$ ($i = 2$, $4 \leq i \leq n$) into Eq. (12.12), we get

$$f(\mathbf{Z}_1, \mathbf{Z}_3) = f(\mathbf{Z}_1, \mathbf{O}) + f(\mathbf{O}, \mathbf{Z}_3).$$

For the sake of convenience we take $\mathbf{Z}_3 = \mathbf{Z}_2$, then

$$f(\mathbf{Z}_1, \mathbf{Z}_2) = f(\mathbf{Z}_1, \mathbf{O}) + f(\mathbf{O}, \mathbf{Z}_2). \tag{12.15}$$

From the equalities (12.14) and (12.15) it follows that

$$f(\mathbf{Z}_1 \cdot \mathbf{Z}_2, \mathbf{O}) + f(\mathbf{O}, \mathbf{Z}_1 \cdot \mathbf{Z}_2) = \mathbf{O},$$

i.e.,

$$f(\mathbf{U}, \mathbf{O}) + f(\mathbf{O}, \mathbf{U}) = \mathbf{O}. \tag{12.16}$$

Therefore, the equality (12.15) becomes

$$f(\mathbf{Z}_1, \mathbf{Z}_2) = f(\mathbf{Z}_1, \mathbf{O}) - f(\mathbf{Z}_2, \mathbf{O}) \tag{12.17}$$

which may be also rewritten in the form Eq. (12.13).

Since any solution of the equation (12.12) has the form Eq. (12.13) and, conversely, the function Eq. (12.13) is a solution of the equation (12.12), it follows that the function Eq. (12.13) is the general solution of the equation (12.12). $\qquad\square$

Now assume that $f : \mathcal{V}^3 \mapsto \mathcal{V}$.

Theorem 12.3 *The general solution of the functional equation*

$$\sum_{i=1}^{5} [f(\mathbf{Z}_i + \mathbf{Z}_{i+1}, \mathbf{Z}_{i+2} \cdot \mathbf{Z}_{i+3}, \mathbf{Z}_{i+4}) + f(\mathbf{Z}_i \cdot \mathbf{Z}_{i+1}, \mathbf{Z}_{i+2}, \mathbf{Z}_{i+3} + \mathbf{Z}_{i+4}) \tag{12.18}$$

$$+ f(\mathbf{Z}_i, \mathbf{Z}_{i+1} + \mathbf{Z}_{i+2}, \mathbf{Z}_{i+3} \cdot \mathbf{Z}_{i+4})] = \mathbf{O} \quad (\mathbf{Z}_{i+5} \equiv \mathbf{Z}_i)$$

is given by

$$f(\mathbf{U}, \mathbf{V}, \mathbf{W}) = F(\mathbf{U}, \mathbf{V}, \mathbf{W}) - F(\mathbf{V}, \mathbf{W}, \mathbf{U}), \tag{12.19}$$

where F is an arbitrary complex vector function with values in $\mathcal{V}$.

Proof. By a direct substitution of the function Eq. (12.19) into Eq. (12.18) we verify that it is a solution of the equation (12.18).

Conversely, if we suppose that the function $f(\mathbf{U}, \mathbf{V}, \mathbf{W})$ is a solution of the equation (12.18), we will prove that its form is given by Eq. (12.19).

From Eq. (12.18) for $\mathbf{Z}_1 = \mathbf{Z}_2 = \mathbf{Z}_3 = \mathbf{Z}_4 = \mathbf{Z}_5 = \mathbf{O}$ we obtain

$$15f(\mathbf{O}, \mathbf{O}, \mathbf{O}) = \mathbf{O},$$

i.e.,

$$f(\mathbf{O}, \mathbf{O}, \mathbf{O}) = \mathbf{O}. \tag{12.20}$$

If we put $\mathbf{Z}_1 = \mathbf{U}$, $\mathbf{Z}_2 = \mathbf{Z}_3 = \mathbf{Z}_4 = \mathbf{Z}_5 = \mathbf{O}$ into Eq. (12.18), in view of Eq. (12.20) we get

$$3[f(\mathbf{U}, \mathbf{O}, \mathbf{O}) + f(\mathbf{O}, \mathbf{U}, \mathbf{O}) + f(\mathbf{O}, \mathbf{O}, \mathbf{U})] = \mathbf{O},$$

i.e.,

$$f(\mathbf{U}, \mathbf{O}, \mathbf{O}) + f(\mathbf{O}, \mathbf{U}, \mathbf{O}) + f(\mathbf{O}, \mathbf{O}, \mathbf{U}) = \mathbf{O}. \tag{12.21}$$

If we substitute $\mathbf{Z}_5 = \mathbf{O}$ in Eq. (12.18), making use of the equality (12.21), we obtain

$$f(\mathbf{Z}_1 \cdot \mathbf{Z}_2, \mathbf{Z}_3, \mathbf{Z}_4) + f(\mathbf{Z}_3, \mathbf{Z}_4, \mathbf{Z}_1 \cdot \mathbf{Z}_2) + f(\mathbf{Z}_4, \mathbf{Z}_1 \cdot \mathbf{Z}_2, \mathbf{Z}_3)$$
$$+ \quad f(\mathbf{Z}_2 \cdot \mathbf{Z}_3, \mathbf{Z}_4, \mathbf{Z}_1) + f(\mathbf{Z}_4, \mathbf{Z}_1, \mathbf{Z}_2 \cdot \mathbf{Z}_3) + f(\mathbf{Z}_1, \mathbf{Z}_2 \cdot \mathbf{Z}_3, \mathbf{Z}_4) = \mathbf{O}.$$

If we put $\mathbf{Z}_2 = \mathbf{I} = (1, 1, \cdots, 1)^T$, then we obtain

$$2[f(\mathbf{Z}_1, \mathbf{Z}_3, \mathbf{Z}_4) + f(\mathbf{Z}_3, \mathbf{Z}_4, \mathbf{Z}_1) + f(\mathbf{Z}_4, \mathbf{Z}_1, \mathbf{Z}_3)] = \mathbf{O}.$$

In other words, we have

$$f(\mathbf{Z}_1, \mathbf{Z}_2, \mathbf{Z}_3) + f(\mathbf{Z}_2, \mathbf{Z}_3, \mathbf{Z}_1) + f(\mathbf{Z}_3, \mathbf{Z}_1, \mathbf{Z}_2) = \mathbf{O}. \tag{12.22}$$

The general solution of the equation (12.22), according to [I. B. Risteski *et al.* (2000A)] is given by Eq. (12.19). $\qquad\square$

Theorem 12.4 *The general solution of the functional equation*

$$\sum_{i=1}^{6}[f(\mathbf{Z}_i + \mathbf{Z}_{i+1}, \mathbf{Z}_{i+2} \cdot \mathbf{Z}_{i+3}, \mathbf{Z}_{i+4}) + f(\mathbf{Z}_i \cdot \mathbf{Z}_{i+1}, \mathbf{Z}_{i+2}, \mathbf{Z}_{i+3} + \mathbf{Z}_{i+4}) \tag{12.23}$$

$$+ f(\mathbf{Z}_i, \mathbf{Z}_{i+1} + \mathbf{Z}_{i+2}, \mathbf{Z}_{i+3} \cdot \mathbf{Z}_{i+4})] = \mathbf{O} \quad (\mathbf{Z}_{i+6} \equiv \mathbf{Z}_i)$$

is given by

$$f(\mathbf{U}, \mathbf{V}, \mathbf{W}) = F(\mathbf{U}, \mathbf{V}) - F(\mathbf{V}, \mathbf{W}), \tag{12.24}$$

where F is an arbitrary complex vector function with values in $\mathcal{V}$.

Proof. If we substitute the function Eq. (12.24) in Eq. (12.23), we obtain that Eq. (12.24) is a solution of the functional equation (12.23).

Now we will prove that the converse is also true, *i.e.*, that any solution of the equation (12.23) has the form Eq. (12.24).

If we put $\mathbf{Z}_i = \mathbf{O}$ $(1 \leq i \leq 6)$ into the equation (12.23), then we get

$$18f(\mathbf{O}, \mathbf{O}, \mathbf{O}) = \mathbf{O},$$

i.e.,

$$f(\mathbf{O}, \mathbf{O}, \mathbf{O}) = \mathbf{O}. \tag{12.25}$$

If we put $\mathbf{Z}_1 = \mathbf{U}$ and $\mathbf{Z}_i = \mathbf{O}$ $(2 \leq i \leq 6)$, we get

$$3[f(\mathbf{U}, \mathbf{O}, \mathbf{O}) + f(\mathbf{O}, \mathbf{U}, \mathbf{O}) + f(\mathbf{O}, \mathbf{O}, \mathbf{U})] = \mathbf{O},$$

i.e.,

$$f(\mathbf{U}, \mathbf{O}, \mathbf{O}) + f(\mathbf{O}, \mathbf{U}, \mathbf{O}) + f(\mathbf{O}, \mathbf{O}, \mathbf{U}) = \mathbf{O}. \tag{12.26}$$

Now, by putting $\mathbf{Z}_3 = \mathbf{Z}_4 = \mathbf{Z}_5 = \mathbf{Z}_6 = \mathbf{O}$ into Eq. (12.23) and taking into account Eqs. (12.25) and (12.26), we find

$$f(\mathbf{Z}_1, \mathbf{Z}_2, \mathbf{O}) + f(\mathbf{O}, \mathbf{Z}_1, \mathbf{Z}_2) + f(\mathbf{O}, \mathbf{O}, \mathbf{Z}_1) + f(\mathbf{Z}_2, \mathbf{O}, \mathbf{O}) = \mathbf{O}. \tag{12.27}$$

If we substitute $\mathbf{Z}_2 = \mathbf{Z}_4 = \mathbf{Z}_5 = \mathbf{Z}_6 = \mathbf{O}$ into Eq. (12.23), according to Eqs. (12.25) and (12.26) we obtain

$$f(\mathbf{Z}_1, \mathbf{Z}_3, \mathbf{O}) + f(\mathbf{Z}_3, \mathbf{O}, \mathbf{Z}_1) + f(\mathbf{O}, \mathbf{Z}_1, \mathbf{Z}_3) = \mathbf{O}.$$

If we replace $\mathbf{Z}_3$ by $\mathbf{Z}_2$ in the above expression, we get

$$f(\mathbf{Z}_1, \mathbf{Z}_2, \mathbf{O}) + f(\mathbf{Z}_2, \mathbf{O}, \mathbf{Z}_1) + f(\mathbf{O}, \mathbf{Z}_1, \mathbf{Z}_2) = \mathbf{O}. \tag{12.28}$$

From the equalities (12.27) and (12.28) it follows that

$$f(\mathbf{Z}_2, \mathbf{O}, \mathbf{Z}_1) = f(\mathbf{Z}_2, \mathbf{O}, \mathbf{O}) + f(\mathbf{O}, \mathbf{O}, \mathbf{Z}_1). \tag{12.29}$$

If $\mathbf{Z}_5 = \mathbf{Z}_6 = \mathbf{O}$ in Eq. (12.23), we obtain

$$f(\mathbf{Z}_1 + \mathbf{Z}_2, \mathbf{Z}_3 \cdot \mathbf{Z}_4, \mathbf{O}) + f(\mathbf{Z}_1 \cdot \mathbf{Z}_2, \mathbf{Z}_3, \mathbf{Z}_4) + f(\mathbf{Z}_1, \mathbf{Z}_2 + \mathbf{Z}_3, \mathbf{O})$$

$$+ \quad f(\mathbf{Z}_2 + \mathbf{Z}_3, \mathbf{O}, \mathbf{O}) + f(\mathbf{Z}_2 \cdot \mathbf{Z}_3, \mathbf{Z}_4, \mathbf{O}) + f(\mathbf{Z}_2, \mathbf{Z}_3 + \mathbf{Z}_4, \mathbf{O})$$

$$+ \quad f(\mathbf{Z}_3 + \mathbf{Z}_4, \mathbf{O}, \mathbf{Z}_1) + f(\mathbf{Z}_3 \cdot \mathbf{Z}_4, \mathbf{O}, \mathbf{Z}_1) + f(\mathbf{Z}_3, \mathbf{Z}_4, \mathbf{O})$$

$$+ \quad f(\mathbf{Z}_4, \mathbf{O}, \mathbf{Z}_2) + f(\mathbf{O}, \mathbf{O}, \mathbf{Z}_1 + \mathbf{Z}_2) + f(\mathbf{Z}_4, \mathbf{O}, \mathbf{Z}_1 \cdot \mathbf{Z}_5)$$

$$+ \quad f(\mathbf{O}, \mathbf{Z}_1 \cdot \mathbf{Z}_2, \mathbf{Z}_3) + f(\mathbf{O}, \mathbf{Z}_1, \mathbf{Z}_2 + \mathbf{Z}_3) + f(\mathbf{O}, \mathbf{Z}_1, \mathbf{Z}_2 \cdot \mathbf{Z}_3)$$

$$+ \quad f(\mathbf{Z}_1, \mathbf{Z}_2 \cdot \mathbf{Z}_3, \mathbf{Z}_4) + f(\mathbf{O}, \mathbf{Z}_2, \mathbf{Z}_3 + \mathbf{Z}_4) + f(\mathbf{O}, \mathbf{Z}_1 + \mathbf{Z}_2, \mathbf{Z}_3 \cdot \mathbf{Z}_4) = \mathbf{O}.$$

By using the equalities (12.28) and (12.29), the above equation reduces to the following simpler form

$$f(\mathbf{Z}_1 \cdot \mathbf{Z}_2, \mathbf{Z}_3, \mathbf{Z}_4) + f(\mathbf{Z}_1, \mathbf{Z}_2 \cdot \mathbf{Z}_3, \mathbf{Z}_4) + f(\mathbf{Z}_2 \cdot \mathbf{Z}_3, \mathbf{Z}_4, \mathbf{O}) + f(\mathbf{Z}_4, \mathbf{O}, \mathbf{Z}_1 \cdot \mathbf{Z}_2)$$

$$+ f(\mathbf{O}, \mathbf{Z}_1, \mathbf{Z}_2 \cdot \mathbf{Z}_3) + f(\mathbf{O}, \mathbf{Z}_1 \cdot \mathbf{Z}_2, \mathbf{Z}_3) + f(\mathbf{Z}_3, \mathbf{Z}_4, \mathbf{O}) + f(\mathbf{Z}_4, \mathbf{O}, \mathbf{Z}_1) = \mathbf{O}.$$

In the above equation we put $\mathbf{Z}_2 = \mathbf{I} = (1, 1, \cdots, 1)^T$, and then we obtain

$$2[f(\mathbf{Z}_1, \mathbf{Z}_3, \mathbf{Z}_4) + f(\mathbf{Z}_3, \mathbf{Z}_4, \mathbf{O}) + f(\mathbf{O}, \mathbf{Z}_1, \mathbf{Z}_3) + f(\mathbf{Z}_4, \mathbf{O}, \mathbf{Z}_1)] = \mathbf{O},$$

i.e., we get

$$f(\mathbf{Z}_1, \mathbf{Z}_2, \mathbf{Z}_3) = -f(\mathbf{Z}_2, \mathbf{Z}_3, \mathbf{O}) - f(\mathbf{O}, \mathbf{Z}_1, \mathbf{Z}_2) - f(\mathbf{Z}_3, \mathbf{O}, \mathbf{Z}_1). \quad (12.30)$$

We may show that the equation (12.23) is a consequence of Eqs. (12.28), (12.29) and (12.30). On the basis of the last three equalities, we obtain

$$f(\mathbf{Z}_1, \mathbf{Z}_2, \mathbf{Z}_3) + f(\mathbf{Z}_2, \mathbf{Z}_3, \mathbf{Z}_1) + f(\mathbf{Z}_3, \mathbf{Z}_1, \mathbf{Z}_2)$$

$$= \quad [-f(\mathbf{Z}_2, \mathbf{Z}_3, \mathbf{O}) - f(\mathbf{O}, \mathbf{Z}_1, \mathbf{Z}_2) - f(\mathbf{Z}_3, \mathbf{O}, \mathbf{Z}_1)]$$

$$+ \quad [-f(\mathbf{Z}_3, \mathbf{Z}_1, \mathbf{O}) - f(\mathbf{O}, \mathbf{Z}_2, \mathbf{Z}_3) - f(\mathbf{Z}_1, \mathbf{O}, \mathbf{Z}_2)]$$

$$+ \quad [-f(\mathbf{Z}_1, \mathbf{Z}_2, \mathbf{O}) - f(\mathbf{O}, \mathbf{Z}_3, \mathbf{Z}_1) - f(\mathbf{Z}_2, \mathbf{O}, \mathbf{Z}_3)]$$

$$= \quad -[f(\mathbf{Z}_2, \mathbf{Z}_3, \mathbf{O}) + f(\mathbf{O}, \mathbf{Z}_2, \mathbf{Z}_3) + f(\mathbf{Z}_3, \mathbf{O}, \mathbf{Z}_2)]$$

$$+ f(\mathbf{Z}_3, \mathbf{O}, \mathbf{Z}_2) - f(\mathbf{Z}_2, \mathbf{O}, \mathbf{Z}_3)$$

$$- \quad [f(\mathbf{Z}_3, \mathbf{Z}_1, \mathbf{O}) + f(\mathbf{O}, \mathbf{Z}_3, \mathbf{Z}_1) + f(\mathbf{Z}_1, \mathbf{O}, \mathbf{Z}_3)]$$

$$+ f(\mathbf{Z}_1, \mathbf{O}, \mathbf{Z}_3) - f(\mathbf{Z}_3, \mathbf{O}, \mathbf{Z}_1)$$

$$- \quad [f(\mathbf{Z}_1, \mathbf{Z}_2, \mathbf{O}) + f(\mathbf{O}, \mathbf{Z}_1, \mathbf{Z}_2) + f(\mathbf{Z}_2, \mathbf{O}, \mathbf{Z}_1)]$$

$$+ f(\mathbf{Z}_2, \mathbf{O}, \mathbf{Z}_1) - f(\mathbf{Z}_1, \mathbf{O}, \mathbf{Z}_2)$$

$$\begin{aligned}
&= \; f(\mathbf{Z}_3,\mathbf{O},\mathbf{O}) + f(\mathbf{O},\mathbf{O},\mathbf{Z}_2) - f(\mathbf{Z}_2,\mathbf{O},\mathbf{O}) + f(\mathbf{O},\mathbf{O},\mathbf{Z}_3) \\
&+ \; f(\mathbf{Z}_1,\mathbf{O},\mathbf{O}) + f(\mathbf{O},\mathbf{O},\mathbf{Z}_3) - f(\mathbf{Z}_3,\mathbf{O},\mathbf{O}) + f(\mathbf{O},\mathbf{O},\mathbf{Z}_1) \\
&+ \; f(\mathbf{Z}_2,\mathbf{O},\mathbf{O}) + f(\mathbf{O},\mathbf{O},\mathbf{Z}_1) - f(\mathbf{Z}_1,\mathbf{O},\mathbf{O}) + f(\mathbf{O},\mathbf{O},\mathbf{Z}_2) = \mathbf{O}.
\end{aligned}$$

From the last equality it follows that there exists a function $G(\mathbf{U},\mathbf{V},\mathbf{W})$ such that

$$f(\mathbf{Z}_1,\mathbf{Z}_2,\mathbf{Z}_3) = G(\mathbf{Z}_1,\mathbf{Z}_2,\mathbf{Z}_3) - G(\mathbf{Z}_2,\mathbf{Z}_3,\mathbf{Z}_1). \tag{12.31}$$

On the basis of the equalities (12.30) and (12.31), we obtain

$$\begin{aligned}
& G(\mathbf{Z}_1,\mathbf{Z}_2,\mathbf{Z}_3) - G(\mathbf{Z}_2,\mathbf{Z}_3,\mathbf{Z}_1) && \tag{12.32} \\
&= \; - \;\; G(\mathbf{Z}_2,\mathbf{Z}_3,\mathbf{O}) + G(\mathbf{Z}_3,\mathbf{O},\mathbf{Z}_2) \\
&\quad - \;\; G(\mathbf{O},\mathbf{Z}_1,\mathbf{Z}_2) + G(\mathbf{Z}_1,\mathbf{Z}_2,\mathbf{O}) \\
&\quad - \;\; G(\mathbf{Z}_3,\mathbf{O},\mathbf{Z}_1) + G(\mathbf{O},\mathbf{Z}_1,\mathbf{Z}_2).
\end{aligned}$$

If we substitute $\mathbf{Z}_3 = \mathbf{O}$ in the above equation, we obtain

$$\begin{aligned}
G(\mathbf{Z}_1,\mathbf{Z}_2,\mathbf{O}) \;\; - \;\; & G(\mathbf{Z}_2,\mathbf{O},\mathbf{Z}_1) = -G(\mathbf{Z}_2,\mathbf{O},\mathbf{O}) + G(\mathbf{O},\mathbf{O},\mathbf{Z}_2) \\
-G(\mathbf{O},\mathbf{Z}_1,\mathbf{Z}_2) \;\; + \;\; & G(\mathbf{Z}_1,\mathbf{Z}_2,\mathbf{O}) - G(\mathbf{O},\mathbf{O},\mathbf{Z}_1) + G(\mathbf{O},\mathbf{Z}_1,\mathbf{O})
\end{aligned}$$

or, written in a more convenient form,

$$\begin{aligned}
G(\mathbf{Z}_2,\mathbf{O},\mathbf{Z}_1) \;\; - \;\; & G(\mathbf{O},\mathbf{Z}_1,\mathbf{Z}_2) = G(\mathbf{Z}_2,\mathbf{O},\mathbf{O}) \\
-G(\mathbf{O},\mathbf{O},\mathbf{Z}_2) \;\; + \;\; & G(\mathbf{O},\mathbf{O},\mathbf{Z}_1) - G(\mathbf{O},\mathbf{Z}_1,\mathbf{O}).
\end{aligned}$$

By using the last equality, the function Eq. (12.31) takes the following form

$$\begin{aligned}
f(\mathbf{Z}_1,\mathbf{Z}_2,\mathbf{Z}_3) \;\; = \;\; & G(\mathbf{Z}_1,\mathbf{Z}_2,\mathbf{O}) - G(\mathbf{Z}_2,\mathbf{Z}_3,\mathbf{O}) \\
+G(\mathbf{Z}_3,\mathbf{O},\mathbf{Z}_2) - G(\mathbf{O},\mathbf{Z}_1,\mathbf{Z}_2) \;\; + \;\; & G(\mathbf{O},\mathbf{Z}_1,\mathbf{Z}_3) - G(\mathbf{Z}_3,\mathbf{O},\mathbf{Z}_1) \\
= [G(\mathbf{Z}_1,\mathbf{Z}_2,\mathbf{O}) - G(\mathbf{Z}_2,\mathbf{Z}_3,\mathbf{O})] \;\; - \;\; & [G(\mathbf{O},\mathbf{Z}_1,\mathbf{Z}_2) - G(\mathbf{O},\mathbf{Z}_2,\mathbf{Z}_3)] \\
+[G(\mathbf{Z}_3,\mathbf{O},\mathbf{Z}_2) - G(\mathbf{O},\mathbf{Z}_2,\mathbf{Z}_3)] \;\; - \;\; & [G(\mathbf{Z}_3,\mathbf{O},\mathbf{Z}_1) - G(\mathbf{O},\mathbf{Z}_1,\mathbf{Z}_3)]
\end{aligned}$$

$$\begin{aligned}
= [G(\mathbf{Z}_1,\mathbf{Z}_2,\mathbf{O}) - G(\mathbf{Z}_2,\mathbf{Z}_3,\mathbf{O})] \;\; - \;\; & [G(\mathbf{O},\mathbf{Z}_1,\mathbf{Z}_2) - G(\mathbf{O},\mathbf{Z}_2,\mathbf{Z}_3)] \\
+[G(\mathbf{Z}_3,\mathbf{O},\mathbf{O}) - G(\mathbf{O},\mathbf{O},\mathbf{Z}_3) \;\; + \;\; & G(\mathbf{O},\mathbf{O},\mathbf{Z}_2) - G(\mathbf{O},\mathbf{Z}_2,\mathbf{O})] \\
-[G(\mathbf{Z}_3,\mathbf{O},\mathbf{O}) - G(\mathbf{O},\mathbf{O},\mathbf{Z}_3) \;\; + \;\; & G(\mathbf{O},\mathbf{O},\mathbf{Z}_1) - G(\mathbf{O},\mathbf{Z}_1,\mathbf{O})] \\
= [G(\mathbf{Z}_1,\mathbf{Z}_2,\mathbf{O}) - G(\mathbf{O},\mathbf{Z}_1,\mathbf{Z}_2) \;\; + \;\; & G(\mathbf{O},\mathbf{Z}_1,\mathbf{O}) - G(\mathbf{O},\mathbf{O},\mathbf{Z}_1)] \\
-[G(\mathbf{Z}_2,\mathbf{Z}_3,\mathbf{O}) - G(\mathbf{O},\mathbf{Z}_2,\mathbf{Z}_3) \;\; + \;\; & G(\mathbf{O},\mathbf{Z}_2,\mathbf{O}) - G(\mathbf{O},\mathbf{O},\mathbf{Z}_2)].
\end{aligned}$$

Finally, if we denote

$$\begin{aligned} F(\mathbf{Z}_1, \mathbf{Z}_2) \;=\;& G(\mathbf{Z}_1, \mathbf{Z}_2, \mathbf{O}) - G(\mathbf{O}, \mathbf{Z}_1, \mathbf{Z}_2) \\ +\;& G(\mathbf{O}, \mathbf{Z}_1, \mathbf{O}) - G(\mathbf{O}, \mathbf{O}, \mathbf{Z}_1), \end{aligned}$$

then the function $f(\mathbf{Z}_1, \mathbf{Z}_2, \mathbf{Z}_3)$ can be rewritten in the following form

$$f(\mathbf{Z}_1, \mathbf{Z}_2, \mathbf{Z}_3) = F(\mathbf{Z}_1, \mathbf{Z}_2) - F(\mathbf{Z}_2, \mathbf{Z}_3),$$

which means that we have obtained the form given by Eq. (12.24). We conclude that the function Eq. (12.24) is the general solution of the equation (12.23). $\qquad\qquad\square$

The above results are obtained in [I. B. Risteski *et al.* (to appear A)].

Chapter 3

Functional Equations with Constant Parameters

Functional equations which contain constant parameters are studied in this chapter. Such equations with constant parameters solved here are the general parametric functional equation, the special parametric functional equation, expanded parametric functional equation and the general expanded parametric functional equation. The results presented here are obtained in [I. B. Risteski et $al.$ (to appear B); I. B. Risteski et $al.$ (2001A)].

First we introduce the following notations. Let $\mathbf{N}$ be the set of all positive integers. Let $\mathcal{V}$, $\mathcal{V}'$ be finite dimensional complex vector spaces and $\mathbf{Z}_i$, $i \in \mathbf{N}$, be vectors in $\mathcal{V}$. We may assume that $\mathbf{Z}_i = (z_{i1}(t), \cdots, z_{in}(t))^T$, where $z_{ij}(t)$ $(1 \le j \le n)$ are complex functions and $\mathbf{O} = (0, \cdots, 0)^T$ is the zero-vector in $\mathcal{V}$ or $\mathcal{V}'$. For a vector $\mathbf{U} \in \mathcal{V}$ denote by Re $\mathbf{U}$ (respectively Im $\mathbf{U}$) the real (respectively imaginary) part of $\mathbf{U}$. Moreover, we denote by $\mathcal{V}^0$ the subspace of all real vectors in $\mathcal{V}$ (thus $\mathcal{V} = \mathcal{V}^0 + i\mathcal{V}^0$), and by $\mathcal{L}(\mathcal{V}^0, \mathcal{V}')$ the space of linear mappings $\mathcal{V}^0 \mapsto \mathcal{V}'$. Let (m,n) be the greatest common divisor of m and n.

13   General Parametric Functional Equation

In the present section our object of investigation will be the following functional equation

$$\sum_{i=1}^{m+n} f_i \left(\sum_{j=0}^{m-1} a^{m-1-j} \mathbf{Z}_{i+j}, \ \sum_{j=0}^{n-1} a^{n-1-j} \mathbf{Z}_{i+m+j} \right) = \mathbf{O} \qquad (13.1)$$

$$(\mathbf{Z}_{m+n+i} \equiv \mathbf{Z}_i),$$

121

where a is a complex number and $f_i : \mathcal{V}^2 \mapsto \mathcal{V}'$ $(1 \leq i \leq m+n)$ are unknown complex vector functions.

The above equation for $a = 1$ was solved in [S. B. Prešić and D. Ž. Djoković (1961)] under the assumption that the functions and variables are real. But the argument given there is valid only if the greatest common divisor of m and n is 1. Also, one special general case is solved in [D. Ž. Djoković (1965B)]. The theorems of [D. Ž. Djoković (1965B)] concerning the cases $m \neq n$ should be modified to give the general continuous solutions. In the more general formulation as given in [D. Ž. Djoković (1965B)] they are invalid.

Now we will give the following result which is obtained in [I. B. Risteski *et al.* (to appear B)].

Theorem 13.1   *If $a = 1$, $(m,n) = 1$ and $m + n > 2$, then the general continuous solution of the functional equation (13.1) is*

$$f_i(\mathbf{U}, \mathbf{V}) = F_1(\mathbf{U} + \mathbf{V})\operatorname{Re}\mathbf{U} + F_2(\mathbf{U} + \mathbf{V})\operatorname{Im}\mathbf{U} + G_i(\mathbf{U} + \mathbf{V}) \qquad (13.2)$$

$$(1 \leq i \leq m + n),$$

so that

$$\sum_{i=1}^{n+m} G_i(\mathbf{U}) = -m[F_1(\mathbf{U})\operatorname{Re}\mathbf{U} + F_2(\mathbf{U})\operatorname{Im}\mathbf{U}],$$

where $F_i : \mathcal{V} \mapsto \mathcal{L}(\mathcal{V}^0, \mathcal{V}')$ $(i = 1, 2)$ and $G_i : \mathcal{V} \mapsto \mathcal{V}'$ $(1 \leq i \leq m + n - 1)$ are arbitrary continuous complex vector functions.

Proof.   We accept the convention to reduce the indices $\pmod{m+n}$. If we set

$$\mathbf{S} = \sum_{i=1}^{m+n} \mathbf{Z}_i, \qquad (13.3)$$

$$\mathbf{T}_i = \mathbf{Z}_i + \mathbf{Z}_{i+1} + \cdots + \mathbf{Z}_{i+m-1} - \frac{m\mathbf{S}}{m+n} \quad (1 \leq i \leq m + n - 1),$$

the vectors $\mathbf{T}_i$ $(1 \leq i \leq m + n - 1)$ and $\mathbf{S}$ are independent since $(m,n) = 1$.

The equation (13.1) becomes

$$\sum_{i=1}^{m+n-1} f_i\left(\mathbf{T}_i + \frac{m\mathbf{S}}{m+n}, \frac{n\mathbf{S}}{m+n} - \mathbf{T}_i\right) \tag{13.4}$$

$$+\ f_{m+n}\left(-\mathbf{T}_1 - \mathbf{T}_2 - \cdots - \mathbf{T}_{m+n-1} + \frac{m\mathbf{S}}{m+n},\right.$$

$$\left.\frac{n\mathbf{S}}{m+n} + \mathbf{T}_1 + \mathbf{T}_2 + \cdots + \mathbf{T}_{m+n-1}\right) = \mathbf{O}.$$

We introduce the new notations

$$f_i\left(\mathbf{U} + \frac{m\mathbf{S}}{m+n}, \frac{n\mathbf{S}}{m+n} - \mathbf{U}\right) = g_i(\mathbf{U}, \mathbf{S}) \quad (1 \le i \le m+n),$$

i.e.,

$$f_i(\mathbf{U}, \mathbf{V}) = g_i\left(\frac{n\mathbf{U} - m\mathbf{V}}{m+n}, \mathbf{U} + \mathbf{V}\right) \quad (1 \le i \le m+n). \tag{13.5}$$

The equation (13.4) is transformed into

$$\sum_{i=1}^{m+n-1} g_i(\mathbf{T}_i, \mathbf{S}) + g_{m+n}(-\mathbf{T}_1 - \mathbf{T}_2 - \cdots - \mathbf{T}_{m+n-1}, \mathbf{S}) = \mathbf{O}. \tag{13.6}$$

By the substitution $\mathbf{T}_1 = \mathbf{T}_2 = \cdots = \mathbf{T}_{r-1} = \mathbf{T}_{r+1} = \cdots = \mathbf{T}_{m+n-1} = \mathbf{O}$, we obtain

$$g_r(\mathbf{T}_r, \mathbf{S}) = -g_{m+n}(-\mathbf{T}_r, \mathbf{S}) - H_r(\mathbf{S}) \quad (1 \le r \le m+n-1). \tag{13.7}$$

Putting the equalities (13.7) into Eq. (13.6), we get

$$g_{m+n}(-\mathbf{T}_1 - \mathbf{T}_2 - \cdots - \mathbf{T}_{m+n-1}, \mathbf{S}) \tag{13.8}$$

$$= \sum_{i=1}^{m+n-1} g_{m+n}(-\mathbf{T}_i, \mathbf{S}) + \sum_{i=1}^{m+n-1} H_i(\mathbf{S}).$$

We conclude that the function

$$K(\mathbf{U}, \mathbf{S}) = g_{m+n}(\mathbf{U}, \mathbf{S}) + \frac{1}{m+n-2} \sum_{i=1}^{m+n-1} H_i(\mathbf{S}) \tag{13.9}$$

satisfies the functional equation

$$K(\mathbf{Z}_1 + \mathbf{Z}_2 + \cdots + \mathbf{Z}_{m+n-1}, \mathbf{S}) = \sum_{i=1}^{m+n-1} K(\mathbf{Z}_i, \mathbf{S}). \tag{13.10}$$

Using the continuity of K, from Eq. (13.10) we deduce that for fixed $\mathbf{S}$ we have

$$K(\mathbf{U}, \mathbf{S}) = c_1 \operatorname{Re} \mathbf{U} + c_2 \operatorname{Im} \mathbf{U}.$$

The mappings $c_1, c_2 \in \mathcal{L}(\mathcal{V}^0, \mathcal{V}')$ may depend upon $\mathbf{S}$. Hence,

$$K(\mathbf{U}, \mathbf{V}) = F_1(\mathbf{V}) \operatorname{Re} \mathbf{U} + F_2(\mathbf{V}) \operatorname{Im} \mathbf{U}, \tag{13.11}$$

where $F_i : \mathcal{V} \mapsto \mathcal{L}(\mathcal{V}^0, \mathcal{V}')$ are continuous functions.

From Eqs. (13.9), (13.11) and (13.7) we obtain

$$
\begin{aligned}
g_{m+n}(\mathbf{U}, \mathbf{V}) &= F_1(\mathbf{V}) \operatorname{Re} \mathbf{U} + f_2(\mathbf{V}) \operatorname{Im} \mathbf{U} - \frac{1}{m+n-2} \sum_{i=1}^{m+n-1} H_i(\mathbf{V}), \\
g_r(\mathbf{U}, \mathbf{V}) &= F_1(\mathbf{V}) \operatorname{Re} \mathbf{U} + F_2(\mathbf{V}) \operatorname{Im} \mathbf{U} - H_r(\mathbf{V}) \\
&\quad + \frac{1}{m+n-2} \sum_{i=1}^{m+n-1} H_i(\mathbf{V}) \quad (1 \le r \le m+n-1).
\end{aligned}
\tag{13.12}
$$

From Eqs. (13.5) and (13.12) we deduce that

$$
\begin{aligned}
f_r(\mathbf{U}, \mathbf{V}) &= F_1(\mathbf{U}+\mathbf{V}) \operatorname{Re} \left(\frac{n\mathbf{U} - m\mathbf{V}}{m+n} \right) \\
&\quad + F_2(\mathbf{U}+\mathbf{V}) \operatorname{Im} \left(\frac{n\mathbf{U} - m\mathbf{V}}{m+n} \right) - H_r(\mathbf{U}+\mathbf{V}) \\
&\quad + \frac{1}{m+n-2} \sum_{i=1}^{m+n-1} H_i(\mathbf{U}+\mathbf{V}) \quad (1 \le r \le m+n-1), \\
f_{m+n}(\mathbf{U}+\mathbf{V}) &= F_1(\mathbf{U}+\mathbf{V}) \operatorname{Re} \left(\frac{n\mathbf{U} - m\mathbf{V}}{m+n} \right) \\
&\quad + F_2(\mathbf{U}+\mathbf{V}) \operatorname{Im} \left(\frac{n\mathbf{U} - m\mathbf{V}}{m+n} \right) \\
&\quad - \frac{1}{m+n-2} \sum_{i=1}^{m+n-1} H_i(\mathbf{U}+\mathbf{V}).
\end{aligned}
\tag{13.13}
$$

By denoting

$$
\begin{aligned}
&-F_1(\mathbf{U}+\mathbf{V}) \operatorname{Re} \left[\frac{m(\mathbf{U}+\mathbf{V})}{m+n} \right] - F_2(\mathbf{U}+\mathbf{V}) \operatorname{Im} \left[\frac{m(\mathbf{U}+\mathbf{V})}{m+n} \right] \\
&+ \frac{1}{m+n-2} \sum_{i=1}^{m+n-1} H_i(\mathbf{U}+\mathbf{V}) - H_r(\mathbf{U}+\mathbf{V}) = G_r(\mathbf{U}+\mathbf{V})
\end{aligned}
$$

$$(1 \le r \le m + n - 1),$$

$$-F_1(\mathbf{U} + \mathbf{V})\mathrm{Re}\left[\frac{m(\mathbf{U} + \mathbf{V})}{m + n}\right] - F_2(\mathbf{U} + \mathbf{V})\mathrm{Im}\left[\frac{m(\mathbf{U} + \mathbf{V})}{m + n}\right]$$

$$-\frac{1}{m + n - 2}\sum_{i=1}^{m+n-1} H_i(\mathbf{U} + \mathbf{V}) = G_{m+n}(\mathbf{U} + \mathbf{V}),$$

from Eq. (13.13) we get Eq. (13.2).

The converse can be established by a straightforward verification. $\square$

Example 13.1 The general continuous solution of the functional equation

$$f_1(\mathbf{Z}_1 + \mathbf{Z}_2,\ \mathbf{Z}_3) + f_2(\mathbf{Z}_2 + \mathbf{Z}_3,\ \mathbf{Z}_1) + f_3(\mathbf{Z}_3 + \mathbf{Z}_1,\ \mathbf{Z}_2) = \mathbf{O}$$

is given by

$$
\begin{aligned}
f_1(\mathbf{U}, \mathbf{V}) &= F_1(\mathbf{U} + \mathbf{V})\mathrm{Re}\,\mathbf{U} + F_2(\mathbf{U} + \mathbf{V})\mathrm{Im}\,\mathbf{U} + G_1(\mathbf{U} + \mathbf{V}), \\
f_2(\mathbf{U}, \mathbf{V}) &= F_1(\mathbf{U} + \mathbf{V})\mathrm{Re}\,\mathbf{U} + F_2(\mathbf{U} + \mathbf{V})\mathrm{Im}\,\mathbf{U} + G_2(\mathbf{U} + \mathbf{V}), \\
f_3(\mathbf{U}, \mathbf{V}) &= -F_1(\mathbf{U} + \mathbf{V})\mathrm{Re}\,(\mathbf{U} + 2\mathbf{V}) - F_2(\mathbf{U} + \mathbf{V})\mathrm{Im}\,(\mathbf{U} + 2\mathbf{V}) \\
&\quad -G_1(\mathbf{U} + \mathbf{V}) - G_2(\mathbf{U} + \mathbf{V}),
\end{aligned}
$$

where $F_1, F_2 : \mathcal{V} \mapsto \mathcal{L}(\mathcal{V}^0, \mathcal{V}')$ and $G_1, G_2 : \mathcal{V} \mapsto \mathcal{V}'$ are arbitrary continuous complex vector functions.

Corollary 13.2 *The general continuous solution of the vector functional equation*

$$\sum_{i=1}^{m+n} g_i(\mathbf{Z}_i + \cdots + \mathbf{Z}_{i+m-1},\ \mathbf{Z}_1 + \mathbf{Z}_2 + \cdots + \mathbf{Z}_{m+n}) = \mathbf{O}$$

if $(m, n) = 1$ and $m + n > 2$ is given by

$$g_i(\mathbf{U}, \mathbf{V}) = F_1(\mathbf{V})\mathrm{Re}\,\mathbf{U} + F_2(\mathbf{V})\mathrm{Im}\,\mathbf{U} + G_i(\mathbf{V}) \quad (1 \le i \le m + n),$$

$$\sum_{i=1}^{m+n} G_i(\mathbf{V}) = -m[F_1(\mathbf{V})\mathrm{Re}\,\mathbf{V} + F_2(\mathbf{V})\mathrm{Im}\,\mathbf{V}],$$

where $F_1, F_2 : \mathcal{V} \mapsto \mathcal{L}(\mathcal{V}^0, \mathcal{V}')$, $G_i : \mathcal{V} \mapsto \mathcal{V}'$ $(1 \le i \le m+n-1)$ are arbitrary continuous complex vector functions.

Proof.   Put $f_i(\mathbf{U}, \mathbf{V}) = g_i(\mathbf{U}, \mathbf{U} + \mathbf{V})$ in Theorem 13.1. $\square$

Theorem 13.3   *The general continuous solution of the complex vector functional equation (13.1) if $a = 1$, $(m, n) = d > 1$, $m/d = p$, $n/d = q$ and $p + q > 2$ is given by*

$$f_{id+j}(\mathbf{U}, \mathbf{V}) = F_{1j}(\mathbf{U} + \mathbf{V})\mathrm{Re}\,\mathbf{U} + F_{2j}(\mathbf{U} + \mathbf{V})\mathrm{Im}\,\mathbf{U} + G_{ij}(\mathbf{U} + \mathbf{V})$$

$$(0 \leq i \leq p + q - 1, \quad 1 \leq j \leq d),$$

$$\sum_{i=0}^{p+q-1} G_{ij}(\mathbf{U}) = H_j(\mathbf{U}) - p[F_{1j}(\mathbf{U})\mathrm{Re}\,\mathbf{U} + F_{2j}(\mathbf{U})\mathrm{Im}\,\mathbf{U}] \qquad (13.14)$$

$$(1 \leq j \leq d),$$

$$\sum_{j=1}^{d} H_j(\mathbf{U}) = \mathbf{O},$$

where

$$F_{ij} : \mathcal{V} \mapsto \mathcal{L}(\mathcal{V}^0, \mathcal{V}') \quad (i = 1, 2; \ 1 \leq j \leq d),$$

$$H_j : \mathcal{V} \mapsto \mathcal{V}' \quad (1 \leq j \leq d - 1),$$

$$G_{ij} : \mathcal{V} \mapsto \mathcal{V}' \quad (0 \leq i \leq p + q - 2; \ 1 \leq j \leq d)$$

are arbitrary continuous complex vector functions.

Proof.   We set

$$f_i(\mathbf{U}, \mathbf{V}) = g_i(\mathbf{U}, \mathbf{U} + \mathbf{V}) \quad (1 \leq i \leq m + n) \qquad (13.15)$$

and we obtain

$$\sum_{i=1}^{m+n} g_i(\mathbf{Z}_i + \mathbf{Z}_{i+1} + \cdots + \mathbf{Z}_{i+m-1}, \ \mathbf{Z}_1 + \mathbf{Z}_2 + \cdots + \mathbf{Z}_{m+n}) = \mathbf{O}. \quad (13.16)$$

Let us introduce the new vectors

$$\mathbf{V}_i = \mathbf{Z}_i + \mathbf{Z}_{i+1} + \cdots + \mathbf{Z}_{i+d-1} \quad (1 \leq i \leq m + n) \qquad (13.17)$$

so that $\mathbf{V}_{i+m+n} = \mathbf{V}_i$, and

$$\mathbf{W} = \mathbf{Z}_1 + \mathbf{Z}_2 + \cdots + \mathbf{Z}_{m+n}. \qquad (13.18)$$

They are not independent because

$$\sum_{i=0}^{p+q-1} \mathbf{V}_{id+j} = \mathbf{W} \quad (1 \le j \le d). \tag{13.19}$$

The vectors $\mathbf{V}_i$ $(1 \le i \le m+n-d)$ and $\mathbf{W}$ are independent because the rank of the matrix of linear forms determining them is $m+n-d+1$, which is easy to verify. In the sequel we will use all vectors given by Eqs. (13.17) and (13.18) but we must have always in mind that Eq. (13.19) holds. The equation (13.16) becomes

$$\sum_{i=1}^{m+n} g_i(\mathbf{V}_i + \mathbf{V}_{i+d} + \cdots + \mathbf{V}_{i+(p-1)d}, \ \mathbf{W}) = \mathbf{O}.$$

It can be written in the following form

$$\sum_{j=1}^{d} \sum_{i=0}^{p+q-1} g_{id+j}(\mathbf{V}_{id+j} + \mathbf{V}_{(i+1)d+j} + \cdots + \mathbf{V}_{(i+p-1)d+j}, \ \mathbf{W}) = \mathbf{O}.$$

If we set here

$$\mathbf{V}_{id+j} = \mathbf{O} \quad (0 \le i \le p+q-2; \ j = 1, 2, \cdots, r-1, r+1, \cdots, d),$$

$$\mathbf{V}_{(p+q-1)d+j} = \mathbf{W} \quad (j = 1, 2, \cdots, r-1, r+1, \cdots, d),$$

we get

$$\sum_{i=0}^{p+q-1} g_{id+r}(\mathbf{V}_{id+r} + \mathbf{V}_{(i+1)d+r} + \cdots + \mathbf{V}_{(i+p-1)d+r}, \ \mathbf{W}) - \frac{H_r(\mathbf{W})}{p+q} = \mathbf{O}$$

$$(1 \le r \le d)$$

and

$$\sum_{r=1}^{d} H_r(\mathbf{W}) = \mathbf{O}.$$

By using the corollary of Theorem 13.1 we get

$$g_{id+r}(\mathbf{U}, \mathbf{V}) = F_{1r}(\mathbf{V})\operatorname{Re}\mathbf{U} + F_{2r}(\mathbf{V})\operatorname{Im}\mathbf{U} + G_{ir}(\mathbf{V})$$

$$(0 \le i \le p+q-1; \ 1 \le r \le d),$$

$$\sum_{i=0}^{p+q-1} G_{ir}(\mathbf{V}) = H_r(\mathbf{V}) - p[F_{1r}(\mathbf{V})\operatorname{Re}\mathbf{V} + F_{2r}(\mathbf{V})\operatorname{Im}\mathbf{V}]$$

$$(1 \le r \le d),$$

where

$$F_{ir} : \mathcal{V} \mapsto \mathcal{L}(\mathcal{V}^0, \mathcal{V}') \quad (i = 1, 2;\ 1 \le r \le d),$$

$$G_{ir} : \mathcal{V} \mapsto \mathcal{V}' \quad (0 \le i \le p+q-2;\ 1 \le r \le d),$$

$$H_r : \mathcal{V} \mapsto \mathcal{V}' \quad (1 \le r \le d-1)$$

are arbitrary continuous complex vector functions. By application of the equalities (13.15) these formulae give Eq. (13.14).

It is easy to prove that the functions $f_i : \mathcal{V}^2 \mapsto \mathcal{V}'$ $(1 \le i \le m+n)$ defined by Eq. (13.15) satisfy the complex vector functional equations (13.1).   □

Example 13.2   The general continuous solution of the functional equation

$$f_1(\mathbf{Z}_1 + \mathbf{Z}_2 + \mathbf{Z}_3 + \mathbf{Z}_4,\ \mathbf{Z}_5 + \mathbf{Z}_6) + f_2(\mathbf{Z}_2 + \mathbf{Z}_3 + \mathbf{Z}_4 + \mathbf{Z}_5,\ \mathbf{Z}_6 + \mathbf{Z}_1)$$

$$+ f_3(\mathbf{Z}_3 + \mathbf{Z}_4 + \mathbf{Z}_5 + \mathbf{Z}_6,\ \mathbf{Z}_1 + \mathbf{Z}_2) + f_4(\mathbf{Z}_4 + \mathbf{Z}_5 + \mathbf{Z}_6 + \mathbf{Z}_1,\ \mathbf{Z}_2 + \mathbf{Z}_3)$$

$$+ f_5(\mathbf{Z}_5 + \mathbf{Z}_6 + \mathbf{Z}_1 + \mathbf{Z}_2,\ \mathbf{Z}_3 + \mathbf{Z}_4) + f_6(\mathbf{Z}_6 + \mathbf{Z}_1 + \mathbf{Z}_2 + \mathbf{Z}_3,\ \mathbf{Z}_4 + \mathbf{Z}_5) = \mathbf{O}$$

is given by

$$f_1(\mathbf{U}, \mathbf{V}) = F_{11}(\mathbf{U} + \mathbf{V})\operatorname{Re}\mathbf{U} + F_{21}(\mathbf{U} + \mathbf{V})\operatorname{Im}\mathbf{U} + G_{01}(\mathbf{U} + \mathbf{V}),$$

$$f_2(\mathbf{U}, \mathbf{V}) = F_{12}(\mathbf{U} + \mathbf{V})\operatorname{Re}\mathbf{U} + F_{22}(\mathbf{U} + \mathbf{V})\operatorname{Im}\mathbf{U} + G_{02}(\mathbf{U} + \mathbf{V}),$$

$$f_3(\mathbf{U}, \mathbf{V}) = F_{11}(\mathbf{U} + \mathbf{V})\operatorname{Re}\mathbf{U} + F_{21}(\mathbf{U} + \mathbf{V})\operatorname{Im}\mathbf{U} + G_{11}(\mathbf{U} + \mathbf{V}),$$

$$f_4(\mathbf{U}, \mathbf{V}) = F_{12}(\mathbf{U} + \mathbf{V})\operatorname{Re}\mathbf{U} + F_{22}(\mathbf{U} + \mathbf{V})\operatorname{Im}\mathbf{U} + G_{12}(\mathbf{U} + \mathbf{V}),$$

$$f_5(\mathbf{U}, \mathbf{V}) = -F_{11}(\mathbf{U} + \mathbf{V})\operatorname{Re}(\mathbf{U} + 2\mathbf{V}) - F_{21}(\mathbf{U} + \mathbf{V})\operatorname{Im}(\mathbf{U} + 2\mathbf{V})$$

$$+ H_1(\mathbf{U} + \mathcal{V}) - G_{01}(\mathbf{U} + \mathbf{V}) - G_{11}(\mathbf{U} + \mathbf{V}),$$

$$f_6(\mathbf{U}, \mathbf{V}) = -F_{12}(\mathbf{U} + \mathbf{V})\mathrm{Re}\,(\mathbf{U} + 2\mathbf{V}) - F_{22}(\mathbf{U} + \mathbf{V})\mathrm{Im}\,(\mathbf{U} + 2\mathbf{V})$$

$$-H_1(\mathbf{U} + \mathbf{V}) - G_{01}(\mathbf{U} + \mathbf{V}) - G_{12}(\mathbf{U} + \mathbf{V}),$$

where

$$F_{ij} : \mathcal{V} \mapsto \mathcal{L}(\mathcal{V}^0, \mathcal{V}') \quad (i = 1, 2),$$

$$G_{ij} : \mathcal{V} \mapsto \mathcal{V}' \quad (i = 0, 1;\ j = 1, 2),$$

$$H_1 : \mathcal{V} \mapsto \mathcal{V}'$$

are arbitrary continuous complex vector functions.

Theorem 13.4 *The most general solution of* (13.1) *if* $a = 1$ *and* $m = n$ *is*

$$
\begin{aligned}
f_i(\mathbf{U}, \mathbf{V}) & \quad (1 \le i \le m) \quad \text{are arbitrary,} \\
f_{m+i}(\mathbf{U}, \mathbf{V}) &= H_i(\mathbf{U} + \mathbf{V}) - f_i(\mathbf{V}, \mathbf{U}) \quad (1 \le i \le m), \quad (13.20) \\
\sum_{i=1}^{m} H_i(\mathbf{U}) &= \mathbf{O},
\end{aligned}
$$

where $H_i : \mathcal{V} \mapsto \mathcal{V}'$ $(1 \le i \le m - 1)$ *are arbitrary functions.*

Proof.   Put $f_i(\mathbf{U}, \mathbf{V}) = G_i(\mathbf{U}, \mathbf{U} + \mathbf{V})$. $\qquad\qquad\qquad\square$

Example 13.3   The most general solution of the equation

$$
\begin{aligned}
f_1(\mathbf{Z}_1 + \mathbf{Z}_2,\ \mathbf{Z}_3 + \mathbf{Z}_4) &+ f_2(\mathbf{Z}_2 + \mathbf{Z}_3,\ \mathbf{Z}_4 + \mathbf{Z}_1) \\
+f_3(\mathbf{Z}_3 + \mathbf{Z}_4,\ \mathbf{Z}_1 + \mathbf{Z}_2) &+ f_4(\mathbf{Z}_4 + \mathbf{Z}_1,\ \mathbf{Z}_2 + \mathbf{Z}_3) = \mathbf{O}
\end{aligned}
$$

is

$$
\begin{aligned}
f_1(\mathbf{U}, \mathbf{V}) &,\quad f_2(\mathbf{U}, \mathbf{V}) \quad \text{are arbitrary,} \\
f_3(\mathbf{U}, \mathbf{V}) &= H_1(\mathbf{U} + \mathbf{V}) - f_1(\mathbf{V}, \mathbf{U}), \\
f_4(\mathbf{U}, \mathbf{V}) &= -H_1(\mathbf{U} + \mathbf{V}) - f_2(\mathbf{V}, \mathbf{U}),
\end{aligned}
$$

where $H_1 : \mathcal{V} \mapsto \mathcal{V}'$ is an arbitrary function.

Theorem 13.5 *If* $a^{m+n} \ne 1$ *and* $m \ne n$, *the general solution of the functional equation* (13.1) *is given by*

$$f_i(\mathbf{U}, \mathbf{V}) = F_i(\mathbf{U} + a^m \mathbf{V}) - F_{i+n}(a^n \mathbf{U} + \mathbf{V}) + A_i \quad (1 \le i \le m+n), \quad (13.21)$$

where $F_i : \mathcal{V} \mapsto \mathcal{V}'$ $(1 \le i \le m+n)$ are arbitrary complex vector functions, and A_i are arbitrary constant complex vectors such that $\sum_{i=1}^{m+n} A_i = \mathbf{O}$.

Proof.   If we introduce new functions g_i by the equation

$$f_i(\mathbf{U}, \mathbf{V}) = g_i(\mathbf{U} + a^m \mathbf{V}, \ a^n \mathbf{U} + \mathbf{V}) \quad (1 \le i \le m+n), \tag{13.22}$$

then Eq. (13.1) becomes

$$\sum_{i=1}^{m+n} g_i \left(\sum_{j=0}^{m-1} a^{m-1-j} \mathbf{Z}_{i+j} + \sum_{j=0}^{n-1} a^{m+n-1-j} \mathbf{Z}_{m+i+j}, \right.$$

$$\left. \sum_{j=0}^{m-1} a^{m+n-1-j} \mathbf{Z}_{i+j} + \sum_{j=0}^{n-1} a^{n-1-j} \mathbf{Z}_{m+i+j} \right) = \mathbf{O},$$

i.e.,

$$\sum_{i=1}^{m+n} g_i \left(\sum_{j=0}^{m+n-1} a^j \mathbf{Z}_{m+i-1-j}, \ \sum_{j=0}^{m+n-1} a^j \mathbf{Z}_{i-1-j} \right) = \mathbf{O}. \tag{13.23}$$

Since $a^{m+n} \ne 1$, this transformation is possible. Also we may introduce new vectors $\mathbf{V}_i$ by

$$\mathbf{V}_i = \sum_{j=0}^{m+n-1} a^j \mathbf{Z}_{m+i-1-j} \quad (1 \le i \le m+n)$$

and then Eq. (13.23) takes the form

$$\sum_{i=0}^{m+n} g_i(\mathbf{V}_i, \mathbf{V}_{i+n}) = \mathbf{O}. \tag{13.24}$$

By putting $\mathbf{V}_j = \mathbf{O}$ $(j = 1, 2, \cdots, i-1, i+1, \cdots, i+n-1, i+n+1, \cdots, m+n)$ we obtain

$$g_i(\mathbf{V}_i, \mathbf{V}_{i+n}) = F_i(\mathbf{V}_i) + G_i(\mathbf{V}_{i+n}) \quad (1 \le i \le m+n). \tag{13.25}$$

On the basis of the expression Eq. (13.25), the equation (13.24) becomes

$$\sum_{i=1}^{m+n} [F_i(\mathbf{V}_i) + G_i(\mathbf{V}_{i+n})] = \mathbf{O},$$

or

$$\sum_{i=1}^{m+n} [F_i(\mathbf{V}_i) + G_{m+i}(\mathbf{V}_i)] = \mathbf{O}. \tag{13.26}$$

From Eq. (13.26) it follows that

$$G_{i+m}(\mathbf{V}_i) = -F_i(\mathbf{V}_i) + A_i \quad (1 \le i \le m+n), \tag{13.27}$$

where A_i are arbitrary constant complex vectors with the property

$$\sum_{i=1}^{m+n} A_i = \mathbf{O}.$$

On the basis of the expression Eq. (13.27), the equality (13.25) has the form

$$g_i(\mathbf{U}, \mathbf{V}) = F_i(\mathbf{U}) + F_{i+n}(\mathbf{V}) + A_i \quad (1 \le i \le m+n), \tag{13.28}$$

where $\sum_{i=1}^{m+n} A_i = \mathbf{O}$.

On the basis of Eqs. (13.28) and (13.22), we obtain Eq. (13.21). $\qquad \square$

Example 13.4   If $a^3 \neq 1$, the general solution of the functional equation

$$f_1(a^2\mathbf{Z}_1 + a\mathbf{Z}_2 + \mathbf{Z}_3, \mathbf{Z}_4) \quad + \quad f_2(a^2\mathbf{Z}_2 + a\mathbf{Z}_3 + \mathbf{Z}_4, \mathbf{Z}_1)$$
$$+f_3(a^2\mathbf{Z}_3 + a\mathbf{Z}_4 + \mathbf{Z}_1, \mathbf{Z}_2) \quad + \quad f_4(a^2\mathbf{Z}_4 + a\mathbf{Z}_1 + \mathbf{Z}_2, \mathbf{Z}_3) = \mathbf{O}$$

is given by

$$\begin{aligned}
f_1(\mathbf{U}, \mathbf{V}) &= F_1(\mathbf{U} + a^3\mathbf{V}) - F_2(a\mathbf{U} + \mathbf{V}) + A_1, \\
f_2(\mathbf{U}, \mathbf{V}) &= F_2(\mathbf{U} + a^3\mathbf{V}) - F_3(a\mathbf{U} + \mathbf{V}) + A_2, \\
f_3(\mathbf{U}, \mathbf{V}) &= F_3(\mathbf{U} + a^3\mathbf{V}) - F_4(a\mathbf{U} + \mathbf{V}) + A_3, \\
f_4(\mathbf{U}, \mathbf{V}) &= F_4(\mathbf{U} + a^3\mathbf{V}) - F_1(a\mathbf{U} + \mathbf{V}) - A_1 - A_2 - A_3,
\end{aligned}$$

where $F_i : \mathcal{V} \mapsto \mathcal{V}'$ $(i = 1, 2, 3, 4)$ are arbitrary complex vector functions, and A_i $(i = 1, 2, 3)$ are arbitrary constant complex vectors.

Theorem 13.6   *If $a^{m+n} \neq 1$ and $m = n$, the most general solution of the functional equation (13.1) is*

$$f_{i+m}(\mathbf{U}, \mathbf{V}) = -f_i(\mathbf{V}, \mathbf{U}) + A_i \quad (1 \le i \le m), \tag{13.29}$$

where $f_i : \mathcal{V}^2 \mapsto \mathcal{V}'$ $(1 \le i \le m)$ and A_i $(1 \le i \le m)$ are arbitrary complex constant vectors such that $\sum_{i=1}^{m} A_i = \mathbf{O}$.

Proof. By the transformations used in the proof of the previous theorem we may bring the equation (13.1) to the form Eq. (13.24).

For $\mathbf{V}_j = \mathbf{O}$ $(j = 1, 2, \cdots, i-1, i+1, \cdots, i+m-1, i+m+1, \cdots, 2m)$ the equation (13.24) becomes

$$g_i(\mathbf{V}_i, \mathbf{V}_{i+m}) + g_{i+m}(\mathbf{V}_{i+m}, \mathbf{V}_i) = A_i \quad (1 \le i \le m), \tag{13.30}$$

where A_i $(1 \le i \le m)$ are arbitrary complex constant vectors. By substituting Eq. (13.30) into Eq. (13.1), we obtain that it must hold

$$\sum_{i=1}^{m} A_i = \mathbf{O}.$$

On the basis of this equality and Eq. (13.30), we obtain Eq. (13.29). $\square$

Example 13.5 If $a^4 \ne 1$, the most general solution of the functional equation

$$\begin{aligned}
f_1(a\mathbf{Z}_1 + \mathbf{Z}_2,\ a\mathbf{Z}_3 + \mathbf{Z}_4) \quad &+ \quad f_2(a\mathbf{Z}_3 + \mathbf{Z}_3,\ a\mathbf{Z}_4 + \mathbf{Z}_1) \\
+ f_3(a\mathbf{Z}_3 + \mathbf{Z}_4,\ a\mathbf{Z}_1 + \mathbf{Z}_2) \quad &+ \quad f_4(a\mathbf{Z}_4 + \mathbf{Z}_1,\ a\mathbf{Z}_2 + \mathbf{Z}_3) = \mathbf{O}
\end{aligned}$$

is given by

$$\begin{aligned}
f_i(\mathbf{U}, \mathbf{V}) \qquad & (i = 1, 2) \quad are \quad arbitrary, \\
f_3(\mathbf{U}, \mathbf{V}) \ &= \ -f_1(\mathbf{U}, \mathbf{V}) + A, \\
f_4(\mathbf{U}, \mathbf{V}) \ &= \ -f_1(\mathbf{U}, \mathbf{V}) - A,
\end{aligned}$$

where A is an arbitrary complex constant vector.

If $a^{m+n} = 1$, then the functional equation (13.1) may be transformed in the following way.

We introduce new vectors by the equality

$$\mathbf{V}_i = a^{1-i}\mathbf{Z}_i, \quad i.e., \quad \mathbf{Z}_i = a^{i-1}\mathbf{V}_i \quad (1 \le i \le m + n).$$

Then the equation (13.1) becomes

$$\sum_{i=1}^{m+n} f_i\left(a^{m-2+i}\sum_{j=0}^{m-1}\mathbf{V}_{i+j},\ a^{m+n-2+i}\sum_{j=0}^{n-1}\mathbf{V}_{m+i+j}\right) = \mathbf{O}. \tag{13.31}$$

Now, if we put

$$g_i(\mathbf{U}, \mathbf{V}) = f_i(a^{m-2+i}\mathbf{U},\ a^{m+n-2+i}\mathbf{V}) \quad (1 \le i \le m+n),$$

i.e.,

$$f_i(\mathbf{U}, \mathbf{V}) = g_i(a^{n+2-i}\mathbf{U},\ a^{m+n+2-i}\mathbf{V}) \quad (1 \le i \le m+n), \tag{13.32}$$

the functional equation (13.31) takes the form

$$\sum_{i=1}^{m+n} g_i \left(\sum_{j=0}^{m-1} \mathbf{V}_{i+j},\ \sum_{j=0}^{n-1} \mathbf{V}_{m+i+j} \right) = \mathbf{O}. \tag{13.33}$$

The equation (13.33) is just Eq. (13.1) for $a = 1$.

Theorem 13.7 *If $a^{m+n} = 1$, $(m,n) = 1$ and $m+n > 2$, then the general continuous solution of the functional equation (13.1) is given by*

$$f_i(\mathbf{U}, \mathbf{V}) = F_1(a^{n+2-i}\mathbf{U} + a^{m+n+2-i}\mathbf{V})\mathrm{Re}\,(a^{n+2-i}\mathbf{U}) \tag{13.34}$$

$$+F_2(a^{n+2-i}\mathbf{U} + a^{m+n+2-i}\mathbf{V})\mathrm{Im}\,(a^{n+2-i}\mathbf{U}) + G_i(a^{n+2-i}\mathbf{U} + a^{m+n+2-i}\mathbf{V})$$

$$(1 \le i \le m+n)$$

so that

$$\sum_{i=1}^{m+n} G_i(\mathbf{U}) = -m[F_1(\mathbf{U})\mathrm{Re}\,\mathbf{U} + F_2(\mathbf{U})\mathrm{Im}\,\mathbf{U}], \tag{13.35}$$

where $F_i : \mathcal{V} \mapsto \mathcal{L}(\mathcal{V}^0, \mathcal{V}')$ $(i = 1,2)$ and $G_i : \mathcal{V} \mapsto \mathcal{V}'$ $(1 \le i \le m+n-1)$ are arbitrary continuous complex vector functions.

Proof. The proof immediately follows from Eqs. (13.33), (13.32) and Theorem 13.1. $\qquad\square$

Theorem 13.8 *If $a^{m+n} = 1$, $(m,n) = d > 1$, $m/d = p$, $n/d = q$ and $p+q > 2$, then the general continuous solution of the functional equation (13.1) is*

$$f_{id+j}(\mathbf{U}, \mathbf{V}) = F_{1j}(a^{n+2-i}\mathbf{U} + a^{m+n+2-i}\mathbf{V})\mathrm{Re}\,(a^{n+2-i}\mathbf{U}) \tag{13.36}$$

$$+F_{2j}(a^{n+2-i}\mathbf{U} + a^{m+n+2-i}\mathbf{V})\mathrm{Im}\,(a^{n+2-i}\mathbf{U}) + G_{ij}(a^{n+2-i}\mathbf{U} + a^{m+n+2-i}\mathbf{V})$$

$$(0 \le i \le p+q-1; \quad 1 \le j \le d)$$

so that

$$\sum_{i=0}^{p+q-1} G_{ij}(\mathbf{U}) = H_j(\mathbf{U}) - p[F_{1j}(\mathbf{U})\operatorname{Re}\mathbf{U} + F_{2j}(\mathbf{U})\operatorname{Im}\mathbf{U}] \qquad (13.37)$$

$$(1 \le j \le d),$$

$$\sum_{j=1}^{d} H_j(\mathbf{U}) = \mathbf{O}, \qquad (13.38)$$

where

$$F_{ij} : \mathcal{V} \mapsto \mathcal{L}(\mathcal{V}^0, \mathcal{V}') \quad (i = 1, 2; \quad 1 \le j \le d),$$

$$G_{ij} : \mathcal{V} \mapsto \mathcal{V}' \quad (0 \le i \le p + q - 2; \quad 1 \le j \le d),$$

$$H_j : \mathcal{V} \mapsto \mathcal{V}' \quad (1 \le j \le d - 1)$$

are arbitrary continuous complex vector functions.

Proof. On the basis of the expressions Eqs. (13.33), (13.32) and Theorem 13.3 we derive the proof of the theorem. $\square$

Theorem 13.9 *If $a^{m+n} = 1$ and $m = n$, then the most general solution of the functional equation (13.1) is given by*

$$\begin{aligned}
f_i(\mathbf{U}, \mathbf{V}) &\quad (1 \le i \le m) \quad \text{are arbitrary,} \\
f_{m+i}(\mathbf{U}, \mathbf{V}) &= H_i(a^{n+2-i}\mathbf{U} + a^{m+n+2-i}\mathbf{V}) \\
&\quad - f_i(a^{n+2-i}\mathbf{U},\, a^{m+n+2-i}\mathbf{V}) \quad (1 \le i \le m),
\end{aligned} \qquad (13.39)$$

where $H_i : \mathcal{V} \mapsto \mathcal{V}'$ are arbitrary complex vector functions such that
$$\sum_{i=1}^{m} H_i(\mathbf{U}) = \mathbf{O}.$$

Proof. The proof immediately follows from Eqs. (13.33), (13.32) and Theorem 13.4. $\square$

14 Special Parametric Functional Equation

All notations for the vectors are the same as in the previous section.

Now, we will solve the following functional equation [I. B. Risteski *et al.* (to appear B)]

$$\sum_{i=1}^{m+n} f\left(\sum_{j=0}^{m-1} a^{m-1-j}\mathbf{Z}_{i+j},\; \sum_{j=0}^{n-1} a^{n-1-j}\mathbf{Z}_{i+m+j}\right) = \mathbf{O}, \qquad (14.1)$$

which is obtained as a special case of the equation (13.1) for $f_i = f$ ($1 \le i \le m + n$).

Theorem 14.1 *If $a^{m+n} \neq 1$, then the most general solution of the complex vector functional equation (14.1) is given by*

$$f(\mathbf{U}, \mathbf{V}) \quad = \qquad\qquad\qquad (14.2)$$

$$\begin{cases} F(\mathbf{U} + a^m\mathbf{V}) - F(a^n\mathbf{U} + \mathbf{V}) & (m \neq n), \\[2mm] G(\mathbf{U} + a^m\mathbf{V},\; a^m\mathbf{U} + \mathbf{V}) - G(a^m\mathbf{U} + \mathbf{V},\; \mathbf{U} + a^m\mathbf{V}) & (m = n), \end{cases}$$

where $F : \mathcal{V} \mapsto \mathcal{V}'$, $G : \mathcal{V}^2 \mapsto \mathcal{V}'$ are arbitrary complex vector functions.

Proof. We set

$$f(\mathbf{U}, \mathbf{V}) = g(\mathbf{U} + a^m\mathbf{V},\; a^n\mathbf{U} + \mathbf{V}) \qquad (14.3)$$

into Eq. (14.1) and deduce that

$$\sum_{i=1}^{m+n} g\left(\sum_{j=0}^{m+1} a^{m-1-j}\mathbf{Z}_{i+j} + \sum_{j=0}^{n-1} a^{m+n-1}\mathbf{Z}_{i+m+j},\right.$$

$$\left.\sum_{j=0}^{m-1} a^{m+n-1-j}\mathbf{Z}_{i+j} + \sum_{j=0}^{n-1} a^{n-1-j}\mathbf{Z}_{i+m+j}\right) = \mathbf{O},$$

i.e.,

$$\sum_{i=1}^{m+n} g\left(\sum_{j=0}^{m+n-1} a^j\mathbf{Z}_{i+m-1-j},\; \sum_{j=0}^{m+n-1} a^j\mathbf{Z}_{i-1-j}\right) = \mathbf{O}. \qquad (14.4)$$

This transformation of the equation (14.1) is possible since $a^{m+n} \neq 1$.

Now we introduce new vectors

$$\mathbf{V}_i = \sum_{j=0}^{m+n+1} a^j \mathbf{Z}_{i-1-j} \quad (1 \le i \le m+n). \tag{14.5}$$

The linear forms Eq. (14.5) are independent since their determinant is $(a^{m+n} - 1)^{m+n-1}$.

Making use of these notations, the equation (14.4) becomes

$$\sum_{i=1}^{m+n} g(\mathbf{V}_i, \mathbf{V}_{i+n}) = \mathbf{O}. \tag{14.6}$$

If $m \ne n$, we set $\mathbf{V}_1 = \mathbf{V}_2 = \cdots = \mathbf{V}_{m-1} = \mathbf{V}_{m+1} = \mathbf{V}_{m+2} = \cdots = \mathbf{V}_{m+n-1} = \mathbf{O}$ and we get

$$g(\mathbf{U}, \mathbf{V}) = F(\mathbf{U}) + F_1(\mathbf{V}). \tag{14.7}$$

We substitute g from Eq. (14.7) into Eq. (14.6) and obtain

$$\sum_{i=1}^{m+n} [F(\mathbf{V}_i) + F_1(\mathbf{V}_i)] = \mathbf{O},$$

which implies that $F_1(\mathbf{V}_i) = -F(\mathbf{V}_i)$. Hence,

$$g(\mathbf{U}, \mathbf{V}) = F(\mathbf{U}) - F(\mathbf{V}). \tag{14.8}$$

If $m = n$, the equation (14.4) yields

$$g(\mathbf{U}, \mathbf{V}) + g(\mathbf{V}, \mathbf{U}) = \mathbf{O},$$

i.e.,

$$g(\mathbf{U}, \mathbf{V}) = G(\mathbf{U}, \mathbf{V}) - G(\mathbf{V}, \mathbf{U}). \tag{14.9}$$

From Eqs. (14.3), (14.8) and (14.9) we conclude that Eq. (14.2) holds. It is easy to verify that Eq. (14.2) satisfies Eq. (14.1).   □

Example 14.1   If $a^3 \ne 1$, then the most general solution of the functional equation

$$f(a\mathbf{Z}_1 + \mathbf{Z}_2, \mathbf{Z}_3) + f(a\mathbf{Z}_2 + \mathbf{Z}_3, \mathbf{Z}_1) + f(a\mathbf{Z}_3 + \mathbf{Z}_1, \mathbf{Z}_2) = \mathbf{O}$$

is given by

$$f(\mathbf{U}, \mathbf{V}) = F(\mathbf{U} + a^2\mathbf{V}) - F(a\mathbf{U} + \mathbf{V}),$$

where $F : \mathcal{V} \mapsto \mathcal{V}'$ is an arbitrary complex vector function.

Example 14.2 If $a^4 \neq 1$, the most general solution of the functional equation

$$f(a\mathbf{Z}_1 + \mathbf{Z}_2, \ a\mathbf{Z}_3 + \mathbf{Z}_4) \quad + \quad f(a\mathbf{Z}_2 + \mathbf{Z}_3, \ a\mathbf{Z}_4 + \mathbf{Z}_1)$$
$$+ f(a\mathbf{Z}_3 + \mathbf{Z}_4, \ a\mathbf{Z}_1 + \mathbf{Z}_2) \quad + \quad f(a\mathbf{Z}_4 + \mathbf{Z}_1, \ a\mathbf{Z}_2 + \mathbf{Z}_3) = \mathbf{O}$$

is given by

$$f(\mathbf{U}, \mathbf{V}) = G(\mathbf{U} + a^2\mathbf{V}, \ a^2\mathbf{U} + \mathbf{V}) - G(a^2\mathbf{U} + \mathbf{V}, \ \mathbf{U} + a^2\mathbf{V}),$$

where $G : \mathcal{V}^2 \mapsto \mathcal{V}'$ is an arbitrary complex vector function.

Theorem 14.2 *If $a^{m+n} = 1$, $(m, n) = 1$ and $m + n > 2$, then the general continuous solution of the functional equation (14.1) is given by*

$$
\begin{aligned}
f(\mathbf{U}, \mathbf{V}) \ = \ & \sum_{i=1}^{m+n} [F_1(a^i\mathbf{U} + a^{i+m}\mathbf{V})\mathrm{Re}\,(a^i\mathbf{U}) \qquad\qquad (14.10) \\
& \qquad\quad + F_2(a^i\mathbf{U} + a^{i+m}\mathbf{V})\mathrm{Im}\,(a^i\mathbf{U})] \\
+ \ & \sum_{i=1}^{m+n-1} [G_i(a^i\mathbf{U} + a^{i+m}\mathbf{V}) - G_i(a^i\mathbf{U} + a^m\mathbf{V})] \\
- \ & m[F_1(\mathbf{U} + a^m\mathbf{V})\mathrm{Re}\,(\mathbf{U} + a^m\mathbf{V}) \\
& \qquad + F_2(\mathbf{U} + a^m\mathbf{V})\mathrm{Im}\,(\mathbf{U} + a^m\mathbf{V})],
\end{aligned}
$$

where $F_i : \mathcal{V} \mapsto \mathcal{L}(\mathcal{V}^0, \mathcal{V}')$ $(i = 1, 2)$ and $G_i : \mathcal{V} \mapsto \mathcal{V}'$ $(1 \leq i \leq m + n - 1)$ are arbitrary complex vector functions.

Proof. Let us put $\mathbf{Z}_i = a^{i-1}\mathbf{T}_i$ $(1 \leq i \leq m + n)$. The equation (14.1) becomes

$$\sum_{i=1}^{m+n} f\left(a^{m+i-2} \sum_{j=0}^{m-1} \mathbf{T}_{i+j}, \ a^{m+n-2+i} \sum_{j=0}^{n-1} \mathbf{T}_{m+i-j}\right) = \mathbf{O}. \qquad (14.11)$$

Now we make the substitutions

$$f(a^{m+i-2}\mathbf{U}, \ a^{m+n-2+i}\mathbf{V}) = f_i(\mathbf{U}, \mathbf{V}) \quad (1 \leq i \leq m + n),$$

i.e.,

$$f(\mathbf{U}, \mathbf{V}) = f_i(a^{n-i+2}\mathbf{U}, \ a^{m+n+2-i}\mathbf{V}) \quad (1 \leq i \leq m + n), \qquad (14.12)$$

and we obtain

$$\sum_{i=1}^{m+n} f_i \left(\sum_{j=0}^{m-1} \mathbf{T}_{i+j}, \ \sum_{j=0}^{n-1} \mathbf{T}_{m+i+j} \right) = \mathbf{O}. \qquad (14.13)$$

The equation (14.13) is just Eq. (13.1) for $a = 1$, and its solution is determined by Theorem 13.1.

By an application of Theorem 13.1, and by the equality (14.12) we get

$$f_i(\mathbf{U}, \mathbf{V}) = P_1(a^{n+2-i}\mathbf{U} + a^{m+n+2-i}\mathbf{V})\mathrm{Re}\,(a^{n+2-i}\mathbf{U}) \qquad (14.14)$$

$$+ P_2(a^{n+2-i}\mathbf{U} + a^{m+n+2-i}\mathbf{V})\mathrm{Im}\,(a^{n+2-i}\mathbf{U}) + Q_i(a^{n+2-i}\mathbf{U} + a^{m+n+2-i}\mathbf{V})$$

$$(1 \le i \le m + n),$$

so that

$$\sum_{i=1}^{m+n} Q_i(\mathbf{U}) = -m[P_1(\mathbf{U})\mathrm{Re}\,\mathbf{U} + P_2(\mathbf{U})\mathrm{Im}\,\mathbf{U}],$$

where $P_i : \mathcal{V} \mapsto \mathcal{L}(\mathcal{V}^0, \mathcal{V}')$ $(i = 1, 2)$ and $Q_i : \mathcal{V} \mapsto \mathcal{V}'$ $(1 \le i \le m + n)$ are continuous complex vector functions. By addition of all equations (14.14) and putting

$$\begin{aligned} P_1(\mathbf{U}) &= (m+n)F_1(\mathbf{U}), \quad P_2(\mathbf{U}) = (m+n)F_2(\mathbf{U}), \\ Q_i(\mathbf{U}) &= (m+n)G_{n+2-i}(\mathbf{U}) \quad (i = 1, 2, \cdots, m+n) \end{aligned}$$

we obtain Eq. (14.10). $\qquad\qquad\qquad\qquad\qquad\qquad\qquad\qquad \square$

Example 14.3 If $a^3 = 1$, the general continuous solution of the functional equation

$$f(a\mathbf{Z}_1 + \mathbf{Z}_2, \ \mathbf{Z}_3) + f(a\mathbf{Z}_2 + \mathbf{Z}_3, \ \mathbf{Z}_1) + f(a\mathbf{Z}_3 + \mathbf{Z}_1, \ \mathbf{Z}_2) = \mathbf{O}$$

is given by

$$\begin{aligned} f(\mathbf{U}, \mathbf{V}) = \ & F_1(a\mathbf{U} + \mathbf{V})\mathrm{Re}\,(a\mathbf{U}) + F_2(a\mathbf{U} + \mathbf{V})\mathrm{Im}\,(a\mathbf{U}) \\ & + \ F_1(a^2\mathbf{U} + \mathbf{V})\mathrm{Re}\,(a^2\mathbf{U}) + F_2(a^2\mathbf{U} + \mathbf{V})\mathrm{Im}\,(a^2\mathbf{U}) \\ & - \ F_1(\mathbf{U} + a^2\mathbf{V})\mathrm{Re}\,(\mathbf{U} + 2a^2\mathbf{V}) - F_2(\mathbf{U} + a^2\mathbf{V})\mathrm{Im}\,(\mathbf{U} + 2a^2\mathbf{V}) \\ & + \ G_1(a\mathbf{U} + \mathbf{V}) - G_1(\mathbf{U} + a^2\mathbf{V}) \\ & + \ G_2(a^2\mathbf{U} + a\mathbf{V}) - G_2(\mathbf{U} + a^2\mathbf{V}), \end{aligned}$$

where $F_i : \mathcal{V} \mapsto \mathcal{L}(\mathcal{V}^0, \mathcal{V}')$ $(i = 1, 2)$ and $G_i : \mathcal{V} \mapsto \mathcal{V}'$ $(i = 1, 2)$ are arbitrary complex vector functions.

Theorem 14.3 *If $a^{m+n} = 1$, $(m, n) = d > 1$, $m/d = p$, $n/d = q$ and $p + q > 2$, then the general continuous solution of the functional equation (14.1) is given by*

$$
\begin{aligned}
f(\mathbf{U}, \mathbf{V}) \;=\; & \sum_{j=-1}^{d-2} \sum_{i=0}^{p+q-1} [F_{1,j+2}(a^{n-id-j}\mathbf{U} + a^{-id-j}\mathbf{V})\mathrm{Re}\,(a^{n-id-j}\mathbf{U}) \\
+\;& F_{2,j+2}(a^{n-id-j}\mathbf{U} + a^{-id-j}\mathbf{V})\mathrm{Im}\,(a^{n-id-j}\mathbf{U}) \qquad (14.15) \\
+\;& G_{i,j+2}(a^{n-id-j}\mathbf{U} + a^{-id-j}\mathbf{V})],
\end{aligned}
$$

so that

$$
\sum_{i=0}^{p+q-1} G_{ij}(\mathbf{U}) = H_j(\mathbf{U}) - p[F_{1j}(\mathbf{U})\mathrm{Re}\,(\mathbf{U}) + F_{2j}(\mathbf{U})\mathrm{Im}\,(\mathbf{U})] \quad (1 \le j \le d)
$$

and

$$
\sum_{j=1}^{d} H_j(\mathbf{U}) = \mathbf{O},
$$

where

$$
F_{ij} : \mathcal{V} \mapsto \mathcal{L}(\mathcal{V}^0, \mathcal{V}') \quad (i = 1, 2; \quad 1 \le j \le d),
$$

$$
G_{ij} : \mathcal{V} \mapsto \mathcal{V}' \quad (0 \le i \le p + q - 2; \quad 1 \le j \le d),
$$

$$
H_j : \mathcal{V} \mapsto \mathcal{V}' \quad (1 \le j \le d - 1)
$$

are arbitrary continuous complex vector functions.

Proof. We can start from equation (14.11). From Eqs. (14.10) and (14.11) on the basis of Theorem 13.3 we get

$$
\begin{aligned}
f(\mathbf{U}, \mathbf{V}) \;=\; & P_{1j}(a^{n-id-j+2}\mathbf{U} + a^{m+n+2-id-j}\mathbf{V})\mathrm{Re}\,(a^{n-id-j+2}\mathbf{U}) \\
+\;& P_{2j}(a^{n-id-j+2}\mathbf{U} + a^{m+n+2-id-j}\mathbf{V})\mathrm{Im}\,(a^{n-id-j+2}\mathbf{U}) \\
+\;& Q_{ij}(a^{n-id-j+2}\mathbf{U} + a^{n+m+2-id-j}\mathbf{V}) \qquad (14.16) \\
& (0 \le i \le p + q - 1;\ 1 \le j \le d),
\end{aligned}
$$

$$\sum_{i=0}^{p+q-1} Q_{ij}(\mathbf{U}) \;=\; K_j(\mathbf{U}) - p[P_{1j}(\mathbf{U})\mathrm{Re}\,(\mathbf{U}) + P_{2j}(\mathbf{U})\mathrm{Im}\,(\mathbf{U})] \quad (14.17)$$

$$(1 \le j \le d),$$

$$\sum_{j=1}^{d} K_j(\mathbf{U}) = \mathbf{O}, \qquad\qquad (14.18)$$

where

$$P_{ij} : \mathcal{V} \mapsto \mathcal{L}(\mathcal{V}^0, \mathcal{V}') \quad (i = 1, 2; \quad 1 \le j \le d),$$

$$Q_{ij} : \mathcal{V} \mapsto \mathcal{V}' \quad (0 \le i \le p + q - 2; \quad 1 \le j \le d),$$

$$K_j : \mathcal{V} \mapsto \mathcal{V}' \quad (1 \le j \le d - 1)$$

are continuous functions.

We take into account Eqs. (14.17) and (14.18) and we add together all equations (14.16). In this way we obtain Eq. (14.16) with

$$P_{1j}(\mathbf{U}) = (m + n)F_{1j}(\mathbf{U}), \quad P_{2j}(\mathbf{U}) = (m + n)F_{2j}(\mathbf{U}),$$

$$Q_{ij}(\mathbf{U}) = (m + n)G_{ij}(\mathbf{U}), \quad K_j(\mathbf{U}) = (m + n)H_j(\mathbf{U})$$

$$(0 \le i \le p + q - 2; \quad 1 \le j \le d). \qquad \square$$

Example 14.4   If $a^6 = 1$, then the general continuous solution of the functional equation

$$f(a^3\mathbf{Z}_1 + a^2\mathbf{Z}_2 + a\mathbf{Z}_3 + \mathbf{Z}_4,\; a\mathbf{Z}_5 + \mathbf{Z}_6) + f(a^3\mathbf{Z}_2 + a^2\mathbf{Z}_3 + a\mathbf{Z}_4 + \mathbf{Z}_5,\; a\mathbf{Z}_6 + \mathbf{Z}_1)$$

$$+ f(a^3\mathbf{Z}_3 + a^2\mathbf{Z}_4 + a\mathbf{Z}_5 + \mathbf{Z}_6,\; a\mathbf{Z}_1 + \mathbf{Z}_2) + f(a^3\mathbf{Z}_4 + a^2\mathbf{Z}_5 + a\mathbf{Z}_6 + \mathbf{Z}_1,\; a\mathbf{Z}_2 + \mathbf{Z}_3)$$

$$+ f(a^3\mathbf{Z}_5 + a^2\mathbf{Z}_6 + a\mathbf{Z}_1 + \mathbf{Z}_2, a\mathbf{Z}_3 + \mathbf{Z}_4) + f(a^3\mathbf{Z}_6 + a^2\mathbf{Z}_1 + a\mathbf{Z}_2 + \mathbf{Z}_3, a\mathbf{Z}_4 + \mathbf{Z}_5) = \mathbf{O}$$

is given by

$$f(\mathbf{U}, \mathbf{V}) = F_{11}(a\mathbf{U} + a^5\mathbf{V})\mathrm{Re}\,(a\mathbf{U}) + F_{21}(a\mathbf{U} + a^5\mathbf{V})\mathrm{Im}\,(a\mathbf{U})$$

$$+ F_{11}(a^3\mathbf{U} + a\mathbf{V})\mathrm{Re}\,(a^3\mathbf{U}) + F_{21}(a^3\mathbf{U} + a\mathbf{V})\mathrm{Im}\,(a^3\mathbf{U})$$

$$-F_{11}(a^5\mathbf{U} + a^3\mathbf{V})\mathrm{Re}\,(a^5\mathbf{U} + 2a^3\mathbf{V}) - F_{21}(a^5\mathbf{U} + a^3\mathbf{V})\mathrm{Im}\,(a^5\mathbf{U} + 2a^3\mathbf{V})$$

$$+F_{12}(\mathbf{U} + a^4\mathbf{V})\mathrm{Re}\,(\mathbf{U}) + F_{22}(\mathbf{U} + a^4\mathbf{V})\mathrm{Im}\,(\mathbf{U})$$

$$+F_{12}(a^2\mathbf{U} + \mathbf{V})\mathrm{Re}\,(a^2\mathbf{U}) + F_{22}(a^2\mathbf{U} + \mathbf{V})\mathrm{Im}\,(a^2\mathbf{U})$$

$$-F_{12}(a^4\mathbf{U} + a^2\mathbf{V})\mathrm{Re}\,(a^4\mathbf{U} + 2a^2\mathbf{V}) - F_{22}(a^4\mathbf{U} + a^2\mathbf{V})\mathrm{Im}\,(a^4\mathbf{U} + 2a^2\mathbf{V})$$

$$+G_{01}(a\mathbf{U} + a^5\mathbf{V}) - G_{01}(a^5\mathbf{U} + a^3\mathbf{V}) + G_{02}(a\mathbf{U} + a^4\mathbf{V}) - G_{02}(a^4\mathbf{U} + a^2\mathbf{V})$$

$$+G_{11}(a^3\mathbf{U} + a\mathbf{V}) - G_{11}(a^5\mathbf{U} + a^3\mathbf{V}) + G_{12}(a^2\mathbf{U} + \mathbf{V}) - G_{12}(a^4\mathbf{U} + a^2\mathbf{V})$$

$$+H_1(a^5\mathbf{U} + a^3\mathbf{V}) - H_1(a^4\mathbf{U} + a^2\mathbf{V}),$$

where $F_{ij} : \mathcal{V} \mapsto \mathcal{L}(\mathcal{V}^0, \mathcal{V}')$ $(i,j = 1,2)$; $G_{ij} : \mathcal{V} \mapsto \mathcal{V}'$ $(i = 0,1;\ j = 1,2)$ and $H_1 : \mathcal{V} \mapsto \mathcal{V}'$ are arbitrary continuous complex vector functions.

Theorem 14.4 *If* $a^{m+n} = 1$ *and* $m = n$, *the most general solution of the functional equation* (14.1) *is given by*

$$f(\mathbf{U}, \mathbf{V}) \;=\; \sum_{i=1}^{m}[F_1(a^i\mathbf{U},\ a^{n+i}\mathbf{V}) - F_i(a^i\mathbf{U},\ a^{n+i}\mathbf{V})$$

$$+ H_i(a^{n+i}\mathbf{V} + a^i\mathbf{U})], \qquad (14.19)$$

$$\sum_{i=1}^{m} H_i(\mathbf{U}) = \mathbf{O},$$

where $F_i : \mathcal{V}^2 \mapsto \mathcal{V}'$ $(1 \le i \le m)$ *and* $H_i : \mathcal{V} \mapsto \mathcal{V}'$ $(1 \le i \le m - 1)$ *are arbitrary complex vector functions.*

Proof. We start again from the equation (14.11). According to Theorem 13.4 and Eq. (14.10) we have

$$f(\mathbf{U}, \mathbf{V}) = P_i(a^{m-i+2}\mathbf{U},\ a^{m+n+2-i}\mathbf{V}) \quad (1 \le i \le m),$$

$$f(\mathbf{U}, \mathbf{V}) \;=\; Q_i(a^{m-i+2}\mathbf{U} + a^{m+n+2-i}\mathbf{V}) \qquad\qquad (14.20)$$

$$-\; P_i(a^{m+n+2-i}\mathbf{V},\ a^{m+2-i}\mathbf{U}) \qquad (1 \le i \le m),$$

$$\sum_{i=1}^{m} Q_i(\mathbf{U}) = \mathbf{O}.$$

By addition we get Eq. (14.19) with

$$P_i(\mathbf{U}, \mathbf{V}) = 2mF_{m-i+2}(\mathbf{U}, \mathbf{V}), \quad Q_i(\mathbf{U}) = 2mH_{m-i+2}(\mathbf{U}).$$

$\square$

Example 14.5 If $a^4 = 1$, the most general solution of the functional equation

$$\begin{aligned}
f(a\mathbf{Z}_1 + \mathbf{Z}_2,\ a\mathbf{Z}_3 + \mathbf{Z}_4) \quad &+\quad f(a\mathbf{Z}_2 + \mathbf{Z}_3,\ a\mathbf{Z}_4 + \mathbf{Z}_1) \\
+ f(a\mathbf{Z}_3 + \mathbf{Z}_4,\ a\mathbf{Z}_1 + \mathbf{Z}_2) \quad &+\quad f(a\mathbf{Z}_4 + \mathbf{Z}_1,\ a\mathbf{Z}_2 + \mathbf{Z}_3) = \mathbf{O}
\end{aligned}$$

is given by

$$\begin{aligned}
f(\mathbf{U}, \mathbf{V}) \quad = \quad & F_1(a\mathbf{U},\ a^3\mathbf{V}) - F_1(a\mathbf{V},\ a^3\mathbf{U}) + F_2(a^2\mathbf{U},\ \mathbf{V}) \\
- \quad & F_2(a^2\mathbf{V},\ \mathbf{U}) + H_1(a^3\mathbf{U} + a\mathbf{V}) - H_1(\mathbf{U} + a^2\mathbf{V}),
\end{aligned}$$

where $F_i : \mathcal{V}^2 \mapsto \mathcal{V}'$ $(i = 1, 2)$ and $H_1 : \mathcal{V} \mapsto \mathcal{V}'$ are arbitrary complex vector functions.

Now, as special cases we obtain the results given in [D. Ž. Djoković *et al.* (1966); R. Ž. Djordjević and P. M. Vasić (1967); D. S. Mitrinović and J. E. Pečarić (1991)].

15 Expanded Parametric Functional Equation

The notations for the vectors in this section are the same as in Sec. 13.

In this section we will solve the following simple complex vector functional equation [I. B. Risteski *et al.* (2001A)]

$$\sum_{i=1}^{m+n+k} f\left(\sum_{j=0}^{m-1} a^{m-1-j}\mathbf{Z}_{i+j},\ \sum_{j=0}^{n-1} a^{n-1-j}\mathbf{Z}_{i+m+j},\ \sum_{j=0}^{k-1} a^{k-1-j}\mathbf{Z}_{i+m+n+j} \right) = \mathbf{O} \tag{15.1}$$

$$(\mathbf{Z}_{m+n+k+i} \equiv \mathbf{Z}_i),$$

where a is a complex number and $f : \mathcal{V}^3 \mapsto \mathcal{V}'$ is an unknown complex vector function.

The above equation for $k = 0$ was solved in the previous section. Also, the functional equation (15.1) for $a = 1$ was solved in [D. Ž. Djoković (1961)] under the hypothesis that the function and variables are real.

Now we will solve the functional equation (15.1) in the following cases.

I. Let $a = 1$. If we introduce the notations

$$\mathbf{S} = \sum_{j=1}^{m+n+k} \mathbf{Z}_j \tag{15.2}$$

and

$$f(\mathbf{U}, \mathbf{V}, \mathbf{S}) = g(\mathbf{U}, \mathbf{V}, \ \mathbf{S} + \mathbf{U} + \mathbf{V}), \tag{15.3}$$

then the equation (15.1) becomes

$$\sum_{j=1}^{m+n+k} g\left(\sum_{j=0}^{m-1} \mathbf{Z}_{i+j}, \ \sum_{j=0}^{n-1} \mathbf{Z}_{m+i+j}, \ \mathbf{S} \right) - \mathbf{O}. \tag{15.4}$$

If we introduce new variables $\mathbf{T}_i$ $(1 \leq i \leq m+n+k-1)$ by the equalities

$$\mathbf{T}_i = \mathbf{Z}_i - \frac{\mathbf{S}}{m+n+k} \quad (1 \leq i \leq m+n+k-1),$$

i.e.,

$$\mathbf{Z}_{m+n+k} = \frac{\mathbf{S}}{m+n+k} - \sum_{j=1}^{m+n+k-1} \mathbf{T}_j,$$

then the equation (15.4) becomes

$$\sum_{i=1}^{k} g\left(\frac{m\mathbf{S}}{m+n+k} + \sum_{j=0}^{m-1} \mathbf{T}_{i+j}, \ \frac{n\mathbf{S}}{m+n+k} + \sum_{j=0}^{n-1} \mathbf{T}_{m+i+j}, \ \mathbf{S} \right) \tag{15.5}$$

$$+ \sum_{i=k+1}^{n+k} g\left(\frac{m\mathbf{S}}{m+n+k} + \sum_{j=0}^{m-1} \mathbf{T}_{i+j}, \ \frac{n\mathbf{S}}{m+n+k} - \sum_{j=0}^{m+k-1} \mathbf{T}_{m+n+i+j}, \ \mathbf{S} \right)$$

$$+ \sum_{i=n+k+1}^{m+n+k} g\left(\frac{m\mathbf{S}}{m+n+k} - \sum_{j=0}^{n+k-1} \mathbf{T}_{m+i+j}, \ \frac{n\mathbf{S}}{m+n+k} + \sum_{j=0}^{n-1} \mathbf{T}_{m+i+j}, \ \mathbf{S} \right) = \mathbf{O}.$$

By putting

$$g\left(\frac{m\mathbf{S}}{m+n+k}+\mathbf{U},\ \frac{n\mathbf{S}}{m+n+k}+\mathbf{V},\ \mathbf{S}\right) = H(\mathbf{U},\mathbf{V},\mathbf{S}), \qquad (15.6)$$

the equation (15.5) becomes

$$\sum_{i=1}^{k} H\left(\sum_{j=0}^{m-1}\mathbf{T}_{i+j},\ \sum_{j=0}^{n-1}\mathbf{T}_{m+i+j},\ \mathbf{S}\right) \qquad (15.7)$$

$$+ \sum_{i=k+1}^{k+n} H\left(\sum_{j=0}^{m-1}\mathbf{T}_{i+j},\ -\sum_{j=0}^{m+k-1}\mathbf{T}_{m+n+i+j},\ \mathbf{S}\right)$$

$$+ \sum_{i=n+k+1}^{m+n+k} H\left(-\sum_{j=0}^{n+k-1}\mathbf{T}_{m+i+j},\ \sum_{j=0}^{n-1}\mathbf{T}_{m+i+j},\ \mathbf{S}\right) = \mathbf{O}.$$

Now, we may suppose that the variable $\mathbf{S}$ is fixed, and we may put

$$H(\mathbf{U},\mathbf{V},\mathbf{S}) = h(\mathbf{U},\mathbf{V}). \qquad (15.8)$$

In this case, the functional equation (15.7) takes the following form

$$\sum_{i=1}^{k} h\left(\sum_{j=0}^{m-1}\mathbf{T}_{i+j},\ \sum_{j=0}^{n-1}\mathbf{T}_{m+i+j}\right) \qquad (15.9)$$

$$+ \sum_{i=k+1}^{m+k} h\left(\sum_{j=0}^{m-1}\mathbf{T}_{i+j},\ -\sum_{j=0}^{m+k-1}\mathbf{T}_{m+n+i+j}\right)$$

$$+ \sum_{i=n+k+1}^{m+n+k} h\left(-\sum_{j=0}^{n+k+1}\mathbf{T}_{m+i+j},\ \sum_{j=0}^{n-1}\mathbf{T}_{m+i+j}\right) = \mathbf{O}.$$

The complex vector function h has the following properties:

1°. If $m < n < k$ and $m + n < k$, then the following relations hold

$$h(\mathbf{O},\mathbf{O}) = \mathbf{O}, \qquad (15.10)$$

$$h(\mathbf{U},-\mathbf{U}) + h(\mathbf{U},\mathbf{O}) + h(\mathbf{O},-\mathbf{U}) = \mathbf{O}, \qquad (15.11)$$

$$h(\mathbf{O},\mathbf{U}) + h(\mathbf{O},-\mathbf{U}) = \mathbf{O}, \qquad (15.12)$$

$$h(\mathbf{U}, \mathbf{V}) = h(\mathbf{U} + \mathbf{V}, \mathbf{O}) - h(\mathbf{V}, \mathbf{O}) + h(\mathbf{O}, \mathbf{V}), \tag{15.13}$$

$$h(\mathbf{U}, \mathbf{O}) + h(-\mathbf{U}, \mathbf{O}) = \mathbf{O}, \tag{15.14}$$

$$h(\mathbf{U}, \mathbf{V}) = h(\mathbf{O}, \mathbf{U} + \mathbf{V}) - h(\mathbf{O}, \mathbf{U}) + h(\mathbf{U}, \mathbf{O}). \tag{15.15}$$

2°. If $m < n < k$ and $m + n = k$, then Eqs. (15.10), (15.11), (15.12), (15.13), (15.14) and (15.15) hold.

3°. If $m < n < k$ and $m + n > k$, then Eqs. (15.10), (15.11), (15.12), (15.13), (15.14) hold and

$$h(\mathbf{U}, \mathbf{V}) + h(\mathbf{V}, -\mathbf{U} - \mathbf{V}) + h(-\mathbf{U} - \mathbf{V}, \mathbf{U}) = \mathbf{O}. \tag{15.16}$$

4°. If $m < n = k$, then Eqs. (15.10), (15.11), (15.12), (15.13) and (15.16) hold.

5°. If $m = n$ and $2m < k$, then Eqs. (15.10), (15.11) hold and

$$h(\mathbf{U}, \mathbf{V}) + h(\mathbf{V}, \mathbf{O}) + h(\mathbf{O}, -\mathbf{U} - \mathbf{V}) + h(-\mathbf{U} - \mathbf{V}, \mathbf{U}) = \mathbf{O}. \tag{15.17}$$

6°. If $m = n$ and $2m = k$, then Eqs. (15.10), (15.11) and (15.17) hold.

7°. If $m = n < k$ and $2m > k$, then Eqs. (15.10), (15.11) and (15.17) hold.

8°. If $n < m < k$ and $m + n < k$, then Eqs. (15.10), (15.11), (15.14), (15.15), (15.12) and (15.13) hold.

9°. If $n < m < k$ and $m + n = k$, then Eqs. (15.10), (15.11), (15.14), (15.15), (15.12) and (15.13) hold.

10°. If $n < m < k$ and $m + n > k$, then Eqs. (15.10), (15.11), (15.14), (15.15), (15.12) and (15.16) hold.

11°. If $n < m = k$, then Eqs. (15.10), (15.11), (15.14), (15.15) and (15.16) hold.

12°. If $m = n = k$, then Eqs. (15.10), (15.11) and (15.16) hold.

Now, we will prove the theorems which treat the functional equation (15.1).

Theorem 15.1   *If $a = 1$, $m, n < k$, $m \neq n$ and $m + n \neq k$, then the* *general continuous solution of the functional equation* (15.1) *is*

$$
\begin{aligned}
f(\mathbf{U},\mathbf{V},\mathbf{W}) \;=\;& G_1(\mathbf{U}+\mathbf{V}+\mathbf{W})\mathrm{Re}\left(\mathbf{U}-m\frac{\mathbf{U}+\mathbf{V}+\mathbf{W}}{m+n+k}\right) \\
+\;& G_2(\mathbf{U}+\mathbf{V}+\mathbf{W})\mathrm{Im}\left(\mathbf{U}-m\frac{\mathbf{U}+\mathbf{V}+\mathbf{W}}{m+n+k}\right) \\
+\;& G_3(\mathbf{U}+\mathbf{V}+\mathbf{W})\mathrm{Re}\left(\mathbf{V}-n\frac{\mathbf{U}+\mathbf{V}+\mathbf{W}}{m+n+k}\right) \\
+\;& G_4(\mathbf{U}+\mathbf{V}+\mathbf{W})\mathrm{Im}\left(\mathbf{V}-n\frac{\mathbf{U}+\mathbf{V}+\mathbf{W}}{m+n+k}\right),
\end{aligned}
$$

where $G_i : \mathcal{V} \mapsto \mathcal{L}(\mathcal{V}^0,\mathcal{V}')$ $(1 \le i \le 4)$ *are arbitrary continuous complex vector functions.*

Proof. Let $m < n < k$ and $m + n < k$. On the basis of the expressions Eqs. (15.13) and (15.15), we obtain that the complex vector function h satisfies the equation

$$
h(\mathbf{U}+\mathbf{V},\mathbf{O}) - h(\mathbf{O},\mathbf{U}+\mathbf{V}) = h(\mathbf{U},\mathbf{O}) - h(\mathbf{O},\mathbf{U}) + h(\mathbf{V},\mathbf{O}) - h(\mathbf{O},\mathbf{V}),
$$

and hence we get

$$
h(\mathbf{U},\mathbf{O}) - h(\mathbf{O},\mathbf{U}) = G'\mathrm{Re}\,\mathbf{U} + G''\mathrm{Im}\,\mathbf{U},
$$

where $G', G'' \in \mathcal{L}(\mathcal{V}^0, \mathcal{V}')$ are arbitrary continuous functions of $\mathbf{S}$.

Thus, the equation (15.15) has the form

$$
h(\mathbf{U},\mathbf{V}) = h(\mathbf{O},\mathbf{U}+\mathbf{V}) + G'\mathrm{Re}\,\mathbf{U} + G''\mathrm{Im}\,\mathbf{U}. \tag{15.18}
$$

If we substitute the function h determined by Eq. (15.18) into Eq. (15.9), on the basis of the expression Eq. (15.12) we obtain

$$
\sum_{i=1}^{k} h\left(\mathbf{O}, \sum_{j=1}^{m+n-1}\mathbf{T}_{i+j}\right) - \sum_{i=k+1}^{k-1} h\left(\mathbf{O}, \sum_{j=0}^{k-1}\mathbf{T}_{m+n+i+j}\right) = \mathbf{O}. \tag{15.19}
$$

If we put $\mathbf{T}_1 = \mathbf{U}$, $\mathbf{T}_k = \mathbf{V}$ and $\mathbf{T}_j = \mathbf{O}$ $(j \ne 1, k)$ into Eq. (15.19), then we obtain

$$
h(\mathbf{O},\mathbf{U}+\mathbf{V}) = h(\mathbf{O},\mathbf{U}) + h(\mathbf{O},\mathbf{V}),
$$

and hence we deduce

$$
h(\mathbf{O},\mathbf{U}) = G_3\mathrm{Re}\,\mathbf{U} + G_4\mathrm{Im}\,\mathbf{U}, \tag{15.20}
$$

where $G_3, G_4 \in \mathcal{L}(\mathcal{V}^0, \mathcal{V}')$ are arbitrary complex vector functions of $\mathbf{S}$.

On the basis of the expressions Eqs. (15.20) and (15.18), we obtain

$$h(\mathbf{U}, \mathbf{V}) = G_1 \operatorname{Re} \mathbf{U} + G_2 \operatorname{Im} \mathbf{U} + G_3 \operatorname{Re} \mathbf{V} + G_4 \operatorname{Im} \mathbf{V}, \tag{15.21}$$

where $G_1 = G' + G_3$ and $G_2 = G'' + G_4$.

Let $m < n < k$ and $m + n > k$. On the basis of the expressions Eqs. (15.16), (15.13) and (15.12), we get

$$h(\mathbf{U} + \mathbf{V}, \mathbf{O}) - h(-\mathbf{U} - \mathbf{V}, \mathbf{O}) - h(\mathbf{O}, \mathbf{U} + \mathbf{V})$$

$$= h(\mathbf{U}, \mathbf{O}) - h(-\mathbf{U}, \mathbf{O}) - h(\mathbf{O}, \mathbf{U}) + h(\mathbf{V}, \mathbf{O}) - h(-\mathbf{V}, \mathbf{O}) - h(\mathbf{O}, \mathbf{V}),$$

and hence we can derive that

$$h(\mathbf{U}, \mathbf{O}) - h(-\mathbf{U}, \mathbf{O}) - h(\mathbf{O}, \mathbf{U}) = G_3 \operatorname{Re} \mathbf{U} + G_4 \operatorname{Im} \mathbf{U}, \tag{15.22}$$

where $G_3, G_4 \in \mathcal{L}(\mathcal{V}^0, \mathcal{V}')$ are arbitrary complex vector functions of $\mathbf{S}$.

On the basis of the identity Eq. (15.22), the equation (15.13) becomes

$$h(\mathbf{U}, \mathbf{V}) = h(\mathbf{U} + \mathbf{V}, \mathbf{O}) - h(-\mathbf{U}, \mathbf{O}) + G_3 \operatorname{Re} \mathbf{V} + G_4 \operatorname{Im} \mathbf{V}. \tag{15.23}$$

If we substitute the function h determined by Eq. (15.23) into Eq. (15.9), we obtain

$$\sum_{i=1}^{k} h\left(\sum_{i=1}^{m+n-1} \mathbf{T}_{i+j}, \ \mathbf{O}\right) + \sum_{i=k+1}^{m+n+k} h\left(-\sum_{j=0}^{k-1} \mathbf{T}_{m+n+i+j}, \ \mathbf{O}\right) \tag{15.24}$$

$$- \sum_{i=k+1}^{m+n+k} h\left(\sum_{j=0}^{m+k-1} \mathbf{T}_{m+n+i+j}, \ \mathbf{O}\right) - \sum_{i=1}^{m+k} h\left(-\sum_{j=0}^{n-1} \mathbf{T}_{i+j}, \ \mathbf{O}\right) = \mathbf{O}.$$

For $\mathbf{T}_1 = \mathbf{U}$, $\mathbf{T}_{m+k} = \mathbf{V}$ and $\mathbf{T}_j = \mathbf{O}$ ($j \neq 1, m + k$) from Eq. (15.24) we have

$$h(\mathbf{U} + \mathbf{U}, \mathbf{O}) = h(\mathbf{U}, \mathbf{O}) + h(\mathbf{V}, \mathbf{O}),$$

where

$$h(\mathbf{U}, \mathbf{O}) = G_1 \operatorname{Re} \mathbf{U} + G_2 \operatorname{Im} \mathbf{U}.$$

On the basis of this, the equation (15.23) becomes Eq. (15.24).

Let $n < m < k$ and $m + n < k$. On the basis of the equations (15.15) and (15.13) we obtain

$$h(\mathbf{U} + \mathbf{V}, \mathbf{O}) - h(\mathbf{O}, \mathbf{U} + \mathbf{V}) = h(\mathbf{U}, \mathbf{O}) - h(\mathbf{O}, \mathbf{U}) + h(\mathbf{V}, \mathbf{O}) - h(\mathbf{O}, \mathbf{V}),$$

from where it follows that

$$h(\mathbf{U}, \mathbf{O}) - h(\mathbf{O}, \mathbf{U}) = G_1' \operatorname{Re} \mathbf{U} + G_2'' \operatorname{Im} \mathbf{U}. \tag{15.25}$$

On the basis of the equality (15.25), the equation (15.15) becomes

$$h(\mathbf{U}, \mathbf{V}) = h(\mathbf{O}, \mathbf{U} + \mathbf{V}) + G' \operatorname{Re} \mathbf{U} + G'' \operatorname{Im} \mathbf{U}. \tag{15.26}$$

If we substitute the function h determined by Eq. (15.26) into Eq. (15.9), we obtain the following equation

$$\sum_{i=1}^{k} h\left(\mathbf{O}, \sum_{j=0}^{m+n+1} \mathbf{T}_{i+j}\right) = \sum_{i=k+1}^{m+n+k} h\left(\mathbf{O}, \sum_{j=0}^{k-1} \mathbf{T}_{m+n+i+j}\right). \tag{15.27}$$

By putting $\mathbf{T}_1 = \mathbf{U}$, $\mathbf{T}_{m+n} = \mathbf{V}$ and $\mathbf{T}_j = \mathbf{O}$ $(j \neq 1, m + n)$ in Eq. (15.27) we have

$$h(\mathbf{O}, \mathbf{U}, \mathbf{V}) = h(\mathbf{O}, \mathbf{U}) + h(\mathbf{O}, \mathbf{V}),$$

from which we conclude that

$$h(\mathbf{O}, \mathbf{U}) = G_3 \operatorname{Re} \mathbf{U} + G_4 \operatorname{Im} \mathbf{U}.$$

On the basis of this, the equation (15.26) becomes Eq. (15.21).

Let $n < m < k$ and $m + n > k$. On the basis of Eqs. (15.15) and (15.16), we obtain that the function h satisfies the functional equation

$$h(\mathbf{U} + \mathbf{V}, \mathbf{O}) + h(\mathbf{O}, -\mathbf{U} - \mathbf{V}) - h(\mathbf{O}, \mathbf{U} + \mathbf{V})$$

$$= h(\mathbf{U}, \mathbf{O}) + h(\mathbf{O}, -\mathbf{U}) - h(\mathbf{O}, \mathbf{U}) + h(\mathbf{V}, \mathbf{O}) + h(\mathbf{O}, -\mathbf{V}) - h(\mathbf{O}, \mathbf{V}).$$

Therefore, the function $h(\mathbf{U}, \mathbf{O}) + h(\mathbf{O}, -\mathbf{U}) - h(\mathbf{O}, \mathbf{U})$ is determined by

$$h(\mathbf{U}, \mathbf{O}) + h(\mathbf{O}, -\mathbf{U}) - h(\mathbf{O}, \mathbf{V}) = G_1 \operatorname{Re} \mathbf{U} + G_2 \operatorname{Im} \mathbf{U}. \tag{15.28}$$

The equation (15.15), on the basis of the equality (15.28), becomes

$$h(\mathbf{U}, \mathbf{V}) = h(\mathbf{O}, \mathbf{U} + \mathbf{V}) - h(\mathbf{O}, -\mathbf{U}) + G' \operatorname{Re} \mathbf{U} + G'' \operatorname{Im} \mathbf{U}. \tag{15.29}$$

The function h needs to satisfy the equation (15.9). If we substitute Eq. (15.29) into Eq. (15.9), we get

$$\sum_{i=1}^{k} h\left(\mathbf{O}, \sum_{j=0}^{m+n-1} \mathbf{T}_{i+j}\right) + \sum_{i=k+1}^{m+n+k} h\left(\mathbf{O}, -\sum_{j=0}^{k-1} \mathbf{T}_{m+n+i+j}\right) \quad (15.30)$$

$$- \sum_{i=1}^{k+1} h\left(\mathbf{O}, -\sum_{j=0}^{m-1} \mathbf{T}_{i+j}\right) - \sum_{i=n+k+1}^{m+n+k} h\left(\mathbf{O}, \sum_{j=0}^{m-1} \mathbf{T}_{m+n+i+j}\right) = \mathbf{O}.$$

If we put $\mathbf{T}_1 = \mathbf{U}$, $\mathbf{T}_{n+k} = \mathbf{V}$ and $\mathbf{T}_j = \mathbf{O}$ $(j \neq 1, n+k)$ into Eq. (15.30), we obtain

$$h(\mathbf{O}, \mathbf{U} + \mathbf{V}) = h(\mathbf{O}, \mathbf{U}) + h(\mathbf{O}, \mathbf{V}),$$

from which it follows that

$$h(\mathbf{O}, \mathbf{U}) = G_3 \mathrm{Re}\, \mathbf{U} + G_4 \mathrm{Im}\, \mathbf{U}.$$

On the basis of the last equality, the equality (15.29) has just the form Eq. (15.21). Consequently, in all four cases the function h is determined by Eq. (15.21).

We denote the arbitrary complex functions G_i $(1 \leq i \leq n)$ of $\mathbf{S}$ by $G_i(\mathbf{S})$ $(1 \leq i \leq 4)$. On the basis of that, the equality (15.21) and the transformations Eqs. (15.8), (15.6), (15.3) and (15.2) there follows the proof of the theorem. $\qquad \square$

Theorem 15.2 *If $a = 1$, $m \neq n$ and $m + n = k$, then the general continuous solution of the functional equation (15.1) is determined by*

$$\begin{aligned} f(\mathbf{U}, \mathbf{V}, \mathbf{W}) \;=\;& F\left(\mathbf{U} + \mathbf{V} - k\frac{\mathbf{U} + \mathbf{V} + \mathbf{W}}{m + n + k}, \, \mathbf{U} + \mathbf{V} + \mathbf{W}\right) \\ -\;& F\left(-\mathbf{U} - \mathbf{V} + k\frac{\mathbf{U} + \mathbf{V} + \mathbf{W}}{m + n + k}, \, \mathbf{U} + \mathbf{V} + \mathbf{W}\right) \\ +\;& G_1(\mathbf{U} + \mathbf{V} + \mathbf{W})\mathrm{Re}\left(\mathbf{U} - m\frac{\mathbf{U} + \mathbf{V} + \mathbf{W}}{m + n + k}\right) \\ +\;& G_2(\mathbf{U} + \mathbf{V} + \mathbf{W})\mathrm{Im}\left(\mathbf{U} - m\frac{\mathbf{U} + \mathbf{V} + \mathbf{W}}{m + n + k}\right), \end{aligned}$$

where $F : \mathcal{V}^2 \mapsto \mathcal{V}'$ and $G_i : \mathcal{V} \mapsto \mathcal{L}(\mathcal{V}^0, \mathcal{V}')$ $(i = 1, 2)$ are arbitrary continuous complex vector functions.

Proof.   Let $m < n$. As in the proof of the previous theorem, on the basis of the expressions Eqs. (15.13) and (15.15) we may show that the function h has the form determined by Eq. (15.18). If we substitute the function h determined as previously into Eq. (15.1), on the basis of the expression Eq. (15.12) we conclude that the equality is valid and, consequently, the function $h(\mathbf{O}, \mathbf{U})$ may be arbitrary. According to Eq. (15.12), for $h(\mathbf{O}, \mathbf{U})$ we can put

$$h(\mathbf{O}, \mathbf{U}) = F(\mathbf{U}) - F(-\mathbf{U}),$$

where F is an arbitrary continuous function.

On the basis of this, the equality (15.18) becomes

$$h(\mathbf{U}, \mathbf{V}) = F(\mathbf{U} + \mathbf{V}) - F(-\mathbf{U} - \mathbf{V}) + G_1 \operatorname{Re} \mathbf{U} + G_2 \operatorname{Im} \mathbf{U}. \qquad (15.31)$$

Let $n < m$. As in this case the equalities (15.13) and (15.15) hold and the function h has the form determined by Eq. (15.18). If we substitute Eq. (15.18) into Eq. (15.9), we obtain that the equation (15.9) holds, which means that $h(\mathbf{O}, \mathbf{U})$ is an arbitrary function for which Eq. (15.12) holds.

On the basis of this, we obtain that the function h in this case has the form determined by Eq. (15.31).

If we take into account that the arbitrary continuous complex vector functions G_1 and G_2 are $G_1(\mathbf{S})$ and $G_2(\mathbf{S})$ respectively, on the basis of the expressions Eqs. (15.8), (15.6), (15.3) and (15.2), there follows the proof of the theorem. $\square$

Theorem 15.3   *If $a = 1$ and $m < n = k$, the functional equation (15.1) has a general continuous solution determined by*

$$
\begin{aligned}
f(\mathbf{U}, \mathbf{V}, \mathbf{W}) \;=\; & F\left(\mathbf{U} + \mathbf{V} - (m+n)\frac{\mathbf{U} + \mathbf{V} + \mathbf{W}}{m+n+k}, \; \mathbf{U} + \mathbf{V} + \mathbf{W}\right) \\
& - \; F\left(-\mathbf{V} + n\frac{\mathbf{U} + \mathbf{V} + \mathbf{W}}{m+n+k}, \; \mathbf{U} + \mathbf{V} + \mathbf{W}\right) \\
& + \; G_1(\mathbf{U} + \mathbf{V} + \mathbf{W})\operatorname{Re}\left(\mathbf{V} - n\frac{\mathbf{U} + \mathbf{V} + \mathbf{W}}{m+n+k}\right) \\
& + \; G_2(\mathbf{U} + \mathbf{V} + \mathbf{W})\operatorname{Im}\left(\mathbf{V} - n\frac{\mathbf{U} + \mathbf{V} + \mathbf{W}}{m+n+k}\right),
\end{aligned}
$$

where $F : \mathcal{V}^2 \mapsto \mathcal{V}'$ and $G_i : \mathcal{V} \mapsto \mathcal{L}(\mathcal{V}^0, \mathcal{V}')$ $(i = 1, 2)$ are arbitrary continuous complex vector functions.

Proof. On the basis of Eqs. (15.13) and (15.16), as in the proof of Theorem 15.1, we conclude that Eq. (15.22) holds, *i.e.*, Eq. (15.23) is valid. If we substitute Eq. (15.23) into Eq. (15.9), since $n = k$, we obtain an identity which means that for $h(\mathbf{U}, \mathbf{O})$ we may take an arbitrary continuous function.

On the basis of this, for the function $f(\mathbf{U}, \mathbf{V})$ we obtain

$$h(\mathbf{U}, \mathbf{V}) = F(\mathbf{U} + \mathbf{V}) - F(-\mathbf{V}) + G_1 \operatorname{Re} \mathbf{V} + G_2 \operatorname{Im} \mathbf{V}. \qquad (15.32)$$

From Eq. (15.32), on the basis of the transformations Eqs. (15.8), (15.6), (15.3) and (15.2) and taking into account that G_1 and G_2 are arbitrary continuous functions of $\mathbf{S}$, we obtain that the function f has the form given in the Theorem 15.3. $\qquad \square$

Theorem 15.4 *If $a = 1$ and $n < m = k$, then the general continuous solution of the functional equation (15.1) is*

$$
\begin{aligned}
f(\mathbf{U}, \mathbf{V}, \mathbf{W}) \;=\; & F\left(\mathbf{U} + \mathbf{V} - (m + n)\frac{\mathbf{U} + \mathbf{V} + \mathbf{W}}{m + n + k}, \; \mathbf{U} + \mathbf{V} + \mathbf{W}\right) \\
& - \; F\left(-\mathbf{U} + m\frac{\mathbf{U} + \mathbf{V} + \mathbf{W}}{m + n + k}, \; \mathbf{U} + \mathbf{V} + \mathbf{W}\right) \\
& + \; G_1(\mathbf{U} + \mathbf{V} + \mathbf{W})\operatorname{Re}\left(\mathbf{U} - m\frac{\mathbf{U} + \mathbf{V} + \mathbf{W}}{m + n + k}\right) \\
& + \; G_2(\mathbf{U} + \mathbf{V} + \mathbf{W})\operatorname{Im}\left(\mathbf{U} - m\frac{\mathbf{U} + \mathbf{V} + \mathbf{W}}{m + n + k}\right),
\end{aligned}
$$

where $F : \mathcal{V}^2 \mapsto \mathcal{V}'$ and $G_i : \mathcal{V} \mapsto \mathcal{L}(\mathcal{V}^0, \mathcal{V}')$ $(i = 1, 2)$ are arbitrary continuous complex vector functions.

Proof. On the basis of Eqs. (15.15), (15.16) and (15.28), we obtain that the function h satisfies the equation (15.29).

If the function h determined by Eq. (15.29) is substituted into Eq. (15.9), then the equation (15.9) becomes an identity, which means that $h(\mathbf{O}, \mathbf{U})$ can be substituted by an arbitrary continuous function $F(\mathbf{U})$.

On the basis of this, the equality (15.29) has the form

$$h(\mathbf{U}, \mathbf{V}) = F(\mathbf{U} + \mathbf{V}) - F(-\mathbf{U}) + G_1 \operatorname{Re} \mathbf{U} + G_2 \operatorname{Im} \mathbf{U}. \qquad (15.33)$$

According to the transformations Eqs. (15.8), (15.6), (15.3) and (15.2), from the equation (15.33), by putting $G_1 = G_1(\mathbf{S})$ and $G_2 = G_2(\mathbf{S})$, it follows that the function f has the form which is given in the statement of this theorem. $\qquad\square$

Theorem 15.5   *If $a = 1$, $m = n < k$ and $2m \neq k$, then the general continuous solution of the functional equation (15.1) is determined by*

$$
\begin{aligned}
f(\mathbf{U}, \mathbf{V}, \mathbf{W}) \;=\;& F\left(\mathbf{U} - m\frac{\mathbf{U}+\mathbf{V}+\mathbf{W}}{m+n+k},\; \mathbf{U}+\mathbf{V}+\mathbf{W}\right) \\
-\;& F\left(\mathbf{V} - n\frac{\mathbf{U}+\mathbf{V}+\mathbf{W}}{m+n+k},\; \mathbf{U}+\mathbf{V}+\mathbf{W}\right) \\
+\;& G_1(\mathbf{U}+\mathbf{V}+\mathbf{W})\mathrm{Re}\left(\mathbf{U} - m\frac{\mathbf{U}+\mathbf{V}+\mathbf{W}}{m+n+k}\right) \\
+\;& G_2(\mathbf{U}+\mathbf{V}+\mathbf{W})\mathrm{Im}\left(\mathbf{U} - m\frac{\mathbf{U}+\mathbf{V}+\mathbf{W}}{m+n+k}\right),
\end{aligned}
$$

where $F : \mathcal{V}^2 \mapsto \mathcal{V}'$ and $G_i : \mathcal{V} \mapsto \mathcal{L}(\mathcal{V}^0, \mathcal{V}')$ $(i = 1, 2)$ are arbitrary continuous complex vector functions.

Proof.   Let $2m < k$. On the basis of the expression Eq. (15.17), we obtain

$$h(\mathbf{U},\mathbf{V}) + h(\mathbf{V},\mathbf{O}) + h(\mathbf{O},\, -\mathbf{U} - \mathbf{V}) + h(-\mathbf{U} - \mathbf{V},\, \mathbf{U}) = \mathbf{O},$$

$$h(-\mathbf{U} - \mathbf{V},\, \mathbf{U}) + h(\mathbf{U},\mathbf{O}) + h(\mathbf{O},\mathbf{V}) + h(\mathbf{V},\, -\mathbf{U} - \mathbf{V}) = \mathbf{O},$$

$$h(\mathbf{V},\, -\mathbf{U} - \mathbf{V}) + h(-\mathbf{U} - \mathbf{V},\, \mathbf{O}) + h(\mathbf{O},\mathbf{U}) + h(\mathbf{U},\mathbf{V}) = \mathbf{O}.$$

If from these three equalities we eliminate $h(-\mathbf{U} - \mathbf{V},\, \mathbf{U})$ and $h(\mathbf{V},\, -\mathbf{U} - \mathbf{V})$, we obtain

$$
\begin{aligned}
2h(\mathbf{U},\mathbf{V}) \;=\;& -h(-\mathbf{U} - \mathbf{V},\, \mathbf{O}) - h(\mathbf{O},\, -\mathbf{U} - \mathbf{V}) + h(\mathbf{U},\mathbf{O}) \\
-\;& h(\mathbf{O},\mathbf{U}) + h(\mathbf{O},\mathbf{V}) - h(\mathbf{V},\mathbf{O}).
\end{aligned}
$$

If we substitute this expression for h into Eq. (15.9), we get

$$
\sum_{i=1}^{k} h\left(-\sum_{j=0}^{m+n-1} \mathbf{T}_{i+j},\; \mathbf{O}\right) + \sum_{i=1}^{k} h\left(\mathbf{O},\; -\sum_{j=0}^{m+n-1} \mathbf{T}_{i+j}\right) \tag{15.34}
$$

$$
+ \sum_{i=k+1}^{m+n+k} h\left(\sum_{j=0}^{k-1} \mathbf{T}_{m+n+i+j},\; \mathbf{O}\right) + \sum_{i=k+1}^{m+n+k} h\left(\mathbf{O},\; \sum_{j=0}^{k-1} \mathbf{T}_{m+n+i+j}\right) = \mathbf{O}.
$$

If we put $\mathbf{T}_1 = \mathbf{U}$, $\mathbf{T}_{m+n} = \mathbf{V}$ and $\mathbf{T}_j = \mathbf{O}$ $(j \neq 1, m+n)$ into Eq. (15.34), we have

$$h(\mathbf{U} + \mathbf{V}, \mathbf{O}) + h(\mathbf{O}, \mathbf{U} + \mathbf{V}) + (k-1)[h(\mathbf{V}, \mathbf{O}) + h(\mathbf{O}, \mathbf{V})]$$
$$+ \quad h(-\mathbf{U}, \mathbf{O}) + h(\mathbf{O}, -\mathbf{U}) + k[h(-\mathbf{V}, \mathbf{O}) + h(\mathbf{O}, -\mathbf{V})] = \mathbf{O}. \quad (15.35)$$

For $\mathbf{V} = \mathbf{O}$, from Eq. (15.35) we obtain

$$h(-\mathbf{U}, \mathbf{O}) + h(\mathbf{O}, -\mathbf{U}) + h(\mathbf{U}, \mathbf{O}) + h(\mathbf{O}, \mathbf{U}) = \mathbf{O}, \qquad (15.36)$$

i.e., the equality (15.35) has the form

$$h(\mathbf{U} + \mathbf{V}, \mathbf{O}) + h(\mathbf{O}, \mathbf{U} + \mathbf{V}) = h(\mathbf{U}, \mathbf{O}) + h(\mathbf{O}, \mathbf{U}) + h(\mathbf{V}, \mathbf{O}) + h(\mathbf{O}, \mathbf{V}).$$

Hence, it follows that

$$h(\mathbf{U}, \mathbf{O}) + h(\mathbf{O}, \mathbf{U}) = G_1 \operatorname{Re} \mathbf{U} + G_2 \operatorname{Im} \mathbf{U}.$$

Thus we obtain that the function $h(\mathbf{U}, \mathbf{V})$ is determined by

$$h(\mathbf{U}, \mathbf{V}) = h(\mathbf{U}, \mathbf{O}) + h(\mathbf{O}, \mathbf{V}),$$

or

$$h(\mathbf{U}, \mathbf{V}) = F(\mathbf{U}) - F(\mathbf{V}) + G_1 \operatorname{Re} \mathbf{U} + G_2 \operatorname{Im} \mathbf{V}, \qquad (15.37)$$

where we have put $F(\mathbf{U}) = -h(\mathbf{O}, \mathbf{U})$.

Let $2m > k$. If we put $\mathbf{T}_1 = \mathbf{U}$, $\mathbf{T}_k = \mathbf{V}$ and $\mathbf{T}_j = \mathbf{O}$ $(j \neq 1, k)$ into Eq. (15.34), we get

$$h(\mathbf{U}, \mathbf{O}) + h(\mathbf{O}, \mathbf{U}) + (m+n)[h(\mathbf{V}, \mathbf{O}) + h(\mathbf{O}, \mathbf{V})] \quad (15.38)$$
$$+ \quad h(-\mathbf{U} - \mathbf{V}, \mathbf{O}) + h(\mathbf{O}, \mathbf{U} - \mathbf{V})$$
$$+ \quad (m+n-1)[h(-\mathbf{V}, \mathbf{O}) + h(\mathbf{O}, -\mathbf{V})] = \mathbf{O},$$

and hence for $\mathbf{V} = \mathbf{O}$ we obtain Eq. (15.36), *i.e.*, the equality (15.38) has the following form

$$h(\mathbf{U} + \mathbf{V}, \mathbf{O}) + h(\mathbf{O}, \mathbf{U} + \mathbf{V}) = h(\mathbf{U}, \mathbf{O}) + h(\mathbf{O}, \mathbf{U}) + h(\mathbf{V}, \mathbf{O}) + h(\mathbf{O}, \mathbf{V}).$$

Hence it follows that the function h has the form Eq. (15.37).

Thus, on the basis of the expressions Eqs. (15.37), (15.8), (15.6), (15.3) and (15.2), we conclude that the function f has the form given in the theorem. $\qquad\square$

Theorem 15.6   *If $a = 1$ and $2m = 2n = k$, then the general continuous solution of the functional equation (15.1) is given by*

$$
\begin{aligned}
f(\mathbf{U},\mathbf{V},\mathbf{W}) \;=\; & F(\mathbf{U},\ \mathbf{V}+\mathbf{W}) - F(\mathbf{V},\ \mathbf{W}+\mathbf{U}) \\
+\; & G(\mathbf{U}+\mathbf{V},\ \mathbf{W}) - G(\mathbf{W},\ \mathbf{U}+\mathbf{V}),
\end{aligned}
$$

where $F, G : \mathcal{V}^2 \mapsto \mathcal{V}'$ are arbitrary complex vector functions.

Proof.   From the equation (15.34), which holds in this case, there immediately follows the equality

$$
h(\mathbf{U},\mathbf{O}) + h(\mathbf{O},\mathbf{U}) + h(-\mathbf{U},\mathbf{O}) + h(\mathbf{O},-\mathbf{U}) = \mathbf{O},
$$

i.e., we may put

$$
h(\mathbf{U},\mathbf{O}) + h(\mathbf{O},\mathbf{U}) = 2G(\mathbf{U}) - 2G(-\mathbf{U}), \tag{15.39}
$$

where G is an arbitrary complex vector function.

Since by virtue of Eq. (15.14) it holds

$$
\begin{aligned}
2h(\mathbf{U},\mathbf{V}) \;=\; & -h(-\mathbf{U}-\mathbf{V},\ \mathbf{O}) - h(\mathbf{O},\ -\mathbf{U}-\mathbf{V}) + h(\mathbf{U},\mathbf{O}) \\
-\; & h(\mathbf{O},\mathbf{U}) + h(\mathbf{O},\mathbf{V}) - h(\mathbf{V},\mathbf{O}),
\end{aligned}
$$

on the basis of the expression Eq. (16.39) we have

$$
h(\mathbf{U},\mathbf{V}) = F(\mathbf{U}) - F(\mathbf{V}) + G(\mathbf{U}+\mathbf{V}) - G(-\mathbf{U}-\mathbf{V}), \tag{15.40}
$$

where we introduced the notation

$$
2F(\mathbf{U}) = h(\mathbf{U},\mathbf{O}) - h(\mathbf{O},\mathbf{U}).
$$

From the equalities (15.40), (15.8), (15.6), (15.3) and (15.2) there follows the proof of the theorem. $\square$

Theorem 15.7   *If $a = 1$ and $m = n = k$, then the general continuous solution of the functional equation (15.1) is given by*

$$
f(\mathbf{U},\mathbf{V},\mathbf{W}) = F(\mathbf{U},\mathbf{V},\mathbf{W}) - F(\mathbf{V},\mathbf{W},\mathbf{U}),
$$

where $F : \mathcal{V}^3 \mapsto \mathcal{V}'$ is an arbitrary complex vector function.

Proof.   If we put $\mathbf{T}_1 = \mathbf{U}$, $\mathbf{T}_{m+1} = \mathbf{V}$ and $\mathbf{T}_j = \mathbf{O}$ $(j \neq 1, m+1)$ into Eq. (15.9), then we obtain

$$
h(\mathbf{U},\mathbf{V}) + h(\mathbf{V},\ -\mathbf{U}-\mathbf{V}) + h(-\mathbf{U}-\mathbf{V},\ \mathbf{U}) = \mathbf{O}.
$$

According to [M. Ghermănescu (1940)] the general solution of this equation is given by

$$h(\mathbf{U}, \mathbf{V}) = F(\mathbf{U}, \mathbf{V}) - F(\mathbf{V}, -\mathbf{U} - \mathbf{V}), \qquad (15.41)$$

where $F : \mathcal{V}^2 \mapsto \mathcal{V}'$ is an arbitrary complex vector function.

On the basis of the expressions Eqs. (15.41), (15.8), (15.6), (15.3) and (15.2), we conclude that the function f is determined by the form given in the theorem. $\qquad\square$

II. Let $a^{m+n+k} \neq 1$. If we put

$$f(\mathbf{U}, \mathbf{V}, \mathbf{W}) = \qquad (15.42)$$

$$g(\mathbf{U} + a^{k+m}\mathbf{V} + a^m\mathbf{W}, \ \mathbf{V} + a^{m+n}\mathbf{W} + a^n\mathbf{U}, \ \mathbf{W} + a^{n+k}\mathbf{U} + a^k\mathbf{V}),$$

then the functional equation (15.1) becomes

$$\sum_{i=1}^{m+n+k} g\left(\sum_{j=0}^{m-1} a^{m-1-j}\mathbf{Z}_{i+j} \right.$$

$$+ \sum_{j=0}^{n-1} a^{m+n+k-1-j}\mathbf{Z}_{m+i+j} + \sum_{j=0}^{k-1} a^{m+k-1-j}\mathbf{Z}_{m+n+i+j},$$

$$\sum_{j=0}^{n-1} a^{n-1-j}\mathbf{Z}_{m+i+j} + \sum_{j=0}^{k-1} a^{m+n+k-1-j}\mathbf{Z}_{m+n+i+j} + \sum_{j=0}^{m-1} a^{m+n-1-j}\mathbf{Z}_{i+j},$$

$$\sum_{j=0}^{k-1} a^{k-1-j}\mathbf{Z}_{m+n+i+j}$$

$$\left. + \sum_{j=0}^{n-1} a^{m+n+k-1-j}\mathbf{Z}_{i+j} + \sum_{j=0}^{n-1} a^{n+k-1-j}\mathbf{Z}_{m+i+j} \right) = \mathbf{O},$$

i.e.,

$$\sum_{i=0}^{m+n+k} g\left(\sum_{j=0}^{m+n+k-1} a^j \mathbf{Z}_{m-1+i-j}, \ \sum_{j=0}^{m+n+k-1} a^j \mathbf{Z}_{m+n-1+i-j}, \right.$$

$$\left. \sum_{j=0}^{m+n+k-1} a^j \mathbf{Z}_{m+n+k-1+i-j} \right) = \mathbf{O}.$$

This transformation of the equation (15.1) is possible since $a^{m+n+k} \neq 1$. If we introduce new variables $\mathbf{T}_i$ by the relations

$$\mathbf{T}_i = \sum_{j=0}^{m+n+k-1} a^j \mathbf{Z}_{m-1+i-j} \quad (1 \leq i \leq m+n+k),$$

then the previous equation becomes

$$\sum_{i=1}^{m+n+k} g(\mathbf{T}_i, \mathbf{T}_{i+n}, \mathbf{T}_{i+n+k}) = \mathbf{O}. \tag{15.43}$$

Now we will give the following results.

Lemma 15.8   *The general solution of the functional equation (15.43) is determined by*

$1^0.$ $g(\mathbf{U}, \mathbf{V}, \mathbf{W}) = F(\mathbf{U}) - F(\mathbf{W}) - H(\mathbf{V}) - H(\mathbf{W})$

$$(m \neq n \neq k \neq m, \quad m \neq n+k, \quad n \neq m+k, \quad k \neq m+n),$$

$2^0.$ $g(\mathbf{U}, \mathbf{V}, \mathbf{W}) = F(\mathbf{U}) - F(\mathbf{V}) - G(\mathbf{U}, \mathbf{W}) - G(\mathbf{W}, \mathbf{U})$

$$(m \neq n \neq k \neq m, \quad m = n+k),$$

$3^0.$ $g(\mathbf{U}, \mathbf{V}, \mathbf{W}) = F(\mathbf{W}) - F(\mathbf{U}) - G(\mathbf{U}, \mathbf{V}) - G(\mathbf{V}, \mathbf{U})$

$$(m \neq n \neq k \neq m, \quad n = m+k),$$

$4^0.$ $g(\mathbf{U}, \mathbf{V}, \mathbf{W}) = F(\mathbf{U}) - F(\mathbf{V}) + G(\mathbf{V}, \mathbf{W}) - G(\mathbf{W}, \mathbf{V})$

$$(m \neq n \neq k \neq m, \quad k = m+n),$$

$5^0.$ $g(\mathbf{U}, \mathbf{V}, \mathbf{W}) = G(\mathbf{U}, \mathbf{V}) - G(\mathbf{V}, \mathbf{W}) \quad\quad (m \neq n = k, \quad m \neq n+k),$

$6^0.$ $g(\mathbf{U}, \mathbf{V}, \mathbf{W}) = K(\mathbf{U}, \mathbf{V}) - K(\mathbf{V}, \mathbf{W}) + G(\mathbf{W}, \mathbf{U}) - G(\mathbf{U}, \mathbf{W})$

$$(m \neq n = k, \quad m = n+k),$$

$7^0.$ $g(\mathbf{U}, \mathbf{V}, \mathbf{W}) = G(\mathbf{U}, \mathbf{V}) - G(\mathbf{W}, \mathbf{U}) \quad\quad (m = n \neq k, \quad k \neq m+n),$

$8^0.$ $g(\mathbf{U}, \mathbf{V}, \mathbf{W}) = K(\mathbf{U}, \mathbf{V}) - K(\mathbf{W}, \mathbf{U}) + G(\mathbf{V}, \mathbf{W}) - G(\mathbf{W}, \mathbf{V})$

$$(m = n \neq k, \quad k = m+n),$$

9^0. $\quad g(\mathbf{U}, \mathbf{V}, \mathbf{W}) = G(\mathbf{V}, \mathbf{W}) - G(\mathbf{W}, \mathbf{U}) \quad (m = k \neq n, \quad n \neq k + m)$,

10^0. $\quad g(\mathbf{U}, \mathbf{V}, \mathbf{W}) = K(\mathbf{U}, \mathbf{V}) - K(\mathbf{V}, \mathbf{U}) + G(\mathbf{V}, \mathbf{W}) - G(\mathbf{W}, \mathbf{U})$

$$(m = k \neq n, \quad n = k + m),$$

11^0. $\quad g(\mathbf{U}, \mathbf{V}, \mathbf{W}) = L(\mathbf{U}, \mathbf{V}, \mathbf{W}) - L(\mathbf{V}, \mathbf{W}, \mathbf{U}) \quad (m = n = k)$,

where $F, H : \mathcal{V} \mapsto \mathcal{V}'$, $G, K : \mathcal{V}^2 \mapsto \mathcal{V}'$ and $L : \mathcal{V}^3 \mapsto \mathcal{V}'$ are arbitrary complex vector functions.

Proof. Since the proofs of the particular cases are similar or completely the same, we will prove the lemma only in the cases 3^0 and 8^0.

3^0. If all variables in Eq. (15.43), except for $\mathbf{Z}_i$, $\mathbf{Z}_{i+n}$ and $\mathbf{Z}_{i+n+k}$, are equal to some constant, we obtain

$$g(\mathbf{Z}_i, \mathbf{Z}_{i+n}, \mathbf{Z}_{i+n+k}) = F(\mathbf{Z}_{i+n+k}) + K(\mathbf{Z}_i, \mathbf{Z}_{i+n}),$$

i.e.,

$$g(\mathbf{U}, \mathbf{V}, \mathbf{W}) = F(\mathbf{W}) + K(\mathbf{U}, \mathbf{V}). \tag{15.44}$$

Hence, the equation (15.43) becomes

$$\sum_{i=1}^{m+n+k} F(\mathbf{Z}_i) + \sum_{i=1}^{m+n+k} K(\mathbf{Z}_i, \mathbf{Z}_{i+n}) = \mathbf{O}.$$

If we put $\mathbf{Z}_r = \mathbf{O}$ $(r \neq i, i + n)$ in the above equation, we have

$$F(\mathbf{Z}_i) + F(\mathbf{Z}_{i+n}) + K(\mathbf{Z}_i, \mathbf{Z}_{i+n}) + K(\mathbf{Z}_{i+n}, \mathbf{Z}_i) = \mathbf{O},$$

and thus it follows that

$$K(\mathbf{U}, \mathbf{V}) = G(\mathbf{U}, \mathbf{V}) - G(\mathbf{V}, \mathbf{U}) - F(\mathbf{U}).$$

Now, the equality (15.44) becomes

$$g(\mathbf{U}, \mathbf{V}, \mathbf{W}) = F(\mathbf{W}) - F(\mathbf{U}) + G(\mathbf{U}, \mathbf{V}) - G(\mathbf{V}, \mathbf{U}).$$

8^0. If we put $\mathbf{Z}_r = \mathbf{O}$ $(r \neq i, i + n, i + n + k)$ into Eq. (15.43), then the equation (15.43) becomes

$$g(\mathbf{Z}_i, \mathbf{Z}_{i+n}, \mathbf{Z}_{i+n+k}) = K(\mathbf{Z}_i, \mathbf{Z}_{i+n}) + K_1(\mathbf{Z}_{i+n}, \mathbf{Z}_{i+n+k}) + K_2(\mathbf{Z}_{i+n+k}, \mathbf{Z}_i),$$

i.e.,

$$g(\mathbf{U}, \mathbf{V}, \mathbf{W}) = K(\mathbf{U}, \mathbf{V}) + K_1(\mathbf{V}, \mathbf{W}) + K_2(\mathbf{W}, \mathbf{U}),$$

where K, K_1, $K_2 : \mathcal{V}^2 \mapsto \mathcal{V}'$ are arbitrary complex vector functions.

By a substitution of the expression obtained for g into Eq. (15.43), we get

$$\sum_{i=1}^{m+n+k} K(\mathbf{Z}_i, \mathbf{Z}_{i+n}) + \sum_{i=1}^{m+n+k} K_1(\mathbf{Z}_i, \mathbf{Z}_{i+n}) + \sum_{i=1}^{m+n+k} K_2(\mathbf{Z}_i, \mathbf{Z}_{i+n}) = \mathbf{O}. \tag{15.45}$$

Hence, for $\mathbf{Z}_r$ $(r \neq i, i + k)$ we obtain

$$K_2(\mathbf{Z}_i, \mathbf{Z}_{i+k}) + K_2(\mathbf{Z}_{i+k}, \mathbf{Z}_i) + H(\mathbf{Z}_i) + H(\mathbf{Z}_{i+k}) = \mathbf{O},$$

where $H : \mathcal{V} \mapsto \mathcal{V}'$ is an arbitrary complex vector function.

From the above equation it follows that the function K_2 has the form

$$K_2(\mathbf{U}, \mathbf{V}) = G(\mathbf{U}, \mathbf{V}) - G(\mathbf{V}, \mathbf{U}) - H(\mathbf{U}), \tag{15.46}$$

where $G : \mathcal{V}^2 \mapsto \mathcal{V}'$ is an arbitrary complex vector function.

If we put Eq. (15.46) into Eq. (15.45), we obtain

$$\sum_{i=1}^{m+n+k} [K(\mathbf{Z}_i, \mathbf{Z}_{i+n}) + K_1(\mathbf{Z}_i, \mathbf{Z}_{i+n}) + H(\mathbf{Z}_i)] = \mathbf{O}. \tag{15.47}$$

For $\mathbf{Z}_r = \mathbf{O}$ $(r \neq i, i + n)$ from the above equation we obtain

$$K_1(\mathbf{U}, \mathbf{V}) = -K(\mathbf{U}, \mathbf{V}) - 2H(\mathbf{U}) - 2H(\mathbf{V}).$$

By putting the above obtained expression for K_1 into Eq. (15.47), we have

$$\sum_{i=1}^{m+n+k} H(\mathbf{Z}_i) = \mathbf{O},$$

i.e.,

$$H(\mathbf{U}) = \mathbf{O}.$$

On the basis of the previous equalities, the function g is determined by

$$g(\mathbf{U}, \mathbf{V}, \mathbf{W}) = K(\mathbf{U}, \mathbf{V}) - K(\mathbf{W}, \mathbf{U}) + G(\mathbf{V}, \mathbf{W}) - G(\mathbf{W}, \mathbf{V}).$$

$\square$

According to the previous lemma and the transformation Eq. (15.42), we obtain the following theorems.

Theorem 15.9 *If $a^{m+n+k} \neq 1$, $m \neq n \neq k \neq m$, $m \neq n+k$, $n \neq m+k$ and $k \neq m+n$, then the general solution of the functional equation (15.1) is*

$$
\begin{aligned}
f(\mathbf{U}, \mathbf{V}, \mathbf{W}) \;=\; & F(\mathbf{U} + a^{k+m}\mathbf{V} + a^{m}\mathbf{W}) - F(\mathbf{W} + a^{n+k}\mathbf{U} + a^{k}\mathbf{V}) \\
+\; & G(\mathbf{V} + a^{m+n}\mathbf{W} + a^{n}\mathbf{U}) - G(\mathbf{W} + a^{n+k}\mathbf{U} + a^{k}\mathbf{V}),
\end{aligned}
$$

where $F, G : \mathcal{V} \mapsto \mathcal{V}'$ are arbitrary complex vector functions.

Theorem 15.10 *If $a^{m+n+k} \neq 1$, $m \neq n \neq k \neq m$ and $m = n + k$, then the general solution of the functional equation (15.1) is*

$$
\begin{aligned}
f(\mathbf{U}, \mathbf{V}, \mathbf{W}) \;=\; & F(\mathbf{U} + a^{k+m}\mathbf{V} + a^{m}\mathbf{W}) - F(\mathbf{V} + a^{m+n}\mathbf{W} + a^{n}\mathbf{U}) \\
+\; & G(\mathbf{U} + a^{k+m}\mathbf{V} + a^{m}\mathbf{W}, \; \mathbf{W} + a^{n+k}\mathbf{U} + a^{k}\mathbf{V}) \\
-\; & G(\mathbf{W} + a^{n+k}\mathbf{U} + a^{k}\mathbf{V}, \; \mathbf{U} + a^{k+n}\mathbf{V} + a^{m}\mathbf{W}),
\end{aligned}
$$

where $F : \mathcal{V} \mapsto \mathcal{V}'$ and $G : \mathcal{V}^2 \mapsto \mathcal{V}'$ are arbitrary complex vector functions.

Theorem 15.11 *If $a^{m+n+k} \neq 1$, $m \neq n \neq k \neq m$ and $n = m + k$, then the general solution of the functional equation (15.1) is*

$$
\begin{aligned}
f(\mathbf{U}, \mathbf{V}, \mathbf{W}) \;=\; & F(\mathbf{W} + a^{n+k}\mathbf{U} + a^{k}\mathbf{V}) - F(\mathbf{U} + a^{k+m}\mathbf{V} + a^{m}\mathbf{W}) \\
+\; & \dot{G}(\mathbf{U} + a^{k+m}\mathbf{V} + a^{m}\mathbf{W}, \; \mathbf{V} + a^{m+n}\mathbf{W} + a^{n}\mathbf{U}) \\
-\; & G(\mathbf{V} + a^{m+n}\mathbf{W} + a^{n}\mathbf{U}, \; \mathbf{U} + a^{k+m}\mathbf{V} + a^{m}\mathbf{W}),
\end{aligned}
$$

where $F : \mathcal{V} \mapsto \mathcal{V}'$ and $G : \mathcal{V}^2 \mapsto \mathcal{V}'$ are arbitrary complex vector functions.

Theorem 15.12 *If $a^{m+n+k} \neq 1$, $m \neq n \neq k \neq m$ and $k = m + n$, then the functional equation (15.1) has a general solution*

$$
\begin{aligned}
f(\mathbf{U}, \mathbf{V}, \mathbf{W}) \;=\; & F(\mathbf{U} + a^{k+m}\mathbf{V} + a^{m}\mathbf{W}) - F(\mathbf{V} + a^{m+n}\mathbf{W} + a^{n}\mathbf{U}) \\
+\; & G(\mathbf{V} + a^{m+n}\mathbf{W} + a^{n}\mathbf{U}, \; \mathbf{W} + a^{n+k}\mathbf{U} + a^{k}\mathbf{V}) \\
-\; & G(\mathbf{W} + a^{n+k}\mathbf{U} + a^{k}\mathbf{V}, \; \mathbf{V} + a^{m+n}\mathbf{W} + a^{n}\mathbf{U}),
\end{aligned}
$$

where $F : \mathcal{V} \mapsto \mathcal{V}'$ and $G : \mathcal{V}^2 \mapsto \mathcal{V}'$ are arbitrary complex vector functions.

Theorem 15.13   *If $a^{m+n+k} \neq 1$ and $m \neq n+k$, then the general solution of the functional equation (15.1) is*

$$
\begin{aligned}
f(\mathbf{U}, \mathbf{V}, \mathbf{W}) \;=\;& F(\mathbf{U} + a^{k+m}\mathbf{V} + a^m\mathbf{W},\; \mathbf{V} + a^{m+n}\mathbf{W} + a^n\mathbf{U}) \\
&- F(\mathbf{V} + a^{m+n}\mathbf{W} + a^n\mathbf{U},\; \mathbf{U} + a^{k+m}\mathbf{V} + a^m\mathbf{W}),
\end{aligned}
$$

where $F : \mathcal{V}^2 \mapsto \mathcal{V}'$ is an arbitrary complex vector function.

Theorem 15.14   *If $a^{m+n+k} \neq 1$, $m \neq n = k$ and $m = 2n$, then the functional equation (15.1) has a general solution determined by*

$$
\begin{aligned}
f(\mathbf{U}, \mathbf{V}, \mathbf{W}) \;=\;& F(\mathbf{U} + a^{k+m}\mathbf{V} + a^m\mathbf{W},\; \mathbf{V} + a^{m+n}\mathbf{W} + a^n\mathbf{U}) \\
&- F(\mathbf{V} + a^{m+n}\mathbf{W} + a^n\mathbf{U},\; \mathbf{W} + a^{n+k}\mathbf{U} + a^k\mathbf{V}) \\
&+ G(\mathbf{W} + a^{n+k}\mathbf{U} + a^k\mathbf{V},\; \mathbf{U} + a^{k+m}\mathbf{V} + a^m\mathbf{W}) \\
&- G(\mathbf{U} + a^{k+m}\mathbf{V} + a^m\mathbf{W},\; \mathbf{W} + a^{n+k}\mathbf{U} + a^k\mathbf{V}),
\end{aligned}
$$

where $F, G : \mathcal{V}^2 \mapsto \mathcal{V}'$ are arbitrary complex vector functions.

Theorem 15.15   *If $a^{m+n+k} \neq 1$, $m = n \neq k$ and $k \neq m + n$, then the functional equation (15.1) has a general solution given by*

$$
\begin{aligned}
f(\mathbf{U}, \mathbf{V}, \mathbf{W}) \;=\;& F(\mathbf{U} + a^{k+m}\mathbf{V} + a^m\mathbf{W},\; \mathbf{V} + a^{m+n}\mathbf{W} + a^n\mathbf{U}) \\
&- F(\mathbf{W} + a^{n+k}\mathbf{U} + a^k\mathbf{V},\; \mathbf{U} + a^{k+m}\mathbf{V} + a^m\mathbf{W}),
\end{aligned}
$$

where $F : \mathcal{V}^2 \mapsto \mathcal{V}'$ is an arbitrary complex vector function.

Theorem 15.16   *If $a^{m+n+k} \neq 1$, $m = n \neq k$ and $k = m + n$, then the functional equation (15.1) has a general solution*

$$
\begin{aligned}
f(\mathbf{U}, \mathbf{V}, \mathbf{W}) \;=\;& F(\mathbf{U} + a^{k+m}\mathbf{V} + a^m\mathbf{W},\; \mathbf{V} + a^{m+n}\mathbf{W} + a^n\mathbf{U}) \\
&- F(\mathbf{W} + a^{n+k}\mathbf{U} + a^k\mathbf{V},\; \mathbf{U} + a^{k+m}\mathbf{V} + a^m\mathbf{W}) \\
&+ G(\mathbf{V} + a^{m+n}\mathbf{W} + a^n\mathbf{U},\; \mathbf{W} + a^{n+k}\mathbf{U} + a^k\mathbf{V}) \\
&- G(\mathbf{W} + a^{n+k}\mathbf{U} + a^k\mathbf{V},\; \mathbf{V} + a^{m+n}\mathbf{W} + a^n\mathbf{U}),
\end{aligned}
$$

where $F, G : \mathcal{V}^2 \mapsto \mathcal{V}'$ are arbitrary complex vector functions.

Theorem 15.17   *If $a^{m+n+k} \neq 1$, $m = k \neq n$ and $n \neq m + k$, then the functional equation (15.1) has a general solution*

$$
\begin{aligned}
f(\mathbf{U}, \mathbf{V}, \mathbf{W}) \;=\;& F(\mathbf{V} + a^{m+n}\mathbf{W} + a^n\mathbf{U},\; \mathbf{W} + a^{n+k}\mathbf{U} + a^k\mathbf{V}) \\
&- F(\mathbf{W} + a^{n+k}\mathbf{U} + a^k\mathbf{V},\; \mathbf{U} + a^{k+m}\mathbf{V} + a^m\mathbf{W}),
\end{aligned}
$$

where $F : \mathcal{V}^2 \mapsto \mathcal{V}'$ is an arbitrary complex vector function.

Theorem 15.18   *If $a^{m+n+k} \neq 1$, $m = k \neq n$ and $n = m + k$, then the functional equation (15.1) has a solution given by*

$$
\begin{aligned}
f(\mathbf{U}, \mathbf{V}, \mathbf{W}) \;=\;& F(\mathbf{U} + a^{k+m}\mathbf{V} + a^m\mathbf{W}, \; \mathbf{V} + a^{m+n}\mathbf{W} + a^n\mathbf{U}) \\
-\;& F(\mathbf{V} + a^{m+n}\mathbf{W} + a^n\mathbf{U}, \; \mathbf{U} + a^{k+m}\mathbf{V} + a^m\mathbf{W}) \\
+\;& G(\mathbf{V} + a^{m+n}\mathbf{W} + a^n\mathbf{U}, \; \mathbf{W} + a^{n+k}\mathbf{U} + a^k\mathbf{V}) \\
-\;& G(\mathbf{W} + a^{n+k}\mathbf{U} + a^k\mathbf{V}, \; \mathbf{U} + a^{m+n}\mathbf{V} + a^n\mathbf{W}),
\end{aligned}
$$

where $F, G : \mathcal{V}^2 \mapsto \mathcal{V}'$ are arbitrary complex vector functions.

Theorem 15.19   *If $a^{m+n+k} \neq 1$, $m = n = k$, then the general solution of the functional equation (15.1) is*

$$
\begin{aligned}
& f(\mathbf{U}, \mathbf{V}, \mathbf{W}) \\
=\;& F(\mathbf{U} + a^{k+m}\mathbf{V} + a^m\mathbf{W}, \; \mathbf{V} + a^{m+n}\mathbf{W} + a^n\mathbf{U}, \; \mathbf{W} + a^{n+k}\mathbf{U} + a^k\mathbf{V}) \\
-\;& F(\mathbf{V} + a^{m+n}\mathbf{W} + a^n\mathbf{U}, \; \mathbf{W} + a^{n+k}\mathbf{U} + a^k\mathbf{V}, \; \mathbf{U} + a^{k+m}\mathbf{V} + a^m\mathbf{W}),
\end{aligned}
$$

where $F : \mathcal{V}^3 \mapsto \mathcal{V}'$ is an arbitrary complex vector function.

III. If $a^{m+n+k} = 1$, this case is very difficult and up to now we are not able to solve the functional equation (15.1).

16   General Expanded Parametric Functional Equation

All notations for the vectors are the same as in Sec. 13.

The general expanded parametric functional equation [I. B. Risteski *et al.* (2001A)]

$$
\sum_{i=1}^{m+n+k} f_i \left(\sum_{j=0}^{m-1} a^{m-1-j} \mathbf{Z}_{i+j}, \; \sum_{j=0}^{n-1} a^{n-1-j} \mathbf{Z}_{i+m+j}, \right. \tag{16.1}
$$

$$
\left. \sum_{j=0}^{k-1} a^{k-1-j} \mathbf{Z}_{i+m+n+j} \right) = \mathbf{O}
$$

$$
(f_{m+n+k+i} \equiv f_i, \; \mathbf{Z}_{m+n+k+i} \equiv \mathbf{Z}_i)
$$

where a is complex number, will be solved here.

Further, we will consider the solving of the functional equation (16.1). For the equation (16.1) we will determine the general solution only if $a^{m+n+k} \neq 1$.

Now, we transform Eq. (16.1). If we put

$$f_i(\mathbf{U}, \mathbf{V}, \mathbf{W}) = \tag{16.2}$$

$$g_i(\mathbf{U} + a^{k+m}\mathbf{V} + a^m\mathbf{W}, \ \mathbf{V} + a^{m+n}\mathbf{W} + a^n\mathbf{U}, \ \mathbf{W} + a^{n+k}\mathbf{U} + a^k\mathbf{V}),$$

then the equation (16.1) becomes

$$\sum_{i=1}^{m+n+k} g_i \left(\sum_{j=0}^{m-1} a^{m-1-j}\mathbf{Z}_{i+j} + \sum_{j=0}^{n-1} a^{m+n+k-1-j}\mathbf{Z}_{m+i+j} \right.$$

$$+ \sum_{j=0}^{k-1} a^{m+k-1-j}\mathbf{Z}_{m+n+i+j},$$

$$\sum_{j=0}^{n-1} a^{n-1-j}\mathbf{Z}_{m+i+j} + \sum_{j=0}^{k-1} a^{m+n+k-1-j}\mathbf{Z}_{m+n+i+j} + \sum_{j=0}^{m-1} a^{m+n-1-j}\mathbf{Z}_{i+j},$$

$$\sum_{j=0}^{k-1} a^{k-1-j}\mathbf{Z}_{m+n+i+j} + \sum_{j=0}^{m-1} a^{m+n+k-1-j}\mathbf{Z}_{i+j}$$

$$\left. + \sum_{j=0}^{n-1} a^{n+k-1-j}\mathbf{Z}_{m+i+j} \right) = \mathbf{O},$$

i.e.,

$$\sum_{i=1}^{m+n+k} g_i \left(\sum_{j=0}^{m+n+k-1} a^j\mathbf{Z}_{m-1+i-j}, \ \sum_{j=0}^{m+n+k-1} a^j\mathbf{Z}_{m+n-1+i-j}, \right. \tag{16.3}$$

$$\left. \sum_{j=0}^{m+n+k-1} a^j\mathbf{Z}_{m+n+k-1+i-j} \right) = \mathbf{O}.$$

Since the linear forms

$$\mathbf{T}_i = \sum_{j=0}^{m+n+k-1} a^j\mathbf{Z}_{m-1+i-j} \quad (1 \leq i \leq m+n+k)$$

are linearly independent, it is possible to introduce new variables $\mathbf{T}_i$ which are determined by the above relations.

Therefore, the functional equation (16.3) takes the following form

$$\sum_{i=1}^{m+n+k} g_i(\mathbf{T}_i, \mathbf{T}_{i+n}, \mathbf{T}_{i+n+k}) = \mathbf{O}. \tag{16.4}$$

For the last equation the following lemma holds.

Lemma 16.1 *The general solution of the functional equation* (16.4) *is given by the equalities*

$1^0.\quad g_i(\mathbf{U}, \mathbf{V}, \mathbf{W}) = F_i(\mathbf{U}) - F_{i+n+k}(\mathbf{W}) - H_i(\mathbf{V}) - H_{i+k}(\mathbf{W}) - A_i,$

$$\sum_{i=1}^{m+n+k} A_i = \mathbf{O}$$

$$(m \neq n \neq k \neq m, \quad m \neq n+k, \quad n \neq m+k, \quad k \neq m+n),$$

$2^0.\quad g_i(\mathbf{U}, \mathbf{V}, \mathbf{W}) = F_i(\mathbf{V}) - F_{i+k}(\mathbf{W}) + G_i(\mathbf{U}, \mathbf{W}) \quad (1 \leq i \leq m),$

$g_i(\mathbf{U}, \mathbf{V}, \mathbf{W}) = F_i(\mathbf{V}) - F_{i+k}(\mathbf{W}) - G_{i+m}(\mathbf{W}, \mathbf{U}) - A_{i+m} \quad (m+1 \leq i \leq 2m),$

$$\sum_{i=1}^{m} A_i = \mathbf{O}$$

$$(m \neq n \neq k \neq m, \quad m \neq n+k),$$

$3^0.\quad g_i(\mathbf{U}, \mathbf{V}, \mathbf{W}) = F_i(\mathbf{W}) - F_{i+m}(\mathbf{U}) + G_i(\mathbf{U}, \mathbf{V}) \quad (1 \leq i \leq n),$

$g_i(\mathbf{U}, \mathbf{V}, \mathbf{W}) = F_i(\mathbf{W}) - F_{i+m}(\mathbf{U}) - G_{i+m}(\mathbf{V}, \mathbf{U}) - A_{i+m} \quad (n+1 \leq i \leq 2n),$

$$\sum_{i=1}^{n} A_i = \mathbf{O}$$

$$(m \neq n \neq k \neq m, \quad n = m+k),$$

$4^0.\quad g_i(\mathbf{U}, \mathbf{V}, \mathbf{W}) = F_i(\mathbf{U}) - F_{i+n}(\mathbf{V}) + G_i(\mathbf{V}, \mathbf{W}) \quad (1 \leq i \leq k),$

$g_i(\mathbf{U}, \mathbf{V}, \mathbf{W}) = F_i(\mathbf{U}) - F_{i+n}(\mathbf{V}) - G_{i+k}(\mathbf{W}, \mathbf{V}) - A_{i+k} \quad (k+1 \leq i \leq 2k),$

$$\sum_{i=1}^{n} A_i = \mathbf{O}$$

$$(m \neq n \neq k \neq m, \quad k = m + n),$$

$5^0.$ $g_i(\mathbf{U}, \mathbf{V}, \mathbf{W}) = G_i(\mathbf{U}, \mathbf{V}) - G_{i+n}(\mathbf{V}, \mathbf{W}) + A_i \quad (1 \leq i \leq m+n+k),$

$$\sum_{i=1}^{m+n+k} A_i = \mathbf{O}$$

$$(m \neq n = k, \quad m \neq n + k),$$

$6^0.$ $g_i(\mathbf{U}, \mathbf{V}, \mathbf{W}) = K_i(\mathbf{U}, \mathbf{V}) - K_{i+n}(\mathbf{V}, \mathbf{W}) + G_i(\mathbf{W}, \mathbf{U}) \quad (1 \leq i \leq m),$

$$g_i(\mathbf{U}, \mathbf{V}, \mathbf{W}) = K_i(\mathbf{U}, \mathbf{V}) - K_{i+n}(\mathbf{V}, \mathbf{W}) - G_{i+m}(\mathbf{U}, \mathbf{W}) \quad (m+1 \leq i \leq 2m),$$

$$(m \neq n = k, \quad m = n + k),$$

$7^0.$ $g_i(\mathbf{U}, \mathbf{V}, \mathbf{W}) = G_i(\mathbf{U}, \mathbf{V}) - G_{i+n+k}(\mathbf{W}, \mathbf{U}) + A_i \quad (1 \leq i \leq m + n + k),$

$$\sum_{i=1}^{m+n+k} A_i = \mathbf{O}$$

$$(m = n \neq k, \quad k \neq m + n),$$

$8^0.$ $g_i(\mathbf{U}, \mathbf{V}, \mathbf{W}) = K_i(\mathbf{U}, \mathbf{V}) - K_{i+n+k}(\mathbf{W}, \mathbf{U}) + G_i(\mathbf{V}, \mathbf{W}) + A_i \quad (1 \leq i \leq k),$

$$\sum_{i=1}^{k} A_i = \mathbf{O},$$

$$g_i(\mathbf{U}, \mathbf{V}, \mathbf{W}) = K_i(\mathbf{U}, \mathbf{V}) - K_{i+n+k}(\mathbf{W}, \mathbf{U}) - G_i(\mathbf{V}, \mathbf{W}) \quad (k+1 \leq i \leq 2k),$$

$$(m = n \neq k, \quad k = m + n),$$

$9^0.$ $g_i(\mathbf{U}, \mathbf{V}, \mathbf{W}) = G_i(\mathbf{U}, \mathbf{V}) - G_{i+m+n}(\mathbf{W}, \mathbf{V}) + A_i \quad (1 \leq i \leq m + n + k),$

$$\sum_{i=1}^{m+n+k} A_i = \mathbf{O}$$

$$(m = k \neq n, \quad n \neq m + k),$$

$10^0.$ $\quad g_i(\mathbf{U}, \mathbf{V}, \mathbf{W}) = K_i(\mathbf{V}, \mathbf{W}) - K_{i+k}(\mathbf{W}, \mathbf{U}) + G_i(\mathbf{U}, \mathbf{V}) + A_i$ $\quad (1 \leq i \leq n),$

$$\sum_{i=1}^{n} A_i = \mathbf{O},$$

$$g_i(\mathbf{U}, \mathbf{V}, \mathbf{W}) = K_i(\mathbf{V}, \mathbf{W}) - K_{i+k}(\mathbf{W}, \mathbf{U}) - G_{i+n}(\mathbf{V}, \mathbf{U}) \quad (n+1 \leq i \leq 2n),$$

$$(m = k \neq n, \quad n = m + k),$$

$11^0.$ $\quad g_i(\mathbf{U}, \mathbf{V}, \mathbf{W}) = L_i(\mathbf{U}, \mathbf{V}, \mathbf{W}) \quad (1 \leq i \leq 2m),$

$$g_i(\mathbf{U}, \mathbf{V}, \mathbf{W}) = -L_i(\mathbf{V}, \mathbf{W}, \mathbf{U}) - L_{i+2m}(\mathbf{W}, \mathbf{U}, \mathbf{V}) \quad (2m + 1 \leq i \leq 3m),$$

$$(m = n = k).$$

In all cases the functions F_i, $H_i : \mathcal{V} \mapsto \mathcal{V}'$, G_i, $K_i : \mathcal{V}^2 \mapsto \mathcal{V}'$ and $L_i : \mathcal{V}^3 \mapsto \mathcal{V}'$ are arbitrary complex vector functions such that $F_{i+n+m+k} = F_i$, $G_{i+n+m+k} = G_i, \cdots$. Also, for the arbitrary constant complex vectors A_i the equalities $A_{i+n+m+k} = A_i$ hold.

Proof. Since the statements of the lemma for particular cases may be proven in a similar way, therefore it is sufficient to prove the lemma only for some cases, for example 4^0 and 5^0.

4^0. If in Eq. (16.4) all variables, except $\mathbf{T}_i$, $\mathbf{T}_{i+n}$ and $\mathbf{T}_{i+n+k}$, are equal to some constant, we obtain

$$g_i(\mathbf{T}_i, \mathbf{T}_{i+n}, \mathbf{T}_{i+n+k}) = F_i(\mathbf{T}_i) + K_i(\mathbf{T}_{i+n}, \mathbf{T}_{i+n+k}),$$

i.e.,

$$g_i(\mathbf{U}, \mathbf{V}, \mathbf{W}) = F_i(\mathbf{U}) + K_i(\mathbf{V}, \mathbf{W}), \tag{16.5}$$

where F_i and K_i $(1 \leq i \leq m+n+k)$ are arbitrary complex vector functions. On the basis of the equation (16.5), Eq. (16.4) becomes

$$\sum_{i=1}^{m+n+k} F_i(\mathbf{T}_i) + \sum_{i=1}^{m+n+k} K_i(\mathbf{T}_i, \mathbf{T}_{i+n+k}) = \mathbf{O},$$

i.e.,

$$\sum_{i=1}^{m+n+k} F_{i+n}(\mathbf{T}_{i+n}) + \sum_{i=1}^{m+n+k} K_i(\mathbf{T}_i, \mathbf{T}_{i+n+k}) = \mathbf{O}.$$

If we put $\mathbf{T}_r = \mathbf{O}$ $(r = i + n, i + n + k)$ into the previous equality, we obtain

$$\begin{aligned}
F_{i+n}(\mathbf{T}_{i+n}) \quad &+ \quad F_{i+n+k}(\mathbf{T}_{i+n+k}) + K_i(\mathbf{T}_{i+n}, \mathbf{T}_{i+n+k}) \qquad (16.6)\\
&+ \quad K_{i+k}(\mathbf{T}_{i+n+k}, \mathbf{T}_{i+n}) + A_{i+k} = \mathbf{O},
\end{aligned}$$

where A_i are arbitrary constant complex vectors.

From Eq. (16.6) we obtain

$$K_i(\mathbf{V}, \mathbf{W}) = -K_{i+m}(\mathbf{W}, \mathbf{V}) - F_{i+n}(\mathbf{V}) - F_{i+n+k}(\mathbf{W}) - A_{i+k},$$

i.e., the equation (16.5) takes the form

$$\begin{aligned}
g_i(\mathbf{U}, \mathbf{V}, \mathbf{W}) \quad &= \quad F_i(\mathbf{U}) + K_i(\mathbf{V}, \mathbf{W}) \quad (1 \le i \le k),\\
g_i(\mathbf{U}, \mathbf{V}, \mathbf{W}) \quad &= \quad F_i(\mathbf{U}) - K_{i+k}(\mathbf{W}, \mathbf{V})\\
&- \quad F_{i+n}(\mathbf{V}) - F_{i+n+k}(\mathbf{W}) - A_{i+k} \quad (k+1 \le i \le 2k),
\end{aligned}$$

such that

$$\sum_{i=1}^{k} A_i = \mathbf{O}.$$

If we introduce new functions G_i by

$$G_i(\mathbf{V}, \mathbf{W}) = K_i(\mathbf{V}, \mathbf{W}) + F_{i+n}(\mathbf{V}),$$

the previous equations take the following form

$$\begin{aligned}
g_i(\mathbf{U}, \mathbf{V}, \mathbf{W}) \quad &= \quad F_i(\mathbf{U}) - F_{i+n}(\mathbf{V}) + G_i(\mathbf{V}, \mathbf{W}) \quad (1 \le i \le k),\\
g_i(\mathbf{U}, \mathbf{V}, \mathbf{W}) \quad &= \quad F_i(\mathbf{U}) - F_{i+n}(\mathbf{V})\\
&- \quad G_{i+k}(\mathbf{W}, \mathbf{V}) - A_{i+k} \quad (k+1 \le i \le 2k),
\end{aligned}$$

where

$$\sum_{i=1}^{k} A_i = \mathbf{O}.$$

5^0. If $\mathbf{T}_j = \mathbf{O}$ $(j \neq i, i+n, i+n+k)$, from the equation (16.4) we obtain

$$g_i(\mathbf{T}_i, \mathbf{T}_{i+n}, \mathbf{T}_{i+n+k}) = K_i(\mathbf{T}_i, \mathbf{T}_{i+n}) + M_i(\mathbf{T}_{i+n}, \mathbf{T}_{i+n+k}),$$

i.e.,

$$g_i(\mathbf{U}, \mathbf{V}, \mathbf{W}) = K_i(\mathbf{U}, \mathbf{V}) + M_i(\mathbf{V}, \mathbf{W}), \tag{16.7}$$

where K_i and M_i are arbitrary complex vector functions.

On the basis of the relation Eq. (16.7), the functional equation Eq. (16.4) becomes

$$\sum_{i=1}^{m+n+k} K_i(\mathbf{T}_i, \mathbf{T}_{i+n}) + \sum_{i=1}^{m+n+k} M_i(\mathbf{T}_{i+n}, \mathbf{T}_{i+n+k}) = \mathbf{O},$$

i.e.,

$$\sum_{i=1}^{m+n+k} K_i(\mathbf{T}_i, \mathbf{T}_{i+n}) + \sum_{i=1}^{m+n+k} M_{i+m+n}(\mathbf{T}_i, \mathbf{T}_{i+n}) = \mathbf{O}. \tag{16.8}$$

If we put $\mathbf{T}_r = \mathbf{O}$ $(r \neq i, i+n)$ into Eq. (16.8), then

$$K_i(\mathbf{T}_i, \mathbf{T}_{i+n}) + M_{i+m+n}(\mathbf{T}_i, \mathbf{T}_{i+n}) + P_i(\mathbf{T}_i) - Q_{i+n}(\mathbf{T}_{i+n}) = \mathbf{O}, \tag{16.9}$$

where P_i and Q_i are arbitrary complex vector functions.

On the basis of the above equation (16.9), Eq. (16.8) becomes

$$\sum_{i=1}^{m+n+k} P_i(\mathbf{T}_i) - \sum_{i=1}^{m+n+k} Q_i(\mathbf{T}_i) = \mathbf{O},$$

from which we obtain

$$Q_i(\mathbf{T}_i) = P_i(\mathbf{T}_i) + B_i, \qquad \sum_{i=1}^{m+n+k} B_i = \mathbf{O}.$$

Therefore, the equality (16.9) takes the form

$$K_i(\mathbf{T}_i, \mathbf{T}_{i+n}) + M_{i+m+k}(\mathbf{T}_i, \mathbf{T}_{i+n}) + P_i(\mathbf{T}_i) - P_{i+n}(\mathbf{T}_{i+n}) - B_{i+n} = \mathbf{O}.$$

On the basis of this equality and Eq. (16.7), we have

$$g_i(\mathbf{U}, \mathbf{V}, \mathbf{W}) = K_i(\mathbf{U}, \mathbf{V}) - K_{i+n}(\mathbf{V}, \mathbf{W}) + P_{i+n}(\mathbf{U}) - P_{i+2n}(\mathbf{V}) + B_{i+2n},$$

i.e.,

$$g_i(\mathbf{U}, \mathbf{V}, \mathbf{W}) = G_i(\mathbf{U}, \mathbf{V}) - G_{i+n}(\mathbf{V}, \mathbf{W}) + A_i, \quad \sum_{i=1}^{m+n+k} A_i = \mathbf{O},$$

where we put

$$G_i(\mathbf{U}, \mathbf{V}) = K_i(\mathbf{U}, \mathbf{V}) + P_{i+n}(\mathbf{U}), \quad A_i = B_{i+2n}. \qquad \square$$

On the basis of Lemma 16.1 and the transformations Eq. (16.2), there follow the theorems which treat the functional equation (16.1).

Theorem 16.2   *If $a^{m+n+k} \neq 1$, $m \neq n \neq k \neq m$, $m \neq n + k$, $n \neq m + k$ and $k \neq m + n$, then the general solution of the functional equation (16.1) is*

$$
\begin{aligned}
f_i(\mathbf{U}, \mathbf{V}, \mathbf{W}) \;=\; & F_i(\mathbf{V} + a^{m+n}\mathbf{W} + a^n\mathbf{U}) - F_{i+k}(\mathbf{W} + a^{n+k}\mathbf{U} + a^k\mathbf{V}) \\
& + \; G_i(\mathbf{U} + a^{k+m}\mathbf{V} + a^m\mathbf{W}, \; \mathbf{W} + a^{n+k}\mathbf{U} + a^k\mathbf{V}) \\
& \qquad\qquad (1 \leq i \leq m),
\end{aligned}
$$

$$
\begin{aligned}
f_i(\mathbf{U}, \mathbf{V}, \mathbf{W}) \;=\; & F_i(\mathbf{V} + a^{m+n}\mathbf{W} + a^n\mathbf{U}) - F_{i+k}(\mathbf{W} + a^{n+k}\mathbf{U} + a^k\mathbf{V}) \\
& - \; G_{i+m}(\mathbf{W} + a^{n+k}\mathbf{U} + a^k\mathbf{V}, \; \mathbf{U} + a^{k+m}\mathbf{V} + a^m\mathbf{W}) \\
& - \; A_{i+m} \qquad (m + 1 \leq i \leq 2m),
\end{aligned}
$$

$$\sum_{i=1}^{m} A_i = \mathbf{O},$$

where $F_i : \mathcal{V} \mapsto \mathcal{V}'$ and $G_i : \mathcal{V}^2 \mapsto \mathcal{V}'$ are arbitrary complex vector functions, and A_i are arbitrary constant complex vectors such that $\displaystyle\sum_{i=1}^{m+n+k} A_i = \mathbf{O}$.

Theorem 16.3   *If $a^{m+n+k} \neq 1$, $m \neq n \neq k \neq m$ and $m = n + k$, then the general solution of the functional equation (16.1) is*

$$
\begin{aligned}
f_i(\mathbf{U}, \mathbf{V}, \mathbf{W}) \;=\; & F_i(\mathbf{V} + a^{m+n}\mathbf{W} + a^n\mathbf{U}) - F_{i+k}(\mathbf{W} + a^{n+k}\mathbf{U} + a^k\mathbf{V}) \\
& + \; G_i(\mathbf{U} + a^{k+m}\mathbf{V} + a^m\mathbf{W}, \; \mathbf{W} + a^{n+k}\mathbf{U} + a^k\mathbf{V}) \\
& \qquad\qquad (1 \leq i \leq m),
\end{aligned}
$$

$$
\begin{aligned}
f_i(\mathbf{U}, \mathbf{V}, \mathbf{W}) \;=\; & F_i(\mathbf{V} + a^{m+n}\mathbf{W} + a^n\mathbf{U}) - F_{i+k}(\mathbf{W} + a^{n+k}\mathbf{U} + a^k\mathbf{V}) \\
& - \; G_{i+m}(\mathbf{W} + a^{n+k}\mathbf{U} + a^k\mathbf{V}, \; \mathbf{U} + a^{k+m}\mathbf{V} + a^m\mathbf{W}) \\
& - \; A_{i+m} \qquad (m + 1 \leq i \leq 2m),
\end{aligned}
$$

where $F_i : \mathcal{V} \mapsto \mathcal{V}'$ and $G_i : \mathcal{V}^2 \mapsto \mathcal{V}'$ are arbitrary complex vector functions, and A_i are arbitrary constant complex vectors such that $\sum_{i=1}^{m} A_i = \mathbf{O}$.

Theorem 16.4 *If $a^{m+n+k} \neq 1$, $m \neq n \neq k \neq m$ and $n = m + k$, then the general solution of the functional equation (16.1) is*

$$
\begin{aligned}
f_i(\mathbf{U}, \mathbf{V}, \mathbf{W}) \;=\;& F_i(\mathbf{W} + a^{n+k}\mathbf{U} + a^k\mathbf{V}) - F_{i+m}(\mathbf{U} + a^{k+m}\mathbf{V} + a^m\mathbf{W}) \\
+\;& G_i(\mathbf{U} + a^{k+m}\mathbf{V} + a^m\mathbf{W},\ \mathbf{V} + a^{m+n}\mathbf{W} + a^n\mathbf{U}) \\
& \hspace{3cm} (1 \leq i \leq n),
\end{aligned}
$$

$$
\begin{aligned}
f_i(\mathbf{U}, \mathbf{V}, \mathbf{W}) \;=\;& F_i(\mathbf{W} + a^{n+k}\mathbf{U} + a^k\mathbf{V}) - F_{i+m}(\mathbf{U} + a^{k+m}\mathbf{V} + a^m\mathbf{W}) \\
-\;& G_{i+n}(\mathbf{V} + a^{m+n}\mathbf{W} + a^n\mathbf{U},\ \mathbf{U} + a^{k+m}\mathbf{V} + a^m\mathbf{W}) \\
-\;& A_{i+n} \hspace{1.5cm} (n + 1 \leq i \leq 2n),
\end{aligned}
$$

where $F_i : \mathcal{V} \mapsto \mathcal{V}'$ and $G_i : \mathcal{V}^2 \mapsto \mathcal{V}'$ are arbitrary complex vector functions, A_i are arbitrary constant complex vectors such that $\sum_{i=1}^{n} A_i = \mathbf{O}$.

Theorem 16.5 *If $a^{m+n+k} \neq 1$, $m \neq n \neq k \neq m$ and $k = m + n$, then the general solution of the functional equation (16.1) is given by*

$$
\begin{aligned}
f_i(\mathbf{U}, \mathbf{V}, \mathbf{W}) \;=\;& F_i(\mathbf{U} + a^{k+m}\mathbf{V} + a^m\mathbf{W}) - F_{i+n}(\mathbf{V} + a^{m+n}\mathbf{W} + a^n\mathbf{U}) \\
+\;& G_i(\mathbf{V} + a^{m+n}\mathbf{W} + a^n\mathbf{U},\ \mathbf{W} + a^{n+k}\mathbf{U} + a^k\mathbf{V}) \\
& \hspace{3cm} (1 \leq i \leq k),
\end{aligned}
$$

$$
\begin{aligned}
f_i(\mathbf{U}, \mathbf{V}, \mathbf{W}) \;=\;& F_i(\mathbf{U} + a^{k+m}\mathbf{V} + a^m\mathbf{W}) - F_{i+n}(\mathbf{V} + a^{m+n}\mathbf{W} + a^n\mathbf{U}) \\
-\;& G_{i+k}(\mathbf{W} + a^{n+k}\mathbf{U} + a^k\mathbf{V},\ \mathbf{V} + a^{m+n}\mathbf{W} + a^n\mathbf{U}) \\
-\;& A_{i+k} \hspace{1.5cm} (k + 1 \leq i \leq 2k),
\end{aligned}
$$

where $F_i : \mathcal{V} \mapsto \mathcal{V}'$ and $G_i : \mathcal{V}^2 \mapsto \mathcal{V}'$ are arbitrary complex vector functions, A_i are arbitrary constant complex vectors such that $\sum_{i=1}^{k} A_i = \mathbf{O}$.

Theorem 16.6 *If $a^{m+n+k} \neq 1$, $m \neq n = k$ and $m \neq n + k$, then the general solution of the functional equation (16.1) is*

$$
\begin{aligned}
f_i(\mathbf{U}, \mathbf{V}, \mathbf{W}) \;=\;& F_i(\mathbf{U} + a^{k+m}\mathbf{V} + a^m\mathbf{W},\ \mathbf{V} + a^{m+n}\mathbf{W} + a^n\mathbf{U}) \\
-\;& F_{i+n}(\mathbf{V} + a^{m+n}\mathbf{W} + a^n\mathbf{U},\ \mathbf{W} + a^{n+k}\mathbf{U} + a^k\mathbf{V}) + A_i \\
& \hspace{3cm} (1 \leq i \leq m + n + k),
\end{aligned}
$$

where $F_i : \mathcal{V}^2 \mapsto \mathcal{V}'$ are arbitrary complex vector functions, A_i are arbitrary constant complex vectors such that $\displaystyle\sum_{i=1}^{m+n+k} A_i = \mathbf{O}$.

Theorem 16.7 *If $a^{m+n+k} \neq 1$, $m \neq n = k$ and $m = n + k$, then the general solution of the functional equation (16.1) is determined by*

$$
\begin{aligned}
f_i(\mathbf{U}, \mathbf{V}, \mathbf{W}) \;=\; & F_i(\mathbf{U} + a^{k+m}\mathbf{V} + a^m\mathbf{W}, \ \mathbf{V} + a^{m+n}\mathbf{W} + a^n\mathbf{U}) \\
- \; & F_{i+n}(\mathbf{V} + a^{m+n}\mathbf{W} + a^n\mathbf{U}, \ \mathbf{W} + a^{n+k}\mathbf{U} + a^k\mathbf{V}) \\
+ \; & G_i(\mathbf{W} + a^{n+k}\mathbf{U} + a^k\mathbf{V}, \ \mathbf{U} + a^{k+m}\mathbf{V} + a^m\mathbf{W}) \\
& (1 \leq i \leq m), \\[6pt]
f_i(\mathbf{U}, \mathbf{V}, \mathbf{W}) \;=\; & F_i(\mathbf{U} + a^{k+m}\mathbf{V} + a^m\mathbf{W}, \ \mathbf{V} + a^{m+n}\mathbf{W} + a^n\mathbf{U}) \\
- \; & F_{i+n}(\mathbf{V} + a^{m+n}\mathbf{W} + a^n\mathbf{U}, \ \mathbf{W} + a^{n+k}\mathbf{U} + a^k\mathbf{V}) \\
- \; & G_{i+m}(\mathbf{U} + a^{k+m}\mathbf{V} + a^m\mathbf{W}, \ \mathbf{W} + a^{n+k}\mathbf{U} + a^k\mathbf{V}) \\
& (m + 1 \leq i \leq 2m),
\end{aligned}
$$

where $F_i, G_i : \mathcal{V}^2 \mapsto \mathcal{V}'$ are arbitrary complex vector functions.

Theorem 16.8 *If $a^{m+n+k} \neq 1$, $m = n \neq k$ and $k \neq m + n$, then the general solution of the functional equation (16.1) is*

$$
\begin{aligned}
f_i(\mathbf{U}, \mathbf{V}, \mathbf{W}) = \; & F_i(\mathbf{U} + a^{k+m}\mathbf{V} + a^m\mathbf{W}, \ \mathbf{V} + a^{m+n}\mathbf{W} + a^n\mathbf{U}) \\
& -F_{i+n+k}(\mathbf{W} + a^{n+k}\mathbf{U} + a^k\mathbf{V}, \ \mathbf{U} + a^{k+m}\mathbf{V} + a^m\mathbf{W}) + A_i \\
& (1 \leq i \leq m + n + k),
\end{aligned}
$$

where $F_i : \mathcal{V}^2 \mapsto \mathcal{V}'$ are arbitrary complex vector functions, A_i are arbitrary constant complex vectors such that $\displaystyle\sum_{i=1}^{m+n+k} A_i = \mathbf{O}$.

Theorem 16.9 *If $a^{m+n+k} \neq 1$ and $k = m + n$, then the general solution of the functional equation (16.1) is*

$$
\begin{aligned}
f_i(\mathbf{U}, \mathbf{V}, \mathbf{W}) \;=\; & F_i(\mathbf{U} + a^{k+m}\mathbf{V} + a^m\mathbf{W}, \ \mathbf{V} + a^{m+n}\mathbf{W} + a^n\mathbf{U}) \\
- \; & F_{i+n+k}(\mathbf{W} + a^{n+k}\mathbf{U} + a^k\mathbf{V}, \ \mathbf{U} + a^{k+m}\mathbf{V} + a^m\mathbf{W}) \\
+ \; & G_i(\mathbf{V} + a^{m+n}\mathbf{W} + a^n\mathbf{U}, \ \mathbf{W} + a^{n+k}\mathbf{U} + a^k\mathbf{V}) + A_i \\
& (1 \leq i \leq k),
\end{aligned}
$$

$$f_i(\mathbf{U}, \mathbf{V}, \mathbf{W}) \;=\; F_i(\mathbf{U} + a^{k+m}\mathbf{V} + a^m\mathbf{W}, \; \mathbf{V} + a^{m+n}\mathbf{W} + a^n\mathbf{U})$$
$$- \; F_{i+n+k}(\mathbf{W} + a^{n+k}\mathbf{U} + a^k\mathbf{V}, \; \mathbf{U} + a^{k+m}\mathbf{V} + a^m\mathbf{W})$$
$$- \; G_{i+k}(\mathbf{W} + a^{n+k}\mathbf{U} + a^k\mathbf{V}, \; \mathbf{V} + a^{m+n}\mathbf{W} + a^n\mathbf{U})$$
$$(k + 1 \le i \le 2k),$$

where F_i, $G_i : \mathcal{V}^2 \mapsto \mathcal{V}'$ are arbitrary complex vector functions, and A_i are arbitrary constant complex vectors such that $\sum_{i=1}^{k} A_i = \mathbf{O}$.

Theorem 16.10 If $a^{m+n+k} \ne 1$, $m = k \ne n$ and $n \ne m + k$, then the general solution of the functional equation (16.1) is

$$f_i(\mathbf{U}, \mathbf{V}, \mathbf{W}) = F_i(\mathbf{U} + a^{k+m}\mathbf{V} + a^m\mathbf{W}, \; \mathbf{V} + a^{m+n}\mathbf{W} + a^n\mathbf{U})$$

$$-F_{i+m+n}(\mathbf{W} + a^{n+k}\mathbf{U} + a^k\mathbf{V}, \; \mathbf{V} + a^{m+n}\mathbf{W} + a^n\mathbf{U}) + A_i$$

$$(1 \le i \le m + n + k),$$

where $F_i : \mathcal{V}^2 \mapsto \mathcal{V}'$ are arbitrary complex vector functions, and A_i are arbitrary constant complex vectors such that $\sum_{i=1}^{m+n+k} A_i = \mathbf{O}$.

Theorem 16.11 If $a^{m+n+k} \ne 1$, $m = k \ne n$ and $n = m + k$, then the general solution of the functional equation (16.1) is determined by

$$f_i(\mathbf{U}, \mathbf{V}, \mathbf{W}) \;=\; F_i(\mathbf{V} + a^{m+n}\mathbf{W} + a^n\mathbf{U}, \; \mathbf{W} + a^{n+k}\mathbf{U} + a^k\mathbf{V})$$
$$- \; F_{i+k}(\mathbf{W} + a^{n+k}\mathbf{U} + a^k\mathbf{V}, \; \mathbf{U} + a^{k+m}\mathbf{V} + a^m\mathbf{W})$$
$$+ \; G_i(\mathbf{U} + a^{k+m}\mathbf{V} + a^m\mathbf{W}, \; \mathbf{V} + a^{m+n}\mathbf{W} + a^n\mathbf{U}) + A_i$$
$$(1 \le i \le n),$$

$$f_i(\mathbf{U}, \mathbf{V}, \mathbf{W}) \;=\; F_i(\mathbf{V} + a^{m+n}\mathbf{W} + a^n\mathbf{U}, \; \mathbf{W} + a^{n+k}\mathbf{U} + a^k\mathbf{V})$$
$$- \; F_{i+k}(\mathbf{W} + a^{n+k}\mathbf{U} + a^k\mathbf{V}, \; \mathbf{U} + a^{k+m}\mathbf{V} + a^m\mathbf{W})$$
$$- \; G_{i+n}(\mathbf{V} + a^{m+n}\mathbf{W} + a^n\mathbf{U}, \; \mathbf{U} + a^{k+m}\mathbf{V} + a^m\mathbf{W})$$
$$(n + 1 \le i \le 2n),$$

where F_i, $G_i : \mathcal{V}^2 \mapsto \mathcal{V}'$ are arbitrary complex vector functions, and A_i are arbitrary constant complex vectors such that $\sum_{i=1}^{n} A_i = \mathbf{O}$.

Theorem 16.12   *If $a^{m+n+k} \neq 1$, $m = n = k$, then the general solution of the functional equation (16.1) is*

$$f_i(\mathbf{U}, \mathbf{V}, \mathbf{W})$$

$$= \; F_i(\mathbf{U} + a^{k+m}\mathbf{V} + a^m\mathbf{W}, \; \mathbf{V} + a^{m+n}\mathbf{W} + a^n\mathbf{U}, \; \mathbf{W} + a^{n+k}\mathbf{U} + a^k\mathbf{V})$$

$$(1 \leq i \leq 2m),$$

$$f_i(\mathbf{U}, \mathbf{V}, \mathbf{W})$$

$$= \; -F_i(\mathbf{V} + a^{m+n}\mathbf{W} + a^n\mathbf{U}, \; \mathbf{W} + a^{n+k}\mathbf{U} + a^k\mathbf{V}, \; \mathbf{U} + a^{k+m}\mathbf{V} + a^m\mathbf{W})$$

$$- \; F_{i+2m}(\mathbf{W} + a^{n+k}\mathbf{U} + a^k\mathbf{V}, \; \mathbf{U} + a^{k+m}\mathbf{V} + a^m\mathbf{W}, \; \mathbf{V} + a^{m+n}\mathbf{W} + a^n\mathbf{U})$$

$$(2m + 1 \leq i \leq 3m),$$

where $F_i : \mathcal{V}^3 \mapsto \mathcal{V}'$ are arbitrary complex vector functions.

If $a = 1$ or $a^{m+n+k} = 1$ $(a \neq 1)$, then the solution of the functional equation (16.1) is very complicated and up to now we have not been able to obtain it.

Chapter 4

Functional Equations with Constant Coefficients

In [I. B. Risteski (to appear A)] the solution of one class of homogeneous complex vector functional equations is obtained by the method of intersections, while in [I. B. Risteski and V. C. Covachev (2000)] several classes of homogeneous complex vector functional equations are solved on the basis of the well-known method of elimination of variables. For the purpose of expanding of our investigations, here we will propose a matrix method suitable for both homogeneous and nonhomogeneous complex vector functional equations with constant complex coefficients. The results presented here are obtained in [I. B. Risteski (to appear B)].

Now we will introduce the following notations. Let $\mathcal{V}$ be a finite dimensional complex vector space and let a mapping $f : \mathcal{V}^n \mapsto \mathcal{V}$ exist. Throughout this chapter $\mathbf{Z}_i$ $(1 \leq i \leq n)$ are vectors in $\mathcal{V}$. We assume that $\mathbf{Z}_i = (z_{i1}(t), \cdots, z_{in}(t))^T$, where $z_{ij}(t)$ $(1 \leq i \leq n)$ are complex functions and $\mathbf{O} = (0, 0, \cdots, 0)^T$ is the zero vector in $\mathcal{V}$.

17 Matrix Equations

Let A be an $n \times n$ matrix. Suppose that by elementary transformations the matrix A is transformed into $A = P_1 D P_2$, where P_1 and P_2 are regular matrices and D is a diagonal matrix with diagonal entries 0 and 1 such that the number of the units is equal to the rank of the matrix A. The matrix $B = P_2^{-1} D P_1^{-1}$ satisfies the equality $ABA = A$. This means that the matrix equation $AXA = A$ has at least one solution for X.

173

If A satisfies the identity

$$A^r + k_1 A^{r-1} + \cdots + k_{r-1}A = O \qquad (k_{r-1} \neq 0),$$

then the matrix

$$X = -\frac{1}{k_{r-1}}(A^{r-2} + k_1 A^{r-3} + \cdots + k_{r-2}I) \qquad (I \text{ is the unit } n \times n \text{ matrix})$$

is also a solution of the equation $AXA = A$.

Now we will prove the following theorem.

Theorem 17.1 *If B satisfies the condition $ABA = A$, then*
$1^o \quad AX = O \quad \Leftrightarrow \quad X = (I - BA)Q, \quad (X \text{ and } Q \text{ are } n \times m \text{ matrices});$
$2^o \quad XA = O \quad \Leftrightarrow \quad X = Q(I - AB), \quad (X \text{ and } Q \text{ are } m \times n \text{ matrices});$
$3^o \quad AXA = A \quad \Leftrightarrow \quad X = B + Q - BAQAB \quad (X \text{ and } Q \text{ are } n \times n$
matrices);
$4^o \quad AX = A \quad \Leftrightarrow \quad X = I + (I - BA)Q;$
$5^o \quad XA = A \quad \Leftrightarrow \quad X = I + Q(I - AB).$

Proof. The theorem will be proved only for the case 3°, because the other cases can be proved analogously.

3°. Let $X = B + Q - BAQAB$. Further it holds

$$AXA = ABA + AQA - ABAQABA, \quad ABA = A$$

$$\Rightarrow \quad AXA = A + AQA - AQA \quad \Rightarrow \quad AXA = A.$$

Conversely, assume that $AXA = A$, then it holds

$$B + (X - B) - BA(X - B)AB$$

$$= X - BAXAB + BABAB = X - BAB + BAB = X.$$

Thus

$$AXA = A \quad \Rightarrow \quad X = B + Q - BAQAB, \quad \text{for} \quad Q = X - B.$$

$\square$

If X is an $n \times k$-matrix, according to 1° all the solutions of the homogeneous system of equations

$$
A \begin{bmatrix} x_1 \\ x_2 \\ \vdots \\ x_n \end{bmatrix} = O
$$

have the following form

$$
\begin{bmatrix} x_1 \\ x_2 \\ \vdots \\ x_n \end{bmatrix} = (I - BA) \begin{bmatrix} u_1 \\ u_2 \\ \vdots \\ u_n \end{bmatrix} \qquad (u_1, \cdots, u_n \text{ are arbitrary}).
$$

18 Homogeneous Functional Equations with Constant Coefficients

Now we will prove the following results.

Theorem 18.1 *The general solution of the basic cyclic complex vector functional equation with complex constant coefficients*

$$
E(f) \equiv \sum_{i=1}^{n} a_i f(\mathbf{Z}_i, \mathbf{Z}_{i+1}, \cdots, \mathbf{Z}_{i+n-1}) = \mathbf{O} \quad (\mathbf{Z}_{n+i} \equiv \mathbf{Z}_i), \tag{18.1}
$$

where a_i $(1 \le i \le n)$ are complex constants, is given by the following formula

$$
\begin{bmatrix} f(\mathbf{Z}_1, \mathbf{Z}_2, \cdots, \mathbf{Z}_n) \\ f(\mathbf{Z}_2, \mathbf{Z}_3, \cdots, \mathbf{Z}_1) \\ \vdots \\ f(\mathbf{Z}_n, \mathbf{Z}_1, \cdots, \mathbf{Z}_{n-1}) \end{bmatrix} = B \begin{bmatrix} h(\mathbf{Z}_1, \mathbf{Z}_2, \cdots, \mathbf{Z}_n) \\ h(\mathbf{Z}_2, \mathbf{Z}_3, \cdots, \mathbf{Z}_1) \\ \vdots \\ h(\mathbf{Z}_n, \mathbf{Z}_1, \cdots, \mathbf{Z}_{n-1}) \end{bmatrix}, \tag{18.2}
$$

where

$$A = \begin{bmatrix} a_1 & a_2 & \cdots & a_n \\ a_n & a_1 & \cdots & a_{n-1} \\ \vdots & & & \\ a_2 & a_3 & \cdots & a_1 \end{bmatrix} \quad and \quad B = \begin{bmatrix} b_1 & b_2 & \cdots & b_n \\ b_n & b_1 & \cdots & b_{n-1} \\ \vdots & & & \\ b_2 & b_3 & \cdots & b_1 \end{bmatrix} \quad (18.3)$$

are nonzero $n \times n$ cyclic matrices with complex constant elements, such that

$$AB = O, \qquad (18.4)$$

O is the $n \times n$ zero matrix and h is an arbitrary complex vector function with values in $\mathcal{V}$.

Proof. If we permute successively the vectors in Eq. (18.1), we get

$$\begin{aligned}
a_1 f(\mathbf{Z}_1, \mathbf{Z}_2, \cdots, \mathbf{Z}_n) \quad &+ \quad a_2 f(\mathbf{Z}_2, \mathbf{Z}_3, \cdots, \mathbf{Z}_n, \mathbf{Z}_1) \\
&+ \quad \cdots + a_n f(\mathbf{Z}_n, \mathbf{Z}_1, \cdots, \mathbf{Z}_{n-1}) = \mathbf{O}, \\
a_n f(\mathbf{Z}_1, \mathbf{Z}_2, \cdots, \mathbf{Z}_n) \quad &+ \quad a_1 f(\mathbf{Z}_2, \mathbf{Z}_3, \cdots, \mathbf{Z}_n, \mathbf{Z}_1) \\
&+ \quad \cdots + a_{n-1} f(\mathbf{Z}_n, \mathbf{Z}_1, \cdots, \mathbf{Z}_{n-1}) = \mathbf{O}, \\
\vdots \quad & \\
a_2 f(\mathbf{Z}_1, \mathbf{Z}_2, \cdots, \mathbf{Z}_n) \quad &+ \quad a_3 f(\mathbf{Z}_2, \mathbf{Z}_3, \cdots, \mathbf{Z}_n, \mathbf{Z}_1) \\
&+ \quad \cdots + a_1 f(\mathbf{Z}_n, \mathbf{Z}_1, \cdots, \mathbf{Z}_{n-1}) = \mathbf{O},
\end{aligned}$$

i.e., in a matrix form

$$AF = \mathcal{O}, \qquad (18.5)$$

where A is given by Eq. (18.3),

$$F = \begin{bmatrix} f(\mathbf{Z}_1, \mathbf{Z}_2, \cdots, \mathbf{Z}_n) \\ f(\mathbf{Z}_2, \mathbf{Z}_3, \cdots, \mathbf{Z}_1) \\ \vdots \\ f(\mathbf{Z}_n, \mathbf{Z}_1, \cdots, \mathbf{Z}_{n-1}) \end{bmatrix} \quad and \quad \mathcal{O} = \begin{bmatrix} \mathbf{O} \\ \mathbf{O} \\ \vdots \\ \mathbf{O} \end{bmatrix}. \qquad (18.6)$$

We shall write $f \in F$ to express the first equality of Eq. (18.6).

The necessary and sufficient condition for the system (18.5) to have a nontrivial solution is

$$\det A = 0. \qquad (18.7)$$

Let

$$f(\mathbf{Z}_1, \mathbf{Z}_2, \cdots, \mathbf{Z}_n) = \sum_{i=1}^{n} b_i h(\mathbf{Z}_i, \mathbf{Z}_{i+1}, \cdots, \mathbf{Z}_{i+n-1}) \quad (\mathbf{Z}_{n+i} \equiv \mathbf{Z}_i), \quad (18.8)$$

where b_i $(1 \leq i \leq n)$ are complex constants such that the matrix B defined by Eq. (18.3) satisfies Eq. (18.4) and h is an arbitrary complex vector function with values in $\mathcal{V}$.

By a cyclic permutation of the vectors in Eq. (18.8), we obtain

$$
\begin{aligned}
f(\mathbf{Z}_1, \mathbf{Z}_2, \cdots, \mathbf{Z}_n) = \quad & b_1 h(\mathbf{Z}_1, \mathbf{Z}_2, \cdots, \mathbf{Z}_n) + b_2 h(\mathbf{Z}_2, \mathbf{Z}_3, \cdots, \mathbf{Z}_1) \\
& + \cdots + b_n h(\mathbf{Z}_n, \mathbf{Z}_1, \cdots, \mathbf{Z}_{n-1}), \\
f(\mathbf{Z}_2, \mathbf{Z}_3, \cdots, \mathbf{Z}_1) = \quad & b_n h(\mathbf{Z}_1, \mathbf{Z}_2, \cdots, \mathbf{Z}_n) + b_1 h(\mathbf{Z}_2, \mathbf{Z}_3, \cdots, \mathbf{Z}_1) \\
& + \cdots + b_{n-1} h(\mathbf{Z}_n, \mathbf{Z}_1, \cdots, \mathbf{Z}_{n-1}), \\
\vdots \quad & \\
f(\mathbf{Z}_n, \mathbf{Z}_1, \cdots, \mathbf{Z}_{n-1}) = \quad & b_2 h(\mathbf{Z}_1, \mathbf{Z}_2, \cdots, \mathbf{Z}_n) + b_3 h(\mathbf{Z}_2, \mathbf{Z}_3, \cdots, \mathbf{Z}_1) \\
& + \cdots + b_1 h(\mathbf{Z}_n, \mathbf{Z}_1, \cdots, \mathbf{Z}_{n-1}),
\end{aligned}
$$

i.e., in a matrix form

$$F = BH \tag{18.9}$$

where

$$H = \begin{bmatrix} h(\mathbf{Z}_1, \mathbf{Z}_2, \cdots, \mathbf{Z}_n) \\ h(\mathbf{Z}_2, \mathbf{Z}_3, \cdots, \mathbf{Z}_1) \\ \vdots \\ h(\mathbf{Z}_n, \mathbf{Z}_1, \cdots, \mathbf{Z}_{n-1}) \end{bmatrix} \tag{18.10}$$

and F is defined in Eq. (18.6).

After a multiplication of Eq. (18.9) by A, we have

$$AF = ABH = OH = \mathcal{O},$$

which means that the function $f \in F$ satisfies the functional equation (18.1) for any $h \in H$.

On the other hand, according to 1° of Theorem 17.1 each solution of equation (18.5) has the form

$$F = (I - \tilde{B}A)H,$$

where $\tilde{B}$ satisfies $A\tilde{B}A = A$. If we put $B = I - \tilde{B}A$, we have $AB = A - A\tilde{B}A = O$ and $B \neq O$ since $\det A = 0$. Thus Eq. (18.9) where B satisfies Eq. (18.4) is the general solution of the system (18.5), *i.e.*, Eq. (18.8) gives the general solution of the functional equation (18.1). $\qquad\square$

Example 18.1   By a cyclic permutation of the variables in the functional equation

$$f(\mathbf{Z}_1, \mathbf{Z}_2, \mathbf{Z}_3, \mathbf{Z}_4) - f(\mathbf{Z}_2, \mathbf{Z}_3, \mathbf{Z}_4, \mathbf{Z}_1)$$
$$+\quad f(\mathbf{Z}_3, \mathbf{Z}_4, \mathbf{Z}_1, \mathbf{Z}_2) - f(\mathbf{Z}_4, \mathbf{Z}_1, \mathbf{Z}_2, \mathbf{Z}_3) = \mathbf{O}$$

we obtain the following system

$$f(\mathbf{Z}_1, \mathbf{Z}_2, \mathbf{Z}_3, \mathbf{Z}_4) - f(\mathbf{Z}_2, \mathbf{Z}_3, \mathbf{Z}_4, \mathbf{Z}_1)$$
$$+\quad f(\mathbf{Z}_3, \mathbf{Z}_4, \mathbf{Z}_1, \mathbf{Z}_2) - f(\mathbf{Z}_4, \mathbf{Z}_1, \mathbf{Z}_2, \mathbf{Z}_3) = \mathbf{O},$$
$$-\quad f(\mathbf{Z}_1, \mathbf{Z}_2, \mathbf{Z}_3, \mathbf{Z}_4) + f(\mathbf{Z}_2, \mathbf{Z}_3, \mathbf{Z}_4, \mathbf{Z}_1)$$
$$-\quad f(\mathbf{Z}_3, \mathbf{Z}_4, \mathbf{Z}_1, \mathbf{Z}_2) + f(\mathbf{Z}_4, \mathbf{Z}_1, \mathbf{Z}_2, \mathbf{Z}_3) = \mathbf{O},$$
$$f(\mathbf{Z}_1, \mathbf{Z}_2, \mathbf{Z}_3, \mathbf{Z}_4) - f(\mathbf{Z}_2, \mathbf{Z}_3, \mathbf{Z}_4, \mathbf{Z}_1)$$
$$+\quad f(\mathbf{Z}_3, \mathbf{Z}_4, \mathbf{Z}_1, \mathbf{Z}_2) - f(\mathbf{Z}_4, \mathbf{Z}_1, \mathbf{Z}_2, \mathbf{Z}_3) = \mathbf{O},$$
$$-\quad f(\mathbf{Z}_1, \mathbf{Z}_2, \mathbf{Z}_3, \mathbf{Z}_4) + f(\mathbf{Z}_2, \mathbf{Z}_3, \mathbf{Z}_4, \mathbf{Z}_1)$$
$$-\quad f(\mathbf{Z}_3, \mathbf{Z}_4, \mathbf{Z}_1, \mathbf{Z}_2) + f(\mathbf{Z}_4, \mathbf{Z}_1, \mathbf{Z}_2, \mathbf{Z}_3) = \mathbf{O}.$$

The matrix of coefficients of this system is

$$A = \begin{bmatrix} 1 & -1 & 1 & -1 \\ -1 & 1 & -1 & 1 \\ 1 & -1 & 1 & -1 \\ -1 & 1 & -1 & 1 \end{bmatrix}.$$

Since $\det A = 0$, the above system has a nontrivial solution.

For the matrix A there exists a nonzero 4×4 cyclic matrix B, such that $AB = O$, *i.e.*,

$$B = \begin{bmatrix} b_1 & b_2 & b_3 & b_1 - b_2 + b_3 \\ b_1 - b_2 + b_3 & b_1 & b_2 & b_3 \\ b_3 & b_1 - b_2 + b_3 & b_1 & b_2 \\ b_2 & b_3 & b_1 - b_2 + b_3 & b_1 \end{bmatrix}.$$

Therefore the general solution of the given functional equation is

$$f(\mathbf{Z}_1, \mathbf{Z}_2, \mathbf{Z}_3, \mathbf{Z}_4) = b_1 h(\mathbf{Z}_1, \mathbf{Z}_2, \mathbf{Z}_3, \mathbf{Z}_4) + b_2 h(\mathbf{Z}_2, \mathbf{Z}_3, \mathbf{Z}_4, \mathbf{Z}_1)$$

$$+ b_3 h(\mathbf{Z}_3, \mathbf{Z}_4, \mathbf{Z}_1, \mathbf{Z}_2) + (b_1 - b_2 + b_3) h(\mathbf{Z}_4, \mathbf{Z}_1, \mathbf{Z}_2, \mathbf{Z}_3),$$

where h is an arbitrary complex vector function with values in $\mathcal{V}$.

Theorem 18.2   *If the matrix A satisfies the condition*

$$A^m + \lambda_1 A^{m-1} + \cdots + \lambda_{m-1} A = O \quad (\lambda_{m-1} \neq 0), \tag{18.11}$$

then the general solution of the functional equation (18.1) is given by

$$F = \frac{1}{\lambda_{m-1}} (A^{m-1} + \lambda_1 A^{m-2} + \cdots + \lambda_{m-1} I) H, \tag{18.12}$$

where I is the $n \times n$ unit matrix and λ_i $(1 \leq i \leq m - 1)$ are complex numbers.

Proof.   The proof of this theorem is very easy.

By multiplication of the formula Eq. (18.12) with A, we obtain

$$AF = \frac{1}{\lambda_{m-1}} (A^m + \lambda_1 A^{m-1} + \cdots + \lambda_{m-1} A) H = OH = \mathcal{O},$$

or in other words the function $f \in F$ satisfies the functional equation (18.1) for any $h \in H$.

Conversely, if F is a solution of $AF = \mathcal{O}$, then obviously F satisfies the identity

$$F = \frac{1}{\lambda_{m-1}} (A^{m-1} + \lambda_1 A^{m-2} + \cdots + \lambda_{m-1} I) F,$$

i.e F can be represented by the formula Eq. (18.12) with $H = F$.

Hence we proved that the function given by Eq. (18.12) is the general solution of Eq. (18.1). $\square$

Example 18.2   For the functional equation

$$2f(\mathbf{Z}_1, \mathbf{Z}_2, \mathbf{Z}_3) - 3f(\mathbf{Z}_2, \mathbf{Z}_2, \mathbf{Z}_3) + f(\mathbf{Z}_3, \mathbf{Z}_3, \mathbf{Z}_3) = \mathbf{O}$$

the matrix of coefficients is

$$A = \begin{bmatrix} 2 & -3 & 1 \\ 0 & -1 & 1 \\ 0 & 0 & 0 \end{bmatrix}.$$

The matrix A satisfies the following equation

$$A^3 - A^2 - 2A = O.$$

The required general solution of the above functional equation is

$$\begin{bmatrix} f(\mathbf{Z}_1, \mathbf{Z}_2, \mathbf{Z}_3) \\ f(\mathbf{Z}_2, \mathbf{Z}_2, \mathbf{Z}_3) \\ f(\mathbf{Z}_3, \mathbf{Z}_3, \mathbf{Z}_3) \end{bmatrix} = -\frac{1}{2}(A^2 - A - 2I) \begin{bmatrix} h(\mathbf{Z}_1, \mathbf{Z}_2, \mathbf{Z}_3) \\ h(\mathbf{Z}_2, \mathbf{Z}_2, \mathbf{Z}_3) \\ h(\mathbf{Z}_3, \mathbf{Z}_3, \mathbf{Z}_3) \end{bmatrix},$$

i.e.,

$$f(\mathbf{Z}_1, \mathbf{Z}_2, \mathbf{Z}_3) = h(\mathbf{Z}_3, \mathbf{Z}_3, \mathbf{Z}_3) \equiv p(\mathbf{Z}_3),$$

where p is an arbitrary complex vector function with values in $\mathcal{V}$.

This example shows that Theorem 18.2 can be also applied to equations not of the form Eq. (18.1).

Theorem 18.3 *The general solution of form*

$$f(\mathbf{Z}_1, \mathbf{Z}_2, \cdots, \mathbf{Z}_n) = R(h(\mathbf{Z}_1, \mathbf{Z}_2, \cdots, \mathbf{Z}_n)) \qquad (18.13)$$

$$= \sum_{i=1}^{n} b_i h(\mathbf{Z}_i, \mathbf{Z}_{i+1}, \cdots, \mathbf{Z}_{i+n-1}) \quad (\mathbf{Z}_{n+i} \equiv \mathbf{Z}_i)$$

of the functional equation (18.1) *is reproductive* $(R(R(h)) = R(h))$ *if and only if the following condition is satisfied*

$$E(f) = \mathbf{O} \quad \Rightarrow \quad f = R(f). \qquad (18.14)$$

Proof. Assume that $R(h)$ is the general solution of the equation $E(f) = \mathbf{O}$.

Let $R(R(h)) = R(h)$ hold for every h. Then, from $E(f) = \mathbf{O}$ it follows that $f = R(h)$ for some h, so that for the same h we have $f = R(R(h)) = R(f)$.

Conversely, let the condition $E(f) = \mathbf{O} \Rightarrow f = R(f)$ be satisfied. Since for every h it holds $E(R(h)) = \mathbf{O}$, then according to the assumption we obtain $R(h) = R(R(h))$ for every h. $\square$

Example 18.3 We will determine the general reproductive solution for the functional equation given in Example 18.1.

On the basis of the general solution, we obtain

$$R(f(\mathbf{Z}_1, \mathbf{Z}_2, \mathbf{Z}_3, \mathbf{Z}_4))$$

$$= b_1 f(\mathbf{Z}_1, \mathbf{Z}_2, \mathbf{Z}_3, \mathbf{Z}_4) + b_2 f(\mathbf{Z}_2, \mathbf{Z}_3, \mathbf{Z}_4, \mathbf{Z}_1)$$

$$+ b_3 f(\mathbf{Z}_3, \mathbf{Z}_4, \mathbf{Z}_1, \mathbf{Z}_2) + (b_1 - b_2 + b_3) f(\mathbf{Z}_4, \mathbf{Z}_1, \mathbf{Z}_2, \mathbf{Z}_3)$$

$$= b_1 f(\mathbf{Z}_1, \mathbf{Z}_2, \mathbf{Z}_3, \mathbf{Z}_4) + b_2 f(\mathbf{Z}_2, \mathbf{Z}_3, \mathbf{Z}_4, \mathbf{Z}_1)$$

$$+ b_3 f(\mathbf{Z}_3, \mathbf{Z}_4, \mathbf{Z}_1, \mathbf{Z}_2) + (b_1 - b_2 + b_3)[f(\mathbf{Z}_1, \mathbf{Z}_2, \mathbf{Z}_3, \mathbf{Z}_4)$$

$$- f(\mathbf{Z}_2, \mathbf{Z}_3, \mathbf{Z}_4, \mathbf{Z}_1) + f(\mathbf{Z}_3, \mathbf{Z}_4, \mathbf{Z}_1, \mathbf{Z}_2)]$$

$$= (2b_1 - b_2 + b_3) f(\mathbf{Z}_1, \mathbf{Z}_2, \mathbf{Z}_3, \mathbf{Z}_4)$$

$$+ (-b_1 + 2b_2 - b_3) f(\mathbf{Z}_2, \mathbf{Z}_3, \mathbf{Z}_4, \mathbf{Z}_1)$$

$$+ (b_1 - b_2 + 2b_3) f(\mathbf{Z}_3, \mathbf{Z}_4, \mathbf{Z}_1, \mathbf{Z}_2).$$

The condition $R(f(\mathbf{Z}_1, \mathbf{Z}_2, \mathbf{Z}_3, \mathbf{Z}_4)) = f(\mathbf{Z}_1, \mathbf{Z}_2, \mathbf{Z}_3, \mathbf{Z}_4)$ holds if

$$2b_1 - b_2 + b_3 = 1, \quad -b_1 + 2b_2 - b_3 = 0, \quad b_1 - b_2 + 2b_3 = 0,$$

i.e.,

$$b_1 = 3/4, \quad b_2 = 1/4 \quad \text{and} \quad b_3 = -1/4.$$

Therefore, the general reproductive solution of the given functional equation is

$$f(\mathbf{Z}_1, \mathbf{Z}_2, \mathbf{Z}_3, \mathbf{Z}_4) \;=\; \frac{3}{4} h(\mathbf{Z}_1, \mathbf{Z}_2, \mathbf{Z}_3, \mathbf{Z}_4) + \frac{1}{4} h(\mathbf{Z}_2, \mathbf{Z}_3, \mathbf{Z}_4, \mathbf{Z}_1)$$
$$- \; \frac{1}{4} h(\mathbf{Z}_3, \mathbf{Z}_4, \mathbf{Z}_1, \mathbf{Z}_2) + \frac{1}{4} h(\mathbf{Z}_4, \mathbf{Z}_1, \mathbf{Z}_2, \mathbf{Z}_3).$$

Next we will give a procedure by which for every functional equation (18.1) in the case $n = 3$ we may determine the canonical equation which is equivalent to it.

For every canonical equation we will determine the general and reproductive solution.

Now we will consider the equation

$$a_1 f(\mathbf{Z}_1, \mathbf{Z}_2, \mathbf{Z}_3) + a_2 f(\mathbf{Z}_2, \mathbf{Z}_3, \mathbf{Z}_1) + a_3 f(\mathbf{Z}_3, \mathbf{Z}_1, \mathbf{Z}_2) = \mathbf{O}. \qquad (18.15)$$

From Eq. (18.15) by a cyclic permutation of the variables we obtain the following system

$$a_1 f(\mathbf{Z}_1, \mathbf{Z}_2, \mathbf{Z}_3) + a_2 f(\mathbf{Z}_2, \mathbf{Z}_3, \mathbf{Z}_1) + a_3 f(\mathbf{Z}_3, \mathbf{Z}_1, \mathbf{Z}_2) = \mathbf{O},$$
$$a_3 f(\mathbf{Z}_1, \mathbf{Z}_2, \mathbf{Z}_3) + a_1 f(\mathbf{Z}_2, \mathbf{Z}_3, \mathbf{Z}_1) + a_2 f(\mathbf{Z}_3, \mathbf{Z}_1, \mathbf{Z}_2) = \mathbf{O}, \qquad (18.16)$$
$$a_2 f(\mathbf{Z}_1, \mathbf{Z}_2, \mathbf{Z}_3) + a_3 f(\mathbf{Z}_2, \mathbf{Z}_3, \mathbf{Z}_1) + a_1 f(\mathbf{Z}_3, \mathbf{Z}_1, \mathbf{Z}_2) = \mathbf{O}.$$

The determinant of the system (18.16) is

$$\Delta = \frac{1}{2}(a_1 + a_2 + a_3)[(a_1 - a_2)^2 + (a_2 - a_3)^2 + (a_3 - a_1)^2].$$

There are four possible cases:
a) $a_1 + a_2 + a_3 \neq 0$ and $(a_1 - a_2)^2 + (a_2 - a_3)^2 + (a_3 - a_1)^2 \neq 0$,
b) $a_1 + a_2 + a_3 \neq 0$ and $(a_1 - a_2)^2 + (a_2 - a_3)^2 + (a_3 - a_1)^2 = 0$,
c) $a_1 + a_2 + a_3 = 0$ and $(a_1 - a_2)^2 + (a_2 - a_3)^2 + (a_3 - a_1)^2 \neq 0$,
d) $a_1 + a_2 + a_3 = 0$ and $(a_1 - a_2)^2 + (a_2 - a_3)^2 + (a_3 - a_1)^2 = 0$.
In the case a) the system (18.16) is obviously equivalent to the equation

$$f(\mathbf{Z}_1, \mathbf{Z}_2, \mathbf{Z}_3) = \mathbf{O}.$$

In the case b) we have

$$a_3^2 - (a_1 + a_2)a_3 + a_1^2 - a_1 a_2 + a_2^2 = 0$$

or

$$a_3 = \left[a_1 + a_2 \pm i(a_1 - a_2)\sqrt{3}\right]/2.$$

If $a_2 = a_1$, then also $a_3 = a_1$ and $a_1 + a_2 + a_3 \neq 0$ implies $a_1 \neq 0$. Thus the system (18.16) is equivalent to the equation

$$f(\mathbf{Z}_1, \mathbf{Z}_2, \mathbf{Z}_3) + f(\mathbf{Z}_2, \mathbf{Z}_3, \mathbf{Z}_1) + f(\mathbf{Z}_3, \mathbf{Z}_1, \mathbf{Z}_2) = \mathbf{O}. \qquad (18.17)$$

In the general case we can write

$$a_3 = \frac{1 \pm i\sqrt{3}}{2}a_1 + \frac{1 \mp i\sqrt{3}}{2}a_2$$

or, if ω_6 and ω_6^{-1} are the primitive 6th roots of 1, then $a_3 = a_1\omega_6 + a_2\omega_6^{-1}$. Now Eq. (18.15) takes the form

$$a_1 f(\mathbf{Z}_1, \mathbf{Z}_2, \mathbf{Z}_3) + a_2 f(\mathbf{Z}_2, \mathbf{Z}_3, \mathbf{Z}_1) \qquad (18.18)$$
$$+ \quad (a_1\omega_6 + a_2\omega_6^{-1})f(\mathbf{Z}_3, \mathbf{Z}_1, \mathbf{Z}_2) = \mathbf{O}.$$

If $a_3 = 0$, we have $a_1 + a_2 \neq 0$, $a_1 \omega_6^2 + a_2 = 0$ and the equation takes the form

$$f(\mathbf{Z}_1, \mathbf{Z}_2, \mathbf{Z}_3) - \omega_3 f(\mathbf{Z}_2, \mathbf{Z}_3, \mathbf{Z}_1) = \mathbf{O}, \tag{18.19}$$

where ω_3 is a primitive third root of 1 (we assume $\omega_3 = \omega_6^2$).

If $a_2 = 0$ or $a_1 = 0$, we obtain respectively the equations

$$f(\mathbf{Z}_1, \mathbf{Z}_2, \mathbf{Z}_3) + \omega_6 f(\mathbf{Z}_3, \mathbf{Z}_1, \mathbf{Z}_2) = \mathbf{O}$$

and

$$f(\mathbf{Z}_2, \mathbf{Z}_3, \mathbf{Z}_1) + \omega_6^{-1} f(\mathbf{Z}_3, \mathbf{Z}_1, \mathbf{Z}_2) = \mathbf{O}$$

which can be reduced to Eq. (18.19) by a cyclic permutation of the vectors.

We will see that equation (18.18) can be always reduced to Eq. (18.17) or Eq. (18.19). To this end we shall use the following lemma.

Lemma 18.4 *Suppose that equation (18.15) can be written in the form*

$$\alpha_1 [c_1 f(\mathbf{Z}_1, \mathbf{Z}_2, \mathbf{Z}_3) + c_2 f(\mathbf{Z}_2, \mathbf{Z}_3, \mathbf{Z}_1)] \tag{18.20}$$
$$+ \quad \alpha_2 [c_1 f(\mathbf{Z}_2, \mathbf{Z}_3, \mathbf{Z}_1) + c_2 f(\mathbf{Z}_3, \mathbf{Z}_1, \mathbf{Z}_2)]$$
$$+ \quad \alpha_3 [c_1 f(\mathbf{Z}_3, \mathbf{Z}_1, \mathbf{Z}_2) + c_2 f(\mathbf{Z}_1, \mathbf{Z}_2, \mathbf{Z}_3)] = \mathbf{O},$$

where

$$\begin{vmatrix} \alpha_1 & \alpha_2 & \alpha_3 \\ \alpha_3 & \alpha_1 & \alpha_2 \\ \alpha_2 & \alpha_3 & \alpha_1 \end{vmatrix} \neq 0. \tag{18.21}$$

Then equation (18.15) is equivalent to

$$c_1 f(\mathbf{Z}_1, \mathbf{Z}_2, \mathbf{Z}_3) + c_2 f(\mathbf{Z}_2, \mathbf{Z}_3, \mathbf{Z}_1) = \mathbf{O}. \tag{18.22}$$

Proof. By a cyclic permutation of Eq. (18.20) we derive the following system

$$\alpha_1 [c_1 f(\mathbf{Z}_1, \mathbf{Z}_2, \mathbf{Z}_3) + c_2 f(\mathbf{Z}_2, \mathbf{Z}_3, \mathbf{Z}_1)]$$
$$+ \quad \alpha_2 [c_1 f(\mathbf{Z}_2, \mathbf{Z}_3, \mathbf{Z}_1) + c_2 f(\mathbf{Z}_3, \mathbf{Z}_1, \mathbf{Z}_2)]$$
$$+ \quad \alpha_3 [c_1 f(\mathbf{Z}_3, \mathbf{Z}_1, \mathbf{Z}_2) + c_2 f(\mathbf{Z}_1, \mathbf{Z}_2, \mathbf{Z}_3)] = \mathbf{O},$$

$$\alpha_3[c_1 f(\mathbf{Z}_1, \mathbf{Z}_2, \mathbf{Z}_3) + c_2 f(\mathbf{Z}_2, \mathbf{Z}_3, \mathbf{Z}_1)]$$
$$+ \quad \alpha_1[c_1 f(\mathbf{Z}_2, \mathbf{Z}_3, \mathbf{Z}_1) + c_2 f(\mathbf{Z}_3, \mathbf{Z}_1, \mathbf{Z}_2)]$$
$$+ \quad \alpha_2[c_1 f(\mathbf{Z}_3, \mathbf{Z}_1, \mathbf{Z}_2) + c_2 f(\mathbf{Z}_1, \mathbf{Z}_2, \mathbf{Z}_3)] = \mathbf{O},$$

$$\alpha_2[c_1 f(\mathbf{Z}_1, \mathbf{Z}_2, \mathbf{Z}_3) + c_2 f(\mathbf{Z}_2, \mathbf{Z}_3, \mathbf{Z}_1)]$$
$$+ \quad \alpha_3[c_1 f(\mathbf{Z}_2, \mathbf{Z}_3, \mathbf{Z}_1) + c_2 f(\mathbf{Z}_3, \mathbf{Z}_1, \mathbf{Z}_2)]$$
$$+ \quad \alpha_1[c_1 f(\mathbf{Z}_3, \mathbf{Z}_1, \mathbf{Z}_2) + c_2 f(\mathbf{Z}_1, \mathbf{Z}_2, \mathbf{Z}_3)] = \mathbf{O}.$$

The determinant of this system is

$$\begin{vmatrix} \alpha_1 & \alpha_2 & \alpha_3 \\ \alpha_3 & \alpha_1 & \alpha_2 \\ \alpha_2 & \alpha_3 & \alpha_1 \end{vmatrix} \neq 0,$$

thus we deduce Eq. (18.22). $\square$

We see that equation (18.18) can be written in the form Eq. (18.20):

$$0 \cdot [f(\mathbf{Z}_1, \mathbf{Z}_2, \mathbf{Z}_3) - \omega_3 f(\mathbf{Z}_2, \mathbf{Z}_3, \mathbf{Z}_1)]$$
$$+ \quad a_2[f(\mathbf{Z}_2, \mathbf{Z}_3, \mathbf{Z}_1) - \omega_3 f(\mathbf{Z}_3, \mathbf{Z}_1, \mathbf{Z}_2)]$$
$$+ \quad a_1\omega_6[f(\mathbf{Z}_3, \mathbf{Z}_1, \mathbf{Z}_2) - \omega_3 f(\mathbf{Z}_1, \mathbf{Z}_2, \mathbf{Z}_3)] = \mathbf{O}.$$

The determinant in condition Eq. (18.21) is

$$\begin{vmatrix} 0 & a_2 & a_1\omega_6 \\ a_1\omega_6 & 0 & a_2 \\ a_2 & a_1\omega_6 & 0 \end{vmatrix} = a_2^3 - a_1^3.$$

If $a_2^3 \neq a_1^3$, then by Lemma 18.4 equation (18.18) is equivalent to Eq. (18.19). If $a_2 = a_1$, then it is equivalent to Eq. (18.17). If $a_2^3 = a_1^3$, but $a_2 \neq a_1$, then $a_2 = a_1\omega_6^2$ or $a_2 = a_1\omega_6^4$. In the latter case we have

$$a_1 + a_2 + a_3 = a_1 + a_1\omega_6^4 + a_1\omega_6 + a_1\omega_6^3 = a_1 - a_1\omega_6 + a_1\omega_6 - a_1 = 0$$

which is a contradiction. Suppose that $a_2 = a_1\omega_6^2$. Then $a_3 = 2a_1\omega_6$ and equation (18.18) becomes

$$f(\mathbf{Z}_1, \mathbf{Z}_2, \mathbf{Z}_3) + \omega_6^2 f(\mathbf{Z}_2, \mathbf{Z}_3, \mathbf{Z}_1) + 2\omega_6 f(\mathbf{Z}_3, \mathbf{Z}_1, \mathbf{Z}_2) = \mathbf{O}$$

or

$$f(\mathbf{Z}_1, \mathbf{Z}_2, \mathbf{Z}_3) - \omega_3 f(\mathbf{Z}_2, \mathbf{Z}_3, \mathbf{Z}_1)$$
$$+ \quad 2\omega_3[f(\mathbf{Z}_2, \mathbf{Z}_3, \mathbf{Z}_1) - \omega_3 f(\mathbf{Z}_3, \mathbf{Z}_1, \mathbf{Z}_2)]$$
$$+ \quad 0 \cdot [f(\mathbf{Z}_3, \mathbf{Z}_1, \mathbf{Z}_2) - \omega_3 f(\mathbf{Z}_1, \mathbf{Z}_2, \mathbf{Z}_3)] = \mathbf{O}.$$

The determinant

$$\begin{vmatrix} 1 & 2\omega_3 & 0 \\ 0 & 1 & 2\omega_3 \\ 2\omega_3 & 0 & 1 \end{vmatrix} = 1 + 8\omega_3^3 = 9 \neq 0,$$

thus by Lemma 18.4 equation (18.18) is equivalent to Eq. (18.19).

In the case c) $a_1 + a_2 + a_3 = 0$ and at least one of the inequalities $a_1 - a_2 \neq 0$, $a_2 - a_3 \neq 0$ and $a_3 - a_1 \neq 0$ is valid. Suppose, for the sake of definiteness, that $a_1 - a_2 \neq 0$. In view of $a_3 = -a_1 - a_2$ we have

$$a_1 f(\mathbf{Z}_1, \mathbf{Z}_2, \mathbf{Z}_3) + a_2 f(\mathbf{Z}_2, \mathbf{Z}_3, \mathbf{Z}_1) + (\ a_1 - a_2) f(\mathbf{Z}_3, \mathbf{Z}_1, \mathbf{Z}_2) - \mathbf{O},$$

or the form Eq. (18.20)

$$0 \cdot [f(\mathbf{Z}_1, \mathbf{Z}_2, \mathbf{Z}_3) - f(\mathbf{Z}_2, \mathbf{Z}_3, \mathbf{Z}_1)]$$
$$+ \quad a_2[f(\mathbf{Z}_2, \mathbf{Z}_3, \mathbf{Z}_1) - f(\mathbf{Z}_3, \mathbf{Z}_1, \mathbf{Z}_2)]$$
$$- \quad a_1[f(\mathbf{Z}_3, \mathbf{Z}_1, \mathbf{Z}_2) - f(\mathbf{Z}_1, \mathbf{Z}_2, \mathbf{Z}_3)] = \mathbf{O}.$$

The determinant in Eq. (18.21) is

$$\begin{vmatrix} 0 & a_2 & -a_1 \\ -a_1 & 0 & a_2 \\ a_2 & -a_1 & 0 \end{vmatrix} = a_2^3 - a_1^3.$$

Since $a_1 - a_2 \neq 0$, the equality $a_2^3 - a_1^3 = 0$ is possible only for $a_2 = a_1\omega_3$ (ω_3 a primitive third root of 1), then $a_3 = -a_1(1 + \omega_3)$ and we are led to a contradiction with $(a_1 - a_2)^2 + (a_2 - a_3)^2 + (a_3 - a_1)^2 \neq 0$. Thus, in this case the equation (18.15) is equivalent to the equation

$$f(\mathbf{Z}_1, \mathbf{Z}_2, \mathbf{Z}_3) - f(\mathbf{Z}_2, \mathbf{Z}_3, \mathbf{Z}_1) = \mathbf{O}. \tag{18.23}$$

If $a_1 - a_2 = 0$, then at least one of the differences $a_2 - a_3$ and $a_3 - a_1$ is not 0 and we come to the same conclusion.

In the case d) we have $a_3 = -a_1 - a_2$, $a_1^2 + a_1 a_2 + a_2^2 = 0$, *i.e.*, either $a_1 = a_2 = a_3 = 0$ and the equation (18.15) reduces to the identity

$$\mathbf{O} = \mathbf{O},$$

or $a_2 = a_1 \omega_3$, $a_3 = a_1 \omega_3^2$, where ω_3 is as above. Now equation (18.15) reduces to

$$f(\mathbf{Z}_1, \mathbf{Z}_2, \mathbf{Z}_3) + \omega_3 f(\mathbf{Z}_2, \mathbf{Z}_3, \mathbf{Z}_1) + \omega_3^2 f(\mathbf{Z}_3, \mathbf{Z}_1, \mathbf{Z}_2) = \mathbf{O}. \qquad (18.24)$$

On the basis of the exposition we conclude that the following lemma holds.

Lemma 18.5   *The functional equation (18.15) is equivalent to the following equations*

I. $f(\mathbf{Z}_1, \mathbf{Z}_2, \mathbf{Z}_3) = \mathbf{O}$ *if*

$$a_1 + a_2 + a_3 \neq 0 \quad and \quad (a_1 - a_2)^2 + (a_2 - a_3)^2 + (a_3 - a_1)^2 \neq 0;$$

II.   *Eq. (18.17) or Eq. (18.19) if*

$$a_1 + a_2 + a_3 \neq 0 \quad and \quad (a_1 - a_2)^2 + (a_2 - a_3)^2 + (a_3 - a_1)^2 = 0;$$

III.   *Eq. (18.23) if*

$$a_1 + a_2 + a_3 = 0 \quad and \quad (a_1 - a_2)^2 + (a_2 - a_3)^2 + (a_3 - a_1)^2 \neq 0;$$

IV. $\mathbf{O} = \mathbf{O}$ *or Eq. (18.24) if*

$$a_1 + a_2 + a_3 = 0 \quad and \quad (a_1 - a_2)^2 + (a_2 - a_3)^2 + (a_3 - a_1)^2 = 0.$$

For any of the above equations we give formulae for the general solutions and formulae for the general reproductive solutions of these equations.

Proposition 18.6   *The equation (18.17)*

$$f(\mathbf{Z}_1, \mathbf{Z}_2, \mathbf{Z}_3) + f(\mathbf{Z}_2, \mathbf{Z}_3, \mathbf{Z}_1) + f(\mathbf{Z}_3, \mathbf{Z}_1, \mathbf{Z}_2) = \mathbf{O}$$

has a general solution given by

$$\begin{aligned}
f(\mathbf{Z}_1, \mathbf{Z}_2, \mathbf{Z}_3) \;=\; & b_1 h(\mathbf{Z}_1, \mathbf{Z}_2, \mathbf{Z}_3) + b_2 h(\mathbf{Z}_2, \mathbf{Z}_3, \mathbf{Z}_1) \qquad (18.25) \\
& -\; (b_1 + b_2) h(\mathbf{Z}_3, \mathbf{Z}_1, \mathbf{Z}_2),
\end{aligned}$$

where h is an arbitrary complex vector function with values in $\mathcal{V}$ and b_1, b_2 are arbitrary complex constants.

Proof. From the given equation we obtain the system

$$f(\mathbf{Z}_1, \mathbf{Z}_2, \mathbf{Z}_3) + f(\mathbf{Z}_2, \mathbf{Z}_3, \mathbf{Z}_1) + f(\mathbf{Z}_3, \mathbf{Z}_1, \mathbf{Z}_2) = \mathbf{O},$$
$$f(\mathbf{Z}_1, \mathbf{Z}_2, \mathbf{Z}_3) + f(\mathbf{Z}_2, \mathbf{Z}_3, \mathbf{Z}_1) + f(\mathbf{Z}_3, \mathbf{Z}_1, \mathbf{Z}_2) = \mathbf{O},$$
$$f(\mathbf{Z}_1, \mathbf{Z}_2, \mathbf{Z}_3) + f(\mathbf{Z}_2, \mathbf{Z}_3, \mathbf{Z}_1) + f(\mathbf{Z}_3, \mathbf{Z}_1, \mathbf{Z}_2) = \mathbf{O},$$

with matrix of coefficients

$$A = \begin{bmatrix} 1 & 1 & 1 \\ 1 & 1 & 1 \\ 1 & 1 & 1 \end{bmatrix}.$$

The general solution of this system, according to Eq. (18.2) is given by

$$\begin{bmatrix} f(\mathbf{Z}_1, \mathbf{Z}_2, \mathbf{Z}_3) \\ f(\mathbf{Z}_2, \mathbf{Z}_3, \mathbf{Z}_1) \\ f(\mathbf{Z}_3, \mathbf{Z}_1, \mathbf{Z}_2) \end{bmatrix} = B \begin{bmatrix} h(\mathbf{Z}_1, \mathbf{Z}_2, \mathbf{Z}_3) \\ h(\mathbf{Z}_2, \mathbf{Z}_3, \mathbf{Z}_1) \\ h(\mathbf{Z}_3, \mathbf{Z}_1, \mathbf{Z}_2) \end{bmatrix},$$

where B is a general cyclic 3×3 matrix which satisfies the condition $AB = O$.

On the basis of the formula Eq. (18.8), we may write

$$f(\mathbf{Z}_1, \mathbf{Z}_2, \mathbf{Z}_3) = b_1 h(\mathbf{Z}_1, \mathbf{Z}_2, \mathbf{Z}_3) + b_2 h(\mathbf{Z}_2, \mathbf{Z}_3, \mathbf{Z}_1) - (b_1 + b_2) h(\mathbf{Z}_3, \mathbf{Z}_1, \mathbf{Z}_2),$$

which proves the proposition. $\square$

Proposition 18.7 *The reproductive solution of the equation* (18.17)

$$f(\mathbf{Z}_1, \mathbf{Z}_2, \mathbf{Z}_3) + f(\mathbf{Z}_2, \mathbf{Z}_3, \mathbf{Z}_1) + f(\mathbf{Z}_3, \mathbf{Z}_1, \mathbf{Z}_2) = \mathbf{O}$$

is

$$f(\mathbf{Z}_1, \mathbf{Z}_2, \mathbf{Z}_3) = \frac{2}{3} h(\mathbf{Z}_1, \mathbf{Z}_2, \mathbf{Z}_3) - \frac{1}{3} h(\mathbf{Z}_2, \mathbf{Z}_3, \mathbf{Z}_1) - \frac{1}{3} h(\mathbf{Z}_3, \mathbf{Z}_1, \mathbf{Z}_2),$$

where h is an arbitrary complex vector function with values in $\mathcal{V}$.

Proof. This statement will be proved in the following way. From the general solution Eq. (18.25) we obtain

$$\begin{aligned} R(f(\mathbf{Z}_1, \mathbf{Z}_2, \mathbf{Z}_3)) &= b_1 f(\mathbf{Z}_1, \mathbf{Z}_2, \mathbf{Z}_3) + b_2 f(\mathbf{Z}_2, \mathbf{Z}_3, \mathbf{Z}_1) \\ &- (b_1 + b_2) f(\mathbf{Z}_3, \mathbf{Z}_1, \mathbf{Z}_2) \\ &= b_1 f(\mathbf{Z}_1, \mathbf{Z}_2, \mathbf{Z}_3) + b_2 f(\mathbf{Z}_2, \mathbf{Z}_3, \mathbf{Z}_1) \\ &+ (b_1 + b_2) [f(\mathbf{Z}_1, \mathbf{Z}_2, \mathbf{Z}_3) + f(\mathbf{Z}_2, \mathbf{Z}_3, \mathbf{Z}_1)] \\ &= (2b_1 + b_2) f(\mathbf{Z}_1, \mathbf{Z}_2, \mathbf{Z}_3) + (b_1 + 2b_2) f(\mathbf{Z}_2, \mathbf{Z}_3, \mathbf{Z}_1). \end{aligned}$$

The condition $R(f(\mathbf{Z}_1, \mathbf{Z}_2, \mathbf{Z}_3)) = f(\mathbf{Z}_1, \mathbf{Z}_2, \mathbf{Z}_3)$ is satisfied if

$$b_1 = 2/3 \quad \text{and} \quad b_2 = -1/3.$$

Thus

$$f(\mathbf{Z}_1, \mathbf{Z}_2, \mathbf{Z}_3) = \frac{2}{3}h(\mathbf{Z}_1, \mathbf{Z}_2, \mathbf{Z}_3) - \frac{1}{3}h(\mathbf{Z}_2, \mathbf{Z}_3, \mathbf{Z}_1) - \frac{1}{3}h(\mathbf{Z}_3, \mathbf{Z}_1, \mathbf{Z}_2)$$

is a reproductive solution of the equation

$$f(\mathbf{Z}_1, \mathbf{Z}_2, \mathbf{Z}_3) + f(\mathbf{Z}_2, \mathbf{Z}_3, \mathbf{Z}_1) + f(\mathbf{Z}_3, \mathbf{Z}_1, \mathbf{Z}_2) = \mathbf{O}.$$

$\square$

In a similar way we can prove that the equation (18.19) has a general solution given by

$$f(\mathbf{Z}_1, \mathbf{Z}_2, \mathbf{Z}_3) = b\left[h(\mathbf{Z}_1, \mathbf{Z}_2, \mathbf{Z}_3) + \omega_3 h(\mathbf{Z}_2, \mathbf{Z}_3, \mathbf{Z}_1) + \omega_3^2 h(\mathbf{Z}_3, \mathbf{Z}_1, \mathbf{Z}_2)\right],$$

where b is a complex constant and h is an arbitrary complex vector function with values in $\mathcal{V}$.

Of course, in the last equality we can put $b = 1$, including the arbitrary constant in the function h. Also, if we put $b_1 = b$, $b_2 = b\omega_3$, then by virtue of the equality $1 + \omega_3 + \omega_3^2 = 0$ we see that the general solution of Eq. (18.19) is of the form Eq. (18.25).

On the other hand, it is sometimes convenient to keep this factor. For instance, we see that the reproductive solution of Eq. (18.19) is obtained for $b = 1/3$, *i.e.*,

$$f(\mathbf{Z}_1, \mathbf{Z}_2, \mathbf{Z}_3) = \frac{1}{3}\left[h(\mathbf{Z}_1, \mathbf{Z}_2, \mathbf{Z}_3) + \omega_3 h(\mathbf{Z}_2, \mathbf{Z}_3, \mathbf{Z}_1) + \omega_3^2 h(\mathbf{Z}_3, \mathbf{Z}_1, \mathbf{Z}_2)\right].$$

Similarly, the equation (18.23) has a general solution given by

$$f(\mathbf{Z}_1, \mathbf{Z}_2, \mathbf{Z}_3) = h(\mathbf{Z}_1, \mathbf{Z}_2, \mathbf{Z}_3) + h(\mathbf{Z}_2, \mathbf{Z}_3, \mathbf{Z}_1) + h(\mathbf{Z}_3, \mathbf{Z}_1, \mathbf{Z}_2)$$

and a reproductive solution given by

$$f(\mathbf{Z}_1, \mathbf{Z}_2, \mathbf{Z}_3) = \frac{1}{3}\left[h(\mathbf{Z}_1, \mathbf{Z}_2, \mathbf{Z}_3) + h(\mathbf{Z}_2, \mathbf{Z}_3, \mathbf{Z}_1) + h(\mathbf{Z}_3, \mathbf{Z}_1, \mathbf{Z}_2)\right],$$

and the equation (18.24) has a general solution given by

$$\begin{aligned}
f(\mathbf{Z}_1, \mathbf{Z}_2, \mathbf{Z}_3) &= b_1 h(\mathbf{Z}_1, \mathbf{Z}_2, \mathbf{Z}_3) + b_2 h(\mathbf{Z}_2, \mathbf{Z}_3, \mathbf{Z}_1) \\
&\quad - \left(\omega_3^2 b_1 + \omega_3 b_2\right) h(\mathbf{Z}_3, \mathbf{Z}_1, \mathbf{Z}_2)
\end{aligned}$$

and a reproductive solution given by

$$f(\mathbf{Z}_1,\mathbf{Z}_2,\mathbf{Z}_3) = \frac{2}{3}h(\mathbf{Z}_1,\mathbf{Z}_2,\mathbf{Z}_3) - \frac{\omega_3}{3}h(\mathbf{Z}_2,\mathbf{Z}_3,\mathbf{Z}_1) - \frac{\omega_3^2}{3}h(\mathbf{Z}_3,\mathbf{Z}_1,\mathbf{Z}_2).$$

On the basis of the previous results the following two theorems hold.

Theorem 18.8 *The general solution of the equation*

$$a_1 f(\mathbf{Z}_1,\mathbf{Z}_2,\mathbf{Z}_3) + a_2 f(\mathbf{Z}_2,\mathbf{Z}_3,\mathbf{Z}_1) + a_3 f(\mathbf{Z}_3,\mathbf{Z}_1,\mathbf{Z}_2) = \mathbf{O}$$

is given by the following formulae

*1°.   If $a_1+a_2+a_3 \neq 0$ and $(a_1-a_2)^2+(a_2-a_3)^2+(a_3-a_1)^2 \neq 0$,
then $f(\mathbf{Z}_1,\mathbf{Z}_2,\mathbf{Z}_3) \equiv \mathbf{O}$;*

*2°.   If $a_1+a_2+a_3 \neq 0$ and $(a_1-a_2)^2+(a_2-a_3)^2+(a_3-a_1)^2 = 0$,
then*

$$f(\mathbf{Z}_1,\mathbf{Z}_2,\mathbf{Z}_3) = b_1 h(\mathbf{Z}_1,\mathbf{Z}_2,\mathbf{Z}_3) + b_2 h(\mathbf{Z}_2,\mathbf{Z}_3,\mathbf{Z}_1) - (b_1+b_2)h(\mathbf{Z}_3,\mathbf{Z}_1,\mathbf{Z}_2)$$

(in particular,

$$f(\mathbf{Z}_1,\mathbf{Z}_2,\mathbf{Z}_3) = h(\mathbf{Z}_1,\mathbf{Z}_2,\mathbf{Z}_3) + \omega_3 h(\mathbf{Z}_2,\mathbf{Z}_3,\mathbf{Z}_1) + \omega_3^2 h(\mathbf{Z}_3,\mathbf{Z}_1,\mathbf{Z}_2)$$

if a_1,a_2,a_3 are distinct nonzero numbers);

*3°.   If $a_1+a_2+a_3 = 0$ and $(a_1-a_2)^2+(a_2-a_3)^2+(a_3-a_1)^2 \neq 0$,
then*

$$f(\mathbf{Z}_1,\mathbf{Z}_2,\mathbf{Z}_3) = h(\mathbf{Z}_1,\mathbf{Z}_2,\mathbf{Z}_3) + h(\mathbf{Z}_2,\mathbf{Z}_3,\mathbf{Z}_1) + h(\mathbf{Z}_3,\mathbf{Z}_1,\mathbf{Z}_2);$$

*4°.   If $a_1+a_2+a_3 = 0$ and $(a_1-a_2)^2+(a_2-a_3)^2+(a_3-a_1)^2 = 0$,
then:*

$$a)\ \ f(\mathbf{Z}_1,\mathbf{Z}_2,\mathbf{Z}_3)\ =\ h(\mathbf{Z}_1,\mathbf{Z}_2,\mathbf{Z}_3)\ \text{if } a_1 = a_2 = a_3 = 0;$$

$$\begin{aligned} b)\ \ f(\mathbf{Z}_1,\mathbf{Z}_2,\mathbf{Z}_3)\ =\ & b_1 h(\mathbf{Z}_1,\mathbf{Z}_2,\mathbf{Z}_3) + b_2 h(\mathbf{Z}_2,\mathbf{Z}_3,\mathbf{Z}_1) \\ & - \left(\omega_3^2 b_1 + \omega_3 b_2\right) h(\mathbf{Z}_3,\mathbf{Z}_1,\mathbf{Z}_2) \end{aligned}$$

if $a_2 = a_1\omega_3$, $a_3 = a_1\omega_3^2$, $a_1 \neq 0$,

where h is an arbitrary complex vector function with values in $\mathcal{V}$.

Theorem 18.9 *Under the assumptions of the previous theorem denoted by $1° - 4°$ the general reproductive solutions are given by the following formulae*

$1^o.$ $f(\mathbf{Z}_1, \mathbf{Z}_2, \mathbf{Z}_3) \equiv \mathbf{O};$

$2^o.$ *a)* $f(\mathbf{Z}_1, \mathbf{Z}_2, \mathbf{Z}_3) = \frac{2}{3}h(\mathbf{Z}_1, \mathbf{Z}_2, \mathbf{Z}_3) - \frac{1}{3}h(\mathbf{Z}_2, \mathbf{Z}_3, \mathbf{Z}_1) - \frac{1}{3}h(\mathbf{Z}_3, \mathbf{Z}_1, \mathbf{Z}_2)$
if $a_1 = a_2 = a_3 \neq 0;$

b) $f(\mathbf{Z}_1, \mathbf{Z}_2, \mathbf{Z}_3) = \frac{2}{3}h(\mathbf{Z}_1, \mathbf{Z}_2, \mathbf{Z}_3) - \frac{\omega_3}{3}h(\mathbf{Z}_2, \mathbf{Z}_3, \mathbf{Z}_1) - \frac{\omega_3^2}{3}h(\mathbf{Z}_3, \mathbf{Z}_1, \mathbf{Z}_2)$
if a_1, a_2, a_3 *are distinct;*

$3^o.$ $f(\mathbf{Z}_1, \mathbf{Z}_2, \mathbf{Z}_3) = \frac{1}{3}[h(\mathbf{Z}_1, \mathbf{Z}_2, \mathbf{Z}_3) + h(\mathbf{Z}_2, \mathbf{Z}_3, \mathbf{Z}_1) + h(\mathbf{Z}_3, \mathbf{Z}_1, \mathbf{Z}_2)];$

$4^o.$ *a)* $f(\mathbf{Z}_1, \mathbf{Z}_2, \mathbf{Z}_3) = h(\mathbf{Z}_1, \mathbf{Z}_2, \mathbf{Z}_3);$

b) $f(\mathbf{Z}_1, \mathbf{Z}_2, \mathbf{Z}_3) = \frac{2}{3}h(\mathbf{Z}_1, \mathbf{Z}_2, \mathbf{Z}_3) - \frac{\omega_3}{3}h(\mathbf{Z}_2, \mathbf{Z}_3, \mathbf{Z}_1) - \frac{\omega_3^2}{3}h(\mathbf{Z}_3, \mathbf{Z}_1, \mathbf{Z}_2).$

19   Nonhomogeneous Functional Equations with Constant Coefficients

Next we will give the following results.

Theorem 19.1   *The basic cyclic complex vector nonhomogeneous functional equation with complex constant coefficients*

$$E(f) \equiv \sum_{i=1}^{n} a_i f(\mathbf{Z}_i, \mathbf{Z}_{i+1}, \cdots, \mathbf{Z}_{i+n-1}) \tag{19.1}$$
$$= g(\mathbf{Z}_1, \mathbf{Z}_2, \cdots, \mathbf{Z}_n) \qquad (\mathbf{Z}_{n+i} \equiv \mathbf{Z}_i)$$

where a_i $(1 \leq i \leq n)$ *are complex constants, has a solution if the right-hand side* g *satisfies*

$$(AC + I) \begin{bmatrix} g(\mathbf{Z}_1, \mathbf{Z}_2, \cdots, \mathbf{Z}_n) \\ g(\mathbf{Z}_2, \mathbf{Z}_3, \cdots, \mathbf{Z}_1) \\ \vdots \\ g(\mathbf{Z}_n, \mathbf{Z}_1, \cdots, \mathbf{Z}_{n-1}) \end{bmatrix} = \mathcal{O}, \tag{19.2}$$

where A *is given by Eq. (18.3),* C *is any nonzero* $n \times n$ *cyclic matrix with complex constant entries satisfying* $ACA + A = O$, O *is the* $n \times n$ *zero matrix,* I *is the* $n \times n$ *unit matrix and* $\mathcal{O}$ *is defined as in Eq. (18.6).*

If Eq. (19.2) holds for some C, *then the general solution of Eq. (19.1) is given by the following formula*

$$\begin{bmatrix} f(\mathbf{Z}_1, \mathbf{Z}_2, \cdots, \mathbf{Z}_n) \\ f(\mathbf{Z}_2, \mathbf{Z}_3, \cdots, \mathbf{Z}_1) \\ \vdots \\ f(\mathbf{Z}_n, \mathbf{Z}_1, \cdots, \mathbf{Z}_{n-1}) \end{bmatrix} \tag{19.3}$$

$$= B \begin{bmatrix} h(\mathbf{Z}_1, \mathbf{Z}_2, \cdots, \mathbf{Z}_n) \\ h(\mathbf{Z}_2, \mathbf{Z}_3, \cdots, \mathbf{Z}_1) \\ \vdots \\ h(\mathbf{Z}_n, \mathbf{Z}_1, \cdots, \mathbf{Z}_{n-1}) \end{bmatrix} - C \begin{bmatrix} g(\mathbf{Z}_1, \mathbf{Z}_2, \cdots, \mathbf{Z}_n) \\ g(\mathbf{Z}_2, \mathbf{Z}_3, \cdots, \mathbf{Z}_1) \\ \vdots \\ g(\mathbf{Z}_n, \mathbf{Z}_1, \cdots, \mathbf{Z}_{n-1}) \end{bmatrix},$$

where the nonzero $n \times n$ cyclic matrix B given by Eq. (18.3) satisfies the condition

$$AB = O \tag{19.4}$$

and h is an arbitrary complex vector function with values in $\mathcal{V}$.

Proof. By a cyclic permutation of the vectors in Eq. (19.1) we get

$$\begin{aligned} & a_1 f(\mathbf{Z}_1, \mathbf{Z}_2, \cdots, \mathbf{Z}_n) + a_2 f(\mathbf{Z}_2, \mathbf{Z}_3, \cdots, \mathbf{Z}_1) \\ + \quad & \cdots + a_n f(\mathbf{Z}_n, \mathbf{Z}_1, \cdots, \mathbf{Z}_{n-1}) = g(\mathbf{Z}_1, \mathbf{Z}_2, \cdots, \mathbf{Z}_n), \\ & a_n f(\mathbf{Z}_1, \mathbf{Z}_2, \cdots, \mathbf{Z}_n) + a_1 f(\mathbf{Z}_2, \mathbf{Z}_3, \cdots, \mathbf{Z}_1) \\ + \quad & \cdots + a_{n-1} f(\mathbf{Z}_n, \mathbf{Z}_1, \cdots, \mathbf{Z}_{n-1}) = g(\mathbf{Z}_2, \mathbf{Z}_3, \cdots, \mathbf{Z}_1), \\ \vdots \quad & \\ & a_2 f(\mathbf{Z}_1, \mathbf{Z}_2, \cdots, \mathbf{Z}_n) + a_3 f(\mathbf{Z}_2, \mathbf{Z}_3, \cdots, \mathbf{Z}_1) \\ + \quad & \cdots + a_1 f(\mathbf{Z}_n, \mathbf{Z}_1, \cdots, \mathbf{Z}_{n-1}) = g(\mathbf{Z}_n, \mathbf{Z}_1, \cdots, \mathbf{Z}_{n-1}), \end{aligned} \tag{19.5}$$

i.e. in a matrix form

$$AF = G, \tag{19.6}$$

where

$$F = \begin{bmatrix} f(\mathbf{Z}_1, \mathbf{Z}_2, \cdots, \mathbf{Z}_n) \\ f(\mathbf{Z}_2, \mathbf{Z}_3, \cdots, \mathbf{Z}_1) \\ \vdots \\ f(\mathbf{Z}_n, \mathbf{Z}_1, \cdots, \mathbf{Z}_{n-1}) \end{bmatrix} \quad \text{and} \quad G = \begin{bmatrix} g(\mathbf{Z}_1, \mathbf{Z}_2, \cdots, \mathbf{Z}_n) \\ g(\mathbf{Z}_2, \mathbf{Z}_3, \cdots, \mathbf{Z}_1) \\ \vdots \\ g(\mathbf{Z}_n, \mathbf{Z}_1, \cdots, \mathbf{Z}_{n-1}) \end{bmatrix}. \tag{19.7}$$

Suppose that equation (19.6) has a solution F and that C satisfies $ACA + A = O$. Then

$$(AC + I)G = (AC + I)AF = (ACA + A)F = \mathcal{O},$$

i.e., Eq. (19.2) must be satisfied. Conversely, let Eq. (19.2) hold for some cyclic matrix C. Then $-CG$ is easily seen to be a solution of Eq. (19.6):

$$A(-CG) = -(AC + I)G + IG = IG = G.$$

Now let us prove that Eq. (19.3) is the general solution of the equation (19.1).

Let f be a solution of the equation (19.1), which we will write in the form

$$E(f) = g. \tag{19.8}$$

We denote by f_h the general solution of the equation $E(f) = \mathbf{O}$, and by f_p we denote a particular solution of the equation (19.8).

Then $f = f_h + f_p$ is the general solution of the equation (19.8). Indeed,

$$E(f_h + f_p) = E(f_h) + E(f_p) = g.$$

On the other hand, let f be an arbitrary solution of the equation (19.8). Then

$$E(f - f_p) = E(f) - E(f_p) = g - g = \mathbf{O},$$

i.e., $f - f_p$ is a solution of the associated homogeneous equation. So there exists a specialization $\bar{f}_h$ of the expression f_h such that

$$f - f_p = \bar{f}_h, \quad i.e., \quad f = \bar{f}_h + f_p.$$

Thus $f_h + f_p$ includes all solutions of the equation (19.8).

The general solution of the homogeneous equation $E(f) = \mathbf{O}$ given in a matrix form according to Theorem 18.1 is BH, where B and H are defined by Eqs. (18.3) and (18.10) respectively, and a particular solution of the equation $E(f) = g$ in a matrix form is $-CG$, then $F = BH - CG$ includes all solutions of the nonhomogeneous equation.

On the other hand, every function of the form Eq. (19.3) satisfies the functional equation (19.1). $\qquad\qquad\qquad\qquad\qquad\qquad\qquad\qquad\qquad\quad\square$

Next we will consider the functional equation

$$a_1 f(\mathbf{Z}_1, \mathbf{Z}_2, \mathbf{Z}_3) + a_2 f(\mathbf{Z}_2, \mathbf{Z}_3, \mathbf{Z}_1) + a_3 f(\mathbf{Z}_3, \mathbf{Z}_1, \mathbf{Z}_2) = g(\mathbf{Z}_1, \mathbf{Z}_2, \mathbf{Z}_3). \tag{19.9}$$

By a similar procedure as in the previous section one can prove the following lemma.

Lemma 19.2 *The functional equation (19.9) is equivalent to*

I. $f(\mathbf{Z}_1,\mathbf{Z}_2,\mathbf{Z}_3) = \frac{1}{\Delta}[(a_1^2-a_2a_3)g(\mathbf{Z}_1,\mathbf{Z}_2,\mathbf{Z}_3)+(a_3^2-a_1a_2)g(\mathbf{Z}_2,\mathbf{Z}_3,\mathbf{Z}_1)$

$$+(a_2^2 - a_1a_3)g(\mathbf{Z}_3,\mathbf{Z}_1,\mathbf{Z}_2)]$$

if

$$a_1 + a_2 + a_3 \neq 0 \quad and \quad (a_1 - a_2)^2 + (a_2 - a_3)^2 + (a_3 - a_1)^2 \neq 0;$$

II. $f(\mathbf{Z}_1,\mathbf{Z}_2,\mathbf{Z}_3) + f(\mathbf{Z}_2,\mathbf{Z}_3,\mathbf{Z}_1) + f(\mathbf{Z}_3,\mathbf{Z}_1,\mathbf{Z}_2) = \frac{1}{a_1}g(\mathbf{Z}_1,\mathbf{Z}_2,\mathbf{Z}_3)$ *if*
$a_1 = a_2 = a_3 \neq 0$
or

$$f(\mathbf{Z}_1,\mathbf{Z}_2,\mathbf{Z}_3) - \omega_3 f(\mathbf{Z}_2,\mathbf{Z}_3,\mathbf{Z}_1)$$

$$= \begin{cases} \dfrac{-a_1a_2\omega_6 g(\mathbf{Z}_1,\mathbf{Z}_2,\mathbf{Z}_3) + a_1^2\omega_6^2 g(\mathbf{Z}_2,\mathbf{Z}_3,\mathbf{Z}_1) + a_2^2 g(\mathbf{Z}_3,\mathbf{Z}_1,\mathbf{Z}_2)}{a_2^3 - a_1^3} \\ \quad if\ a_2^3 \neq a_1^3, \\[2mm] \dfrac{g(\mathbf{Z}_1,\mathbf{Z}_2,\mathbf{Z}_3) - 2\omega_3 g(\mathbf{Z}_2,\mathbf{Z}_3,\mathbf{Z}_1) + 4\omega_3^2 g(\mathbf{Z}_3,\mathbf{Z}_1,\mathbf{Z}_2)}{9a_1} \\ \quad if\ a_2 = a_1\omega_3 \end{cases}$$

if

$$a_1 + a_2 + a_3 \neq 0 \quad and \quad (a_1 - a_2)^2 + (a_2 - a_3)^2 + (a_3 - a_1)^2 = 0;$$

III. $f(\mathbf{Z}_1,\mathbf{Z}_2,\mathbf{Z}_3) - f(\mathbf{Z}_2,\mathbf{Z}_3,\mathbf{Z}_1)$

$$= \frac{1}{a_2^3 - a_1^3}[a_1a_2 g(\mathbf{Z}_1,\mathbf{Z}_2,\mathbf{Z}_3) + a_1^2 g(\mathbf{Z}_2,\mathbf{Z}_3,\mathbf{Z}_1) + a_2^2 g(\mathbf{Z}_3,\mathbf{Z}_1,\mathbf{Z}_2)]$$

if

$$a_1 + a_2 + a_3 = 0 \quad and \quad (a_1 - a_2)^2 + (a_2 - a_3)^2 + (a_3 - a_1)^2 \neq 0;$$

IV. $\mathbf{O} = g(\mathbf{Z}_1,\mathbf{Z}_2,\mathbf{Z}_3)$ *or*

$$f(\mathbf{Z}_1,\mathbf{Z}_2,\mathbf{Z}_3) + \omega_3 f(\mathbf{Z}_2,\mathbf{Z}_3,\mathbf{Z}_1) + \omega_3^2 f(\mathbf{Z}_3,\mathbf{Z}_1,\mathbf{Z}_2) = \frac{1}{a_1}g(\mathbf{Z}_1,\mathbf{Z}_2,\mathbf{Z}_3)$$

if

$$a_1 + a_2 + a_3 = 0 \quad and \quad (a_1 - a_2)^2 + (a_2 - a_3)^2 + (a_3 - a_1)^2 = 0.$$

For each of the above equations we will determine the conditions which must be satisfied by the function g so that the equation should have a solution.

Proposition 19.3 *The equation*

$$a_1 f(\mathbf{Z}_1, \mathbf{Z}_2, \mathbf{Z}_3) + a_2 f(\mathbf{Z}_2, \mathbf{Z}_3, \mathbf{Z}_1) + a_3 f(\mathbf{Z}_3, \mathbf{Z}_1, \mathbf{Z}_2) = g(\mathbf{Z}_1, \mathbf{Z}_2, \mathbf{Z}_3)$$

whose coefficients satisfy the conditions

$$a_1 = a_2 = a_3 \neq 0$$

has a solution if and only if the function g satisfies the condition

$$g(\mathbf{Z}_1, \mathbf{Z}_2, \mathbf{Z}_3) - g(\mathbf{Z}_2, \mathbf{Z}_3, \mathbf{Z}_1) = \mathbf{O}.$$

Proof. In this case the equation considered is equivalent to the equation

$$a_1 \Big[f(\mathbf{Z}_1, \mathbf{Z}_2, \mathbf{Z}_3) + f(\mathbf{Z}_2, \mathbf{Z}_3, \mathbf{Z}_1) + f(\mathbf{Z}_3, \mathbf{Z}_1, \mathbf{Z}_2) \Big] = g(\mathbf{Z}_1, \mathbf{Z}_2, \mathbf{Z}_3).$$

Since

$$A = a_1 \begin{bmatrix} 1 & 1 & 1 \\ 1 & 1 & 1 \\ 1 & 1 & 1 \end{bmatrix},$$

then $A^2 = 3a_1 A$, i.e. $A\left(\frac{-I}{3a_1}\right) A + A = O$, and hence we obtain $C = \frac{-I}{3a_1}$. The condition $ACG + G = O$ reduces to

$$\frac{2}{3} g(\mathbf{Z}_1, \mathbf{Z}_2, \mathbf{Z}_3) - \frac{1}{3} g(\mathbf{Z}_2, \mathbf{Z}_3, \mathbf{Z}_1) - \frac{1}{3} g(\mathbf{Z}_3, \mathbf{Z}_1, \mathbf{Z}_2) = \mathbf{O}$$

and (see Lemma 18.4) eventually to

$$g(\mathbf{Z}_1, \mathbf{Z}_2, \mathbf{Z}_3) - g(\mathbf{Z}_2, \mathbf{Z}_3, \mathbf{Z}_1) = \mathbf{O}.$$

$\square$

In a similar way, we obtain the necessary and sufficient conditions for solvability of the other equations. Thus, we obtain the following result.

Theorem 19.4 *The functional equation*

$$a_1 f(\mathbf{Z}_1, \mathbf{Z}_2, \mathbf{Z}_3) + a_2 f(\mathbf{Z}_2, \mathbf{Z}_3, \mathbf{Z}_1) + a_3 f(\mathbf{Z}_3, \mathbf{Z}_1, \mathbf{Z}_2) = g(\mathbf{Z}_1, \mathbf{Z}_2, \mathbf{Z}_3)$$

has a solution if and only if the function g satisfies the following conditions

1°. $\quad g(\mathbf{Z}_1, \mathbf{Z}_2, \mathbf{Z}_3)$ *is arbitrary if*

$$a_1 + a_2 + a_3 \neq 0 \quad and \quad (a_1 - a_2)^2 + (a_2 - a_3)^2 + (a_3 - a_1)^2 \neq 0;$$

2°. $\quad g(\mathbf{Z}_1, \mathbf{Z}_2, \mathbf{Z}_3) - g(\mathbf{Z}_2, \mathbf{Z}_3, \mathbf{Z}_1) = \mathbf{O}$ *or*

$$g(\mathbf{Z}_1, \mathbf{Z}_2, \mathbf{Z}_3) + \omega_3 g(\mathbf{Z}_2, \mathbf{Z}_3, \mathbf{Z}_1) + \omega_3^2 g(\mathbf{Z}_3, \mathbf{Z}_1, \mathbf{Z}_2) = \mathbf{O}$$

if

$$a_1 + a_2 + a_3 \neq 0 \quad and \quad (a_1 - a_2)^2 + (a_2 - a_3)^2 + (a_3 - a_1)^2 = 0;$$

3°. $\quad g(\mathbf{Z}_1, \mathbf{Z}_2, \mathbf{Z}_3) + g(\mathbf{Z}_2, \mathbf{Z}_3, \mathbf{Z}_1) + g(\mathbf{Z}_3, \mathbf{Z}_1, \mathbf{Z}_2) = \mathbf{O}$ *if*

$$a_1 + a_2 + a_3 = 0 \quad and \quad (a_1 - a_2)^2 + (a_2 - a_3)^2 + (a_3 - a_1)^2 \neq 0;$$

4°. $\quad g = \mathbf{O}$ *or*

$$g(\mathbf{Z}_1, \mathbf{Z}_2, \mathbf{Z}_3) - \omega_3 g(\mathbf{Z}_2, \mathbf{Z}_3, \mathbf{Z}_1) = \mathbf{O}$$

if

$$a_1 + a_2 + a_3 = 0 \quad and \quad (a_1 - a_2)^2 + (a_2 - a_3)^2 + (a_3 - a_1)^2 = 0.$$

Now we will find the general solution for any equation of Lemma 19.2. We will illustrate this only for the second equation.

Proposition 19.5 *The general solution of the equation*

$$f(\mathbf{Z}_1, \mathbf{Z}_2, \mathbf{Z}_3) + f(\mathbf{Z}_2, \mathbf{Z}_3, \mathbf{Z}_1) + f(\mathbf{Z}_3, \mathbf{Z}_1, \mathbf{Z}_2) = \frac{1}{a_1} g(\mathbf{Z}_1, \mathbf{Z}_2, \mathbf{Z}_3)$$

is given by

$$f(\mathbf{Z}_1, \mathbf{Z}_2, \mathbf{Z}_3) = b_1 h(\mathbf{Z}_1, \mathbf{Z}_2, \mathbf{Z}_3) + b_2 h(\mathbf{Z}_2, \mathbf{Z}_3, \mathbf{Z}_1)$$

$$- (b_1 + b_2) h(\mathbf{Z}_3, \mathbf{Z}_1, \mathbf{Z}_2) + \frac{1}{3a_1} g(\mathbf{Z}_1, \mathbf{Z}_2, \mathbf{Z}_3),$$

where h is an arbitrary complex vector function with values in $\mathcal{V}$.

Proof. The general solution of the corresponding homogeneous equation is

$$f(\mathbf{Z}_1, \mathbf{Z}_2, \mathbf{Z}_3) = b_1 h(\mathbf{Z}_1, \mathbf{Z}_2, \mathbf{Z}_3) + b_2 h(\mathbf{Z}_2, \mathbf{Z}_3, \mathbf{Z}_1) - (b_1 + b_2) h(\mathbf{Z}_3, \mathbf{Z}_1, \mathbf{Z}_2).$$

The particular solution of the considered nonhomogeneous equation is

$$f(\mathbf{Z}_1, \mathbf{Z}_2, \mathbf{Z}_3) = \frac{1}{3a_1} g(\mathbf{Z}_1, \mathbf{Z}_2, \mathbf{Z}_3).$$

Thus, the general solution of the given nonhomogeneous equation is

$$f(\mathbf{Z}_1, \mathbf{Z}_2, \mathbf{Z}_3) = b_1 h(\mathbf{Z}_1, \mathbf{Z}_2, \mathbf{Z}_3) + b_2 h(\mathbf{Z}_2, \mathbf{Z}_3, \mathbf{Z}_1)$$

$$-(b_1 + b_2) h(\mathbf{Z}_3, \mathbf{Z}_1, \mathbf{Z}_2) + \frac{1}{3a_1} g(\mathbf{Z}_1, \mathbf{Z}_2, \mathbf{Z}_3).$$

$\square$

Theorem 19.6　　*The general solution of the equation*

$$a_1 f(\mathbf{Z}_1, \mathbf{Z}_2, \mathbf{Z}_3) + a_2 f(\mathbf{Z}_2, \mathbf{Z}_3, \mathbf{Z}_1) + a_3 f(\mathbf{Z}_3, \mathbf{Z}_1, \mathbf{Z}_2) = g(\mathbf{Z}_1, \mathbf{Z}_2, \mathbf{Z}_3)$$

in the cases given in the Theorem 19.4 is

$1^o.$　$f(\mathbf{Z}_1, \mathbf{Z}_2, \mathbf{Z}_3) = \frac{1}{\Delta}[(a_1^2 - a_2 a_3) g(\mathbf{Z}_1, \mathbf{Z}_2, \mathbf{Z}_3)$

$$+(a_3^2 - a_1 a_2) g(\mathbf{Z}_2, \mathbf{Z}_3, \mathbf{Z}_1) + (a_2^2 - a_1 a_3) g(\mathbf{Z}_3, \mathbf{Z}_1, \mathbf{Z}_2)]$$

if

$$a_1 + a_2 + a_3 \neq 0 \quad and \quad (a_1 - a_2)^2 + (a_2 - a_3)^2 + (a_3 - a_1)^2 \neq 0;$$

$2^o.$　$f(\mathbf{Z}_1, \mathbf{Z}_2, \mathbf{Z}_3) = b_1 h(\mathbf{Z}_1, \mathbf{Z}_2, \mathbf{Z}_3) + b_2 h(\mathbf{Z}_2, \mathbf{Z}_3, \mathbf{Z}_1)$

$$-(b_1 + b_2) h(\mathbf{Z}_3, \mathbf{Z}_1, \mathbf{Z}_2) + \frac{1}{3a_1} g(\mathbf{Z}_1, \mathbf{Z}_2, \mathbf{Z}_3) \quad if \quad a_1 = a_2 = a_3 \neq 0$$

or

$$f(\mathbf{Z}_1, \mathbf{Z}_2, \mathbf{Z}_3) = h(\mathbf{Z}_1, \mathbf{Z}_2, \mathbf{Z}_3) + \omega_3 h(\mathbf{Z}_2, \mathbf{Z}_3, \mathbf{Z}_1) + \omega_3^2 h(\mathbf{Z}_3, \mathbf{Z}_1, \mathbf{Z}_2) +$$

$$\begin{cases} \dfrac{(a_1^2 - 2a_1 a_2 \omega_6 - a_2^2 \omega_6^2) g(\mathbf{Z}_1, \mathbf{Z}_2, \mathbf{Z}_3) + \omega_3 (a_1^2 - a_1 a_2 \omega_6 - 2a_2^2 \omega_6^2) g(\mathbf{Z}_2, \mathbf{Z}_3, \mathbf{Z}_1)}{3(a_2^3 - a_1^3)} \\ \qquad if\ a_2^3 \neq a_1^3, \\ -\dfrac{\omega_3}{3a_1} g(\mathbf{Z}_2, \mathbf{Z}_3, \mathbf{Z}_1) \quad if\ a_2 = a_1 \omega_3 \end{cases}$$

if

$$a_1 + a_2 + a_3 \neq 0 \quad and \quad (a_1 - a_2)^2 + (a_2 - a_3)^2 + (a_3 - a_1)^2 = 0;$$

$3^o.\quad f(\mathbf{Z}_1, \mathbf{Z}_2, \mathbf{Z}_3) = b_1[h(\mathbf{Z}_1, \mathbf{Z}_2, \mathbf{Z}_3) + h(\mathbf{Z}_2, \mathbf{Z}_3, \mathbf{Z}_1) + h(\mathbf{Z}_3, \mathbf{Z}_1, \mathbf{Z}_2)]$

$$+ \frac{(a_2 - a_1)g(\mathbf{Z}_1, \mathbf{Z}_2, \mathbf{Z}_3) + (a_1 + 2a_2)g(\mathbf{Z}_2, \mathbf{Z}_3, \mathbf{Z}_1)}{3(a_1^2 + a_1 a_2 + a_2^2)}$$

if

$$a_1 + a_2 + a_3 = 0 \quad and \quad (a_1 - a_2)^2 + (a_2 - a_3)^2 + (a_3 - a_1)^2 \neq 0;$$

$4^o.\quad f(\mathbf{Z}_1, \mathbf{Z}_2, \mathbf{Z}_3) = h(\mathbf{Z}_1, \mathbf{Z}_2, \mathbf{Z}_3) \ or$

$$\begin{aligned}
f(\mathbf{Z}_1, \mathbf{Z}_2, \mathbf{Z}_3) \ &= \ b_1 h(\mathbf{Z}_1, \mathbf{Z}_2, \mathbf{Z}_3) + b_2 h(\mathbf{Z}_2, \mathbf{Z}_3, \mathbf{Z}_1) \\
&- \ (\omega_3^2 b_1 + \omega_3 b_2)h(\mathbf{Z}_3, \mathbf{Z}_1, \mathbf{Z}_2) + \frac{1}{3a_1}g(\mathbf{Z}_1, \mathbf{Z}_2, \mathbf{Z}_3)
\end{aligned}$$

if

$$a_1 + a_2 + a_3 = 0 \quad and \ (a_1 - a_2)^2 + (a_2 - a_3)^2 + (a_3 - a_1)^2 = 0.$$

20 Paracyclic Functional Equations with Constant Coefficients

Let $\mathcal{V}$ be a complex vector space with complex dimension n, and let the complex vectors $\mathbf{X}_i$, $\mathbf{Y}_j \in \mathcal{V}$ $(1 \leq i, j \leq n)$ be given as above. Throughout this section $\mathcal{C}_i$ are constant complex vectors in $\mathcal{V}$ and let $f : \mathcal{V}^{n+k} \mapsto \mathcal{V}$.

Now we will consider the following paracyclic complex vector functional equation of the first kind

$$\sum_{i=1}^{n} a_i f(\mathbf{X}_i, \mathbf{X}_{i+1}, \cdots, \mathbf{X}_{i+n-1}, \mathbf{Y}_i, \mathbf{Y}_{i+1}, \cdots, \mathbf{Y}_{i+k-1}) = \mathbf{O} \qquad (20.1)$$

$$(\mathbf{X}_{n+i} \equiv \mathbf{X}_i, \quad \mathbf{Y}_{n+i} \equiv \mathbf{Y}_i)$$

where a_i $(1 \leq i \leq n)$ are complex constants.

First, we will consider two particular cases for $k = 1$ $(n > 1)$ and $k = n$. Let us determine the general solution of the equation

$$\sum_{i=1}^{n} a_i f(\mathbf{X}_i, \mathbf{X}_{i+1}, \cdots, \mathbf{X}_{i+n-1}, \mathbf{Y}_i) = \mathbf{O}. \qquad (20.2)$$

By a cyclic permutation of the vectors in Eq. (20.2), we obtain the matrix system

$$AF = \mathcal{O},\qquad\qquad(20.3)$$

where

$$A = \begin{bmatrix} a_1 & a_2 & \cdots & a_n \\ a_n & a_1 & \cdots & a_{n-1} \\ \vdots & & & \\ a_2 & a_3 & \cdots & a_1 \end{bmatrix},\qquad\qquad(20.4)$$

$$F = \begin{bmatrix} f(\mathbf{X}_1,\cdots,\mathbf{X}_n,\mathbf{Y}_1) \\ f(\mathbf{X}_2,\cdots,\mathbf{X}_1,\mathbf{Y}_2) \\ \vdots \\ f(\mathbf{X}_n,\cdots,\mathbf{X}_{n-1},\mathbf{Y}_n) \end{bmatrix} \quad \text{and} \quad \mathcal{O} = \begin{bmatrix} \mathbf{O} \\ \mathbf{O} \\ \vdots \\ \mathbf{O} \end{bmatrix}.$$

For the system (20.3) the following theorem holds.

Theorem 20.1 *The general solution of the functional equation (20.2) is given by the formula*

$$F = BH,\qquad\qquad(20.5)$$

if

$$AB = O,\qquad\qquad(20.6)$$

where A and B are nonzero $n \times n$ cyclic matrices given by Eq. (18.3), O is the $n \times n$ zero matrix and

$$H = \begin{bmatrix} h(\mathbf{X}_1,\mathbf{X}_2,\cdots,\mathbf{X}_n) \\ h(\mathbf{X}_2,\mathbf{X}_3,\cdots,\mathbf{X}_1) \\ \vdots \\ h(\mathbf{X}_n,\mathbf{X}_1,\cdots,\mathbf{X}_{n-1}) \end{bmatrix}\qquad\qquad(20.7)$$

where h is an arbitrary complex vector function with values in $\mathcal{V}$.

Proof.   If not all coefficients a_i $(1 \leq i \leq n)$ are 0, we can suppose without loss of generality that $a_1 \neq 0$. Then equation (20.2) is equivalent to the equation

$$f(\mathbf{X}_1,\cdots,\mathbf{X}_n,\mathbf{Y}_1) = -\frac{a_2}{a_1}f(\mathbf{X}_2,\cdots,\mathbf{X}_n,\mathbf{X}_1,\mathbf{Y}_2) - \cdots$$
$$- \frac{a_n}{a_1}f(\mathbf{X}_n,\mathbf{X}_1,\cdots,\mathbf{X}_{n-1},\mathbf{Y}_n).$$

By putting $\mathbf{Y}_i = \mathcal{C}_i$ $(2 \le i \le n)$ where $\mathcal{C}_i$ are arbitrary complex constant vectors from $\mathcal{V}$, we obtain

$$f(\mathbf{X}_1,\cdots,\mathbf{X}_n,\mathbf{Y}_1) = -\frac{a_2}{a_1}f(\mathbf{X}_2,\cdots,\mathbf{X}_n,\mathbf{X}_1,\mathcal{C}_2) - \cdots \qquad (20.8)$$
$$- \frac{a_n}{a_1}f(\mathbf{X}_n,\mathbf{X}_1,\cdots,\mathbf{X}_{n-1},\mathcal{C}_n).$$

The right-hand side of the last equation depends on $\mathbf{X}_1,\cdots,\mathbf{X}_n$ only. Denote this expression by $h(\mathbf{X}_1,\cdots,\mathbf{X}_n)$.

Therefore, Eq. (20.8) obtains the following form

$$f(\mathbf{X}_1,\cdots,\mathbf{X}_n,\mathbf{Y}_1) = h(\mathbf{X}_1,\cdots,\mathbf{X}_n). \qquad (20.9)$$

The formula Eq. (20.9) is the general solution of the equation (20.2) if and only if it holds

$$\sum_{i=1}^{n} a_i h(\mathbf{X}_i,\mathbf{X}_{i+1},\cdots,\mathbf{X}_{i+n-1}) = \mathbf{O}. \qquad (20.10)$$

The above equation is equivalent to the functional equation (18.1), and therefore Theorem 18.1 holds. Thus Eq. (20.5) is true. $\square$

Now, we will solve the functional equation (20.1) if $k = n$.

By denoting the pairs $(\mathbf{X}_i,\mathbf{Y}_i) = \mathbf{Z}_i$ $(1 \le i \le n)$, the functional equation (20.1) takes the form Eq. (18.1), then Theorem 18.1 holds, *i.e.*, the general solution is given by

$$F = BH,$$

where

$$F = \begin{bmatrix} f(\mathbf{Z}_1,\mathbf{Z}_2,\cdots,\mathbf{Z}_n) \\ f(\mathbf{Z}_2,\mathbf{Z}_3,\cdots,\mathbf{Z}_1) \\ \vdots \\ f(\mathbf{Z}_n,\mathbf{Z}_1,\cdots,\mathbf{Z}_{n-1}) \end{bmatrix}$$

$$
= \begin{bmatrix}
f(\mathbf{X}_1, \mathbf{X}_2, \cdots, \mathbf{X}_n, \mathbf{Y}_1, \mathbf{Y}_2, \cdots, \mathbf{Y}_n) \\
f(\mathbf{X}_2, \mathbf{X}_3, \cdots, \mathbf{X}_1, \mathbf{Y}_2, \mathbf{Y}_3, \cdots, \mathbf{Y}_1) \\
\vdots \\
f(\mathbf{X}_n, \mathbf{X}_1, \cdots, \mathbf{X}_{n-1}, \mathbf{Y}_n, \mathbf{Y}_1, \cdots, \mathbf{Y}_{n-1})
\end{bmatrix},
$$

$$
H = \begin{bmatrix}
h(\mathbf{X}_1, \mathbf{X}_2, \cdots, \mathbf{X}_n, \mathbf{Y}_1, \mathbf{Y}_2, \cdots, \mathbf{Y}_n) \\
h(\mathbf{X}_2, \mathbf{X}_3, \cdots, \mathbf{X}_1, \mathbf{Y}_2, \mathbf{Y}_3, \cdots, \mathbf{Y}_1) \\
\vdots \\
h(\mathbf{X}_n, \mathbf{X}_1, \cdots, \mathbf{X}_{n-1}, \mathbf{Y}_n, \mathbf{Y}_1, \cdots, \mathbf{Y}_{n-1})
\end{bmatrix}
$$

and B is given by Eq. (18.3).

Next, we will consider the case $1 < k < n$. To this end, instead of the equation (20.1) we will consider the equation

$$
a_1 f(\mathbf{X}_1, \cdots, \mathbf{X}_n, \mathbf{Y}_1, \cdots, \mathbf{Y}_k, \mathbf{Y}_{k+1}, \cdots, \mathbf{Y}_n) \tag{20.11}
$$
$$
+ \quad a_2 f(\mathbf{X}_2, \cdots, \mathbf{X}_n, \mathbf{X}_1, \mathbf{Y}_2, \cdots, \mathbf{Y}_k, \mathbf{Y}_{k+1}, \cdots, \mathbf{Y}_n, \mathbf{Y}_1) + \cdots
$$

$$
+ a_n f(\mathbf{X}_n, \mathbf{X}_1, \cdots, \mathbf{X}_{n-1}, \mathbf{Y}_n, \mathbf{Y}_1, \cdots, \mathbf{Y}_{n+k-1}, \mathbf{Y}_{n+k}, \cdots, \mathbf{Y}_{n-1}) = \mathbf{O}.
$$

By a cyclic permutation of the vectors in the last equation, we obtain the matrix system (20.3), where

$$
F = \tag{20.12}
$$

$$
\begin{bmatrix}
f(\mathbf{X}_1, \cdots, \mathbf{X}_n, \mathbf{Y}_1, \cdots, \mathbf{Y}_k, \mathbf{Y}_{k+1}, \cdots, \mathbf{Y}_n) \\
f(\mathbf{X}_2, \cdots, \mathbf{X}_n, \mathbf{X}_1, \mathbf{Y}_2, \cdots, \mathbf{Y}_k, \mathbf{Y}_{k+1}, \cdots, \mathbf{Y}_n, \mathbf{Y}_1) \\
\vdots \\
f(\mathbf{X}_n, \mathbf{X}_1, \cdots, \mathbf{X}_{n-1}, \mathbf{Y}_n, \mathbf{Y}_1, \cdots, \mathbf{Y}_{n+k-1}, \mathbf{Y}_{n+k}, \cdots, \mathbf{Y}_{n-1})
\end{bmatrix},
$$

A and $\mathcal{O}$ are as in Eq. (20.4).

The necessary and sufficient condition for the system (20.3) with Eq. (20.12) to have nontrivial solution is $\det A = 0$. Since $\det A$ is cyclic, then its value is

$$
\det A = \prod_{i=0}^{n-1} E(\varepsilon_i),
$$

where ε_i $(0 \leq i \leq n-1)$ are distinct roots of the binomial equation

$$
b(x) \equiv 1 - x^n = 0.
$$

Therefore, the equation (20.11) has nontrivial solutions if and only if the characteristic equation $E(x) \equiv a_1 + a_2 x + \cdots + a_n x^{n-1} = 0$ has common roots with the binomial equation $b(x) \equiv 1 - x^n = 0$. If this is so, we can write $E(x) = P(x)D(x)$, $b(x) = D(x)F(x)$, where $D(x)$ is the greatest common divisor of the polynomials $E(x)$ and $b(x)$.

The general solution of the equation (20.11) is given by the formula

$$
\begin{aligned}
& f(\mathbf{X}_1, \cdots, \mathbf{X}_n, \mathbf{Y}_1, \cdots, \mathbf{Y}_k, \mathbf{Y}_{k+1}, \cdots, \mathbf{Y}_n) \\
= \ & b_1 h(\mathbf{X}_1, \cdots, \mathbf{X}_n, \mathbf{Y}_1, \cdots, \mathbf{Y}_m, \mathbf{Y}_{m+1}, \cdots, \mathbf{Y}_n) \\
+ \ & b_2 h(\mathbf{X}_2, \cdots, \mathbf{X}_n, \mathbf{X}_1, \mathbf{Y}_2, \cdots, \mathbf{Y}_m, \mathbf{Y}_{m+1}, \cdots, \mathbf{Y}_n, \mathbf{Y}_1) + \cdots \\
+ \ & b_{s+1} h(\mathbf{X}_{s+1}, \cdots, \mathbf{X}_n, \mathbf{X}_1, \cdots, \mathbf{X}_s, \\
& \qquad \mathbf{Y}_{s+1}, \cdots, \mathbf{Y}_{m+s}, \mathbf{Y}_{m+s+1}, \cdots, \mathbf{Y}_n, \mathbf{Y}_1, \cdots, \mathbf{Y}_s),
\end{aligned}
$$

where the complex numbers b_i $(1 \leq i \leq s+1)$ are coefficients of the polynomial

$$
b_1 + b_2 x + \cdots + b_{s+1} x^s = F(x).
$$

The functional equation (20.1) will be called *reduced* equation of the equation (20.11).

Now we will prove the following result.

Theorem 20.2 *Every function f given by*

$$
\begin{aligned}
& f(\mathbf{X}_1, \cdots, \mathbf{X}_n, \mathbf{Y}_1, \cdots, \mathbf{Y}_n) \qquad\qquad (20.13) \\
= \ & b_1 h(\mathbf{X}_1, \cdots, \mathbf{X}_n, \mathbf{Y}_1, \cdots, \mathbf{Y}_m) \\
+ \ & b_2 h(\mathbf{X}_2, \cdots, \mathbf{X}_n, \mathbf{X}_1, \mathbf{Y}_2, \cdots, \mathbf{Y}_{m+1}) + \cdots \\
+ \ & b_{s+1} h(\mathbf{X}_{s+1}, \cdots, \mathbf{X}_n, \mathbf{X}_1, \cdots, \mathbf{X}_s, \mathbf{Y}_{s+1}, \cdots, \mathbf{Y}_{m+s}),
\end{aligned}
$$

satisfies the equation (20.1), where $m = k - s$ for $k > s$, $h(\mathbf{X}_1, \cdots, \mathbf{X}_n, \mathbf{Y}_s, \cdots, \mathbf{Y}_m)$ is an arbitrary complex vector function with values in $\mathcal{V}$ and if $k - s \leq 0$, then h is an arbitrary complex vector function only of $\mathbf{X}_1, \cdots, \mathbf{X}_n$ with values in the same space $\mathcal{V}$.

Proof. We should prove that $f = F(h)$ is solution of the equation (20.1), where h is an arbitrary complex vector function with values in $\mathcal{V}$.

Indeed, we have

$$
D(f) = D(F(h)) = b(h) = 0,
$$

from where it follows that $E(f) = P(D(f)) = \mathbf{O}$, which we were required to prove. $\qquad\square$

Next, we will solve the functional equation

$$a_1 f(\mathbf{X}_1, \mathbf{X}_2, \mathbf{X}_3, \mathbf{Y}_1, \mathbf{Y}_2) + a_2 f(\mathbf{X}_2, \mathbf{X}_3, \mathbf{X}_1, \mathbf{Y}_2, \mathbf{Y}_3) \quad (20.14)$$
$$+ \quad a_3 f(\mathbf{X}_3, \mathbf{X}_1, \mathbf{X}_2, \mathbf{Y}_3, \mathbf{Y}_1) = \mathbf{O}.$$

By a procedure similar to that in the first section, we may prove the following lemma.

Lemma 20.3 *The functional equation (20.14) is equivalent to the equation*
I. $f(\mathbf{X}_1, \mathbf{X}_2, \mathbf{X}_3, \mathbf{Y}_1, \mathbf{Y}_2) = \mathbf{O}$ *if*

$$a_1 + a_2 + a_3 \neq 0 \quad and \quad (a_1 - a_2)^2 + (a_2 - a_3)^2 + (a_3 - a_1)^2 \neq 0;$$

II. $f(\mathbf{X}_1, \mathbf{X}_2, \mathbf{X}_3, \mathbf{Y}_1, \mathbf{Y}_2) + f(\mathbf{X}_2, \mathbf{X}_3, \mathbf{X}_1, \mathbf{Y}_2, \mathbf{Y}_3) + f(\mathbf{X}_3, \mathbf{X}_1, \mathbf{X}_2, \mathbf{Y}_3, \mathbf{Y}_1)$
$= \mathbf{O}$ *or*

$$f(\mathbf{X}_1, \mathbf{X}_2\mathbf{X}_3, \mathbf{Y}_1, \mathbf{Y}_2) - \omega_3 f(\mathbf{X}_2, \mathbf{X}_3, \mathbf{X}_1, \mathbf{Y}_2, \mathbf{Y}_3) = \mathbf{O}$$

if

$$a_1 + a_2 + a_3 \neq 0 \quad and \quad (a_1 - a_2)^2 + (a_2 - a_3)^2 + (a_3 - a_1)^2 = 0;$$

III. $f(\mathbf{X}_1, \mathbf{X}_2, \mathbf{X}_3, \mathbf{Y}_1, \mathbf{Y}_2) - f(\mathbf{X}_2, \mathbf{X}_3, \mathbf{X}_1, \mathbf{Y}_2, \mathbf{Y}_3) = \mathbf{O}$ *if*

$$a_1 + a_2 + a_3 = 0 \quad and \quad (a_1 - a_2)^2 + (a_2 - a_3)^2 + (a_3 - a_1)^2 \neq 0;$$

IV. $\mathbf{O} = \mathbf{O}$ *or*

$$f(\mathbf{X}_1, \mathbf{X}_2, \mathbf{X}_3, \mathbf{Y}_1, \mathbf{Y}_2) + \omega_3 f(\mathbf{X}_2, \mathbf{X}_3, \mathbf{X}_1, \mathbf{Y}_2, \mathbf{Y}_3) + \omega_3^2 f(\mathbf{X}_3, \mathbf{X}_1, \mathbf{X}_2, \mathbf{Y}_3, \mathbf{Y}_1)$$

$= \mathbf{O}$ *if*

$$a_1 + a_2 + a_3 = 0 \quad and \quad (a_1 - a_2)^2 + (a_2 - a_3)^2 + (a_3 - a_1)^2 = 0.$$

Proposition 20.4 *The functional equation*

$$f(\mathbf{X}_1, \mathbf{X}_2, \mathbf{X}_3, \mathbf{Y}_1, \mathbf{Y}_2) - f(\mathbf{X}_2, \mathbf{X}_3, \mathbf{X}_1, \mathbf{Y}_2, \mathbf{Y}_3) = \mathbf{O},$$

has a general solution

$$f(\mathbf{X}_1, \mathbf{X}_2, \mathbf{X}_3, \mathbf{Y}_1, \mathbf{Y}_2) = h(\mathbf{X}_1, \mathbf{X}_2, \mathbf{X}_3) + h(\mathbf{X}_2, \mathbf{X}_3, \mathbf{X}_1) + h(\mathbf{X}_3, \mathbf{X}_1, \mathbf{X}_2),$$

where h is an arbitrary complex vector function of the variables $\mathbf{X}_1, \mathbf{X}_2, \mathbf{X}_3$ *with values in* $\mathcal{V}$.

Proof.   The given equation may be written in the following form

$$f(\mathbf{X}_1, \mathbf{X}_2, \mathbf{X}_3, \mathbf{Y}_1, \mathbf{Y}_2) = f(\mathbf{X}_2, \mathbf{X}_3, \mathbf{X}_1, \mathbf{Y}_2, \mathbf{Y}_3).$$

The left-hand side of the equation is independent of $\mathbf{Y}_3$ and the right-hand side is independent of $\mathbf{Y}_1$, so we have

$$f(\mathbf{X}_1, \mathbf{X}_2, \mathbf{X}_3, \mathbf{Y}_1, \mathbf{Y}_2) = F(\mathbf{X}_1, \mathbf{X}_2, \mathbf{X}_3, \mathbf{Y}_2) \tag{20.15}$$

and

$$f(\mathbf{X}_2, \mathbf{X}_3, \mathbf{X}_1, \mathbf{Y}_2, \mathbf{Y}_3) = F(\mathbf{X}_1, \mathbf{X}_2, \mathbf{X}_3, \mathbf{Y}_2). \tag{20.16}$$

On the other hand, from Eq. (20.15) we find

$$f(\mathbf{X}_2, \mathbf{X}_3, \mathbf{X}_1, \mathbf{Y}_2, \mathbf{Y}_3) = F(\mathbf{X}_2, \mathbf{X}_3, \mathbf{X}_1, \mathbf{Y}_3),$$

thus we have

$$F(\mathbf{X}_1, \mathbf{X}_2, \mathbf{X}_3, \mathbf{Y}_2) = F(\mathbf{X}_2, \mathbf{X}_3, \mathbf{X}_1, \mathbf{Y}_3). \tag{20.17}$$

Since the left-hand side of the above equation is independent of $\mathbf{Y}_3$ and the right-hand side is independent of $\mathbf{Y}_2$, we obtain

$$F(\mathbf{X}_1, \mathbf{X}_2, \mathbf{X}_3, \mathbf{Y}_2) = G(\mathbf{X}_1, \mathbf{X}_2, \mathbf{X}_3). \tag{20.18}$$

On the basis of the equality (20.18), the formula Eq. (20.15) becomes

$$f(\mathbf{X}_1, \mathbf{X}_2, \mathbf{X}_3, \mathbf{Y}_1, \mathbf{Y}_2) = G(\mathbf{X}_1, \mathbf{X}_2, \mathbf{X}_3). \tag{20.19}$$

The formula Eq. (20.19) gives a solution of the equation if and only if

$$G(\mathbf{X}_1, \mathbf{X}_2, \mathbf{X}_3) - G(\mathbf{X}_2, \mathbf{X}_3, \mathbf{X}_1) = \mathbf{O},$$

whose general solution is given by

$$G(\mathbf{X}_1, \mathbf{X}_2, \mathbf{X}_3) = h(\mathbf{X}_1, \mathbf{X}_2, \mathbf{X}_3) + h(\mathbf{X}_2, \mathbf{X}_3, \mathbf{X}_1) + h(\mathbf{X}_3, \mathbf{X}_1, \mathbf{X}_2).$$

Therefore, the general solution of the equation is

$$
\begin{aligned}
f(\mathbf{X}_1, \mathbf{X}_2, \mathbf{X}_3, \mathbf{Y}_1, \mathbf{Y}_2) &= h(\mathbf{X}_1, \mathbf{X}_2, \mathbf{X}_3) + h(\mathbf{X}_2, \mathbf{X}_3, \mathbf{X}_1) \\
&\quad + h(\mathbf{X}_3, \mathbf{X}_1, \mathbf{X}_2).
\end{aligned}
$$

$\square$

On the basis of the previous results, the following theorem holds.

Theorem 20.5   *The general solution of the equation*

$$a_1 f(\mathbf{X}_1, \mathbf{X}_2, \mathbf{X}_3, \mathbf{Y}_1, \mathbf{Y}_2) + a_2 f(\mathbf{X}_2, \mathbf{X}_3, \mathbf{X}_1, \mathbf{Y}_2, \mathbf{Y}_3)$$
$$+\ \ a_3 f(\mathbf{X}_3, \mathbf{X}_1, \mathbf{X}_2, \mathbf{Y}_3, \mathbf{Y}_1) = \mathbf{O}$$

is given by the formulae

$1^o.$ $f(\mathbf{X}_1, \mathbf{X}_2, \mathbf{X}_3, \mathbf{Y}_1, \mathbf{Y}_2) \equiv \mathbf{O}$ *if*

$$a_1 + a_2 + a_3 \neq 0 \quad and \quad (a_1 - a_2)^2 + (a_2 - a_3)^2 + (a_3 - a_1)^2 \neq 0;$$

$2^o.$ $f(\mathbf{X}_1, \mathbf{X}_2, \mathbf{X}_3, \mathbf{Y}_1, \mathbf{Y}_2) = h(\mathbf{X}_1, \mathbf{X}_2, \mathbf{X}_3, \mathbf{Y}_1) - h(\mathbf{X}_2, \mathbf{X}_3, \mathbf{X}_1, \mathbf{Y}_2)$ *or*

$$f(\mathbf{X}_1, \mathbf{X}_2, \mathbf{X}_3, \mathbf{Y}_1, \mathbf{Y}_2) = h(\mathbf{X}_1, \mathbf{X}_2, \mathbf{X}_3) + \omega_3 h(\mathbf{X}_2, \mathbf{X}_3, \mathbf{X}_1)$$

$+\omega_3^2 h(\mathbf{X}_3, \mathbf{X}_1, \mathbf{X}_2)$ *if*

$$a_1 + a_2 + a_3 \neq 0 \quad and \quad (a_1 - a_2)^2 + (a_2 - a_3)^2 + (a_3 - a_1)^2 = 0;$$

$3^o.$ $f(\mathbf{X}_1, \mathbf{X}_2, \mathbf{X}_3, \mathbf{Y}_1, \mathbf{Y}_2) = h(\mathbf{X}_1, \mathbf{X}_2, \mathbf{X}_3) + h(\mathbf{X}_2, \mathbf{X}_3, \mathbf{X}_1)$
$+h(\mathbf{X}_3, \mathbf{X}_1, \mathbf{X}_2)$ *if*

$$a_1 + a_2 + a_3 = 0 \quad and \quad (a_1 - a_2)^2 + (a_2 - a_3)^2 + (a_3 - a_1)^2 \neq 0;$$

$4^o.$ $f(\mathbf{X}_1, \mathbf{X}_2, \mathbf{X}_3, \mathbf{Y}_1, \mathbf{Y}_2) = h(\mathbf{X}_1, \mathbf{X}_2, \mathbf{X}_3, \mathbf{Y}_1, \mathbf{Y}_2)$ *or*

$$f(\mathbf{X}_1, \mathbf{X}_2, \mathbf{X}_3, \mathbf{Y}_1, \mathbf{Y}_2) = h(\mathbf{X}_1, \mathbf{X}_2, \mathbf{X}_3, \mathbf{Y}_1) - \omega_3 h(\mathbf{X}_2, \mathbf{X}_3, \mathbf{X}_1, \mathbf{Y}_2)$$

if

$$a_1 + a_2 + a_3 = 0 \quad and \quad (a_1 - a_2)^2 + (a_2 - a_3)^2 + (a_3 - a_1)^2 = 0,$$

where h is an arbitrary complex vector function with values in V.

Chapter 5

Systems of Linear Functional Equations

In this chapter two types of systems of linear functional equations are solved, namely systems in which each equation contains all the unknown functions and systems in which not all equations contain all the unknown functions.

The results presented here are obtained in [I. B. Risteski *et al.* (submitted)] (see also [I. B. Risteski *et al.* (2001B)]).

21   Systems in Which Each Equation Contains All Unknown Functions

Now we prove the following results.

Theorem 21.1   *The general solution of the system of functional equations*

$$
\begin{aligned}
f_0(\mathbf{Z}_1, \mathbf{Z}_2) \;+\;& f_1(\mathbf{Z}_2, \mathbf{Z}_3, \mathbf{Z}_4) + g_1(\mathbf{Z}_1, \mathbf{Z}_3, \mathbf{Z}_4) \\
+\;& f_2(\mathbf{Z}_3, \mathbf{Z}_4, \mathbf{Z}_5, \mathbf{Z}_6) + g_2(\mathbf{Z}_1, \mathbf{Z}_3, \mathbf{Z}_5, \mathbf{Z}_6) = \mathbf{O},
\end{aligned} \tag{21.1}
$$

$$
\begin{aligned}
f_0(\mathbf{Z}_1, \mathbf{Z}_2) \;+\;& f_1(\mathbf{Z}_2, \mathbf{Z}_3, \mathbf{Z}_4) + g_1(\mathbf{Z}_1, \mathbf{Z}_3, \mathbf{Z}_4) \\
+\;& g_2(\mathbf{Z}_3, \mathbf{Z}_4, \mathbf{Z}_5, \mathbf{Z}_6) + f_2(\mathbf{Z}_1, \mathbf{Z}_3, \mathbf{Z}_5, \mathbf{Z}_6) = \mathbf{O},
\end{aligned} \tag{21.2}
$$

is determined by

$$
f_0(\mathbf{Z}_1, \mathbf{Z}_2) = F_1(\mathbf{Z}_1) - F_2(\mathbf{Z}_2),
$$

$$
f_1(\mathbf{Z}_1, \mathbf{Z}_2, \mathbf{Z}_3) = F_2(\mathbf{Z}_1) + G_1(\mathbf{Z}_2, \mathbf{Z}_3),
$$

$$
\begin{aligned}
g_1(\mathbf{Z}_1, \mathbf{Z}_2, \mathbf{Z}_3) \;=\;& -F_1(\mathbf{Z}_1) + G_2(\mathbf{Z}_1, \mathbf{Z}_3) \\
& -G_1(\mathbf{Z}_2, \mathbf{Z}_3) + G_2(\mathbf{Z}_2, \mathbf{Z}_3) + A,
\end{aligned} \tag{21.3}
$$

205

$$f_2(\mathbf{Z}_1, \mathbf{Z}_2, \mathbf{Z}_3, \mathbf{Z}_4) = -G_2(\mathbf{Z}_1, \mathbf{Z}_2) + G_3(\mathbf{Z}_3, \mathbf{Z}_4),$$

$$g_2(\mathbf{Z}_1, \mathbf{Z}_2, \mathbf{Z}_3, \mathbf{Z}_4) = -G_2(\mathbf{Z}_1, \mathbf{Z}_2) - G_3(\mathbf{Z}_3, \mathbf{Z}_4) - A,$$

where F_1, F_2, G_i $(i = 1, 2, 3)$ are arbitrary functions with values in $\mathcal{V}'$, and A is an arbitrary constant vector from $\mathcal{V}'$.

Proof.   By putting $\mathbf{Z}_i = C_i$ $(i = 3, 4, 5, 6)$ into Eq. (21.1), we obtain

$$f_0(\mathbf{Z}_1, \mathbf{Z}_2) = F_1(\mathbf{Z}_1) - F_2(\mathbf{Z}_2), \tag{21.4}$$

where we introduced the notations

$$\begin{aligned}
F_1(\mathbf{Z}_1) &= -g_1(\mathbf{Z}_1, C_3, C_4) - g_2(\mathbf{Z}_1, C_3, C_5, C_6), \\
F_2(\mathbf{Z}_2) &= f_1(\mathbf{Z}_2, C_3, C_4) + f_2(C_3, C_4, C_5, C_6).
\end{aligned}$$

By virtue of the expression Eq. (21.4), by putting $\mathbf{Z}_i = C_i$ $(i = 1, 5, 6)$ into Eq. (21.1) we get

$$f_1(\mathbf{Z}_2, \mathbf{Z}_3, \mathbf{Z}_4) = F_2(\mathbf{Z}_2) + G_1(\mathbf{Z}_3, \mathbf{Z}_4), \tag{21.5}$$

where

$$\begin{aligned}
G_1(\mathbf{Z}_3, \mathbf{Z}_4) &= -F_1(C_1) - g_1(C_1, \mathbf{Z}_3, \mathbf{Z}_4) \\
&\quad - f_2(\mathbf{Z}_3, \mathbf{Z}_4, C_5, C_6) - g_2(C_1, \mathbf{Z}_3, C_5, C_6).
\end{aligned}$$

If we put $\mathbf{Z}_i = C_i$ $(i = 5, 6)$ into Eq. (21.1), in view of Eqs. (21.4), (21.5) we have

$$g_1(\mathbf{Z}_1, \mathbf{Z}_3, \mathbf{Z}_4) = -F_1(\mathbf{Z}_1) - G_1(\mathbf{Z}_3, \mathbf{Z}_4) + G_2'(\mathbf{Z}_3, \mathbf{Z}_4) - H(\mathbf{Z}_1, \mathbf{Z}_3), \tag{21.6}$$

where we introduced the notations

$$G_2'(\mathbf{Z}_3, \mathbf{Z}_4) = -f_2(\mathbf{Z}_3, \mathbf{Z}_4, C_5, C_6), \quad H(\mathbf{Z}_1, \mathbf{Z}_3) = g_2(\mathbf{Z}_1, \mathbf{Z}_3, C_5, C_6).$$

By substituting Eqs. (21.4), (21.5) and (21.6) into Eqs. (21.1) and (21.2), the system becomes

$$G_2'(\mathbf{Z}_{73}, \mathbf{Z}_4) - H(\mathbf{Z}_1, \mathbf{Z}_3) \tag{21.7}$$
$$+ \ f_2(\mathbf{Z}_3, \mathbf{Z}_4, \mathbf{Z}_5, \mathbf{Z}_6) + g_2(\mathbf{Z}_1, \mathbf{Z}_3, \mathbf{Z}_5, \mathbf{Z}_6) = \mathbf{O},$$

$$G_2'(\mathbf{Z}_3, \mathbf{Z}_4) - H(\mathbf{Z}_1, \mathbf{Z}_3) \tag{21.8}$$
$$+ \ g_2(\mathbf{Z}_3, \mathbf{Z}_4, \mathbf{Z}_5, \mathbf{Z}_6) + f_2(\mathbf{Z}_1, \mathbf{Z}_3, \mathbf{Z}_5, \mathbf{Z}_6) = \mathbf{O}.$$

By putting $Z_i = C_1$ $(i = 1, 3)$ into Eq. (21.8), after replacing Z_4 by Z_3 and putting $Z_1 = C_1$ into Eq. (21.7), the equations (21.7) and (21.8) become

$$G_2'(Z_3, Z_4) - H(C_1, Z_3) \tag{21.9}$$
$$+ \quad f_2(Z_3, Z_4, Z_5, Z_6) + g_2(C_1, Z_3, Z_5, Z_6) = O,$$
$$G_2'(C_1, Z_3) - H(C_1, C_1) \tag{21.10}$$
$$+ \quad g_2(C_1, Z_3, Z_5, Z_6) + f_2(C_1, C_1, Z_5, Z_6) = O.$$

By subtracting Eq. (21.10) from Eq. (21.9), we obtain

$$f_2(Z_3, Z_4, Z_5, Z_6) = F_3(Z_3) - G_2'(Z_3, Z_4) + G_3(Z_5, Z_6), \tag{21.11}$$

where

$$F_3(Z_3) \quad = \quad G_2'(C_1, Z_3) - H(C_1, C_1) + H(C_1, Z_3),$$
$$G_3(Z_5, Z_6) \quad = \quad f_2(C_1, C_1, Z_5, Z_6).$$

From Eqs. (21.7) and (21.11) we obtain

$$g_2(Z_1, Z_3, Z_5, Z_6) = H(Z_1, Z_3) - F_3(Z_3) - G_3(Z_5, Z_6). \tag{21.12}$$

We substitute the functions f_2 and g_2 determined by Eqs. (21.11) and (21.12) (with Z_3, Z_4 replaced by Z_1, Z_3 and *vice versa*) into the equation (21.8) and obtain

$$G_2'(Z_3, Z_4) + H(Z_3, Z_4) - F_3(Z_4) \tag{21.13}$$
$$= \quad G_2'(Z_1, Z_3) + H(Z_1, Z_3) - F_3(Z_1).$$

It is clear that both sides of this equality are a function just of Z_3, say, $F_4(Z_3)$. Then we have

$$G_2'(Z_3, Z_4) + H(Z_3, Z_4) = F_3(Z_4) + F_4(Z_3). \tag{21.14}$$

If in Eq. (21.14) we replace Z_3, Z_4 by Z_1, Z_3, we obtain

$$G_2'(Z_1, Z_3) + H(Z_1, Z_3) = F_3(Z_3) + F_4(Z_1).$$

The last two equalities, together with Eq. (21.13), yield

$$F_3(Z_1) - F_4(Z_1) = F_3(Z_3) - F_4(Z_3). \tag{21.15}$$

It is clear that both sides of this equality are equal to a complex constant vector A, and

$$F_4(\mathbf{Z}_3) = F_3(\mathbf{Z}_3) - A.$$

On the basis of this, the equation (21.14) takes on the form

$$H(\mathbf{Z}_3, \mathbf{Z}_4) = -G_2(\mathbf{Z}_3, \mathbf{Z}_4) + F_3(\mathbf{Z}_4) - A, \qquad (21.16)$$

where we introduced a new function

$$G_2(\mathbf{Z}_3, \mathbf{Z}_4) = G_2'(\mathbf{Z}_3, \mathbf{Z}_4) - F_3(\mathbf{Z}_3). \qquad (21.17)$$

From Eqs. (21.16), (21.15), (21.12), (21.11), (21.6), (21.5) and (21.4) there follows the result Eq. (21.3). $\qquad\qquad\square$

This theorem generalizes the result given in [R. Ž. Djordjević (1965B)].

Theorem 21.2   *The general solution of the system of functional equations*

$$\begin{aligned}
f_0(\mathbf{Z}_1, \mathbf{Z}_2) \;&+\; f_1(\mathbf{Z}_2, \mathbf{Z}_3, \mathbf{Z}_4) + g_1(\mathbf{Z}_1, \mathbf{Z}_3, \mathbf{Z}_4) && (21.18) \\
&+\; f_2(\mathbf{Z}_3, \mathbf{Z}_4, \mathbf{Z}_5, \mathbf{Z}_6) + g_2(\mathbf{Z}_1, \mathbf{Z}_3, \mathbf{Z}_5, \mathbf{Z}_6) = \mathbf{O}, \\
f_0(\mathbf{Z}_1, \mathbf{Z}_2) \;&+\; g_1(\mathbf{Z}_2, \mathbf{Z}_3, \mathbf{Z}_4) + f_1(\mathbf{Z}_1, \mathbf{Z}_3, \mathbf{Z}_4) && (21.19) \\
&+\; f_2(\mathbf{Z}_3, \mathbf{Z}_4, \mathbf{Z}_5, \mathbf{Z}_6) + g_2(\mathbf{Z}_1, \mathbf{Z}_3, \mathbf{Z}_5, \mathbf{Z}_6) = \mathbf{O}
\end{aligned}$$

is determined by

$$f_0(\mathbf{Z}_1, \mathbf{Z}_2) = F_1(\mathbf{Z}_1) - F_2(\mathbf{Z}_2),$$

$$f_1(\mathbf{Z}_1, \mathbf{Z}_2, \mathbf{Z}_3) = F_2(\mathbf{Z}_1) + F_3(\mathbf{Z}_2) + G_2(\mathbf{Z}_2, \mathbf{Z}_3),$$

$$g_1(\mathbf{Z}_1, \mathbf{Z}_2, \mathbf{Z}_3) = F_2(\mathbf{Z}_1) - G_1(\mathbf{Z}_2, \mathbf{Z}_3) - G_2(\mathbf{Z}_2, \mathbf{Z}_3), \qquad (21.20)$$

$$f_2(\mathbf{Z}_1, \mathbf{Z}_2, \mathbf{Z}_3, \mathbf{Z}_4) = G_1(\mathbf{Z}_1, \mathbf{Z}_2) - H(\mathbf{Z}_1, \mathbf{Z}_3, \mathbf{Z}_4),$$

$$g_2(\mathbf{Z}_1, \mathbf{Z}_2, \mathbf{Z}_3, \mathbf{Z}_4) = -F_1(\mathbf{Z}_1) - F_2(\mathbf{Z}_1) - F_3(\mathbf{Z}_2) + H(\mathbf{Z}_2, \mathbf{Z}_3, \mathbf{Z}_4),$$

where F_i $(i = 1, 2, 3)$, G_j $(j = 1, 2)$ and H are arbitrary functions with values in $\mathcal{V}'$.

Proof. If we put $\mathbf{Z}_i = C_i$ $(i = 3, 4, 5, 6)$ into Eq. (21.18), we obtain

$$f_0(\mathbf{Z}_1, \mathbf{Z}_2) = F_1(\mathbf{Z}_1) - F_2(\mathbf{Z}_2), \tag{21.21}$$

where

$$\begin{aligned}
F_1(\mathbf{Z}_1) &= -g_1(\mathbf{Z}_1, C_3, C_4) - g_2(\mathbf{Z}_1, C_3, C_5, C_6), \\
F_2(\mathbf{Z}_2) &= f_1(\mathbf{Z}_2, C_3, C_4) + f_2(C_3, C_4, C_5, C_6).
\end{aligned}$$

By putting $\mathbf{Z}_i = C_i$ $(i = 2, 4)$ into Eq. (21.18), by virtue of the expression Eq. (21.21) we get

$$g_2(\mathbf{Z}_1, \mathbf{Z}_3, \mathbf{Z}_5, \mathbf{Z}_6) = -F_1(\mathbf{Z}_1) - G(\mathbf{Z}_1, \mathbf{Z}_3) + H(\mathbf{Z}_3, \mathbf{Z}_5, \mathbf{Z}_6), \tag{21.22}$$

where

$$H(\mathbf{Z}_3, \mathbf{Z}_5, \mathbf{Z}_6) = -f_1(C_2, \mathbf{Z}_3, C_4) - f_2(\mathbf{Z}_3, C_4, \mathbf{Z}_5, \mathbf{Z}_6) + F_2(C_2),$$

$$G(\mathbf{Z}_1, \mathbf{Z}_3) = -g_1(\mathbf{Z}_1, \mathbf{Z}_3, C_4).$$

By virtue of the expressions Eqs. (21.21) and (21.22), if we put $\mathbf{Z}_i = C_i$ $(i = 1, 2)$ into Eq. (21.18), we get

$$f_2(\mathbf{Z}_3, \mathbf{Z}_4, \mathbf{Z}_5, \mathbf{Z}_6) = -H(\mathbf{Z}_3, \mathbf{Z}_5, \mathbf{Z}_6) + G_1(\mathbf{Z}_3, \mathbf{Z}_4), \tag{21.23}$$

where we have put

$$G_1(\mathbf{Z}_3, \mathbf{Z}_4) = -f_1(C_2, \mathbf{Z}_3, \mathbf{Z}_4) - g_1(C_1, \mathbf{Z}_3, \mathbf{Z}_4) + G(C_1, \mathbf{Z}_3) + F_2(C_2).$$

In view of the expressions Eqs. (21.21), (21.22) and (21.23), the system of functional equations (21.18) and (21.19) becomes

$$\begin{aligned}
f_1(\mathbf{Z}_2, \mathbf{Z}_3, \mathbf{Z}_4) &+ g_1(\mathbf{Z}_1, \mathbf{Z}_3, \mathbf{Z}_4) \\
- \ F_2(\mathbf{Z}_2) &- G(\mathbf{Z}_1, \mathbf{Z}_3) + G_1(\mathbf{Z}_3, \mathbf{Z}_4) = 0,
\end{aligned} \tag{21.24}$$

$$\begin{aligned}
g_1(\mathbf{Z}_2, \mathbf{Z}_3, \mathbf{Z}_4) &+ f_1(\mathbf{Z}_1, \mathbf{Z}_3, \mathbf{Z}_4) \\
- \ F_2(\mathbf{Z}_2) &- G(\mathbf{Z}_1, \mathbf{Z}_3) + G_1(\mathbf{Z}_3, \mathbf{Z}_4) = 0.
\end{aligned} \tag{21.25}$$

By putting $\mathbf{Z}_2 = C_2$ into Eq. (21.24), after exchanging the roles of $\mathbf{Z}_1$ and $\mathbf{Z}_2$ in Eq. (21.25), subtracting Eq. (21.24) from Eq. (21.25), we get

$$f_1(\mathbf{Z}_2, \mathbf{Z}_3, \mathbf{Z}_4) = F_2(\mathbf{Z}_1) + G(\mathbf{Z}_2, \mathbf{Z}_3) - G(\mathbf{Z}_1, \mathbf{Z}_3) + f_1(C_2, \mathbf{Z}_3, \mathbf{Z}_4) - F_2(C_2),$$

or if we put $\mathbf{Z}_1 = C_2$, we obtain

$$f_1(\mathbf{Z}_2, \mathbf{Z}_3, \mathbf{Z}_4) = G(\mathbf{Z}_2, \mathbf{Z}_3) + G_2(\mathbf{Z}_3, \mathbf{Z}_4), \tag{21.26}$$

where

$$G_2(\mathbf{Z}_3, \mathbf{Z}_4) = f_1(C_2, \mathbf{Z}_3, \mathbf{Z}_4) - G(C_2, \mathbf{Z}_3).$$

From Eqs. (21.25) and (21.26) it follows that

$$g_1(\mathbf{Z}_2, \mathbf{Z}_3, \mathbf{Z}_4) = F_2(\mathbf{Z}_2) - G_1(\mathbf{Z}_3, \mathbf{Z}_4) - G_2(\mathbf{Z}_3, \mathbf{Z}_4). \tag{21.27}$$

We substitute the functions f_1 and g_1 determined by Eqs. (21.26) and (21.27) into the equation (21.24) and find

$$G(\mathbf{Z}_2, \mathbf{Z}_3) - F_2(\mathbf{Z}_2) = G(\mathbf{Z}_1, \mathbf{Z}_3) - F_2(\mathbf{Z}_1).$$

It is clear that both sides of this equality are equal to a function $F_3(\mathbf{Z}_3)$ (independent both of $\mathbf{Z}_1$ and $\mathbf{Z}_2$), thus

$$G(\mathbf{Z}_2, \mathbf{Z}_3) = F_2(\mathbf{Z}_2) + F_3(\mathbf{Z}_3). \tag{21.28}$$

Thus, on the basis of Eqs. (21.28), (21.27), (21.26), (21.23), (21.22) and (21.21), we obtain Eq. (21.20). $\qquad\square$

Theorem 21.3 *The general solution of the system of functional equations*

$$\begin{aligned}
f_0(\mathbf{Z}_1, \mathbf{Z}_2) \quad &+ \quad f_1(\mathbf{Z}_2, \mathbf{Z}_3, \mathbf{Z}_4) + g_1(\mathbf{Z}_1, \mathbf{Z}_3, \mathbf{Z}_4) \\
&+ \quad f_2(\mathbf{Z}_3, \mathbf{Z}_4, \mathbf{Z}_5, \mathbf{Z}_6) + g_2(\mathbf{Z}_1, \mathbf{Z}_3, \mathbf{Z}_5, \mathbf{Z}_6) = \mathbf{O},
\end{aligned} \tag{21.29}$$

$$\begin{aligned}
f_0(\mathbf{Z}_1, \mathbf{Z}_2) \quad &+ \quad g_1(\mathbf{Z}_2, \mathbf{Z}_3, \mathbf{Z}_4) + f_1(\mathbf{Z}_1, \mathbf{Z}_3, \mathbf{Z}_4) \\
&+ \quad g_2(\mathbf{Z}_3, \mathbf{Z}_4, \mathbf{Z}_5, \mathbf{Z}_6) + f_2(\mathbf{Z}_1, \mathbf{Z}_3, \mathbf{Z}_5, \mathbf{Z}_6) = \mathbf{O}
\end{aligned} \tag{21.30}$$

is determined by

$$f_0(\mathbf{Z}_1, \mathbf{Z}_2) = F_1(\mathbf{Z}_1) - F_2(\mathbf{Z}_2),$$

$$f_1(\mathbf{Z}_1, \mathbf{Z}_2, \mathbf{Z}_3) = F_2(\mathbf{Z}_1) + G_1(\mathbf{Z}_2, \mathbf{Z}_3),$$

$$g_1(\mathbf{Z}_1, \mathbf{Z}_2, \mathbf{Z}_3) = F_2(\mathbf{Z}_1) + F(\mathbf{Z}_2) + F_3(\mathbf{Z}_3) - G_1(\mathbf{Z}_2, \mathbf{Z}_3) - A,$$

$$\tag{21.31}$$

$$f_2(\mathbf{Z}_1, \mathbf{Z}_2, \mathbf{Z}_3, \mathbf{Z}_4) = F_3(\mathbf{Z}_1) - F(\mathbf{Z}_1) - F_3(\mathbf{Z}_2) + G_2(\mathbf{Z}_3, \mathbf{Z}_4) + A,$$

$$g_2(\mathbf{Z}_1, \mathbf{Z}_2, \mathbf{Z}_3, \mathbf{Z}_4) = F_3(\mathbf{Z}_1) - F(\mathbf{Z}_1) - F_3(\mathbf{Z}_2) - G_2(\mathbf{Z}_3, \mathbf{Z}_4),$$

where F_i $(i = 1, 2, 3)$, G_j $(j = 1, 2)$ and H are arbitrary functions with values in $\mathcal{V}'$, $F = F_1 + F_2 + F_3$ and A is an arbitrary constant vector from $\mathcal{V}'$.

Proof. By putting $\mathbf{Z}_i = C_i$ $(i = 3, 4, 5, 6)$ into Eq. (21.29), we obtain

$$f_0(\mathbf{Z}_1, \mathbf{Z}_2) = F_1(\mathbf{Z}_1) - F_2(\mathbf{Z}_2), \tag{21.32}$$

where

$$\begin{aligned}
F_1(\mathbf{Z}_1) &= -g_1(\mathbf{Z}_1, C_3, C_4) - g_2(\mathbf{Z}_1, C_3, C_5, C_6), \\
F_2(\mathbf{Z}_2) &= f_1(\mathbf{Z}_2, C_3, C_4) + f_2(C_3, C_4, C_5, C_6).
\end{aligned}$$

If we put $\mathbf{Z}_i = C_i$ $(i = 1, 5, 6)$ into Eq. (21.29), in view of the expression Eq. (21.32) we get

$$f_1(\mathbf{Z}_2, \mathbf{Z}_3, \mathbf{Z}_4) = F_2(\mathbf{Z}_2) + G_1(\mathbf{Z}_3, \mathbf{Z}_4), \tag{21.33}$$

where

$$G_1(\mathbf{Z}_3, \mathbf{Z}_4) = -F_1(C_1) - g_1(C_1, \mathbf{Z}_3, \mathbf{Z}_4) - f_2(\mathbf{Z}_3, \mathbf{Z}_4, C_5, C_6) - g_2(C_1, \mathbf{Z}_3, C_5, C_6).$$

By putting $\mathbf{Z}_i = C_i$ $(i = 1, 5, 6)$ into Eq. (21.30), by virtue of the expression Eq. (21.32) we obtain

$$g_1(\mathbf{Z}_2, \mathbf{Z}_3, \mathbf{Z}_4) = F_2(\mathbf{Z}_2) + G(\mathbf{Z}_3, \mathbf{Z}_4), \tag{21.34}$$

where we have put

$$G(\mathbf{Z}_3, \mathbf{Z}_4) = -F_1(C_1) - f_1(C_1, \mathbf{Z}_3, \mathbf{Z}_4) - g_2(\mathbf{Z}_3, \mathbf{Z}_4, C_5, C_6) - f_2(C_1, \mathbf{Z}_3, C_5, C_6).$$

On the basis of the expressions Eqs. (21.32), (21.33) and (21.34), the system of functional equations (21.29) and (21.30) becomes

$$F_1(\mathbf{Z}_1) + F_2(\mathbf{Z}_1) + G_1(\mathbf{Z}_3, \mathbf{Z}_4) \tag{21.35}$$
$$+ G(\mathbf{Z}_3, \mathbf{Z}_4) + f_2(\mathbf{Z}_3, \mathbf{Z}_4, \mathbf{Z}_5, \mathbf{Z}_6) + g_2(\mathbf{Z}_1, \mathbf{Z}_3, \mathbf{Z}_5, \mathbf{Z}_6) = \mathbf{O},$$

$$F_1(\mathbf{Z}_1) + F_2(\mathbf{Z}_1) + G_1(\mathbf{Z}_3, \mathbf{Z}_4) \tag{21.36}$$
$$+ G(\mathbf{Z}_3, \mathbf{Z}_4) + g_2(\mathbf{Z}_3, \mathbf{Z}_4, \mathbf{Z}_5, \mathbf{Z}_6) + f_2(\mathbf{Z}_1, \mathbf{Z}_3, \mathbf{Z}_5, \mathbf{Z}_6) = \mathbf{O}.$$

We put $\mathbf{Z}_i = C_1$ $(i = 1, 3)$ into Eq. (21.36), and further we replace $\mathbf{Z}_4$ by $\mathbf{Z}_3$. Also, we put $\mathbf{Z}_1 = C_1$ into Eq. (21.35). By subtracting the equations obtained, we get

$$f_2(\mathbf{Z}_3, \mathbf{Z}_4, \mathbf{Z}_5, \mathbf{Z}_6) = F_3(\mathbf{Z}_3) - G_1(\mathbf{Z}_3, \mathbf{Z}_4) + G_2(\mathbf{Z}_5, \mathbf{Z}_6) - G(\mathbf{Z}_3, \mathbf{Z}_4),$$

$$(21.37)$$

where we introduced the notations

$$F_3(\mathbf{Z}_3) = G_1(C_1, \mathbf{Z}_3) + G(C_1, \mathbf{Z}_3), \quad G_2(\mathbf{Z}_5, \mathbf{Z}_6) = f_2(C_1, C_1, \mathbf{Z}_5, \mathbf{Z}_6).$$

From the equation (21.35), on the basis of the expression Eq. (21.37), we get

$$g_2(\mathbf{Z}_1, \mathbf{Z}_3, \mathbf{Z}_5, \mathbf{Z}_6) = -F_1(\mathbf{Z}_1) - F_2(\mathbf{Z}_1) - F_3(\mathbf{Z}_3) - G_2(\mathbf{Z}_5, \mathbf{Z}_6). \quad (21.38)$$

We substitute the functions f_2 and g_2 determined by Eqs. (21.37) and (21.38) into the equation (21.36) and find

$$F_1(\mathbf{Z}_1) + F_2(\mathbf{Z}_1) + F_3(\mathbf{Z}_1) - F_1(\mathbf{Z}_3) - F_2(\mathbf{Z}_3) - F_3(\mathbf{Z}_4) \quad (21.39)$$
$$+ \quad G_1(\mathbf{Z}_3, \mathbf{Z}_4) + G(\mathbf{Z}_3, \mathbf{Z}_4) - G_1(\mathbf{Z}_1, \mathbf{Z}_3) - G(\mathbf{Z}_1, \mathbf{Z}_3) = \mathbf{O}.$$

If we put $\mathbf{Z}_1 = C_1$, the equation (21.39) becomes

$$G_1(\mathbf{Z}_3, \mathbf{Z}_4) + G(\mathbf{Z}_3, \mathbf{Z}_4) = F_1(\mathbf{Z}_3) + F_2(\mathbf{Z}_3) + F_3(\mathbf{Z}_3) + F_3(\mathbf{Z}_4) - A,$$

$$(21.40)$$

since

$$G_1(C_1, \mathbf{Z}_3) + G(C_1, \mathbf{Z}_3) = F_3(\mathbf{Z}_3),$$

where we have put $A = \sum_{i=1}^{3} F_i(C_1)$.

On the basis of the expression Eq. (21.40), the equalities (21.34) and (21.37) obtain respectively the forms

$$\begin{aligned}
g_1(\mathbf{Z}_2, \mathbf{Z}_3, \mathbf{Z}_4) &= F_2(\mathbf{Z}_2) + F(\mathbf{Z}_3) & (21.41)\\
&+ F_3(\mathbf{Z}_4) - G_1(\mathbf{Z}_3, \mathbf{Z}_4) - A,\\
f_2(\mathbf{Z}_3, \mathbf{Z}_4, \mathbf{Z}_5, \mathbf{Z}_6) &= F_3(\mathbf{Z}_3) - F(\mathbf{Z}_3) & (21.42)\\
&- F_3(\mathbf{Z}_4) + G_2(\mathbf{Z}_5, \mathbf{Z}_6) + A,
\end{aligned}$$

where we introduced the new function

$$F(\mathbf{Z}_3) = \sum_{i=1}^{3} F_i(\mathbf{Z}_3). \quad (21.43)$$

On the basis of Eqs. (21.43), (21.42), (21.41), (21.38), (21.33) and (21.32), we obtain the equalities (21.31). $\qquad\square$

Theorem 21.4 *The general solution of the system of functional equations*

$$f_0(\mathbf{Z}_1, \mathbf{Z}_2) \; + \; f_1(\mathbf{Z}_2, \mathbf{Z}_3, \mathbf{Z}_4) + g_1(\mathbf{Z}_1, \mathbf{Z}_3, \mathbf{Z}_4) \qquad (21.44)$$
$$+ \; f_2(\mathbf{Z}_3, \mathbf{Z}_4, \mathbf{Z}_5, \mathbf{Z}_6) + g_2(\mathbf{Z}_1, \mathbf{Z}_3, \mathbf{Z}_5, \mathbf{Z}_6) = \mathbf{O},$$

$$f_0(\mathbf{Z}_1, \mathbf{Z}_2) \; + \; f_1(\mathbf{Z}_2, \mathbf{Z}_3, \mathbf{Z}_4) + g_1(\mathbf{Z}_1, \mathbf{Z}_3, \mathbf{Z}_4) \qquad (21.45)$$
$$+ \; g_2(\mathbf{Z}_3, \mathbf{Z}_4, \mathbf{Z}_5, \mathbf{Z}_6) + f_2(\mathbf{Z}_1, \mathbf{Z}_3, \mathbf{Z}_5, \mathbf{Z}_6) = \mathbf{O},$$

$$f_0(\mathbf{Z}_1, \mathbf{Z}_2) \; + \; g_1(\mathbf{Z}_2, \mathbf{Z}_3, \mathbf{Z}_4) + f_1(\mathbf{Z}_1, \mathbf{Z}_3, \mathbf{Z}_4) \qquad (21.46)$$
$$+ \; f_2(\mathbf{Z}_3, \mathbf{Z}_4, \mathbf{Z}_5, \mathbf{Z}_6) + g_2(\mathbf{Z}_1, \mathbf{Z}_3, \mathbf{Z}_5, \mathbf{Z}_6) = \mathbf{O}$$

is determined by the equalities

$$f_0(\mathbf{Z}_1, \mathbf{Z}_2) = F_1(\mathbf{Z}_1) - F_2(\mathbf{Z}_2),$$

$$f_1(\mathbf{Z}_1, \mathbf{Z}_2, \mathbf{Z}_3) = F_2(\mathbf{Z}_1) + G_1(\mathbf{Z}_2, \mathbf{Z}_3),$$

$$g_1(\mathbf{Z}_1, \mathbf{Z}_2, \mathbf{Z}_3) = F_2(\mathbf{Z}_1) + F_1(\mathbf{Z}_2) + F_2(\mathbf{Z}_2) - G_1(\mathbf{Z}_2, \mathbf{Z}_3) + K, \quad (21.47)$$

$$f_2(\mathbf{Z}_1, \mathbf{Z}_2, \mathbf{Z}_3, \mathbf{Z}_4) = -F_1(\mathbf{Z}_1) - F_2(\mathbf{Z}_1) + G(\mathbf{Z}_3, \mathbf{Z}_4),$$

$$g_2(\mathbf{Z}_1, \mathbf{Z}_2, \mathbf{Z}_3, \mathbf{Z}_4) = F_1(\mathbf{Z}_1) - F_2(\mathbf{Z}_1) - G(\mathbf{Z}_3, \mathbf{Z}_4) - K,$$

where F_1, F_2, G, G_1 are arbitrary functions with values in $\mathcal{V}'$, and K is an arbitrary constant vector from $\mathcal{V}'$.

Proof. The equations (21.44) and (21.45) form the system of functional equations given by Eqs. (21.1) and (21.2). Consequently, the functions determined by the equalities (21.3) satisfy Eqs. (21.44) and (21.45).

In order for the functions Eq. (21.3) to satisfy the equation (21.46) we must have

$$F_1(\mathbf{Z}_1) + F_2(\mathbf{Z}_1) - G_2(\mathbf{Z}_1, \mathbf{Z}_3) \qquad (21.48)$$
$$= \; F_1(\mathbf{Z}_2) + F_2(\mathbf{Z}_2) - G_2(\mathbf{Z}_2, \mathbf{Z}_4) = \mathbf{O}.$$

It is clear that both sides of Eq. (21.48) are equal to a complex constant vector M, thus

$$G_2(\mathbf{Z}_1, \mathbf{Z}_3) = F_1(\mathbf{Z}_1) + F_2(\mathbf{Z}_1) - M. \qquad (21.49)$$

If we introduce a new function G by

$$G(\mathbf{Z}_1, \mathbf{Z}_2) = G_3(\mathbf{Z}_1, \mathbf{Z}_2) + M$$

and denote $K = A - 2M$, from Eqs. (21.49) and (21.3) there follows Eq. (21.47). $\square$

22 Systems in Which Not All Equations Contain All Unknown Functions

Theorem 22.1   *The general solution of the system of functional equations*

$$f_0(\mathbf{Z}_1, \mathbf{Z}_2) \quad + \quad f_1(\mathbf{Z}_2, \mathbf{Z}_3, \mathbf{Z}_4) \tag{22.1}$$
$$+ \quad g_1(\mathbf{Z}_1, \mathbf{Z}_3, \mathbf{Z}_4) = \mathbf{O},$$

$$f_0(\mathbf{Z}_1, \mathbf{Z}_2) \quad + \quad g_1(\mathbf{Z}_2, \mathbf{Z}_3, \mathbf{Z}_4) \tag{22.2}$$
$$+ f_1(\mathbf{Z}_1, \mathbf{Z}_3, \mathbf{Z}_4) \quad + \quad f_2(\mathbf{Z}_3, \mathbf{Z}_4, \mathbf{Z}_5, \mathbf{Z}_6) = \mathbf{O},$$

is given by the equalities

$$
\begin{aligned}
f_0(\mathbf{Z}_1, \mathbf{Z}_2) &= -F(\mathbf{Z}_1) - F(\mathbf{Z}_2), \\
f_1(\mathbf{Z}_1, \mathbf{Z}_2, \mathbf{Z}_3) &= F(\mathbf{Z}_1) - G(\mathbf{Z}_2, \mathbf{Z}_3), \\
g_1(\mathbf{Z}_1, \mathbf{Z}_2, \mathbf{Z}_3) &= F(\mathbf{Z}_1) + G(\mathbf{Z}_2, \mathbf{Z}_3),
\end{aligned}
\tag{22.3}
$$

$$f_2(\mathbf{Z}_1, \mathbf{Z}_2, \mathbf{Z}_3, \mathbf{Z}_4) = \mathbf{O},$$

where F and G are arbitrary functions with values in $\mathcal{V}'$.

Proof.   According to [D. S. Mitrinović (1963A)] the general solution of the functional equation (22.1) is

$$
\begin{aligned}
f_0(\mathbf{Z}_1, \mathbf{Z}_2) &= H_1(\mathbf{Z}_1) - F_1(\mathbf{Z}_2), \\
f_1(\mathbf{Z}_1, \mathbf{Z}_2, \mathbf{Z}_3) &= F_1(\mathbf{Z}_1) - G_1(\mathbf{Z}_2, \mathbf{Z}_3), \\
g_1(\mathbf{Z}_1, \mathbf{Z}_2, \mathbf{Z}_3) &= -H_1(\mathbf{Z}_1) + G_1(\mathbf{Z}_2, \mathbf{Z}_3),
\end{aligned}
\tag{22.4}
$$

where F_1, H_1 and G_1 are arbitrary functions with values in $\mathcal{V}'$.

We substitute the functions f_0, f_1 and g_1 determined by the equalities (22.4) into the equation (22.2). We must have

$$F_1(\mathbf{Z}_1) + H_1(\mathbf{Z}_1) - F_1(\mathbf{Z}_2) - H_1(\mathbf{Z}_2) = -f_2(\mathbf{Z}_3, \mathbf{Z}_4, \mathbf{Z}_5, \mathbf{Z}_6). \tag{22.5}$$

It is obvious that both sides of this equality are constant. By putting $\mathbf{Z}_1 = \mathbf{Z}_2$ into Eq. (22.5), we get

$$f_2(\mathbf{Z}_3, \mathbf{Z}_4, \mathbf{Z}_5, \mathbf{Z}_6) = \mathbf{O}, \tag{22.6}$$

and further

$$F_1(\mathbf{Z}_1) + H_1(\mathbf{Z}_1) = F_1(\mathbf{Z}_2) + H_1(\mathbf{Z}_2) = 2A, \tag{22.7}$$

where A is a constant vector from $\mathcal{V}'$.

By introducing new functions F and G by the following equalities

$$\begin{aligned}
F(\mathbf{Z}_1) &= F_1(\mathbf{Z}_1) - A, \\
G(\mathbf{Z}_1, \mathbf{Z}_2) &= G_1(\mathbf{Z}_1, \mathbf{Z}_2) - A,
\end{aligned}$$

on the basis of the expressions Eqs. (22.7), (22.6) and (22.4), we obtain Eq. (22.3). $\qquad\square$

Theorem 22.2   *The general solution of the system of functional equations*

$$f_0(\mathbf{Z}_1, \mathbf{Z}_2) + f_1(\mathbf{Z}_2, \mathbf{Z}_3, \mathbf{Z}_4) + g_1(\mathbf{Z}_1, \mathbf{Z}_3, \mathbf{Z}_4) = \mathbf{O},$$

$$f_0(\mathbf{Z}_1, \mathbf{Z}_2) + g_1(\mathbf{Z}_2, \mathbf{Z}_3, \mathbf{Z}_4) + f_1(\mathbf{Z}_1, \mathbf{Z}_3, \mathbf{Z}_4) + f_2(\mathbf{Z}_3, \mathbf{Z}_4, \mathbf{Z}_5, \mathbf{Z}_6) = \mathbf{O},$$

$$\begin{aligned}
f_0(\mathbf{Z}_1, \mathbf{Z}_2) \quad &+ \quad f_1(\mathbf{Z}_2, \mathbf{Z}_3, \mathbf{Z}_4) + g_1(\mathbf{Z}_1, \mathbf{Z}_3, \mathbf{Z}_4) \\
&+ \quad f_2(\mathbf{Z}_3, \mathbf{Z}_4, \mathbf{Z}_5, \mathbf{Z}_6) + g_2(\mathbf{Z}_1, \mathbf{Z}_3, \mathbf{Z}_5, \mathbf{Z}_6) = \mathbf{O},
\end{aligned}$$

$$\cdots$$

$$\begin{aligned}
f_0(\mathbf{Z}_1, \mathbf{Z}_2) &+ \sum_{i=1}^{k-1} g_i(\mathbf{Z}_{i+1}, \mathbf{Z}_{i+2}, \cdots, \mathbf{Z}_{2i+2}) \\
+ \sum_{i=1}^{k-1} & f_i(\mathbf{Z}_1, \mathbf{Z}_3, \cdots, \mathbf{Z}_{2i+1}, \mathbf{Z}_{2i+2}) + f_k(\mathbf{Z}_{k+1}, \mathbf{Z}_{k+2}, \cdots, \mathbf{Z}_{2k+2}) = \mathbf{O},
\end{aligned}$$

$$\begin{aligned}
f_0(\mathbf{Z}_1, \mathbf{Z}_2) &+ \sum_{i=1}^{k} f_i(\mathbf{Z}_{i+1}, \mathbf{Z}_{i+2}, \cdots, \mathbf{Z}_{2i+2}) \\
&+ \sum_{i=1}^{k} g_i(\mathbf{Z}_1, \mathbf{Z}_3, \cdots, \mathbf{Z}_{2i+1}, \mathbf{Z}_{2i+2}) = \mathbf{O},
\end{aligned}$$

$$\cdots$$

$$f_0(\mathbf{Z}_1, \mathbf{Z}_2) + \sum_{i=1}^{m-1} g_i(\mathbf{Z}_{i+1}, \mathbf{Z}_{i+2}, \cdots, \mathbf{Z}_{2i+2})$$

$$+ \sum_{i=1}^{m-1} f_i(\mathbf{Z}_1, \mathbf{Z}_3, \cdots, \mathbf{Z}_{2i+1}, \mathbf{Z}_{2i+2}) + f_m(\mathbf{Z}_{m+1}, \mathbf{Z}_{m+2}, \cdots, \mathbf{Z}_{2m+2}) = \mathbf{O},$$

$$f_0(\mathbf{Z}_1, \mathbf{Z}_2) + \sum_{i=1}^{m} f_i(\mathbf{Z}_{i+1}, \mathbf{Z}_{i+2}, \cdots, \mathbf{Z}_{2i+2})$$

$$+ \sum_{i=1}^{m} g_i(\mathbf{Z}_1, \mathbf{Z}_3, \cdots, \mathbf{Z}_{2i+1}, \mathbf{Z}_{2i+2}) = \mathbf{O},$$

is given by

$$
\begin{aligned}
f_0(\mathbf{Z}_1, \mathbf{Z}_2) &= -F(\mathbf{Z}_1) - F(\mathbf{Z}_2), \\
f_1(\mathbf{Z}_1, \mathbf{Z}_2, \mathbf{Z}_3) &= F(\mathbf{Z}_1) - G(\mathbf{Z}_2, \mathbf{Z}_3), \\
g_1(\mathbf{Z}_1, \mathbf{Z}_2, \mathbf{Z}_3) &= F(\mathbf{Z}_1) + G(\mathbf{Z}_2, \mathbf{Z}_3), \\
f_i = g_i &= \mathbf{O} \qquad (2 \le i \le m),
\end{aligned}
$$

where F and G are arbitrary functions with values in $\mathcal{V}'$.

Proof. The proof of this theorem follows from Theorem 22.1. $\square$

We have noticed that this approach to the solution of systems of linear complex vector functional equations is not considered in the references [W. Eichhorn (1963); M. Ghermănescu (1953); M. Ghermănescu (1955)].

PART 2

Nonlinear Complex Vector Functional Equations

Chapter 6

Quadratic Functional Equations

In this chapter two vector functional equations are considered. First one simple quadratic complex vector functional equation is solved, and after that investigations are made when the special real quadratic functional equation has nontrivial solutions. The results presented here were obtained in [I. B. Risteski and V. C. Covachev (submitted A)].

Now we will introduce the following notations.

Let $\mathcal{V}$ be a finite dimensional complex vector space and let there exist a mapping $f : \mathcal{V}^2 \mapsto \mathcal{V}$. In the next section $\mathbf{Z}_i$ $(0 \leq i \leq n)$ will denote vectors in $\mathcal{V}$. We assume that $\mathbf{Z}_i = (z_{i1}(t), \cdots, z_{in}(t))^T$, where the components $z_{ij}(t)$ $(0 \leq i \leq n; 1 \leq j \leq n)$ are complex functions and $\mathbf{O} = (0, \cdots, 0)^T$ is the zero vector in $\mathcal{V}$. We define multiplication of arbitrary two vectors $\mathbf{U} = (u_1(t), \cdots, u_n(t))^T$ and $\mathbf{V} = (v_1(t), \cdots, v_n(t))^T$ in $\mathcal{V}$ as $\mathbf{UV} = (u_1(t)v_1(t), \cdots, u_n(t)v_n(t))^T$.

23 Simple Quadratic Functional Equation

Now we will prove the following results.

Theorem 23.1 *The general solution of the functional equation*

$$F(\mathbf{Z}_0, \mathbf{Z}_1, \mathbf{Z}_2, \cdots, \mathbf{Z}_n) \ + \ F(\mathbf{Z}_0, \mathbf{Z}_2, \mathbf{Z}_3, \cdots, \mathbf{Z}_n, \mathbf{Z}_1) + \cdots \qquad (23.1)$$

$$+ \ F(\mathbf{Z}_0, \mathbf{Z}_n, \mathbf{Z}_1, \mathbf{Z}_2, \cdots, \mathbf{Z}_{n-2}, \mathbf{Z}_{n-1}) = \mathbf{O},$$

where

$$F(\mathbf{Z}_0, \mathbf{Z}_1, \mathbf{Z}_2, \cdots, \mathbf{Z}_n) = f(\mathbf{Z}_0, \mathbf{Z}_p) f(\mathbf{Z}_q, \mathbf{Z}_r) \qquad (f : \mathcal{V}^2 \mapsto \mathcal{V}) \qquad (23.2)$$

219

is given by

$$f(\mathbf{U}, \mathbf{V}) = g(\mathbf{U})h(\mathbf{V}) - g(\mathbf{V})h(\mathbf{U}) \qquad (23.3)$$

($g, h : \mathcal{V} \mapsto \mathcal{V}$ are arbitrary complex vector functions) in the case when p, q, r, n satisfy one of the following six conditions:

$$
\begin{aligned}
&1'. &&2p = q + r - n, &&2n = 3r - 3p &&(p < q < r),\\
&2'. &&2p = q + r + n, &&n = 3p - 3r &&(q < r < p),\\
&3'. &&2p = q + r, &&n = 3p - 3r &&(r < p < q),\\
&4'. &&2p = q + r - n, &&n = 3r - 3p &&(p < r < q),\\
&5'. &&2p = q + r, &&n = 3r - 3p &&(q < p < r),\\
&6'. &&2p = q + r + n, &&2n = 3p - 3r &&(r < q < p);
\end{aligned}
$$

$$f(\mathbf{U}, \mathbf{V}) = g(\mathbf{V}) - g(\mathbf{U}) \qquad (23.4)$$

($g : \mathcal{V} \mapsto \mathcal{V}$ is an arbitrary complex vector function) when one of the following six conditions is satisfied:

$$
\begin{aligned}
&1''. &&2p = q + r - n, &&2n \neq 3r - 3p &&(p < q < r),\\
&2''. &&2p = q + r + n, &&n \neq 3p - 3r &&(q < r < p),\\
&3''. &&2p = q + r, &&n \neq 3p - 3r &&(r < p < q),\\
&4''. &&2p = q + r - n, &&n \neq 3r - 3p &&(p < r < q),\\
&5''. &&2p = q + r, &&n \neq 3r - 3p &&(q < p < r),\\
&6''. &&2p = q + r + n, &&2n \neq 3p - 3r &&(r < q < p);
\end{aligned}
$$

$$f(\mathbf{U}, \mathbf{V}) \equiv \mathbf{O} \qquad (23.5)$$

in all other cases.

Proof.   Between the parameters p, q, r there may exist one of the following relations:

$1^0. \; p < q < r, \quad 2^0. \; q < r < p, \quad 3^0. \; r < p < q, \quad 4^0. \; p < r < q, \quad 5^0. \; q < p < r,$
$6^0. \; r < q < p.$

In the cases 1^0, 2^0, 3^0 we put

$$F(\mathbf{Z}_0, \mathbf{Z}_1, \mathbf{Z}_2, \cdots, \mathbf{Z}_{n-1}, \mathbf{Z}_n) = G(\mathbf{Z}_0, \mathbf{Z}_{1+r}, \cdots, \mathbf{Z}_n, \mathbf{Z}_1, \cdots, \mathbf{Z}_r), \quad (23.6)$$

$$\mathbf{Z}_0 = \mathbf{U}_0, \qquad \mathbf{Z}_i = \mathbf{U}_{n-r+i} \qquad (1 \leq i \leq n)$$

so that $\mathbf{Z}_{j+n} = \mathbf{Z}_j \;\; (1 \leq j \leq n)$. Then from Eq. (23.1) we obtain

$$
\begin{aligned}
G(\mathbf{U}_0, \mathbf{U}_1, \mathbf{U}_2, \cdots, \mathbf{U}_n) \;\; &+ \;\; G(\mathbf{U}_0, \mathbf{U}_2, \mathbf{U}_3, \cdots, \mathbf{U}_n, \mathbf{U}_1) + \cdots \quad (23.7)\\
&+ \;\; G(\mathbf{U}_0, \mathbf{U}_n, \mathbf{U}_1, \cdots, \mathbf{U}_{n-2}, \mathbf{U}_{n-1}) = \mathbf{O},
\end{aligned}
$$

but from Eq. (23.2) it follows that

$$G(\mathbf{U}_0, \mathbf{U}_1, \cdots, \mathbf{U}_n) \tag{23.8}$$

$$= \begin{cases} f(\mathbf{U}_0, \mathbf{U}_{n+p-r})f(\mathbf{U}_{n+q-r}, \mathbf{U}_n) & (p < q < r), \\ f(\mathbf{U}_0, \mathbf{U}_{p-r})f(\mathbf{U}_{n+q-r}, \mathbf{U}_n) & (q < r < p), \\ f(\mathbf{U}_0, \mathbf{U}_{p-r})f(\mathbf{U}_{q-r}, \mathbf{U}_n) & (r < p < q). \end{cases}$$

In the cases 4^0, 5^0, 6^0 we put

$$F(\mathbf{Z}_0, \mathbf{Z}_1, \mathbf{Z}_2, \cdots, \mathbf{Z}_{n-1}, \mathbf{Z}_n) \tag{23.9}$$
$$= G(\mathbf{Z}_0, \mathbf{Z}_{r-1}, \mathbf{Z}_{r-2}, \cdots, \mathbf{Z}_1, \mathbf{Z}_n, \cdots, \mathbf{Z}_r),$$

$$\mathbf{Z}_0 = \mathbf{U}_0, \qquad \mathbf{Z}_i = \mathbf{U}_{n-r+i} \qquad (1 \le i \le n)$$

so that $\mathbf{Z}_{j+n} = \mathbf{Z}_j$ $(1 \le j \le n)$. Then the equation (23.1) reduces to the equation (23.7), but the function Eq. (23.2) becomes

$$G(\mathbf{U}_0, \mathbf{U}_1, \cdots, \mathbf{U}_n) \tag{23.10}$$

$$= \begin{cases} f(\mathbf{U}_0, \mathbf{U}_{r-p})f(\mathbf{U}_{n+r-q}, \mathbf{U}_n) & (p < r < q), \\ f(\mathbf{U}_0, \mathbf{U}_{r-p})f(\mathbf{U}_{r-q}, \mathbf{U}_n) & (q < p < r), \\ f(\mathbf{U}_0, \mathbf{U}_{n+r-p})f(\mathbf{U}_{n+r-q}, \mathbf{U}_n) & (r < q < p). \end{cases}$$

In the functions given by Eqs. (23.8) and (23.10) the vector variables $\mathbf{U}_i$ $(0 \le i \le n)$ are arranged by increasing indices. Consequently, it suffices to consider the functional equation

$$G(\mathbf{U}_0, \mathbf{U}_1, \mathbf{U}_2, \cdots, \mathbf{U}_n) + G(\mathbf{U}_0, \mathbf{U}_2, \mathbf{U}_3, \cdots, \mathbf{U}_n, \mathbf{U}_1) + \cdots \tag{23.11}$$
$$+ G(\mathbf{U}_0, \mathbf{U}_n, \mathbf{U}_1, \cdots, \mathbf{U}_{n-2}, \mathbf{U}_{n-1}) = \mathbf{O},$$

where

$$G(\mathbf{U}_0, \mathbf{U}_1, \mathbf{U}_2, \cdots, \mathbf{U}_n) = f(\mathbf{U}_0, \mathbf{U}_k)f(\mathbf{U}_m, \mathbf{U}_n) \tag{23.12}$$
$$(0 < k < m < n).$$

We will call the equation (23.11) with the condition Eq. (23.12) *transformed equation*.

By putting $\mathbf{U}_j = \mathbf{U}$ $(0 \le j \le n)$ the transformed equation reduces to

$$f(\mathbf{U}, \mathbf{U}) \equiv \mathbf{O}. \tag{23.13}$$

If we put $\mathbf{U}_j = \mathbf{X}$ for all indices j distinct from k, m, n and

$$\mathbf{U}_k = \mathbf{Y}, \qquad \mathbf{U}_m = \mathbf{U}, \qquad \mathbf{U}_n = \mathbf{V},$$

then by virtue of Eq. (23.13) the transformed equation becomes

$$f(\mathbf{X}, \mathbf{Y})f(\mathbf{U}, \mathbf{V}) + f(\mathbf{X}, \mathbf{U})f(\mathbf{U}_{2m-k}, \mathbf{U}_{m-k}) \qquad (23.14)$$
$$+ \quad f(\mathbf{X}, \mathbf{V})f(\mathbf{U}_{m-k}, \mathbf{U}_{n-k}) = \mathbf{O},$$

where $\mathbf{U}_j = \mathbf{U}_{j-n}$ for $j > n$. In the equation (23.14) the vectors $\mathbf{U}_{2m-k}$, $\mathbf{U}_{m-k}$, $\mathbf{U}_{n-k}$ can be all equal to $\mathbf{X}$, or they can take values among $\mathbf{X}$, $\mathbf{Y}$, $\mathbf{U}$, $\mathbf{V}$ according to the following table:

$\mathbf{U}_{2m-k}$	$\mathbf{Y}$	$\mathbf{Y}$	$\mathbf{V}$	$\mathbf{V}$	$\mathbf{X}$	$\mathbf{X}$
$\mathbf{U}_{m-k}$	$\mathbf{X}$ or $\mathbf{V}$	$\mathbf{X}$	$\mathbf{Y}$	$\mathbf{X}$	$\mathbf{Y}$	$\mathbf{X}$
$\mathbf{U}_{n-k}$	$\mathbf{U}$	$\mathbf{X}$	$\mathbf{U}$	$\mathbf{X}$ or $\mathbf{Y}$	$\mathbf{X}$	$\mathbf{Y}$ or $\mathbf{U}$

For instance, it cannot be $\mathbf{U}_{2m-k} = \mathbf{U}$, because in such a case the equation (23.14) would contain a term $f(\mathbf{X}, \mathbf{U})f(\mathbf{U}, \mathbf{U}_{m-k})$ which is impossible in view of the substitution made.

If $\mathbf{U}_{2m-k} = \mathbf{U}_{m-k} = \mathbf{U}_{n-k} = \mathbf{X}$, by virtue of Eq. (23.13) the equation (23.14) reduces to

$$f(\mathbf{X}, \mathbf{Y})f(\mathbf{U}, \mathbf{V}) = \mathbf{O}.$$

From this equation for $\mathbf{X} = \mathbf{U}$, $\mathbf{Y} = \mathbf{V}$ we obtain $f^2(\mathbf{U}, \mathbf{V}) = \mathbf{O}$, so in this case there exists only the trivial solution

$$f(\mathbf{U}, \mathbf{V}) \equiv \mathbf{O}. \qquad (23.15)$$

Now we will consider all cases when in equation (23.14) besides $f(\mathbf{X}, \mathbf{U})f(\mathbf{U}, \mathbf{V})$ at least one more addend is not $\mathbf{O}$.

a) We will have $\mathbf{U}_{2m-k} = \mathbf{Y}$ if and only if $n = 2m - 2k$. The equation (23.14) then becomes

$$f(\mathbf{X}, \mathbf{Y})f(\mathbf{U}, \mathbf{V}) + f(\mathbf{X}, \mathbf{U})f(\mathbf{Y}, \mathbf{U}_{m-k}) \qquad (23.16)$$
$$+ \quad f(\mathbf{X}, \mathbf{V})f(\mathbf{U}_{m-k}, \mathbf{U}_{2m-3k}) = \mathbf{O}.$$

On the basis of the second addend of this equation, taking into consideration the condition $0 < k < m < 2m - 2k$, we can conclude that $\mathbf{U}_{m-k} = \mathbf{X}$. Consequently, the equation (23.16) takes on the form

$$f(\mathbf{X}, \mathbf{Y})f(\mathbf{U}, \mathbf{V}) + f(\mathbf{X}, \mathbf{U})f(\mathbf{Y}, \mathbf{X}) + f(\mathbf{X}, \mathbf{V})f(\mathbf{X}, \mathbf{U}_{2m-3k}) = \mathbf{O}. \quad (23.17)$$

The only possible values of $\mathbf{U}_{2m-3k}$ are

   (i) $\mathbf{U}_{2m-3k} = \mathbf{Y} \quad \Longrightarrow \quad m = n = 2k$;

   (ii) $\mathbf{U}_{2m-3k} = \mathbf{U} \quad \Longrightarrow \quad m = 3k, \ n = 4k$.

However, the case (i) cannot occur because of the assumption $0 < k < m < n$. In the case $m = 3k$, $n = 4k$ the equation (23.17) takes the form

$$f(\mathbf{X},\mathbf{Y})f(\mathbf{U},\mathbf{V}) + f(\mathbf{X},\mathbf{U})f(\mathbf{Y},\mathbf{X}) + f(\mathbf{X},\mathbf{V})f(\mathbf{X},\mathbf{U}) = \mathbf{O}. \qquad (23.18)$$

If $n = 2m - 2k$ and $m \neq 3k$, then the equation (23.17) becomes

$$f(\mathbf{X},\mathbf{Y})f(\mathbf{U},\mathbf{V}) + f(\mathbf{X},\mathbf{U})f(\mathbf{Y},\mathbf{X}) = \mathbf{O}. \qquad (23.19)$$

b) $\mathbf{U}_{2m-k} = \mathbf{V}$ can appear if and only if $n = 2m - k$, and in this case the equation (23.14) takes the form

$$f(\mathbf{X},\mathbf{Y})f(\mathbf{U},\mathbf{V}) + f(\mathbf{X},\mathbf{U})f(\mathbf{V},\mathbf{U}_{m-k}) \qquad (23.20)$$
$$+ \ f(\mathbf{X},\mathbf{V})f(\mathbf{U}_{m-k},\mathbf{U}_{2m-2k}) = \mathbf{O}.$$

Then the only possible values of $\mathbf{U}_{2m-2k}$ are

   (iii) $\mathbf{U}_{2m-2k} = \mathbf{U} \quad \Longrightarrow \quad m = 2k, \ n = 3k$;

   (iv) $\mathbf{U}_{2m-2k} = \mathbf{Y} \quad \Longrightarrow \quad m = \dfrac{3}{2}k, \ n = 2k$.

In the case (iii) we have $\mathbf{U}_{m-k} = \mathbf{Y}$ and equation (23.20) reduces to the following equation

$$f(\mathbf{X},\mathbf{Y})f(\mathbf{U},\mathbf{V}) + f(\mathbf{X},\mathbf{U})f(\mathbf{V},\mathbf{Y}) + f(\mathbf{X},\mathbf{V})f(\mathbf{Y},\mathbf{U}) = \mathbf{O}. \qquad (23.21)$$

In the case (iv), since k, m, n are natural numbers, there must exist a natural number ν such that $k = 2\nu$, $m = 3\nu$, $n = 4\nu$. Then $\mathbf{U}_\nu = \mathbf{X}$ because otherwise one of the equalities $\nu = 2\nu$, $\nu = 3\nu$, $\nu = 4\nu$ should hold, which is impossible.

Consequently, in this case the equation (23.20) takes the form

$$f(\mathbf{X},\mathbf{Y})f(\mathbf{U},\mathbf{V}) + f(\mathbf{X},\mathbf{U})f(\mathbf{V},\mathbf{X}) + f(\mathbf{X},\mathbf{V})f(\mathbf{X},\mathbf{Y}) = \mathbf{O}. \qquad (23.22)$$

If $n = 2m - k$ and $n \neq 2k$, $n \neq 3k$, then the equation (23.20) becomes

$$f(\mathbf{X},\mathbf{Y})f(\mathbf{U},\mathbf{V}) + f(\mathbf{X},\mathbf{U})f(\mathbf{V},\mathbf{X}) = \mathbf{O}. \qquad (23.23)$$

c) $\mathbf{U}_{m-k} = \mathbf{Y}$ implies $m = 2k$, and the equation (23.14) gets the form

$$f(\mathbf{X},\mathbf{Y})f(\mathbf{U},\mathbf{V})+f(\mathbf{X},\mathbf{U})f(\mathbf{U}_{3k},\mathbf{Y})+f(\mathbf{X},\mathbf{V})f(\mathbf{Y},\mathbf{U}_{n-k}) = \mathbf{O}. \quad (23.24)$$

From this equation we see that neither of $\mathbf{U}_{3k}$ and $\mathbf{U}_{n-k}$ can be equal to $\mathbf{Y}$. We can have $\mathbf{U}_{3k} = \mathbf{V}$ and $\mathbf{U}_{n-k} = \mathbf{U}$ if $m = 2k$, $n = 3k$. But this is just case (iii) which leads to equation (23.21).

If $m = 2k$, $n \neq 3k$, the equation (23.24) has the form

$$f(\mathbf{X}, \mathbf{Y})f(\mathbf{U}, \mathbf{V}) + f(\mathbf{X}, \mathbf{U})f(\mathbf{X}, \mathbf{Y}) + f(\mathbf{X}, \mathbf{V})f(\mathbf{Y}, \mathbf{X}) = \mathbf{O}. \qquad (23.25)$$

d) $\mathbf{U}_{n-k} = \mathbf{Y}$ appears only if $n = 2k$ and equation (23.14) becomes

$$f(\mathbf{X}, \mathbf{Y})f(\mathbf{U}, \mathbf{V}) + f(\mathbf{X}, \mathbf{U})f(\mathbf{U}_{2m-k}, \mathbf{U}_{m-k}) \qquad (23.26)$$
$$+ \ f(\mathbf{X}, \mathbf{V})f(\mathbf{U}_{m-k}, \mathbf{Y}) = \mathbf{O}.$$

The equality $\mathbf{U}_{2m-k} = \mathbf{Y}$ would imply $m = n = 2k$ which is impossible. If $\mathbf{U}_{2m-k} = \mathbf{V}$, then $m = \dfrac{3}{2}k$, $n = 2k$. But this is just the case (iv) which leads to equation (23.22).

If $n = 2k$ and $m \neq \dfrac{3}{2}k$, then the equation (23.26) gets the form

$$f(\mathbf{X}, \mathbf{Y})f(\mathbf{U}, \mathbf{V}) + f(\mathbf{X}, \mathbf{V})f(\mathbf{X}, \mathbf{Y}) = \mathbf{O}. \qquad (23.27)$$

e) $\mathbf{U}_{n-k} = \mathbf{U}$ appears only if $n = k + m$. Then from Eq. (23.14) we obtain

$$f(\mathbf{X}, \mathbf{Y})f(\mathbf{U}, \mathbf{V}) + f(\mathbf{X}, \mathbf{U})f(\mathbf{U}_{2m-k}, \mathbf{U}_{m-k}) \qquad (23.28)$$
$$+ \ f(\mathbf{X}, \mathbf{V})f(\mathbf{U}_{m-k}, \mathbf{U}) = \mathbf{O}.$$

If $\mathbf{U}_{2m-k} = \mathbf{Y}$, then $m = 3k$, $n = 4k$ and this is case (ii) which leads to equation (23.18). If $\mathbf{U}_{2m-k} = \mathbf{V}$, then $m = 2k$, $n = 3k$ and $\mathbf{U}_{m-k} = \mathbf{Y}$. This is case (iii) which leads to equation (23.21).

When $n = k + m$ and $m \neq 3k$, $m \neq 2k$, then the equation (23.28) becomes

$$f(\mathbf{X}, \mathbf{Y})f(\mathbf{U}, \mathbf{V}) + f(\mathbf{X}, \mathbf{V})f(\mathbf{X}, \mathbf{U}) = \mathbf{O}. \qquad (23.29)$$

Therefore, to solve the *transformed equation* for distinct values of k, m, n we must solve the following Eqs. (23.18), (23.19), (23.21), (23.22), (23.23), (23.25), (23.27) and (23.29).

The equations (23.18), (23.19), (23.22), (23.23), (23.27) and (23.29) for $\mathbf{X} = \mathbf{U}$ and $\mathbf{Y} = \mathbf{V}$ in accordance with Eq. (23.13) get the form

$$f^2(\mathbf{U}, \mathbf{V}) = \mathbf{O}$$

which means that the general solution in these cases is

$$f(\mathbf{U}, \mathbf{V}) \equiv \mathbf{O}. \tag{23.30}$$

The equation (23.21) has the general solution

$$f(\mathbf{U}, \mathbf{V}) = g(\mathbf{U})h(\mathbf{V}) - g(\mathbf{V})h(\mathbf{U}), \tag{23.31}$$

where $g, h : \mathcal{V} \mapsto \mathcal{V}$ are arbitrary functions.

In order to find the general solution of the vector equation (23.25), we write the scalar equation for the j-th component, $j = 1, \cdots, n$:

$$f_j(\mathbf{X}, \mathbf{Y})f_j(\mathbf{U}, \mathbf{V}) + f_j(\mathbf{X}, \mathbf{U})f_j(\mathbf{X}, \mathbf{Y}) + f_j(\mathbf{X}, \mathbf{V})f_j(\mathbf{Y}, \mathbf{X}) = 0. \tag{23.25$_j$}$$

For each nontrivial solution f_j of the scalar equation (23.25$_j$) there exists at least one pair of complex vectors $(\mathbf{A}_j, \mathbf{B}_j)$ such that $f_j(\mathbf{A}_j, \mathbf{B}_j) \neq 0$. For $\mathbf{X} = \mathbf{V} = \mathbf{A}_j$ and $\mathbf{Y} = \mathbf{B}_j$ from Eq. (23.25$_j$) it follows that

$$f_j(\mathbf{A}_j, \mathbf{B}_j)[f_j(\mathbf{U}, \mathbf{A}_j) + f_j(\mathbf{A}_j, \mathbf{U})] = 0$$

or

$$f_j(\mathbf{A}_j, \mathbf{U}) = -f_j(\mathbf{U}, \mathbf{A}_j). \tag{23.32}$$

By putting $\mathbf{X} = \mathbf{A}_j$ and $\mathbf{Y} = \mathbf{B}_j$ from Eq. (23.25$_j$), on the basis of Eq. (23.32) we obtain

$$f_j(\mathbf{U}, \mathbf{V}) = f_j(\mathbf{A}_j, \mathbf{V}) - f_j(\mathbf{A}_j, \mathbf{U}). \tag{23.33}$$

Introducing the notation $g_j(\mathbf{U}) = f_j(\mathbf{A}_j, \mathbf{U})$, the equality (23.33) becomes

$$f_j(\mathbf{U}, \mathbf{V}) = g_j(\mathbf{V}) - g_j(\mathbf{U}). \tag{23.34$_j$}$$

On the other hand, if $f_j \equiv 0$ is a trivial solution of Eq. (23.25$_j$), we can still write it in the form Eq. (23.34$_j$), where g_j is an arbitrary constant. So if we define $g(\mathbf{U}) = (g_1(\mathbf{U}), \cdots, g_n(\mathbf{U}))^T$, we have

$$f(\mathbf{U}, \mathbf{V}) = g(\mathbf{V}) - g(\mathbf{U}). \tag{23.34}$$

Since the function Eq. (23.34) is really a solution of the equation (23.25), the general solution of this equation has the form Eq. (23.34), where $g : \mathcal{V} \mapsto \mathcal{V}$ is an arbitrary function.

On the basis of the above there follows the result:

The general solution of the *transformed equation* is

$$f(\mathbf{U},\mathbf{V}) = \begin{cases} g(\mathbf{U})h(\mathbf{V}) - g(\mathbf{V})h(\mathbf{U}) & (m = 2k,\ n = 3k), \\ g(\mathbf{V}) - g(\mathbf{U}) & (m = 2k,\ n \neq 3k), \\ 0 & \text{in all other cases,} \end{cases} \tag{23.35}$$

where $g, h : \mathcal{V} \mapsto \mathcal{V}$ are arbitrary complex vector functions.

Theorem 23.1 immediately follows from the result obtained above and the changes Eqs. (23.6) and (23.9). $\hfill\square$

For illustration of the above theorem we will consider some examples.

Example 23.1   The functional equation (23.1) with the condition Eq. (23.2) for $p = 4$, $q = 6$, $r = 2$ and $n = 6$ becomes

$$f(\mathbf{Z}_0,\mathbf{Z}_4)f(\mathbf{Z}_6,\mathbf{Z}_2) + f(\mathbf{Z}_0,\mathbf{Z}_5)f(\mathbf{Z}_1,\mathbf{Z}_3) + f(\mathbf{Z}_0,\mathbf{Z}_6)f(\mathbf{Z}_2,\mathbf{Z}_4)$$
$$+ \quad f(\mathbf{Z}_0,\mathbf{Z}_1)f(\mathbf{Z}_3,\mathbf{Z}_5) + f(\mathbf{Z}_0,\mathbf{Z}_2)f(\mathbf{Z}_4,\mathbf{Z}_6) + f(\mathbf{Z}_0,\mathbf{Z}_3)f(\mathbf{Z}_5,\mathbf{Z}_1) = \mathbf{O}.$$

By putting $\mathbf{Z}_4 = \mathbf{B}$, $\mathbf{Z}_6 = \mathbf{U}$ and $\mathbf{Z}_2 = \mathbf{V}$ and by replacing all the other variables by $\mathbf{A}$, from the last equation we derive

$$f(\mathbf{A},\mathbf{B})f(\mathbf{U},\mathbf{V}) + f(\mathbf{A},\mathbf{U})f(\mathbf{V},\mathbf{B}) + f(\mathbf{A},\mathbf{V})f(\mathbf{B},\mathbf{U}) = \mathbf{O}$$

whose general solution is given by the formula

$$f(\mathbf{U},\mathbf{V}) = g(\mathbf{U})h(\mathbf{V}) - g(\mathbf{V})h(\mathbf{U}).$$

Example 23.2   The equation (23.1) with Eq. (23.2) for $p = 5$, $q = 3$, $r = 2$ and $n = 5$ takes the form

$$f(\mathbf{Z}_0,\mathbf{Z}_5)f(\mathbf{Z}_3,\mathbf{Z}_2) + f(\mathbf{Z}_0,\mathbf{Z}_1)f(\mathbf{Z}_4,\mathbf{Z}_3) + f(\mathbf{Z}_0,\mathbf{Z}_2)f(\mathbf{Z}_5,\mathbf{Z}_4)$$
$$+ f(\mathbf{Z}_0,\mathbf{Z}_3)f(\mathbf{Z}_1,\mathbf{Z}_5) + f(\mathbf{Z}_0,\mathbf{Z}_4)f(\mathbf{Z}_2,\mathbf{Z}_1) = \mathbf{O}.$$

If we put $\mathbf{Z}_5 = \mathbf{B}$, $\mathbf{Z}_3 = \mathbf{U}$ and $\mathbf{Z}_2 = \mathbf{V}$, after we replace all the other variables by $\mathbf{A}$, we obtain the following equation

$$f(\mathbf{A},\mathbf{B})f(\mathbf{U},\mathbf{V}) + f(\mathbf{A},\mathbf{V})f(\mathbf{B},\mathbf{A}) + f(\mathbf{A},\mathbf{U})f(\mathbf{A},\mathbf{B}) = \mathbf{O}.$$

For $\mathbf{U} = \mathbf{A}$ and $\mathbf{V} = \mathbf{B}$ from the above equation we obtain $f(\mathbf{A},\mathbf{B}) = -f(\mathbf{B},\mathbf{A})$, so that the general solution is the function

$$f(\mathbf{U},\mathbf{V}) = g(\mathbf{V}) - g(\mathbf{U}),$$

where $f(\mathbf{A},\mathbf{U}) = g(\mathbf{U})$.

Example 23.3 The equation (23.1) with Eq. (23.2) for $p = 1$, $q = 2$, $r = 4$ and $n = 5$ becomes

$$f(\mathbf{Z}_0, \mathbf{Z}_1)f(\mathbf{Z}_2, \mathbf{Z}_4) + f(\mathbf{Z}_0, \mathbf{Z}_2)f(\mathbf{Z}_3, \mathbf{Z}_5) + f(\mathbf{Z}_0, \mathbf{Z}_3)f(\mathbf{Z}_4, \mathbf{Z}_1)$$
$$+ f(\mathbf{Z}_0, \mathbf{Z}_4)f(\mathbf{Z}_5, \mathbf{Z}_2) + f(\mathbf{Z}_0, \mathbf{Z}_5)f(\mathbf{Z}_1, \mathbf{Z}_3) = \mathbf{O}.$$

If we put $\mathbf{Z}_0 = \mathbf{Z}_1 = \mathbf{Z}_3 = \mathbf{Z}_4 = \mathbf{U}$ and $\mathbf{Z}_2 = \mathbf{Z}_5 = \mathbf{V}$, we obtain $f^2(\mathbf{U}, \mathbf{V}) = \mathbf{O} \implies f(\mathbf{U}, \mathbf{V}) = \mathbf{O}$, which means that the general solution of the above functional equation is the trivial solution.

24 Special Quadratic Functional Equation

We will call *special equation* the functional equation (23.1) in the special case when the function F is given by the following formula

$$F(\mathbf{Z}_0, \mathbf{Z}_1, \cdots, \mathbf{Z}_n) = f(\mathbf{Z}_1, \mathbf{Z}_p)f(\mathbf{Z}_q, \mathbf{Z}_r), \tag{24.1}$$

where $f : \mathbf{R}^2 \mapsto \mathbf{R}$.

We do not solve the special equation in the general case, but we will investigate when this equation has nontrivial solutions. At the same time, we will give some necessary conditions for this equation to have other than trivial solutions. After that, we will give one hypothesis when this equation has nontrivial solutions.

By putting $\mathbf{Z}_p = \mathbf{Z}_r = \mathbf{V}$ and replacing all the other variables by $\mathbf{U}$, in view of $f(\mathbf{U}, \mathbf{U}) \equiv 0$ the special equation becomes

$$f^2(\mathbf{U}, \mathbf{U}) + f(\mathbf{V}, \mathbf{Z}_{2p-1})f(\mathbf{Z}_{p+q-1}, \mathbf{Z}_{p+r-1}) \tag{24.2}$$
$$+ \quad f(\mathbf{V}, \mathbf{Z}_{p+r-1})f(\mathbf{Z}_{q+r-1}, \mathbf{Z}_{2r-1}) + f(\mathbf{Z}_{\alpha_1}, \mathbf{V})f(\mathbf{Z}_{\alpha_2}, \mathbf{Z}_{\alpha_3}) = 0,$$

where

$$\alpha_1 = \begin{cases} r - p + 1 & (r - p + 1 > 0), \\ n + r - p + 1 & (r - p + 1 \le 0), \end{cases}$$

$$\alpha_2 = \begin{cases} r + q - p & (r + q - p > 0), \\ n + q + r - p & (r + q - p \le 0), \end{cases}$$

$$\alpha_3 = \begin{cases} 2r - p & (2r - p > 0), \\ n + 2r - p & (2r - p \le 0). \end{cases}$$

We can write equation (24.2) in the following form

$$af^2(\mathbf{U}, \mathbf{V}) + bf(\mathbf{U}, \mathbf{V})f(\mathbf{V}, \mathbf{U}) + cf^2(\mathbf{V}, \mathbf{U}) = 0, \qquad (24.3)$$

where $a, b, c \ (a \neq 0)$ are nonnegative integers such that $1 \leq a + b + c \leq 4$.

By permuting $\mathbf{U}$ and $\mathbf{V}$ in equation (24.3) we obtain

$$af^2(\mathbf{V}, \mathbf{U}) + bf(\mathbf{V}, \mathbf{U})f(\mathbf{U}, \mathbf{V}) + cf^2(\mathbf{U}, \mathbf{V}) = 0. \qquad (24.4)$$

If $b = 0$, by virtue of $a > 0$, $c \geq 0$ and $f : \mathbf{R}^2 \mapsto \mathbf{R}$ from Eq. (24.3) we get

$$f(\mathbf{U}, \mathbf{V}) \equiv 0. \qquad (24.5)$$

If $c = 0$, $b \neq 0$ $(b \neq a)$, assuming the existence of a pair of real numbers (A, B) such that $f(A, B) \neq 0$, from Eqs. (24.3) and (24.4) we obtain the following system

$$af^2(A, B) + bf(A, B)f(B, A) = 0,$$
$$af^2(B, A) + bf(A, B)f(B, A) = 0.$$

From this system we obtain $f(A, B) = f(B, A) = 0$ which contradicts the assumption $f(A, B) \neq 0$. Consequently, in this case we have Eq. (24.5).

Now, we will investigate the case $a, b, c \neq 0$. If $a = b = c = 1$, assuming that $f(A, B) \neq 0$, from Eqs. (24.3) and (24.4) we get

$$f^2(A, B) + f(A, B)f(B, A) + f^2(B, A) = 0,$$

but this equation has no nontrivial real solutions, so that $f(A, B) = f(B, A) = 0$ and again we have Eq. (24.5). We obtain the same result in the cases when $a = 2$, $b = c = 1$ and $a = b = 1$, $c = 2$.

Now, we will analyze the case when $a = b \neq 0$ and $c = 0$. From Eqs. (24.3) and (24.4) we have

$$[f(\mathbf{U}, \mathbf{V}) + f(\mathbf{V}, \mathbf{U})]^2 = 0,$$

i.e.,

$$f(\mathbf{U}, \mathbf{V}) + f(\mathbf{V}, \mathbf{U}) = 0. \qquad (24.6)$$

The remaining possibility is $a = c = 1$ and $b = 2$. Then from Eq. (24.3) we again obtain Eq. (24.6).

Consequently, a necessary condition for the existence of nontrivial solutions of the special equation is

$$c = 0, \qquad a = b \,(= 1,\, 2) \tag{24.7}$$

or

$$b = 2, \qquad a = c = 1. \tag{24.8}$$

The relations Eq. (24.7) for the case $a = b = 1$ will hold if and only if one of the following conditions is satisfied:

$$2r - 1 \equiv p, \quad p + q - 1 \not\equiv r, \quad r + q - p \not\equiv p \quad (\text{mod } n), \tag{24.9}$$

$$r + q - p \equiv p, \quad p + q - 1 \not\equiv r, \quad 2r - 1 \not\equiv p \quad (\text{mod } n). \tag{24.10}$$

The case $a = b = 2$, $c = 0$ is impossible because then $p + r - 1 \equiv r$ (mod n) must hold, but this relation is also impossible.

The relations Eq. (24.8) will hold if and only if

$$p + q - 1 \equiv r, \quad 2r - 1 \equiv p, \quad r + q - p \equiv p \quad (\text{mod } n). \tag{24.11}$$

On the basis of this there follows the result:

Theorem 24.1 *For the special equation to have nontrivial solutions one of the conditions Eqs. (24.9), (24.10), (24.11) must be satisfied.*

If there exist nontrivial solutions of the special equation, they have the property Eq. (24.6) so that all solutions must be of the following form

$$f(\mathbf{U}, \mathbf{V}) = G(\mathbf{U}, \mathbf{V}) - G(\mathbf{V}, \mathbf{U}), \tag{24.12}$$

where $G : \mathbf{R}^2 \mapsto \mathbf{R}$ is an appropriately chosen function.

The simplest function of the form Eq. (24.12), except for the function $f(\mathbf{U}, \mathbf{V}) \equiv 0$, is the following function

$$f(\mathbf{U}, \mathbf{V}) = g(\mathbf{U}) - g(\mathbf{V}). \tag{24.13}$$

Let us investigate, when the function given by Eq. (24.13) is a solution of the special equation for any function $g : \mathbf{R} \mapsto \mathbf{R}$. Then we must have

$$
\begin{aligned}
g(\mathbf{Z}_1)g(\mathbf{Z}_q) &\;-\; g(\mathbf{Z}_1)g(\mathbf{Z}_r) \\
-g(\mathbf{Z}_p)g(\mathbf{Z}_q) &\;+\; g(\mathbf{Z}_p)g(\mathbf{Z}_r) \\
+g(\mathbf{Z}_2)g(\mathbf{Z}_{q+1}) &\;-\; g(\mathbf{Z}_2)g(\mathbf{Z}_{r+1}) \\
-g(\mathbf{Z}_{p+1})g(\mathbf{Z}_{q+1}) &\;+\; g(\mathbf{Z}_{p+1})g(\mathbf{Z}_{r+1})
\end{aligned}
$$

$$+ \cdots$$

$$+g(\mathbf{Z}_n)g(\mathbf{Z}_{q-1}) \quad - \quad g(\mathbf{Z}_n)g(\mathbf{Z}_{r-1})$$
$$-g(\mathbf{Z}_{p-1})g(\mathbf{Z}_{q-1}) \quad + \quad g(\mathbf{Z}_{p-1})g(\mathbf{Z}_{r-1}) = 0.$$

The last equation will be an identity for all g in two cases:

1^0. The terms of the first and third column of the above equation mutually cancel, and so do the terms of the second and fourth column.

2^0. The terms of the first and second column mutually cancel, and so do the terms of the third and fourth column.

This will happen if and only if

1^0. $2r - 1 \equiv p, \qquad 2q - 1 \equiv p \pmod{n}$

or

2^0. $q + r - p \equiv p, \qquad r + q - 1 \equiv 1 \pmod{n}$.

The conditions 1^0 and 2^0 indeed present sufficient conditions for the existence of solutions other than the trivial one of the special equation.

Now, we will give the following hypothesis:

Hypothesis 24.1   The special equation has nontrivial solutions if and only if one of the following two conditions holds:

$$2r - 1 \equiv p, \qquad 2q - 1 \equiv p \pmod{n}, \tag{24.14}$$

$$q + r - p \equiv p, \qquad r + q - 1 \equiv 0 \pmod{n}. \tag{24.15}$$

If these conditions are satisfied, simultaneously a necessary condition for existence of nontrivial solutions holds.

Example 24.1   For the equation

$$f(\mathbf{Z}_1, \mathbf{Z}_3)f(\mathbf{Z}_5, \mathbf{Z}_2) + f(\mathbf{Z}_2, \mathbf{Z}_4)f(\mathbf{Z}_6, \mathbf{Z}_3) + f(\mathbf{Z}_3, \mathbf{Z}_5)f(\mathbf{Z}_7, \mathbf{Z}_4)$$
$$+ \quad f(\mathbf{Z}_4, \mathbf{Z}_6)f(\mathbf{Z}_1, \mathbf{Z}_5) + f(\mathbf{Z}_5, \mathbf{Z}_7)f(\mathbf{Z}_2, \mathbf{Z}_6) + f(\mathbf{Z}_6, \mathbf{Z}_1)f(\mathbf{Z}_3, \mathbf{Z}_7)$$
$$+ \quad f(\mathbf{Z}_7, \mathbf{Z}_2)f(\mathbf{Z}_4, \mathbf{Z}_1) = 0$$

the necessary conditions are satisfied, but the conditions of the hypothesis are not. The above equation has only the trivial solution.

Chapter 7

Modified Quadratic Functional Equations

In this chapter the basic quadratic complex vector functional equation, permuted quadratic functional equation, and three modified quadratic complex vector functional equations are solved. The results presented here were obtained in [I. B. Risteski and V. C. Covachev (submitted B)].

The notation used is as in Chapter 6.

25 Basic Quadratic Functional Equation

Now the following result will be proved.

Theorem 25.1 *The general solution of the functional equation*

$$\sum_{i=1}^{2n-1} F(\mathbf{Z}_1, \mathbf{Z}_{i+1}, \mathbf{Z}_{i+2}\cdots, \mathbf{Z}_{2n+i-1}) = \mathbf{O} \tag{25.1}$$

$$(\mathbf{Z}_{2n+i-1} \equiv \mathbf{Z}_i,\ 2 \le i \le 2n - 1),$$

where

$$F(\mathbf{Z}_1, \mathbf{Z}_2, \cdots, \mathbf{Z}_{2n}) \tag{25.2}$$

$$= \left[\sum_{i=1}^{k} f(\mathbf{Z}_{2i-1}, \mathbf{Z}_{2i})\right] \cdot \left[\sum_{i=1}^{n-k} f(\mathbf{Z}_{2k+2i-1}, \mathbf{Z}_{2k+2i})\right]$$

$$(f : \mathcal{V}^2 \mapsto \mathcal{V})$$

is given by the following formula

$$f(\mathbf{U}, \mathbf{V}) = \begin{cases} g(\mathbf{U})h(\mathbf{V}) - g(\mathbf{V})h(\mathbf{U}), & n = 2, \\ h(\mathbf{V}) - h(\mathbf{U}), & n > 2, \end{cases} \qquad (25.3)$$

where $g, h : \mathcal{V} \mapsto \mathcal{V}$ are arbitrary functions.

Proof. The function given by Eq. (25.3) is a solution of the equation (25.1).

Conversely, we will prove that every solution of the equation (25.1) has the form Eq. (25.3).

If we substitute $\mathbf{Z}_i = \mathbf{U}$ $(1 \le i \le 2n)$ into Eq. (25.1), we obtain

$$f(\mathbf{U}, \mathbf{U}) \equiv \mathbf{O}. \qquad (25.4)$$

The trivial solution $f(\mathbf{U}, \mathbf{V}) = \mathbf{O}$ of the equation (25.1) is included in Eq. (25.3). Moreover, a zero component of a nontrivial solution can be represented in this form. Further we will seek only nontrivial solutions.

If we put $\mathbf{Z}_{2k+2} = \mathbf{Z}_2 = \mathbf{V}$ and $\mathbf{Z}_3 = \mathbf{Z}_4 = \cdots = \mathbf{Z}_{2k+1} = \mathbf{Z}_{2k+3} = \cdots = \mathbf{Z}_{2n} = \mathbf{Z}_1 = \mathbf{U}$ into the equation (25.1), on the basis of Eq. (25.4) we have

$$pf^2(\mathbf{U}, \mathbf{V}) + qf(\mathbf{U}, \mathbf{V})f(\mathbf{V}, \mathbf{U}) + rf^2(\mathbf{V}, \mathbf{U}) = \mathbf{O}, \qquad (25.5)$$

where $p > 0$, $q \ge 0$, $r > 0$ are integers.

Since the function $f(\mathbf{U}, \mathbf{V}) = h(\mathbf{V}) - h(\mathbf{U})$ must satisfy the equation (25.5) for every function h, we get the condition $p + r = q$.

By a permutation of the variables $\mathbf{U}$ and $\mathbf{V}$ in Eq. (25.5) we obtain

$$pf^2(\mathbf{V}, \mathbf{U}) + qf(\mathbf{V}, \mathbf{U})f(\mathbf{U}, \mathbf{V}) + rf^2(\mathbf{U}, \mathbf{V}) = \mathbf{O}. \qquad (25.6)$$

By adding together the equalities (25.5) and (25.6), and by virtue of the condition $p + r = q > 0$ we have

$$f(\mathbf{U}, \mathbf{V}) + f(\mathbf{V}, \mathbf{U}) = \mathbf{O}. \qquad (25.7)$$

If $n = 2$, then the equation (25.1) according to Eq. (25.2) becomes

$$f(\mathbf{Z}_1, \mathbf{Z}_2)f(\mathbf{Z}_3, \mathbf{Z}_4) + f(\mathbf{Z}_1, \mathbf{Z}_3)f(\mathbf{Z}_4, \mathbf{Z}_2) + f(\mathbf{Z}_1, \mathbf{Z}_4)f(\mathbf{Z}_2, \mathbf{Z}_3) = \mathbf{O}. \quad (25.8)$$

We denote any nonzero component of a nontrivial solution again by f, but now f will be a scalar function of two vector arguments. For such a component there exists at least one pair of constant complex vectors

$(\mathbf{A}, \mathbf{B})$ $(\mathbf{A}, \mathbf{B} \in \mathcal{V})$ such that $f(\mathbf{A}, \mathbf{B}) \neq 0$. By putting $\mathbf{Z}_1 = \mathbf{A}$, $\mathbf{Z}_2 = \mathbf{B}$ into the *scalar* equation (25.8), we obtain

$$f(\mathbf{Z}_3, \mathbf{Z}_4) = -\frac{f(\mathbf{A}, \mathbf{Z}_3)}{f(\mathbf{A}, \mathbf{B})} f(\mathbf{Z}_4, \mathbf{B}) - \frac{f(\mathbf{A}, \mathbf{Z}_4)}{f(\mathbf{A}, \mathbf{B})} f(\mathbf{B}, \mathbf{Z}_3). \qquad (25.9)$$

If we denote $g(\mathbf{Z}) = f(\mathbf{A}, \mathbf{Z})/f(\mathbf{A}, \mathbf{B})$ and $h(\mathbf{Z}) = f(\mathbf{B}, \mathbf{Z})$, by virtue of the property Eq. (25.7) the equality (25.9) becomes

$$f(\mathbf{Z}_3, \mathbf{Z}_4) = g(\mathbf{Z}_3)h(\mathbf{Z}_4) - g(\mathbf{Z}_4)h(\mathbf{Z}_3).$$

Thus for $n = 2$ the theorem is proved.

Now, we will consider the case when $k = 1$ and $n > 2$.

If we put

$$\mathbf{Z}_3 = \mathbf{Z}_4 = \cdots = \mathbf{Z}_{2n-2} = \mathbf{Z}_1, \qquad \mathbf{Z}_{2n-1} = \mathbf{U}, \qquad \mathbf{Z}_{2n} = \mathbf{V}, \qquad (25.10)$$

then the equation (25.1) takes the following form

$$f(\mathbf{Z}_1, \mathbf{Z}_2)f(\mathbf{U}, \mathbf{V}) + f(\mathbf{Z}_1, \mathbf{U})f(\mathbf{V}, \mathbf{Z}_2)$$
$$+ \quad f(\mathbf{Z}_1, \mathbf{V})f(\mathbf{Z}_2, \mathbf{Z}_1) + f(\mathbf{Z}_1, \mathbf{V})f(\mathbf{Z}_1, \mathbf{U}) = \mathbf{O}.$$

Again we denote by f any nonzero component of the solution f. By putting $\mathbf{Z}_1 = \mathbf{A}$, $\mathbf{Z}_2 = \mathbf{B}$ $(f(\mathbf{A}, \mathbf{B}) \neq 0)$ and applying the property Eq. (25.7), from the previous equation we find

$$f(\mathbf{U}, \mathbf{V}) = \frac{f(\mathbf{A}, \mathbf{U})}{f(\mathbf{A}, \mathbf{B})} [f(\mathbf{B}, \mathbf{V}) - f(\mathbf{A}, \mathbf{V})] + f(\mathbf{A}, \mathbf{V}). \qquad (25.11)$$

For $\mathbf{V} = \mathbf{B}$, from the above equation there follows the equality

$$f(\mathbf{B}, \mathbf{U}) - f(\mathbf{A}, \mathbf{U}) = -f(\mathbf{A}, \mathbf{B}) \qquad (25.12)$$

which holds for every $\mathbf{U}$.

If we take into account Eq. (25.12), from Eq. (25.11) it follows that

$$f(\mathbf{U}, \mathbf{V}) = f(\mathbf{A}, \mathbf{V}) - f(\mathbf{A}, \mathbf{U}) = h(\mathbf{V}) - h(\mathbf{U}),$$

which means that the theorem for $k = 1$ and $n > 2$ is proved.

For $k > 1$, the equation Eq. (25.1) with the substitutions given by Eq. (25.10) takes the form

$$f(\mathbf{Z}_1, \mathbf{U})f(\mathbf{V}, \mathbf{Z}_2) + f(\mathbf{Z}_1, \mathbf{V})f(\mathbf{Z}_1, \mathbf{U}) + f(\mathbf{Z}_2, \mathbf{Z}_1)f(\mathbf{Z}_1, \mathbf{U}) = \mathbf{O}.$$

If we put $\mathbf{Z}_1 = \mathbf{A}$ and $\mathbf{U} = \mathbf{B}$ into any nontrivial scalar equation in the last vector equation, we obtain

$$f(\mathbf{V}, \mathbf{Z}_2) = f(\mathbf{A}, \mathbf{Z}_2) - f(\mathbf{A}, \mathbf{V}),$$

i.e.,

$$f(\mathbf{U}, \mathbf{V}) = h(\mathbf{V}) - h(\mathbf{U}).$$

This completes the proof of Theorem 25.1. $\qquad\qquad\square$

26   Permuted Quadratic Functional Equation

The functional equations considered in the next sections will be solved only for $n > 2$.

Now we will solve the permuted quadratic complex vector functional equation.

Theorem 26.1   *The general solution of the functional equation*

$$\sum_{i=1}^{2n-1} F(\mathbf{Z}_1, \mathbf{Z}_{i+1}, \mathbf{Z}_{i+2} \cdots, \mathbf{Z}_{2n+i-1}) = \mathbf{O} \tag{26.1}$$

$$(\mathbf{Z}_{2n+i-1} \equiv \mathbf{Z}_i, \ 2 \le i \le 2n - 1),$$

where

$$F(\mathbf{Z}_1, \mathbf{Z}_2, \cdots, \mathbf{Z}_{2n}) \tag{26.2}$$

$$= \left[\sum_{i=1}^{k} f(\mathbf{Z}_{2i-1}, \mathbf{U}_i) \right] \cdot \left[\sum_{i=1}^{n-k} f(\mathbf{Z}_{2k+2i-1}, \mathbf{V}_i) \right]$$

$$(f : \mathcal{V}^2 \mapsto \mathcal{V}; \ n > 2)$$

$$\mathbf{U}_i \ \in \ \{\mathbf{Z}_2, \mathbf{Z}_4, \cdots, \mathbf{Z}_{2k}\} \qquad (1 \le i \le k),$$

$$\mathbf{V}_j \ \in \ \{\mathbf{Z}_{2k+2}, \mathbf{Z}_{2k+4}, \cdots, \mathbf{Z}_{2n}\} \quad (1 \le j \le n - k),$$

$$\mathbf{U}_s \ne \mathbf{U}_t \ \text{and} \ \mathbf{V}_s \ne \mathbf{V}_t \ \text{if} \ s \ne t,$$

is given by the following formula

$$f(\mathbf{U}, \mathbf{V}) = g(\mathbf{V}) - g(\mathbf{U}), \tag{26.3}$$

where $g : \mathcal{V} \mapsto \mathcal{V}$ *is an arbitrary function.*

Proof. The function given by Eq. (26.3) is a solution of the equation (26.1). Really, by a substitution of Eq. (26.3) into Eq. (26.2), after a rearrangement of the terms we obtain

$$F(\mathbf{Z}_1, \mathbf{Z}_2, \cdots, \mathbf{Z}_{2n})$$
$$= \; [g(\mathbf{Z}_1) - g(\mathbf{Z}_2) + g(\mathbf{Z}_3) - \cdots + g(\mathbf{Z}_{2k-1}) - g(\mathbf{Z}_{2k})] \times$$
$$\times \; [g(\mathbf{Z}_{2k+1}) - g(\mathbf{Z}_{2k+2}) + g(\mathbf{Z}_{2k+3}) - \cdots + g(\mathbf{Z}_{2n-1}) - g(\mathbf{Z}_{2n})],$$

i.e., the same result as by the substitution of Eq. (26.3) into Eq. (25.2). Since the function Eq. (26.3) is a solution of the equation (25.1), it follows that Eq. (26.3) is a solution of the equation (26.1).

Conversely, we will prove that every solution of the equation (26.1) has the form Eq. (26.3).

If we substitute $\mathbf{Z}_i = \mathbf{U}$ $(1 \le i \le 2n)$, the equation (25.1) becomes

$$f(\mathbf{U}, \mathbf{U}) \equiv \mathbf{O}. \tag{26.4}$$

The trivial solution $f(\mathbf{U}, \mathbf{V}) = \mathbf{O}$ of the equation (26.1) is included in Eq. (26.3). Moreover, a zero component of a nontrivial solution can be represented in this form. Further we will seek only nontrivial solutions.

If we put $\mathbf{Z}_2 = \mathbf{Z}_{2n} = \mathbf{V}$ and $\mathbf{Z}_3 = \mathbf{Z}_4 = \cdots = \mathbf{Z}_{2n-1} = \mathbf{Z}_1 = \mathbf{U}$ into the equation (26.1), on the basis of the above expression Eq. (26.4) we obtain

$$pf^2(\mathbf{U}, \mathbf{V}) + qf(\mathbf{U}, \mathbf{V})f(\mathbf{V}, \mathbf{U}) + rf^2(\mathbf{V}, \mathbf{U}) = \mathbf{O}, \tag{26.5}$$

where $p > 0$, $q \ge 0$, $r > 0$ are integers.

Since the function Eq. (26.3) must satisfy the equation (26.5) for every function g, by a substitution of Eq. (26.3) into Eq. (26.5) we get the condition $p + r = q$.

By a permutation of the variables $\mathbf{U}$ and $\mathbf{V}$, from Eq. (26.5) we obtain

$$pf^2(\mathbf{V}, \mathbf{U}) + qf(\mathbf{V}, \mathbf{U})f(\mathbf{U}, \mathbf{V}) + rf^2(\mathbf{U}, \mathbf{V}) = \mathbf{O}. \tag{26.6}$$

By adding together the equalities (26.5) and (26.6), and by virtue of the condition $p + r = q > 0$ we get

$$f(\mathbf{U}, \mathbf{V}) + f(\mathbf{V}, \mathbf{U}) = \mathbf{O}. \tag{26.7}$$

Now we will distinguish the following two cases:

1^0. The case when $k = 1$. In this case the function Eq. (26.2) becomes

$$F(\mathbf{Z}_1, \mathbf{Z}_2, \cdots, \mathbf{Z}_{2n}) = f(\mathbf{Z}_1, \mathbf{Z}_2)[f(\mathbf{Z}_3, \mathbf{V}_1)$$

$$+ f(\mathbf{Z}_4, \mathbf{V}_2) + \cdots + f(\mathbf{Z}_{2n-1}, \mathbf{V}_{n-1})],$$

$$\mathbf{V}_i \in \{\mathbf{Z}_4, \mathbf{Z}_6, \cdots, \mathbf{Z}_{2n}\} \qquad (1 \le i \le n-1),$$

$$\mathbf{V}_i \ne \mathbf{V}_j \qquad \text{if} \qquad i \ne j.$$

It is necessary to separately investigate the cases $\mathbf{V}_{n-1} \ne \mathbf{Z}_{2n}$ and $\mathbf{V}_{n-1} = \mathbf{Z}_{2n}$. We denote any nonzero component of a nontrivial solution again by f, but now f will be a scalar function of two vector arguments. For such a component there exists at least one pair of constant complex vectors $(\mathbf{A}, \mathbf{B})$ $(\mathbf{A}, \mathbf{B} \in \mathcal{V})$ such that $f(\mathbf{A}, \mathbf{B}) \ne 0$.

In the case $\mathbf{V}_{n-1} \ne \mathbf{Z}_{2n}$ the substitutions

$$\mathbf{Z}_1 = \mathbf{A}, \quad \mathbf{Z}_2 = \mathbf{B}, \quad \mathbf{Z}_{2n-1} = \mathbf{U}, \quad \mathbf{V}_{n-1} = \mathbf{V}, \quad (26.8)$$
$$\mathbf{Z}_3 = \mathbf{Z}_5 = \cdots = \mathbf{Z}_{2n-3} = \mathbf{A}, \quad \mathbf{V}_1 = \mathbf{V}_2 = \cdots = \mathbf{V}_{n-2} = \mathbf{A}$$

transform any nontrivial scalar equation in Eq. (26.1) to the following form

$$\begin{aligned}
f(\mathbf{A}, \mathbf{B}) f(\mathbf{U}, \mathbf{V}) \quad &+ \quad f(\mathbf{A}, \mathbf{U})[f(\mathbf{A}, \mathbf{B}) + f(\mathbf{A}, \mathbf{V})] \\
&+ \quad f(\mathbf{A}, \mathbf{V})[f(\mathbf{B}, \mathbf{A}) + f(\mathbf{U}, \mathbf{A})] = 0.
\end{aligned}$$

From the above equation, on the basis of the expression Eq. (26.7) we obtain

$$f(\mathbf{U}, \mathbf{V}) = f(\mathbf{A}, \mathbf{V}) - f(\mathbf{A}, \mathbf{U}).$$

By introducing the notation $f(\mathbf{A}, \mathbf{U}) = g(\mathbf{U})$, we obtain that the function f really is of the form Eq. (26.3).

In the case when $\mathbf{V}_{n-1} = \mathbf{Z}_{2n}$ the substitutions Eq. (26.8) transform any nontrivial scalar equation in Eq. (26.1) to the following equation

$$\begin{aligned}
f(\mathbf{A}, \mathbf{B}) f(\mathbf{U}, \mathbf{V}) \quad &+ \quad f(\mathbf{A}, \mathbf{U})[f(\mathbf{V}, \mathbf{A}) + f(\mathbf{A}, \mathbf{B})] \qquad (26.9) \\
&+ \quad f(\mathbf{A}, \mathbf{V})[f(\mathbf{B}, \mathbf{A}) + f(\mathbf{A}, \mathbf{U})] = 0
\end{aligned}$$

when $\mathbf{V}_1 \ne \mathbf{Z}_4$, and to the equation

$$\begin{aligned}
f(\mathbf{A}, \mathbf{B}) f(\mathbf{U}, \mathbf{V}) + f(\mathbf{A}, \mathbf{U}) f(\mathbf{V}, \mathbf{B}) \qquad\qquad (26.10) \\
+ \quad f(\mathbf{A}, \mathbf{V}) f(\mathbf{B}, \mathbf{A}) + f(\mathbf{A}, \mathbf{V}) f(\mathbf{A}, \mathbf{U}) = 0
\end{aligned}$$

when $\mathbf{V}_1 = \mathbf{Z}_4$.

The equation (26.9) on the basis of the expression Eq. (26.7) yields

$$f(\mathbf{U}, \mathbf{V}) = g(\mathbf{V}) - g(\mathbf{U}), \qquad (26.11)$$

where the notation $f(\mathbf{A}, \mathbf{U}) = g(\mathbf{U})$ is used.

From Eq. (26.10), for $\mathbf{V} = \mathbf{B}$, we obtain

$$f(\mathbf{U}, \mathbf{B}) = f(\mathbf{A}, \mathbf{B}) - f(\mathbf{A}, \mathbf{U}).$$

By using this equality and by introducing the notation $f(\mathbf{A}, \mathbf{U}) = g(\mathbf{U})$, equation (26.10) becomes Eq. (26.11).

2^0. The case when $k > 1$. Let $\mathbf{U}_1 = \mathbf{Z}_{2p}$ $(1 < p < k)$. Now we will distinguish two cases: $1 < 2p \le k$ and $k < 2p \le 2k$.

For the case $1 < 2p \le k$, we put $\mathbf{Z}_1 = \mathbf{U}$, $\mathbf{Z}_{2p} = \mathbf{V}$, $\mathbf{Z}_{2n} = \mathbf{Y}$ and $\mathbf{Z}_2 = \mathbf{Z}_3 = \cdots = \mathbf{Z}_{2p-1} = \mathbf{Z}_{2p+1} = \cdots = \mathbf{Z}_{2n-1} = \mathbf{Z}$, so that the equation (26.1), on the basis of the expressions Eqs. (26.4) and (26.7), takes the following form

$$
\begin{aligned}
f(\mathbf{Z}, \mathbf{Y})f(\mathbf{U}, \mathbf{V}) \;+\;& af(\mathbf{U}, \mathbf{Z})f(\mathbf{V}, \mathbf{Z}) + bf(\mathbf{U}, \mathbf{Z})f(\mathbf{Y}, \mathbf{Z}) \qquad (26.12) \\
+\, cf(\mathbf{Z}, \mathbf{Y})f(\mathbf{V}, \mathbf{Z}) \;+\;& df(\mathbf{U}, \mathbf{Z})f(\mathbf{Y}, \mathbf{V}) + \varepsilon f(\mathbf{U}, \mathbf{Y})f(\mathbf{V}, \mathbf{Z}) = \mathbf{O},
\end{aligned}
$$

where a, b, c, d, ε are integers.

We notice that in this equation $\varepsilon = 0$. Indeed, the term $f(\mathbf{U}, \mathbf{Y})f(\mathbf{V}, \mathbf{Z})$ in the equation (26.1) can only appear in the item $F(\mathbf{Z}_1, \mathbf{Z}_{2n}, \mathbf{Z}_2, \mathbf{Z}_3, \cdots, \mathbf{Z}_{2n-1})$. The variable $\mathbf{Z}_{2p}$ appears among the arguments of this function in $(2p+2)$-nd place, so that by the above substitution we obtain

$$f(\mathbf{Z}_1, \mathbf{Z}_{2n}, \mathbf{Z}_2, \mathbf{Z}_3, \cdots, \mathbf{Z}_{2n-1}) = [f(\mathbf{U}, \mathbf{Y}) + f(\mathbf{V}, \mathbf{Z})](n - k)f(\mathbf{Z}, \mathbf{Z}) \equiv \mathbf{O}.$$

The function $f(\mathbf{U}, \mathbf{V}) = g(\mathbf{V}) - g(\mathbf{U})$ must satisfy the equation (26.12) for every function g. By a substitution of Eq. (26.3) into Eq. (26.12) we come to the following system of equations

$$
\begin{aligned}
a + b = 1, \quad a - c - d = -1, \quad c = 1, \\
b - c + d = 0, \quad a - d = 0, \quad a + b - c = 0.
\end{aligned}
$$

From this system we obtain

$$b = 1 - a, \qquad c = 1, \qquad d = a. \qquad (26.13)$$

Then the equation (26.1) has the form

$$
\begin{aligned}
f(\mathbf{Z},\mathbf{Y})f(\mathbf{U},\mathbf{V}) \;+\; & af(\mathbf{U},\mathbf{Z})f(\mathbf{V},\mathbf{Z}) + (1-a)f(\mathbf{U},\mathbf{Z})f(\mathbf{Y},\mathbf{Z}) \\
+\, f(\mathbf{Z},\mathbf{Y})f(\mathbf{V},\mathbf{Z}) \;+\; & af(\mathbf{U},\mathbf{Z})f(\mathbf{Y},\mathbf{V}) = \mathbf{O}.
\end{aligned}
\tag{26.14}
$$

For $\mathbf{Z} = \mathbf{A}$, $\mathbf{Y} = \mathbf{B}$ from any nontrivial scalar equation in Eq. (26.14) we obtain

$$
\begin{aligned}
f(\mathbf{A},\mathbf{B})f(\mathbf{U},\mathbf{V}) \;+\; & af(\mathbf{U},\mathbf{A})f(\mathbf{V},\mathbf{A}) + (1-a)f(\mathbf{U},\mathbf{A})f(\mathbf{B},\mathbf{A}) \\
+\, f(\mathbf{A},\mathbf{B})f(\mathbf{V},\mathbf{A}) \;+\; & af(\mathbf{U},\mathbf{A})f(\mathbf{B},\mathbf{V}) = 0.
\end{aligned}
\tag{26.15}
$$

For $\mathbf{U} = \mathbf{B}$ from this equality we obtain

$$
f(\mathbf{B},\mathbf{V}) + f(\mathbf{V},\mathbf{A}) + f(\mathbf{A},\mathbf{B}) = 0.
\tag{26.16}
$$

By virtue of (26.16) from (26.15) we get

$$
f(\mathbf{U},\mathbf{V}) = f(\mathbf{A},\mathbf{V}) - f(\mathbf{A},\mathbf{U}).
$$

If we put $g(\mathbf{U}) = f(\mathbf{A},\mathbf{U})$, from the last relation we obtain

$$
f(\mathbf{U},\mathbf{V}) = g(\mathbf{V}) - g(\mathbf{U}).
$$

For the case $k < 2p \leq 2k$, we put $\mathbf{Z}_1 = \mathbf{U}$, $\mathbf{Z}_{2p} = \mathbf{V}, \mathbf{Z}_{2n} = \mathbf{Y}$ and $\mathbf{Z}_2 = \mathbf{Z}_3 = \cdots = \mathbf{Z}_{2p-1} = \mathbf{Z}_{2p+1} = \cdots = \mathbf{Z}_{2n-1} = \mathbf{Z}$, so that the equation (26.1) becomes

$$
\begin{aligned}
f(\mathbf{Z},\mathbf{Y})f(\mathbf{U},\mathbf{V}) \;+\; & a_1 f(\mathbf{U},\mathbf{Z})f(\mathbf{V},\mathbf{Z}) + b_1 f(\mathbf{U},\mathbf{Z})f(\mathbf{Y},\mathbf{Z}) \\
+\, c_1 f(\mathbf{Z},\mathbf{Y})f(\mathbf{V},\mathbf{Z}) \;+\; & d_1 f(\mathbf{U},\mathbf{Z})f(\mathbf{Y},\mathbf{V}) + \varepsilon_1 f(\mathbf{U},\mathbf{Y})f(\mathbf{V},\mathbf{Z}) = \mathbf{O}.
\end{aligned}
\tag{26.17}
$$

Similar to the case $1 < 2p \leq k$, we come to the conclusions that $\varepsilon_1 = 1$ and

$$
b_1 = 1 - d_1, \qquad c_1 = 0, \qquad a_1 = -b_1.
$$

Thus the equation obtained implies the equation (26.15). Consequently, in this case every solution has the form Eq. (26.3). $\qquad\qquad\square$

27   First Modified Quadratic Functional Equation

In this section the first modified equation will be solved.

Theorem 27.1 *The general solution of the first modified quadratic functional equation*

$$\sum_{i=1}^{2n-1} F(\mathbf{Z}_1, \mathbf{Z}_{i+1}, \mathbf{Z}_{i+2}, \cdots, \mathbf{Z}_{2n+i-1}) = \mathbf{O} \qquad (27.1)$$

$$(\mathbf{Z}_{2n+i-1} \equiv \mathbf{Z}_i, \ 2 \le i \le 2n - 1),$$

where

$$F(\mathbf{Z}_1, \mathbf{Z}_2, \cdots, \mathbf{Z}_{2n}) \qquad (27.2)$$

$$= \ [f(\mathbf{Z}_1, \mathbf{U}_1) - f(\mathbf{U}_2, \mathbf{V}_1) + f(\mathbf{V}_2, \mathbf{U}_3) - f(\mathbf{U}_4, \mathbf{V}_3)$$

$$+ \cdots + f(\mathbf{V}_{k-2}, \mathbf{U}_{k-1}) - f(\mathbf{U}_k, \mathbf{V}_{k-1})] \times$$

$$\times \ [f(\mathbf{Z}_{2k+1}, \mathbf{W}_1) + f(\mathbf{Z}_{2k+3}, \mathbf{W}_2) + \cdots + f(\mathbf{Z}_{2n-1}, \mathbf{W}_{n-k})] \quad \textit{for k even,}$$

$$= \ [f(\mathbf{Z}_1, \mathbf{U}_1) - f(\mathbf{U}_2, \mathbf{V}_1) + f(\mathbf{V}_2, \mathbf{U}_3)$$

$$- \cdots - f(\mathbf{U}_{k-1}, \mathbf{V}_{k-2}) + f(\mathbf{V}_{k-1}, \mathbf{U}_k)] \times$$

$$\times \ [f(\mathbf{Z}_{2k+1}, \mathbf{W}_1) + f(\mathbf{Z}_{2k+3}, \mathbf{W}_2) + \cdots + f(\mathbf{Z}_{2n}, \mathbf{W}_{n-k})] \quad \textit{for k odd}$$

$$(f : \ \mathcal{V}^2 \mapsto \mathcal{V}; \quad n > 2),$$

$$\mathbf{U}_i \in \{\mathbf{Z}_2, \mathbf{Z}_4, \cdots, \mathbf{Z}_{2k}\} \qquad (1 \le i \le k),$$

$$\mathbf{V}_j \in \{\mathbf{Z}_3, \mathbf{Z}_5, \cdots, \mathbf{Z}_{2k-1}\} \qquad (1 \le j \le k - 1),$$

$$\mathbf{W}_r \in \{\mathbf{Z}_{2k+2}, \mathbf{Z}_{2k+4}, \cdots, \mathbf{Z}_{2n}\} \qquad (1 \le r \le n - k),$$

$$\mathbf{U}_s \ne \mathbf{U}_t, \quad \mathbf{V}_s \ne \mathbf{V}_t, \quad \mathbf{W}_s \ne \mathbf{W}_t \quad \textit{for s} \ne t,$$

is given by

$$f(\mathbf{U}, \mathbf{V}) = g(\mathbf{V}) - g(\mathbf{U}) \qquad (27.3)$$

for k odd, where $g : \ \mathcal{V} \mapsto \mathcal{V}$ *is an arbitrary function, while for k even the components of the general solution of Eq. (27.1) are given by Eq. (27.3) or by*

$$f(\mathbf{U}, \mathbf{V}) = h(\mathbf{V}), \qquad (27.4)$$

where h is a suitably chosen function.

Proof. First we will find the general solution of the functional equation (27.1) for k odd.

By putting $\mathbf{Z}_i = \mathbf{U}$ $(1 \le i \le 2n)$ into Eq. (27.1) we obtain

$$f(\mathbf{U}, \mathbf{U}) \equiv \mathbf{O}. \tag{27.5}$$

For $\mathbf{Z}_1 = \mathbf{Z}_{2n} = \mathbf{U}$ and $\mathbf{Z}_i = \mathbf{V}$ $(2 \le i \le 2n - 1)$, on the basis of the expression Eq. (27.5), from the equation (27.1) we get

$$f^2(\mathbf{U}, \mathbf{V}) + f(\mathbf{U}, \mathbf{V})f(\mathbf{V}, \mathbf{U}) = \mathbf{O}. \tag{27.6}$$

By a permutation of $\mathbf{U}$ and $\mathbf{V}$, the last equality becomes

$$f^2(\mathbf{V}, \mathbf{U}) + f(\mathbf{V}, \mathbf{U})f(\mathbf{U}, \mathbf{V}) = \mathbf{O}. \tag{27.7}$$

If we add together Eqs. (27.6) and (27.7), we obtain

$$[f(\mathbf{U}, \mathbf{V}) + f(\mathbf{V}, \mathbf{U})]^2 = \mathbf{O},$$

i.e.,

$$f(\mathbf{U}, \mathbf{V}) = -f(\mathbf{V}, \mathbf{U}). \tag{27.8}$$

By using the relation Eq. (27.8), the equation (27.1) can be transformed into the equation (26.1), whose solution according to Theorem 26.1 is given by

$$f(\mathbf{U}, \mathbf{V}) = g(\mathbf{V}) - g(\mathbf{U}). \tag{27.9}$$

Since the last function in Eq. (27.9) satisfies the equation (27.1), Theorem 27.1 is proved for the case k odd.

Now, we will prove Theorem 27.1 for the case k even. For the components of the function $f : \mathcal{V}^2 \mapsto \mathcal{V}$ (denoted again by f) there exist two possibilities.

1^0. $f(\mathbf{U}, \mathbf{U}) \equiv 0$. By putting $\mathbf{Z}_1 = \mathbf{Z}_{2n} = \mathbf{U}$ and $\mathbf{Z}_i = \mathbf{V}$ $(2 \le i \le 2n - 1)$, on the basis of the assumption $f(\mathbf{U}, \mathbf{U}) \equiv 0$, from Eq. (27.1) we obtain the relation Eq. (27.6). As above, we deduce Eq. (27.8). Thus in this case the equation (27.1) becomes the equation (26.1). Consequently, for k even the components of the general solution of the equation (27.1) satisfying $f(\mathbf{U}, \mathbf{U}) \equiv 0$ are given by the formula Eq. (27.9).

2^0. $f(\mathbf{U}, \mathbf{U}) \not\equiv 0$. In this case there exists at least one vector $C \in \mathcal{V}$ such that $f(C, C) \ne 0$.

By putting $\mathbf{Z}_1 = \mathbf{U}$, $\mathbf{Z}_2 = \mathbf{V}$ and $\mathbf{Z}_i = C$ $(3 \le i \le 2n)$, from the equation (27.1) we obtain

$$(n - k)\{[f(\mathbf{U},\mathbf{V}) - K + 2(k - 1)f(\mathbf{U},C) \tag{27.10}$$
$$-2(k - 2)K - f(C,\mathbf{V}) - f(\mathbf{V},C)]K$$
$$+ \quad [f(\mathbf{U},C) - K][f(C,\mathbf{V}) + f(\mathbf{V},C) + 2(n - k - 1)K]\} = 0,$$

where we have put $f(C,C) = K$. From the last equation, for $\mathbf{V} = C$, we obtain

$$f(\mathbf{U},C) = K. \tag{27.11}$$

By using Eq. (27.11), from Eq. (27.10) we get

$$f(\mathbf{U},\mathbf{V}) = h(\mathbf{V}) \qquad (h(\mathbf{U}) = f(C,\mathbf{U})). \tag{27.12}$$

However, in the general case the function Eq. (27.12) is not a solution of the equation (27.1) for an arbitrary function h. $\qquad\square$

In the next section we will solve the second modified equation.

28 Second Modified Quadratic Functional Equation

Now, we will give the following result.

Theorem 28.1 *The functional equation*

$$\sum_{i=1}^{2n-1} F(\mathbf{Z}_1, \mathbf{Z}_{i+1}, \mathbf{Z}_{i+2}, \cdots, \mathbf{Z}_{2n+i-1}) = \mathbf{O} \tag{28.1}$$
$$(\mathbf{Z}_{2n+i-1} \equiv \mathbf{Z}_i,\ 2 \le i \le 2n - 1),$$

where

$$F(\mathbf{Z}_1, \mathbf{Z}_2, \cdots, \mathbf{Z}_{2n}) \tag{28.2}$$
$$= \quad [f(\mathbf{Z}_1, \mathbf{U}_1) + f(\mathbf{Z}_3, \mathbf{U}_2) + \cdots + f(\mathbf{Z}_{2k-1}, \mathbf{U}_k)] \times$$
$$\times \quad [f(\mathbf{V}_1, \mathbf{W}_1) - f(\mathbf{W}_2, \mathbf{V}_2) + f(\mathbf{V}_3, \mathbf{W}_3) - \cdots - f(\mathbf{W}_{n-k-1}, \mathbf{V}_{n-k-1})$$
$$+ f(\mathbf{V}_{n-k}, \mathbf{W}_{n-k})] \quad \textit{for } n - k \textit{ odd,}$$
$$= \quad [f(\mathbf{Z}_1, \mathbf{U}_1) + f(\mathbf{Z}_3, \mathbf{U}_2) + \cdots + f(\mathbf{Z}_{2k-1}, \mathbf{U}_k)] \times$$
$$\times \quad [f(\mathbf{V}_1, \mathbf{W}_1) - f(\mathbf{W}_2, \mathbf{V}_2) + f(\mathbf{V}_3, \mathbf{W}_3) - \cdots + f(\mathbf{W}_{n-k-1}, \mathbf{V}_{n-k-1})$$
$$- f(\mathbf{V}_{n-k}, \mathbf{W}_{n-k})] \quad \textit{for } n - k \textit{ even}$$

$$(f : \mathcal{V}^2 \mapsto \mathcal{V}; \quad n > 2),$$

$$\mathbf{U}_i \in \{\mathbf{Z}_2, \mathbf{Z}_4, \cdots, \mathbf{Z}_{2k}\} \qquad (1 \le i \le k),$$

$$\mathbf{V}_j \in \{\mathbf{Z}_{2k+1}, \mathbf{Z}_{2k+3}, \cdots, \mathbf{Z}_{2n-1}\} \qquad (1 \le j \le n - k),$$

$$\mathbf{W}_r \in \{\mathbf{Z}_{2k+2}, \mathbf{Z}_{2k+4}, \cdots, \mathbf{Z}_{2n}\} \qquad (1 \le r \le n - k),$$

$$\mathbf{U}_s \ne \mathbf{U}_t, \quad \mathbf{V}_s \ne \mathbf{V}_t, \quad \mathbf{W}_s \ne \mathbf{W}_t \quad \text{for } s \ne t,$$

1^0. *has a general solution given by*

$$f(\mathbf{U}, \mathbf{V}) = g(\mathbf{U}) - g(\mathbf{V}), \tag{28.3}$$

where $g : \mathcal{V} \mapsto \mathcal{V}$ is an arbitrary function, if $n - k$ is odd;

　　2^0. *If $n - k$ is even and $\mathbf{V}_i = \mathbf{Z}_{2k+2i+1}$, $\mathbf{W}_j = \mathbf{Z}_{2k+2j+2}$ $(1 \le i, j \le n - k)$, then the function*

$$f(\mathbf{U}, \mathbf{V}) = K_1 g(\mathbf{U}) - K_2 g(\mathbf{V}) \tag{28.4}$$

is a solution of the equation (28.1).

　　For $k > 1$ and $K_1 = K_2$ the function Eq. (28.4) is a general solution of the equation (28.1) if $f(\mathbf{U}, \mathbf{U}) \equiv \mathbf{O}$ holds;

　　3^0. *If in Eq. (28.2) the variable $\mathbf{Z}_{2n}$ appears as an argument of a function f before which the $+$ sign stands, then the general solution of the equation (28.1) is the function (28.3) if $f(\mathbf{U}, \mathbf{U}) = \mathbf{O}$ holds.*

Proof. 1^0. By putting $\mathbf{Z}_i = \mathbf{U}$ $(1 \le i \le 2n)$, from the equation (28.1) we obtain

$$f(\mathbf{U}, \mathbf{U}) \equiv \mathbf{O}. \tag{28.5}$$

According to this relation, for $\mathbf{Z}_1 = \mathbf{Z}_{2n} = \mathbf{V}$ and $\mathbf{Z}_i = \mathbf{U}$ $(2 \le i \le 2n - 1)$, from the equation (28.1) it follows that

$$[f(\mathbf{U}, \mathbf{V}) + f(\mathbf{V}, \mathbf{U})]f(\mathbf{V}, \mathbf{U}) = \mathbf{O}. \tag{28.6}$$

By a permutation of the variables $\mathbf{U}$ and $\mathbf{V}$ in Eq. (28.6) and adding together the relation obtained and Eq. (28.6), we deduce

$$f(\mathbf{U}, \mathbf{V}) + f(\mathbf{V}, \mathbf{U}) = \mathbf{O}.$$

On the basis of this equality, we can transform the equation (28.1) into the equation (26.1), whose general solution is given by

$$f(\mathbf{U}, \mathbf{V}) = g(\mathbf{V}) - g(\mathbf{U}).$$

Since this function satisfies the equation (28.1) if $n - k$ is odd, this means that Theorem 28.1 is proved.

2^0. It is not difficult to check that Eq. (28.4) is indeed a solution of the equation (28.1) if $n - k$ is even and when the variables $\mathbf{V}_j$ and $\mathbf{W}_r$ ($1 \leq j, r \leq n - k$) have values specified in the theorem.

If $n - k$ is even ($k > 1$), by putting $\mathbf{Z}_2 = \mathbf{Z}_3 = \mathbf{Z}_4 = \mathbf{U}$ and $\mathbf{Z}_5 = \mathbf{Z}_6 = \cdots = \mathbf{Z}_{2n} = \mathbf{Z}_1 = \mathbf{V}$, from Eq. (28.1) we obtain

$$[f(\mathbf{U}, \mathbf{V}) + f(\mathbf{V}, \mathbf{U}) - f(\mathbf{U}, \mathbf{U}) - f(\mathbf{V}, \mathbf{V})][f(\mathbf{U}, \mathbf{V}) - f(\mathbf{U}, \mathbf{U})] = \mathbf{O}. \quad (28.7)$$

By a permutation of the variables, we have

$$[f(\mathbf{V}, \mathbf{U}) + f(\mathbf{U}, \mathbf{V}) - f(\mathbf{V}, \mathbf{V}) - f(\mathbf{U}, \mathbf{U})][f(\mathbf{V}, \mathbf{U}) - f(\mathbf{V}, \mathbf{V})] = \mathbf{O}. \quad (28.8)$$

If we add together Eqs. (28.7) and (28.8), we obtain

$$f(\mathbf{U}, \mathbf{V}) + f(\mathbf{V}, \mathbf{U}) - f(\mathbf{U}, \mathbf{U}) - f(\mathbf{V}, \mathbf{V}) = \mathbf{O}.$$

Since $f(\mathbf{U}, \mathbf{U}) \equiv \mathbf{O}$, the last equation gives

$$f(\mathbf{U}, \mathbf{V}) + f(\mathbf{V}, \mathbf{U}) = \mathbf{O},$$

which means that equation (28.1) in the case considered becomes Eq. (26.1).

3^0. Under the assumptions of the theorem the substitutions $\mathbf{Z}_2 = \mathbf{Z}_{2n} = \mathbf{U}$, $\mathbf{Z}_i = \mathbf{V}$ ($3 \leq i \leq 2n - 1$) and $\mathbf{Z}_1 = \mathbf{V}$ transform the equation (28.1) into the following form

$$f(\mathbf{V}, \mathbf{U})[f(\mathbf{U}, \mathbf{V}) + f(\mathbf{V}, \mathbf{U})] = \mathbf{O}. \quad (28.9)$$

By a permutation of the variables in this relation and adding together the relation obtained and Eq. (28.9), we get

$$f(\mathbf{U}, \mathbf{V}) + f(\mathbf{V}, \mathbf{U}) = \mathbf{O}.$$

Consequently, in this case the equation (28.1) becomes Eq. (26.1), whose general solution is the function Eq. (28.3). $\qquad\qquad\square$

We have not been able to obtain the general solution of the equation (28.1) if $n - k$ is even.

In the last section we will solve the third modified equation.

29 Third Modified Quadratic Functional Equation

We will finish this chapter with the following result.

Theorem 29.1 *The general solution of the functional equation*

$$\sum_{i=1}^{4m-1} F(\mathbf{Z}_1, \mathbf{Z}_{i+1}, \mathbf{Z}_{i+2}, \cdots, \mathbf{Z}_{4m+i-1}) = \mathbf{O} \qquad (29.1)$$

$$(\mathbf{Z}_{4m+i-1} \equiv \mathbf{Z}_i; \quad 2 \le i \le 4m - 1),$$

where

$$F(\mathbf{Z}_1, \mathbf{Z}_2, \mathbf{Z}_3, \cdots, \mathbf{Z}_{4m}) \qquad (29.2)$$

$$= [f(\mathbf{Z}_1, \mathbf{Z}_2) - f(\mathbf{Z}_4, \mathbf{Z}_3) + f(\mathbf{Z}_5, \mathbf{Z}_6)$$

$$- \cdots + f(\mathbf{Z}_{4r-3}, \mathbf{Z}_{4r-2}) - f(\mathbf{Z}_{4r}, \mathbf{Z}_{4r-1})] \times$$

$$\times \ [f(\mathbf{Z}_{4r+1}, \mathbf{Z}_{4r+2}) - f(\mathbf{Z}_{4r+4}, \mathbf{Z}_{4r+3})$$

$$+ \cdots + f(\mathbf{Z}_{4m-3}, \mathbf{Z}_{4m-2}) - f(\mathbf{Z}_{4m}, \mathbf{Z}_{4m-1})]$$

$$(f : \mathcal{V}^2 \mapsto \mathcal{V}; \quad m, r > 1)$$

has components which are either constants or functions of the form

$$f(\mathbf{U}, \mathbf{V}) = g(\mathbf{V}) - g(\mathbf{U}), \qquad (29.3)$$

where g is an arbitrary function defined in $\mathcal{V}$.

Proof. For the components of the function $f : \mathcal{V}^2 \mapsto \mathcal{V}$ (denoted again by f) we have two possibilities:

1^0. If $f(\mathbf{U}, \mathbf{U}) \equiv 0$, then for every nontrivial solution f there exists at least one pair of constant complex vectors $(\mathbf{A}, \mathbf{B})$ $(\mathbf{A}, \mathbf{B} \in \mathcal{V})$ such that $f(\mathbf{A}, \mathbf{B}) \ne 0$. By putting $\mathbf{Z}_1 = \mathbf{U}$, $\mathbf{Z}_2 = \mathbf{V}$, $\mathbf{Z}_{4m-1} = \mathbf{B}$ and $\mathbf{Z}_i = \mathbf{A}$ $(3 \le i \le 4m - 2)$, we obtain

$$f(\mathbf{A}, \mathbf{B})f(\mathbf{U}, \mathbf{V}) - f(\mathbf{A}, \mathbf{B})f(\mathbf{U}, \mathbf{A}) + f(\mathbf{A}, \mathbf{B})f(\mathbf{V}, \mathbf{A}) = 0,$$

i.e., by introduction of the notation $f(\mathbf{U}, \mathbf{A}) = -g(\mathbf{U})$ we have Eq. (29.3).

2^0. If $f(\mathbf{U}, \mathbf{U}) \not\equiv 0$, then there exists at least one constant vector $C \in \mathcal{V}$ such that $f(C, C) \ne 0$. By the substitution $\mathbf{Z}_2 = \mathbf{U}$, $\mathbf{Z}_{4r+2} = \mathbf{U}$ and $\mathbf{Z}_3 = \mathbf{Z}_4 = \cdots = \mathbf{Z}_{4r+1} = \mathbf{Z}_{4r+3} = \cdots = \mathbf{Z}_{4m} = \mathbf{Z}_1 = C$, from the equation (29.1) we obtain

$$[f(C, \mathbf{U}) - f(C, C)]^2 + 2[f(C, C) - f(\mathbf{U}, C)][f(C, C) - f(C, \mathbf{U})]$$

$$+[f(C,C) - f(\mathbf{U},C)]^2 = 0,$$

i.e.,

$$f(C,\mathbf{U}) + f(\mathbf{U},C) = 2K, \tag{29.4}$$

where we have put $f(C,C) = K$.

If we substitute $\mathbf{Z}_{4r-1} = \mathbf{Z}_{4r} = \mathbf{Z}_{4r+1} = \mathbf{Z}_{4r+2} = \mathbf{U}$ and the remaining variables by C, then from Eq. (29.1) it follows that

$$-[f(\mathbf{U},\mathbf{U}) - K]^2 + [f(C,\mathbf{U}) - f(\mathbf{U},\mathbf{U})][f(C,\mathbf{U}) + f(\mathbf{U},C) - 2K] = 0,$$

and according to Eq. (29.4) we find

$$f(\mathbf{U},\mathbf{U}) = K. \tag{29.5}$$

By substituting $\mathbf{Z}_2 = \mathbf{Z}_4 = \mathbf{Z}_{4r+2} = \mathbf{Z}_{4r+4} = C$ and the rest of the variables by $\mathbf{U}$, by virtue of Eq. (29.5) from the equation (29.1) we obtain

$$f(\mathbf{U},C) = f(C,\mathbf{U}),$$

so that from Eq. (29.4) it follows that

$$f(\mathbf{U},C) = f(C,\mathbf{U}) = K.$$

By putting $\mathbf{Z}_4 = \mathbf{Z}_{4r+4} = \mathbf{U}$, $\mathbf{Z}_3 = \mathbf{Z}_{4r+3} = \mathbf{V}$ and substituting all the other variables by C, according to the equality $f(\mathbf{U},C) = f(\mathbf{V},C) = K$ from the equation considered we obtain

$$[f(\mathbf{U},\mathbf{V}) - f(C,C)]^2 = 0,$$

i.e.,

$$f(\mathbf{U},\mathbf{V}) = K \qquad \text{(a constant).} \tag{29.6}$$

Since the functions given by Eqs. (29.3) and (29.6) are indeed solutions of the equation (29.1), the proof of the theorem is complete. $\qquad\square$

Chapter 8

Expanded Quadratic Functional Equations

In this chapter expanded quadratic functional equations with functional arguments, with the same signs or alternating signs between the functions, and a generalized expanded quadratic functional equation are solved. The results presented here were obtained in [I. B. Risteski and V. C. Covachev (submitted C)].

The notation used is as in Chapters 6 and 7. We denote by $\mathbf{I} = (1, 1, \cdots, 1)^T$ the unit vector in $\mathcal{V}$.

30 Expanded Quadratic Functional Equations with Functional Arguments

Now the following results will be proved.

Theorem 30.1 *The general solution of the functional equation*

$$\sum_{i=1}^{3k} F(\mathbf{Z}_1, \mathbf{Z}_{i+1}, \mathbf{Z}_{i+2} \cdots, \mathbf{Z}_{3k+i}) = \mathbf{O} \qquad (30.1)$$

$$(\mathbf{Z}_{3k+i} \equiv \mathbf{Z}_i,\ 2 \leq i \leq 3k),$$

where

$$F(\mathbf{U}_1, \mathbf{U}_2, \cdots, \mathbf{U}_{3k+1}) = f(\mathbf{U}_1, g(\mathbf{U}_2, \mathbf{U}_3, \cdots, \mathbf{U}_{k+1})) \times \qquad (30.2)$$

$$\times f(g(\mathbf{U}_{k+2}, \mathbf{U}_{k+3}, \cdots, \mathbf{U}_{2k+1}), g(\mathbf{U}_{2k+2}, \mathbf{U}_{2k+3}, \cdots, \mathbf{U}_{3k+1})),$$

247

and the function $g : \mathcal{V}^k \mapsto \mathcal{V}$ *satisfies the condition*

$$g(\mathbf{U}, \mathbf{I}, \mathbf{I}, \cdots, \mathbf{I}) = g(\mathbf{I}, \mathbf{U}, \mathbf{I}, \cdots, \mathbf{I}) = \cdots = g(\mathbf{I}, \mathbf{I}, \cdots, \mathbf{I}, \mathbf{U}) = \mathbf{U}, \quad (30.3)$$

is given by the following formula

$$f(\mathbf{U}, \mathbf{V}) = G(\mathbf{U})H(\mathbf{V}) - G(\mathbf{V})H(\mathbf{U}), \qquad (30.4)$$

where $G, H : \mathcal{V} \mapsto \mathcal{V}$ *are arbitrary functions.*

Proof.   Indeed, a straightforward calculation shows that the function given by Eq. (30.4) satisfies the functional equation (30.1) for arbitrary functions G and H.

Conversely, we will prove that every solution of the functional equation (30.1) has the form Eq. (30.4).

The trivial solution $f(\mathbf{U}, \mathbf{V}) = \mathbf{O}$ of the equation (30.1) is included in Eq. (30.4). Moreover, a zero component of a nontrivial solution can be represented in this form. Further we will seek only nontrivial solutions.

If we put

$$\mathbf{Z}_3 = \mathbf{Z}_4 = \cdots = \mathbf{Z}_k = \mathbf{Z}_{k+2} = \cdots = \mathbf{Z}_{2k+1} = \mathbf{Z}_{2k+3} = \cdots = \mathbf{Z}_{3k+1} = \mathbf{I}$$

into the equation (30.1), according to the condition Eq (30.3) we obtain

$$f(\mathbf{Z}_1, \mathbf{Z}_2)f(\mathbf{Z}_{k+1}, \mathbf{Z}_{2k+2}) \qquad (30.5)$$
$$+ \quad f(\mathbf{Z}_1, \mathbf{Z}_{k+1})f(\mathbf{Z}_{2k+2}, \mathbf{Z}_2) + f(\mathbf{Z}_1, \mathbf{Z}_{2k+2})f(\mathbf{Z}_2, \mathbf{Z}_{k+1}) = \mathbf{O}.$$

From the last equation it follows that

$$f(\mathbf{U}, \mathbf{U}) \equiv \mathbf{O}.$$

We denote any nonzero component of a nontrivial solution again by f, but now f will be a scalar function of two vector arguments. For such a component there exists at least one pair of constant complex vectors $(\mathbf{A}, \mathbf{B})$ $(\mathbf{A}, \mathbf{B} \in \mathcal{V})$ such that $f(\mathbf{A}, \mathbf{B}) \neq 0$. By putting $\mathbf{Z}_1 = \mathbf{A}$, $\mathbf{Z}_2 = \mathbf{B}$, $\mathbf{Z}_{k+1} = \mathbf{U}$ and $\mathbf{Z}_{2k+2} = \mathbf{V}$, the *scalar* equation (30.5) takes the form

$$f(\mathbf{U}, \mathbf{V}) = -\frac{f(\mathbf{A}, \mathbf{U})}{f(\mathbf{A}, \mathbf{B})} f(\mathbf{V}, \mathbf{B}) - \frac{f(\mathbf{A}, \mathbf{V})}{f(\mathbf{A}, \mathbf{B})} f(\mathbf{B}, \mathbf{U}). \qquad (30.6)$$

If we substitute $\mathbf{U} = \mathbf{B}$ and if we take into consideration the condition $f(\mathbf{U}, \mathbf{U}) \equiv 0$, from the above equation (30.6) it follows that

$$f(\mathbf{B}, \mathbf{V}) = -f(\mathbf{V}, \mathbf{B}). \qquad (30.7)$$

According to Eq. (30.7), the equation (30.6) takes the form

$$f(\mathbf{U}, \mathbf{V}) = \frac{f(\mathbf{A}, \mathbf{U})}{f(\mathbf{A}, \mathbf{B})} f(\mathbf{B}, \mathbf{V}) - \frac{f(\mathbf{A}, \mathbf{V})}{f(\mathbf{A}, \mathbf{B})} f(\mathbf{B}, \mathbf{U}). \tag{30.8}$$

If we denote

$$\frac{f(\mathbf{A}, \mathbf{U})}{f(\mathbf{A}, \mathbf{B})} = G(\mathbf{U}), \qquad f(\mathbf{B}, \mathbf{U}) = H(\mathbf{U}),$$

then we get

$$f(\mathbf{U}, \mathbf{V}) = G(\mathbf{U})H(\mathbf{V}) - G(\mathbf{V})H(\mathbf{U}),$$

which means that Eq. (30.4) is the general solution of Eq. (30.1). $\qquad \square$

Theorem 30.2 *The general solution of the equation*

$$\sum_{i=1}^{n-1} F(\mathbf{Z}_1, \mathbf{Z}_{i+1}, \mathbf{Z}_{i+2} \cdots, \mathbf{Z}_{n+i-1}) = \mathbf{O} \tag{30.9}$$

$$(\mathbf{Z}_{n+i-1} \equiv \mathbf{Z}_i, \ 2 \le i \le n - 1),$$

where

$$F(\mathbf{U}_1, \mathbf{U}_2, \cdots, \mathbf{U}_n) \tag{30.10}$$

$$= f(\mathbf{U}_1, g_1(\mathbf{U}_2, \mathbf{U}_3, \cdots \mathbf{U}_{k+1})) \times$$

$$\times \ [f(g_2(\mathbf{U}_{k+2}, \mathbf{U}_{k+3}, \cdots \mathbf{U}_{k+\ell+1}), g_3(\mathbf{U}_{k+\ell+2}, \mathbf{U}_{k+\ell+3}, \cdots, \mathbf{U}_n))$$

$$+ \ f(g_3(\mathbf{U}_{k+2}, \mathbf{U}_{k+3}, \cdots, \mathbf{U}_{k+m+1}), g_2(\mathbf{U}_{k+m+2}, \mathbf{U}_{k+m+3}, \cdots, \mathbf{U}_n))]$$

$$+ f(\mathbf{U}_1, g_2(\mathbf{U}_2, \mathbf{U}_3, \cdots \mathbf{U}_{\ell+1})) \times$$

$$\times \ [f(g_1(\mathbf{U}_{\ell+2}, \mathbf{U}_{\ell+3}, \cdots \mathbf{U}_{k+\ell+1}), g_3(\mathbf{U}_{k+\ell+2}, \mathbf{U}_{k+\ell+3}, \cdots, \mathbf{U}_n))$$

$$+ \ f(g_3(\mathbf{U}_{\ell+2}, \mathbf{U}_{\ell+3}, \cdots, \mathbf{U}_{\ell+m+1}), g_1(\mathbf{U}_{\ell+m+2}, \mathbf{U}_{\ell+m+3}, \cdots, \mathbf{U}_n))]$$

$$+ f(\mathbf{U}_1, g_3(\mathbf{U}_2, \mathbf{U}_3, \cdots \mathbf{U}_{m+1})) \times$$

$$\times \ [f(g_1(\mathbf{U}_{m+2}, \mathbf{U}_{m+3}, \mathbf{U}_{k+m+1}), g_2(\mathbf{U}_{k+m+2}, \mathbf{U}_{k+m+3}, \cdots, \mathbf{U}_n))$$

$$+ \ f(g_2(\mathbf{U}_{m+2}, \mathbf{U}_{m+3}, \cdots, \mathbf{U}_{\ell+m+1}), g_1(\mathbf{U}_{\ell+m+2}, \mathbf{U}_{\ell+m+3}, \cdots, \mathbf{U}_n))]$$

$$(k + \ell + m + 1 = n)$$

and the functions g_i $(i = 1, 2, 3)$ satisfy the conditions

$$g_i(\mathbf{U}, \mathbf{I}, \mathbf{I}, \cdots, \mathbf{I}) = g_i(\mathbf{I}, \mathbf{U}, \mathbf{I}, \cdots, \mathbf{I}) = \cdots = g_i(\mathbf{I}, \mathbf{I}, \cdots, \mathbf{I}, \mathbf{U}) = \mathbf{U} \tag{30.11}$$

$$(i = 1, 2, 3),$$

is given by

$$f(\mathbf{U}, \mathbf{V}) = G(\mathbf{U})H(\mathbf{V}) - G(\mathbf{V})H(\mathbf{U}), \qquad (30.12)$$

where $G, H : \mathcal{V} \mapsto \mathcal{V}$ *are arbitrary functions.*

Proof. It is easy to check that the function given by Eq. (30.12) is a solution of the equation (30.9). To this end it suffices to put Eq. (30.12) into Eq. (30.9).

Conversely, we will prove that every solution of the equation (30.9), if the conditions Eq. (30.11) are satisfied, is given by Eq. (30.12).

Without loss of generality of the proof, we may assume that $k \le \ell \le m$.

If we substitute $\mathbf{Z}_i = \mathbf{I}$ $(1 \le i \le 2n)$ and if we apply the properties Eq. (30.11), then the equation (30.9) takes the form $f(\mathbf{I}, \mathbf{I}) = \mathbf{O}$.

The trivial solution $f(\mathbf{U}, \mathbf{V}) = \mathbf{O}$ of the equation (30.9) is included in Eq. (30.12). Moreover, a zero component of a nontrivial solution can be represented in this form. Further we will seek only nontrivial solutions.

First, we will find all solutions of the equation (30.9) for which the following condition holds:

$$f(\mathbf{Z}, \mathbf{I}) \equiv \mathbf{O}. \qquad (30.13)$$

We denote any nonzero component of a nontrivial solution again by f, but now f will be a scalar function of two vector arguments. For such a component there exists at least one pair of constant complex vectors $(\mathbf{A}, \mathbf{B})$ $(\mathbf{A}, \mathbf{B} \in \mathcal{V})$ such that $f(\mathbf{A}, \mathbf{B}) \ne 0$.

If we substitute $\mathbf{Z}_1 = \mathbf{A}$, $\mathbf{Z}_{k+2} = \mathbf{U}$, $\mathbf{Z}_{k+3} = \mathbf{B}$ and $\mathbf{Z}_2 = \mathbf{Z}_3 = \cdots = \mathbf{Z}_{k+1} = \mathbf{Z}_{k+4} = \cdots = \mathbf{Z}_n = \mathbf{I}$, according to Eqs. (30.11) and (30.13) from the equation (30.9) it follows that

$$f(\mathbf{I}, \mathbf{U}) \equiv 0. \qquad (30.14)$$

Now, if we substitute

$$\mathbf{Z}_3 = \mathbf{Z}_4 = \cdots = \mathbf{Z}_{k+1} = \mathbf{Z}_{k+4} = \cdots = \mathbf{Z}_n = \mathbf{I},$$

according to the equalities (30.13) and (30.14) the equation (30.9) becomes

$$f(\mathbf{Z}_1, \mathbf{Z}_2)f(\mathbf{Z}_{k+2}, \mathbf{Z}_{k+3})$$
$$+ \quad f(\mathbf{Z}_1, \mathbf{Z}_{k+2})f(\mathbf{Z}_{k+3}, \mathbf{Z}_2) + f(\mathbf{Z}_1, \mathbf{Z}_{k+3})f(\mathbf{Z}_2, \mathbf{Z}_{k+2}) = 0.$$

The general solution of the above equation is given by Eq. (30.12), where G and H are arbitrary functions which according to Eq. (30.13) satisfy the condition

$$G(\mathbf{U})H(\mathbf{I}) - G(\mathbf{I})H(\mathbf{U}) = \mathbf{O}.$$

Now, we find all solutions of the equation (30.9) for which $f(\mathbf{Z}, \mathbf{I}) \not\equiv \mathbf{O}$.

In this case, for each component of such a solution (denoted again by f) for which $f(\mathbf{Z}, \mathbf{I}) \not\equiv 0$ there exists at least one vector $C \neq \mathbf{I}$ such that $f(C, \mathbf{I}) \neq 0$.

For $\mathbf{Z}_1 = C$, $\mathbf{Z}_2 = \mathbf{U}$ and $\mathbf{Z}_3 = \mathbf{Z}_4 = \cdots = \mathbf{Z}_n = \mathbf{I}$, according to Eq. (30.11) the equation (30.9) takes the form

$$f(C, \mathbf{I})[f(\mathbf{U}, \mathbf{I}) + f(\mathbf{I}, \mathbf{U})] = 0,$$

which implies that

$$f(\mathbf{U}, \mathbf{I}) = -f(\mathbf{I}, \mathbf{U}). \tag{30.15}$$

If we substitute

$$\mathbf{Z}_1 = C, \ \mathbf{Z}_2 = \mathbf{U}, \ \mathbf{Z}_3 = \mathbf{V}, \ \mathbf{Z}_4 = \mathbf{Z}_5 = \cdots = \mathbf{Z}_n = \mathbf{I},$$

and if we take into consideration the property Eq. (30.15), then the equation (30.9) reduces to the following form

$$f(C, \mathbf{I})f(\mathbf{U}, \mathbf{V}) - f(C, \mathbf{U})f(\mathbf{I}, \mathbf{V}) + f(C, \mathbf{V})f(\mathbf{I}, \mathbf{U}) = 0. \tag{30.16}$$

If we denote

$$\frac{f(C, \mathbf{U})}{f(C, \mathbf{I})} = G(\mathbf{U}) \qquad \text{and} \qquad f(\mathbf{I}, \mathbf{U}) = H(\mathbf{U}),$$

we obtain

$$f(\mathbf{U}, \mathbf{V}) = G(\mathbf{U})H(\mathbf{V}) - G(\mathbf{V})H(\mathbf{U}),$$

which represents the general solution Eq. (30.12) according to the hypothesis $f(\mathbf{Z}, \mathbf{I}) \not\equiv \mathbf{O}$. $\qquad\qquad\square$

31 Expanded Quadratic Functional Equations with the Same Signs between the Functions

In this section we will prove the following results.

Theorem 31.1   *The general solution of the functional equation*

$$\sum_{i=1}^{2n-1} F(\mathbf{Z}_1, \mathbf{Z}_{i+1}, \mathbf{Z}_{i+2}, \cdots, \mathbf{Z}_{2n+i-1}) = \mathbf{O} \qquad (31.1)$$

$$(\mathbf{Z}_{2n+i-1} \equiv \mathbf{Z}_i,\ 2 \le i \le 2n - 1),$$

where

$$F(\mathbf{Z}_1, \mathbf{Z}_2, \cdots, \mathbf{Z}_{2n-1}, \mathbf{Z}_{2n}) \qquad (31.2)$$

$$= f(\mathbf{Z}_1, \mathbf{Z}_2) \sum_{j=0}^{n-3} a_j \left[f(\mathbf{Z}_{j+3}, \mathbf{Z}_{j+4}) + f(\mathbf{Z}_{2n-j-1}, \mathbf{Z}_{2n-j}) \right]$$

$$+ a_{n-2} f(\mathbf{Z}_1, \mathbf{Z}_2) f(\mathbf{Z}_{n+1}, \mathbf{Z}_{n+2}) \qquad (f : \mathcal{V}^2 \mapsto \mathcal{V})$$

and a_j $(0 \le j \le n - 2)$ are complex constants such that

$$\sum_{j=0}^{n-2} a_j \ne 0, \qquad (31.3)$$

is given by

$$f(\mathbf{U}, \mathbf{V}) = \begin{cases} G(\mathbf{U}) - G(\mathbf{V}), & n > 2, \\ G(\mathbf{U})H(\mathbf{V}) - G(\mathbf{V})H(\mathbf{U}), & n = 2, \end{cases} \qquad (31.4)$$

where $G, H : \mathcal{V} \mapsto \mathcal{V}$ are arbitrary functions.

Proof.   Next, we will prove the theorem for $n > 2$. If we put $\mathbf{Z}_1 = \mathbf{Z}_2 = \cdots = \mathbf{Z}_{2n} = \mathbf{U}$, then from Eq. (31.1) we obtain $f(\mathbf{U}, \mathbf{U}) = \mathbf{O}$ for each $\mathbf{U}$.

The trivial solution $f(\mathbf{U}, \mathbf{V}) = \mathbf{O}$ of the equation (31.1) is included in Eq. (31.4). Moreover, a zero component of a nontrivial solution can be represented in this form. Further we will seek only nontrivial solutions.

We denote any nonzero component of a nontrivial solution again by f, but now f will be a scalar function of two vector arguments. For such a component there exists at least one pair of constant complex vectors $(\mathbf{A}, \mathbf{B})$ $(\mathbf{A}, \mathbf{B} \in \mathcal{V})$ such that $f(\mathbf{A}, \mathbf{B}) \ne 0$.

According to the condition Eq. (31.3) there exists an index $k \in \{0, 1, \cdots, n - 2\}$ such that $a_k \ne 0$. We will distinguish the following cases:

1^0. Let $k = 0$. If we substitute

$$\mathbf{Z}_5 = \mathbf{Z}_6 = \cdots = \mathbf{Z}_{2n} = \mathbf{Z}_1 = \mathbf{A},\ \mathbf{Z}_2 = \mathbf{B},\ \mathbf{Z}_3 = \mathbf{U} \qquad \text{and} \qquad \mathbf{Z}_4 = \mathbf{V}$$

into Eq. (31.1), we get

$$a_0\left[f(\mathbf{A},\mathbf{B})f(\mathbf{U},\mathbf{V})+f(\mathbf{A},\mathbf{U})f(\mathbf{V},\mathbf{A})\right. \tag{31.5}$$
$$\left.+\,f(\mathbf{A},\mathbf{B})f(\mathbf{A},\mathbf{U})+f(\mathbf{A},\mathbf{V})f(\mathbf{B},\mathbf{U})\right]$$
$$+\;a_1\left[f(\mathbf{A},\mathbf{B})f(\mathbf{V},\mathbf{A})+f(\mathbf{A},\mathbf{B})f(\mathbf{A},\mathbf{V})\right]=0.$$

For $\mathbf{V}=\mathbf{A}$, from the above relation it follows that $f(\mathbf{U},\mathbf{A})=-f(\mathbf{A},\mathbf{U})$. If we put $\mathbf{U}=\mathbf{B}$ into the equation (31.5), then we obtain

$$f(\mathbf{B},\mathbf{V})=-f(\mathbf{V},\mathbf{A})-f(\mathbf{A},\mathbf{B}).$$

If we substitute $G(\mathbf{U})=f(\mathbf{U},\mathbf{A})$, from Eq. (31.5) it follows that

$$f(\mathbf{U},\mathbf{V})=G(\mathbf{U})-G(\mathbf{V}).$$

2^0. Let $0<k<n-2$. If we put

$$\mathbf{Z}_1=\mathbf{Z}_3=\cdots=\mathbf{Z}_{k+2}=\mathbf{Z}_{k+5}=\cdots=\mathbf{Z}_{2n}=\mathbf{A},$$

$$\mathbf{Z}_2=\mathbf{B},\;\mathbf{Z}_{k+3}=\mathbf{U},\;\mathbf{Z}_{k+4}=\mathbf{V},$$

then the equation (31.1) takes the form

$$a_{k+1}\left\{f(\mathbf{A},\mathbf{B})\left[f(\mathbf{A},\mathbf{V})+f(\mathbf{V},\mathbf{A})\right]\right\} \tag{31.6}$$
$$+\;a_k\left[f(\mathbf{A},\mathbf{B})f(\mathbf{U},\mathbf{V})+f(\mathbf{A},\mathbf{B})f(\mathbf{A},\mathbf{U})+f(\mathbf{B},\mathbf{A})f(\mathbf{A},\mathbf{V})\right]=0.$$

For $\mathbf{V}=\mathbf{A}$ from (31.6) it follows that $f(\mathbf{A},\mathbf{U})=-f(\mathbf{U},\mathbf{A})$, *i.e.*,

$$f(\mathbf{U},\mathbf{V})=G(\mathbf{U})-G(\mathbf{V}),$$

where $G(\mathbf{U})=f(\mathbf{U},\mathbf{A})$.

3^0. Let $k=n-2$. If we substitute

$$\mathbf{Z}_1=\mathbf{Z}_3=\cdots=\mathbf{Z}_n=\mathbf{Z}_{n+3}=\cdots=\mathbf{Z}_{2n}=\mathbf{A},$$

$$\mathbf{Z}_2=\mathbf{B},\quad \mathbf{Z}_{n+1}=\mathbf{U},\quad \mathbf{Z}_{n+2}=\mathbf{V}$$

into (31.1), we get

$$a_{n-2}\left[f(\mathbf{A},\mathbf{B})f(\mathbf{U},\mathbf{V})+f(\mathbf{A},\mathbf{B})f(\mathbf{A},\mathbf{U})+f(\mathbf{B},\mathbf{A})f(\mathbf{A},\mathbf{V})\right]=0. \tag{31.7}$$

For $\mathbf{V}=\mathbf{A}$ this equation yields the equality

$$f(\mathbf{A},\mathbf{U})=-f(\mathbf{U},\mathbf{A}).$$

According to this equality, the equation (31.7) becomes

$$f(\mathbf{U}, \mathbf{V}) = G(\mathbf{U}) - G(\mathbf{V}),$$

where $G(\mathbf{U}) = f(\mathbf{U}, \mathbf{A})$.

We have shown that for $n > 2$ $f(\mathbf{U}, \mathbf{V})$ has the form

$$f(\mathbf{U}, \mathbf{V}) = G(\mathbf{U}) - G(\mathbf{V}), \tag{31.8}$$

where G is an arbitrary function from $\mathcal{V}$.

The proof of the theorem for $n = 2$ is very easy. In fact, for $n = 2$ from Eq. (31.1) there follows the equation

$$f(\mathbf{Z}_1, \mathbf{Z}_2) f(\mathbf{Z}_3, \mathbf{Z}_4) \tag{31.9}$$
$$+ \quad f(\mathbf{Z}_1, \mathbf{Z}_3) f(\mathbf{Z}_4, \mathbf{Z}_2) + f(\mathbf{Z}_1, \mathbf{Z}_4) f(\mathbf{Z}_2, \mathbf{Z}_3) = \mathbf{O}.$$

The above equation is the same as equation (30.5) (see also Eq. (25.8)), and according to Theorem 30.1 (or Theorem 25.1) it follows that

$$f(\mathbf{U}, \mathbf{V}) = G(\mathbf{U}) H(\mathbf{V}) - G(\mathbf{V}) H(\mathbf{U}),$$

where $G, H : \mathcal{V} \mapsto \mathcal{V}$ are arbitrary functions. $\square$

Theorem 31.2   *The general solution of the functional equation*

$$\sum_{i=1}^{2n-1} F(\mathbf{Z}_1, \mathbf{Z}_{i+1}, \mathbf{Z}_{i+2}, \cdots, \mathbf{Z}_{2n+i-1}) = \mathbf{O} \tag{31.10}$$

$$(\mathbf{Z}_{2n+i-1} \equiv \mathbf{Z}_i : \ 2 \le i \le 2n - 1)$$

where

$$F(\mathbf{Z}_1, \mathbf{Z}_2, \cdots, \mathbf{Z}_{2n}) \tag{31.11}$$

$$= \left[\sum_{j=0}^{k-1} f(\mathbf{Z}_{2j+1}, \mathbf{Z}_{2j+2}) \right] \cdot \left[\sum_{j=0}^{n-k-1} a_j f(\mathbf{Z}_{2k+j+1}, \mathbf{Z}_{2n-j}) \right],$$

$f : \mathcal{V}^2 \mapsto \mathcal{V}$, a_j $(0 \le j \le n - k - 1)$ *are complex constants such that*

$$\sum_{j=0}^{n-k-1} a_j \ne 0, \tag{31.12}$$

for $n > 2$ is given by the formulae

$$f(\mathbf{U}, \mathbf{V}) = G(\mathbf{U}) - G(\mathbf{V}) \tag{31.13}$$

or

$$f(\mathbf{U}, \mathbf{V}) = G(\mathbf{U})H(\mathbf{V}) - G(\mathbf{V})H(\mathbf{U}), \qquad (31.14)$$

where $G, H : \mathcal{V} \mapsto \mathcal{V}$ are arbitrary functions.

Proof.   In the previous theorem we proved that the function given by Eq. (31.13) is one solution of the equation (31.10).

Now, we will prove the converse, *i.e.*, that every solution of the equation (31.10) has the form Eq. (31.13) or Eq. (31.14).

If we put $\mathbf{Z}_i = \mathbf{U}$ ($1 \leq i \leq 2n$) into equation (31.10), by virtue of the relation Eq. (31.12) we obtain

$$f(\mathbf{U}, \mathbf{U}) = \mathbf{O}. \qquad (31.15)$$

The trivial solution $f(\mathbf{U}, \mathbf{V}) = \mathbf{O}$ of the equation Eq. (31.10) is included in both Eqs. (31.13) and (31.14). Moreover, a zero component of a nontrivial solution can be represented in each of these forms. Further we will seek only nontrivial solutions.

We denote any nonzero component of a nontrivial solution of equation (31.10) again by f, but now f will be a scalar function of two vector arguments. For such a component there exists at least one pair of constant complex vectors $(\mathbf{A}, \mathbf{B})$ $(\mathbf{A}, \mathbf{B} \in \mathcal{V})$ such that

$$f(\mathbf{A}, \mathbf{B}) \neq 0. \qquad (31.16)$$

Since $n > 2$, we will distinguish the following three cases:

1^0. Let $k = 1$. On the basis of the condition Eq. (31.12) there exists at least one coefficient a_j, $j \in \{0, 1, \cdots, n - 2\}$ which is distinct from 0. Let a_r be the nonzero coefficient with the smallest index.

In the case when $0 \leq 2r \leq n - 3$, if we substitute $\mathbf{Z}_2 = \mathbf{B}$, $\mathbf{Z}_{r+3} = \mathbf{U}$, $\mathbf{Z}_{2n-r} = \mathbf{V}$,

$$\mathbf{Z}_1 = \mathbf{Z}_3 = \cdots = \mathbf{Z}_{r+2} = \mathbf{Z}_{r+4}$$

$$= \cdots = \mathbf{Z}_{2n-r-1} = \mathbf{Z}_{2n-r+1} = \cdots = \mathbf{Z}_{2n} = \mathbf{A}$$

into Eq. (31.10), then we obtain

$$a_r \left[f(\mathbf{A}, \mathbf{B})f(\mathbf{U}, \mathbf{V}) + f(\mathbf{A}, \mathbf{B})f(\mathbf{A}, \mathbf{U}) + f(\mathbf{B}, \mathbf{A})f(\mathbf{A}, \mathbf{V}) \right]$$
$$+ \quad a_{2r+1} \left[f(\mathbf{A}, \mathbf{U})f(\mathbf{A}, \mathbf{V}) + f(\mathbf{U}, \mathbf{A})f(\mathbf{A}, \mathbf{V}) \right] = 0. \qquad (31.17)$$

If we put into Eq. (31.17) $\mathbf{V} = \mathbf{A}$, then we get $f(\mathbf{A}, \mathbf{U}) = -f(\mathbf{U}, \mathbf{A})$, and on the basis of this Eq. (31.17) becomes

$$f(\mathbf{U}, \mathbf{V}) = G(\mathbf{U}) - G(\mathbf{V}),$$

where we put $G(\mathbf{U}) = f(\mathbf{U}, \mathbf{A})$.

In the case when $2r > n-3$ and $3r \neq 2n-4$, if we put $\mathbf{Z}_2 = \mathbf{B}$, $\quad \mathbf{Z}_{r+3} = \mathbf{U}$, $\quad \mathbf{Z}_{2n-r} = \mathbf{V}$,

$$\mathbf{Z}_1 = \mathbf{Z}_3 = \cdots = \mathbf{Z}_{r+2} = \mathbf{Z}_{r+4}$$

$$= \cdots = \mathbf{Z}_{2n-r-1} = \mathbf{Z}_{2n-r+1} = \cdots = \mathbf{Z}_{2n} = \mathbf{A}$$

into Eq. (31.10), we get

$$a_r \left[f(\mathbf{A}, \mathbf{B})f(\mathbf{U}, \mathbf{V}) + f(\mathbf{A}, \mathbf{B})f(\mathbf{A}, \mathbf{U}) + f(\mathbf{B}, \mathbf{A})f(\mathbf{A}, \mathbf{V}) \right]$$
$$+ \quad a_{2n-2r-4} \left[f(\mathbf{A}, \mathbf{U})f(\mathbf{A}, \mathbf{V}) + f(\mathbf{A}, \mathbf{U})f(\mathbf{V}, \mathbf{A}) \right] = 0. \qquad (31.18)$$

From the above equation for $\mathbf{V} = \mathbf{A}$ we obtain $f(\mathbf{A}, \mathbf{U}) = -f(\mathbf{U}, \mathbf{A})$, and on the basis of this (31.18) becomes

$$f(\mathbf{U}, \mathbf{V}) = G(\mathbf{U}) - G(\mathbf{V}),$$

where $G(\mathbf{U}) = f(\mathbf{U}, \mathbf{A})$.

Now, let $3r = 2n - 4$ and let besides the coefficient a_r which is distinct from zero, there exist a coefficient a_s $(s > r)$ which is also different from zero. Then, by putting $\mathbf{Z}_2 = \mathbf{B}$, $\mathbf{Z}_{s+3} = \mathbf{U}$, $\mathbf{Z}_{2n-s} = \mathbf{V}$ and by substituting the other variables by $\mathbf{A}$, the equation (31.10) becomes

$$a_s \left[f(\mathbf{A}, \mathbf{B})f(\mathbf{U}, \mathbf{V}) + f(\mathbf{A}, \mathbf{B})f(\mathbf{A}, \mathbf{U}) + f(\mathbf{B}, \mathbf{A})f(\mathbf{A}, \mathbf{V}) \right]$$
$$+ \quad a_{2n-2s-4} \left[f(\mathbf{A}, \mathbf{U})f(\mathbf{A}, \mathbf{V}) + f(\mathbf{A}, \mathbf{U})f(\mathbf{V}, \mathbf{A}) \right] = 0. \qquad (31.19)$$

If we put $\mathbf{V} = \mathbf{A}$, from the above equation we obtain $f(\mathbf{A}, \mathbf{U}) = -f(\mathbf{U}, \mathbf{A})$, and on the basis of this (31.19) becomes

$$f(\mathbf{U}, \mathbf{V}) = G(\mathbf{U}) - G(\mathbf{V}),$$

where $G(\mathbf{U}) = f(\mathbf{U}, \mathbf{A})$.

However, if all the coefficients a_0, $a_1, \cdots, a_{n-2}$ except a_r are equal to zero, by substituting $\mathbf{Z}_2 = \mathbf{B}$, $\mathbf{Z}_{r+3} = \mathbf{U}$, $\mathbf{Z}_{2n-r} = \mathbf{V}$ and the other variables by $\mathbf{A}$, from Eqs. (31.10) and (31.11) we obtain

$$a_r \left[f(\mathbf{A}, \mathbf{B})f(\mathbf{U}, \mathbf{V}) + f(\mathbf{A}, \mathbf{U})f(\mathbf{V}, \mathbf{B}) + f(\mathbf{A}, \mathbf{V})f(\mathbf{B}, \mathbf{U}) \right] = 0. \qquad (31.20)$$

For $\mathbf{V} = \mathbf{B}$, from Eq. (31.20) it follows that

$$f(\mathbf{B}, \mathbf{U}) = -f(\mathbf{U}, \mathbf{B}). \qquad (31.21)$$

On the basis of the equalities (31.20) and (31.21), we get

$$f(\mathbf{U}, \mathbf{V}) = G(\mathbf{U})H(\mathbf{V}) - G(\mathbf{V})H(\mathbf{U}),$$

where we put $G(\mathbf{U}) = f(\mathbf{A}, \mathbf{U})$ and $H(\mathbf{U}) = f(\mathbf{B}, \mathbf{U})/f(\mathbf{A}, \mathbf{B})$.

2^0. Let $1 < k < n - 1$. Also in this case, from the condition Eq. (31.12) we conclude that at least one of the coefficients a_j $(0 \leq j \leq n - k - 1)$ must be different from zero. Let again a_r be the nonzero coefficient with the smallest index. We will distinguish three possibilities:

a) For $r = 0$, *i.e.*, $a_0 \neq 0$, if we put $\mathbf{Z}_4 = \mathbf{Z}_5 = \cdots = \mathbf{Z}_{2n} = \mathbf{Z}_1 = \mathbf{A}$, $\mathbf{Z}_2 = \mathbf{B}$, $\mathbf{Z}_3 = \mathbf{U}$, from the equation (31.10) we obtain $f(\mathbf{A}, \mathbf{U}) = -f(\mathbf{U}, \mathbf{A})$. By putting $\mathbf{Z}_5 = \mathbf{Z}_6 = \cdots = \mathbf{Z}_{2n} = \mathbf{Z}_1 = \mathbf{A}$, $\mathbf{Z}_2 = \mathbf{U}$, $\mathbf{Z}_3 = \mathbf{V}$, $\mathbf{Z}_4 = \mathbf{B}$, the equation (31.10) becomes

$$
\begin{aligned}
& a_0 \left[f(\mathbf{B}, \mathbf{A})f(\mathbf{U}, \mathbf{V}) + f(\mathbf{B}, \mathbf{A})f(\mathbf{A}, \mathbf{U}) + f(\mathbf{A}, \mathbf{B})f(\mathbf{A}, \mathbf{V}) \right] \\
+ \ & a_0 \left[f(\mathbf{A}, \mathbf{U})f(\mathbf{A}, \mathbf{V}) + f(\mathbf{A}, \mathbf{U})f(\mathbf{V}, \mathbf{A}) \right] \qquad (31.22) \\
+ \ & a_1 \left[f(\mathbf{B}, \mathbf{A})f(\mathbf{A}, \mathbf{U}) + f(\mathbf{A}, \mathbf{B})f(\mathbf{A}, \mathbf{U}) \right] = 0.
\end{aligned}
$$

On the basis of the equality $f(\mathbf{A}, \mathbf{U}) = -f(\mathbf{U}, \mathbf{A})$, the above equality (31.22) becomes

$$f(\mathbf{U}, \mathbf{V}) = G(\mathbf{U}) - G(\mathbf{V}),$$

where $G(\mathbf{U}) = f(\mathbf{U}, \mathbf{A})$.

b) For $0 < r < n - k - 1$, by putting $\mathbf{Z}_2 = \mathbf{B}$, $\mathbf{Z}_{r+3} = \mathbf{U}$ and by substituting the other variables by $\mathbf{A}$, we deduce from Eq. (31.10) the following equation

$$a_r f(\mathbf{A}, \mathbf{B})[f(\mathbf{A}, \mathbf{U}) + f(\mathbf{U}, \mathbf{A})] = 0,$$

from which we obtain

$$f(\mathbf{A}, \mathbf{U}) = -f(\mathbf{U}, \mathbf{A}). \qquad (31.23)$$

Now, if we put $\mathbf{Z}_2 = \mathbf{U}$, $\mathbf{Z}_3 = \mathbf{V}$, $\mathbf{Z}_{r+4} = \mathbf{B}$ and if we substitute the other variables by $\mathbf{A}$, then from Eq. (31.10) it follows that

$$
\begin{aligned}
& a_{r+1} \left[f(\mathbf{A}, \mathbf{B})f(\mathbf{A}, \mathbf{U}) + f(\mathbf{B}, \mathbf{A})f(\mathbf{A}, \mathbf{U}) \right] \qquad (31.24) \\
+ \ & a_r \left[f(\mathbf{B}, \mathbf{A})f(\mathbf{U}, \mathbf{V}) + f(\mathbf{B}, \mathbf{A})f(\mathbf{A}, \mathbf{U}) + f(\mathbf{A}, \mathbf{B})f(\mathbf{A}, \mathbf{V}) \right] = 0.
\end{aligned}
$$

On the basis of the equalities (31.23) and (31.24) we obtain

$$f(\mathbf{U}, \mathbf{V}) = G(\mathbf{U}) - G(\mathbf{V}),$$

where we introduced the notation $G(\mathbf{U}) = f(\mathbf{U}, \mathbf{A})$.

c) For $r = n - k - 1$, by putting $\mathbf{Z}_2 = \mathbf{B}$, $\mathbf{Z}_{n-k+3} = \mathbf{V}$, $\mathbf{Z}_{n-k+2} = \mathbf{U}$ and by substituting the other variables by $\mathbf{A}$, from Eq. (31.10) we obtain

$$a_{n-k-1}f(\mathbf{A}, \mathbf{B})\left[f(\mathbf{U}, \mathbf{V}) + f(\mathbf{A}, \mathbf{U}) + f(\mathbf{V}, \mathbf{A})\right] \qquad (31.25)$$
$$+ \quad a_{n-k-1}f(\mathbf{B}, \mathbf{A})\left[f(\mathbf{A}, \mathbf{V}) + f(\mathbf{V}, \mathbf{A})\right] = 0.$$

From Eq. (31.25), for $\mathbf{V} = \mathbf{A}$ we get $f(\mathbf{A}, \mathbf{U}) = -f(\mathbf{U}, \mathbf{A})$, and on the basis of this Eq. (31.25) becomes

$$f(\mathbf{U}, \mathbf{V}) = G(\mathbf{U}) - G(\mathbf{V}),$$

where $G(\mathbf{U}) = f(\mathbf{U}, \mathbf{A})$.

3^0. Let $k = n - 1$. In this case we have

$$F(\mathbf{Z}_1, \mathbf{Z}_2, \cdots, \mathbf{Z}_{2n})$$
$$= \left[f(\mathbf{Z}_1, \mathbf{Z}_2) + f(\mathbf{Z}_3, \mathbf{Z}_4) + \cdots + f(\mathbf{Z}_{2n-3}, \mathbf{Z}_{2n-2})\right] a_0 f(\mathbf{Z}_{2n-1}, \mathbf{Z}_{2n}).$$

From the condition Eq. (31.12) it follows that $a_0 \neq 0$. Then we may divide the equation (31.10) by a_0, but in this case we obtain a functional equation whose general solution according to [I. B. Risteski and V. C. Covachev (submitted B)] is given by

$$f(\mathbf{U}, \mathbf{V}) = G(\mathbf{U}) - G(\mathbf{V}),$$

where $G : \mathcal{V} \mapsto \mathcal{V}$ is an arbitrary function. $\qquad\qquad \Box$

Remark 31.1.   If $\displaystyle\sum_{j=0}^{n-k-1} a_j = 0$, the function Eq. (31.13) is a solution of the equation (31.10), but the question of generality of this solution remains open.

In the next section we will consider an expanded quadratic functional equation with alternating signs between the functions.

32   Expanded Quadratic Functional Equation with Alternating Signs between the Functions

Here the following result will be proved.

Theorem 32.1   *The general solution of the functional equation*

$$\sum_{i=1}^{2n-1} F(\mathbf{Z}_1, \mathbf{Z}_{i+1}, \mathbf{Z}_{i+2}, \cdots, \mathbf{Z}_{2n-1}, \mathbf{Z}_{2n}) = \mathbf{O} \tag{32.1}$$

$$(\mathbf{Z}_{2n+i-1} \equiv \mathbf{Z}_i; \quad 2 \le i \le 2n - 1),$$

where

$$F(\mathbf{Z}_1, \mathbf{Z}_2, \cdots, \mathbf{Z}_{2n-1}, \mathbf{Z}_{2n}) \tag{32.2}$$

$$= \begin{cases} [f(\mathbf{Z}_1, \mathbf{Z}_2) - f(\mathbf{Z}_4, \mathbf{Z}_3) + \cdots - f(\mathbf{Z}_{2k-2}, \mathbf{Z}_{2k-3}) + f(\mathbf{Z}_{2k-1}, \mathbf{Z}_{2k})] \\ \quad \times \displaystyle\sum_{j=0}^{n-k-1} a_j f(\mathbf{Z}_{2k+j+1}, \mathbf{Z}_{2n-j}) \qquad (k \ even), \\[2mm] [f(\mathbf{Z}_1, \mathbf{Z}_2) - f(\mathbf{Z}_4, \mathbf{Z}_3) + \cdots + f(\mathbf{Z}_{2k-3}, \mathbf{Z}_{2k-2}) - f(\mathbf{Z}_{2k}, \mathbf{Z}_{2k-1})] \\ \quad \times \displaystyle\sum_{j=0}^{n-k-1} a_j f(\mathbf{Z}_{2k+j+1}, \mathbf{Z}_{2n-j}) \qquad (k \ odd) \end{cases}$$

$$(f : \mathcal{V}^2 \mapsto \mathcal{V}),$$

and a_j $(0 \le j \le n - k - 1)$ are complex constants such that

$$\sum_{j=0}^{n-k-1} a_j \ne 0, \tag{32.3}$$

for k odd is given by the formulae

$$f(\mathbf{U}, \mathbf{V}) = G(\mathbf{U}) - G(\mathbf{V}) \tag{32.4}$$

or

$$f(\mathbf{U}, \mathbf{V}) = G(\mathbf{U})H(\mathbf{V}) - G(\mathbf{V})H(\mathbf{U}), \tag{32.5}$$

where $G, H : \mathcal{V} \mapsto \mathcal{V}$ are arbitrary functions, while for k even it has components which are either constants or functions of the form Eq. (32.4).

Proof.   Now, we will find the general solution of the equation (32.1) for k odd.

By putting $\mathbf{Z}_i = \mathbf{U}$ $(1 \le i \le 2n)$ from the equation (32.1) we obtain

$$f(\mathbf{U}, \mathbf{U}) = \mathbf{O}. \tag{32.6}$$

For $\mathbf{Z}_1 = \mathbf{Z}_{2n} = \mathbf{U}$, $\mathbf{Z}_j = \mathbf{V}$ $(2 \le j \le 2n - 1)$, on the basis of the equality (32.6), the equation (32.1) becomes

$$\sum_{j=0}^{n-k-1} a_j \left[f^2(\mathbf{U}, \mathbf{V}) + f(\mathbf{U}, \mathbf{V})f(\mathbf{V}, \mathbf{U}) \right] = \mathbf{O}, \tag{32.7}$$

or, on the basis of the condition Eq. (32.3), we get

$$f^2(\mathbf{U}, \mathbf{V}) + f(\mathbf{U}, \mathbf{V})f(\mathbf{V}, \mathbf{U}) = \mathbf{O}. \tag{32.8}$$

By a permutation of the variables $\mathbf{U}$ and $\mathbf{V}$, the last equation becomes

$$f^2(\mathbf{V}, \mathbf{U}) + f(\mathbf{V}, \mathbf{U})f(\mathbf{U}, \mathbf{V}) = \mathbf{O}. \tag{32.9}$$

By addition of Eq. (32.8) and (32.9), we obtain

$$[f(\mathbf{U}, \mathbf{V}) + f(\mathbf{V}, \mathbf{U})]^2 = \mathbf{O},$$

i.e.,

$$f(\mathbf{V}, \mathbf{U}) = -f(\mathbf{U}, \mathbf{V}). \tag{32.10}$$

Making use of the relation Eq. (32.10), we can reduce the equation (32.1) to the equation (31.10), whose general solution according to Theorem 31.2 is given by

$$f(\mathbf{U}, \mathbf{V}) = G(\mathbf{U}) - G(\mathbf{V}) \tag{32.11}$$

or

$$f(\mathbf{U}, \mathbf{V}) = G(\mathbf{U})H(\mathbf{V}) - G(\mathbf{V})H(\mathbf{U}). \tag{32.12}$$

Since the functions given by Eqs. (32.11) and (32.12) satisfy the equation (32.1), Theorem 32.1 is proved for the case of odd k.

Now, we will pass to the proof of Theorem 32.1 for the case of even k.

First, we will assume that a component of f (denoted again by f) satisfies $f(\mathbf{U}, \mathbf{U}) \equiv 0$. By putting $\mathbf{Z}_1 = \mathbf{Z}_{2n} = \mathbf{U}$ and $\mathbf{Z}_j = \mathbf{V}$ $(2 \le j \le 2n - 1)$, on the basis of the assumption $f(\mathbf{U}, \mathbf{U}) \equiv 0$, from the equalities (32.1) and (32.2) we obtain the relation Eq. (32.7) from which, by virtue of the condition Eq. (32.3) we get Eq. (32.8). By a permutation of the variables $\mathbf{U}$ and

$\mathbf{V}$, from Eq. (32.8) we deduce Eq. (32.9). Adding together Eq. (32.8) and (32.9) we obtain Eq. (32.10), so that in this case the equation (32.1) with Eq. (32.2) reduces to the equation (31.10) with Eq. (31.11). Therefore, the general solution of the functional equation (32.1) with Eq. (32.2) for k even is given by Eq. (32.11) if $f(\mathbf{U}, \mathbf{U}) \equiv 0$.

Now, we assume that $f(\mathbf{U}, \mathbf{U}) \not\equiv 0$. According to this assumption, there exists at least one complex vector C such that $f(C, C) \neq 0$.

For $\mathbf{Z}_1 = \mathbf{U}$, $\mathbf{Z}_i = C$ $(2 \leq i \leq 2n)$ the equation (32.1) reduces to

$$f(\mathbf{U}, C) = \mathcal{A}, \tag{32.13}$$

where we put $f(C, C) = \mathcal{A}$.

By putting $\mathbf{Z}_1 = \mathbf{U}$, $\mathbf{Z}_2 = \mathbf{V}$, $\mathbf{Z}_i = C$ $(3 \leq i \leq 2n)$ and by using Eq. (32.13), from the equation (32.1) we obtain

$$\sum_{j=0}^{n-k-1} a_j f(C, C) \left[f(\mathbf{U}, \mathbf{V}) - f(C, \mathbf{V}) \right] = 0,$$

from which on the basis of the condition Eq. (32.3) it follows that

$$f(\mathbf{U}, \mathbf{V}) = f(C, \mathbf{V}). \tag{32.14}$$

If we put $\mathbf{U} = \mathbf{V}$ into the above equality, we obtain $f(\mathbf{V}, \mathbf{V}) = f(C, \mathbf{V})$, *i.e.*, we have

$$f(\mathbf{U}, \mathbf{V}) = f(\mathbf{V}, \mathbf{V}) = f(C, \mathbf{V}). \tag{32.15}$$

By virtue of the condition Eq. (32.3) at least one of the coefficients $a_0, a_1, \cdots, a_{n-k-1}$ is different from zero. Let a_r be the nonzero coefficient with the smallest index.

If we substitute $\mathbf{Z}_1 = \mathbf{Z}_3 = \cdots = \mathbf{Z}_{r+2} = \mathbf{Z}_{r+4} = \cdots = \mathbf{Z}_{2n} = C$ and $\mathbf{Z}_2 = \mathbf{Z}_{r+3} = \mathbf{U}$, from Eqs. (32.1) and (32.2) on the basis of the equalities (32.13) and (32.15) we get

$$a_r \left[f(C, \mathbf{U}) - f(C, C) \right]^2 = 0,$$

from which it follows that

$$f(C, \mathbf{U}) = \mathcal{A}. \tag{32.16}$$

By virtue of the equalities (32.14) and (32.16) we have

$$f(\mathbf{U}, \mathbf{V}) = \mathcal{A}. \tag{32.17}$$

This completes the proof of Theorem 32.1.   □

Remark 32.1.   The functions given by Eqs. (32.4) and (32.17) are solutions of the equation (32.1) if $\sum_{j=0}^{n-k-1} a_j = 0$, but the question of generality of these solutions remains open.

33   Generalized Quadratic Functional Equation

In this section we will solve the functional equation

$$f(\mathbf{Z}_1, \mathbf{Z}_2)g(\mathbf{Z}_3, \mathbf{Z}_4) + f(\mathbf{Z}_1, \mathbf{Z}_3)g(\mathbf{Z}_4, \mathbf{Z}_2) \qquad (33.1)$$
$$+ \quad f(\mathbf{Z}_1, \mathbf{Z}_4)g(\mathbf{Z}_2, \mathbf{Z}_3) = \mathbf{O} \qquad (f, g : \mathcal{V}^2 \mapsto \mathcal{V})$$

which is more general than the equation (31.9).

It is easy to see that if a component of f is identically 0, then the corresponding component of g may be arbitrary. Similarly, if a component of g is identically 0, then the corresponding component of f may be arbitrary. So in the next theorem we consider only solutions (f, g) of Eq. (33.1) for which no component of f or g is identically 0.

Theorem 33.1   *The general solution of the functional equation* (33.1) *is given by the formulae*

$$f(\mathbf{U}, \mathbf{V}) \;=\; K_1(\mathbf{U})H_2(\mathbf{V}) - K_2(\mathbf{U})H_1(\mathbf{V}), \qquad (33.2)$$
$$g(\mathbf{U}, \mathbf{V}) \;=\; H_1(\mathbf{U})H_2(\mathbf{V}) - H_2(\mathbf{U})H_1(\mathbf{V}), \qquad (33.3)$$

where H_1, H_2, K_1, $K_2 : \mathcal{V} \mapsto \mathcal{V}$ *are arbitrary functions.*

Proof.   Let (f, g) be a solution of Eq. (33.1) such that no component of f or g is identically zero. Then for any pair of components of f and g (denoted again by (f, g)) there exist at least two pairs of complex constant vectors $(\mathcal{A}, \mathcal{B})$ and $(\mathcal{C}, \mathcal{D})$ (which may coincide) such that the components $f(\mathcal{A}, \mathcal{B})$ and $g(\mathcal{C}, \mathcal{D})$ are both not zero.

If we put $\mathbf{Z}_1 = \mathcal{A}$, $\mathbf{Z}_2 = \mathbf{Z}_3 = \mathbf{Z}_4 = \mathcal{B}$ into equation (33.1), we obtain $f(\mathcal{A}, \mathcal{B})g(\mathcal{B}, \mathcal{B}) = 0$, *i.e.*,

$$g(\mathcal{B}, \mathcal{B}) = 0. \qquad (33.4)$$

By putting $\mathbf{Z}_1 = \mathcal{A}$, $\mathbf{Z}_2 = \mathbf{Z}_3 = \mathcal{B}$, $\mathbf{Z}_4 = \mathbf{U}$, the equation (33.1) on the basis of the equality (33.4) becomes

$$g(\mathcal{B}, \mathbf{U}) + g(\mathbf{U}, \mathcal{B}) = 0,$$

from which it follows that

$$g(\mathcal{B}, \mathbf{U}) = -g(\mathbf{U}, \mathcal{B}). \tag{33.5}$$

For $\mathbf{Z}_1 = \mathcal{A}$, $\mathbf{Z}_2 = \mathcal{B}$, $\mathbf{Z}_3 = \mathbf{U}$, $\mathbf{Z}_4 = \mathbf{V}$, the equation (33.1) becomes

$$f(\mathcal{A}, \mathcal{B})g(\mathbf{U}, \mathbf{V}) + f(\mathcal{A}, \mathbf{U})g(\mathbf{V}, \mathcal{B}) + f(\mathcal{A}, \mathbf{V})g(\mathcal{B}, \mathbf{U}) = 0. \tag{33.6}$$

If we introduce the notation $H_1(\mathbf{U}) = f(\mathcal{A}, \mathbf{U})/f(\mathcal{A}, \mathcal{B})$ and $H_2(\mathbf{U}) = g(\mathcal{B}, \mathbf{U})$, by virtue of the equalities (33.5) and (33.6) we obtain

$$g(\mathbf{U}, \mathbf{V}) = H_1(\mathbf{U})H_2(\mathbf{V}) - H_1(\mathbf{V})H_2(\mathbf{U}). \tag{33.7}$$

Now, if we put $\mathbf{Z}_1 = \mathbf{U}$, $\mathbf{Z}_2 = \mathbf{V}$, $\mathbf{Z}_3 = \mathcal{C}$, $\mathbf{Z}_4 = \mathcal{D}$, from the equation (33.1), on the basis of the equality (33.7), we have

$$f(\mathbf{U}, \mathbf{V}) = K_1(\mathbf{U})H_2(\mathbf{V}) - H_1(\mathbf{V})K_2(\mathbf{U}), \tag{33.8}$$

where we introduced the notations

$$K_1(\mathbf{U}) = \frac{H_1(\mathcal{C})f(\mathbf{U}, \mathcal{D}) - H_1(\mathcal{D})f(\mathbf{U}, \mathcal{C})}{g(\mathcal{C}, \mathcal{D})},$$

$$K_2(\mathbf{U}) = \frac{H_2(\mathcal{C})f(\mathbf{U}, \mathcal{D}) - H_2(\mathcal{D})f(\mathbf{U}, \mathcal{C})}{g(\mathcal{C}, \mathcal{D})}.$$

This completes the proof of Theorem 33.1. $\qquad\qquad\square$

Remark 33.1. If (f, g) is a solution of Eq. (33.1), for the corresponding components of f and g one of the following three possibilities may occur:

a) The component of f is identically 0, the component of g is arbitrary;

a) The component of g is identically 0, the component of f is arbitrary;

c) The components of f and g are given by formulae of the form Eqs. (33.2), (33.3).

Chapter 9

Higher Order Functional Equations

In this chapter some complex vector functional equations of higher order without parameters and with complex parameters are solved. The results presented here were obtained in [I. B. Risteski (submitted)].

The notation used is as in the previous chapters.

34 Higher Order Functional Equation without Parameters

In this section the following result will be proved.

Theorem 34.1 *The general solution of the functional equation*

$$\sum_{i=1}^{2n-1} F(\mathbf{Z}_1, \mathbf{Z}_{i+1}, \mathbf{Z}_{i+2}, \cdots, \mathbf{Z}_{2n+i-1}) = \mathbf{O} \qquad (34.1)$$

$$(\mathbf{Z}_{2n+i-1} \equiv \mathbf{Z}_i,\ 2 \le i \le 2n - 1),$$

where

$$F(\mathbf{Z}_1, \mathbf{Z}_2, \mathbf{Z}_3, \cdots, \mathbf{Z}_{2n}) = \prod_{k=1}^{n} f(\mathbf{Z}_{2k-1}, \mathbf{Z}_{2k}) \qquad (34.2)$$

$$(f:\ \mathcal{V}^2 \mapsto \mathcal{V};\quad n > 1),$$

$$(34.3)$$

is given by

$$f(\mathbf{U}, \mathbf{V}) \ = \ g(\mathbf{U})h(\mathbf{V}) - g(\mathbf{V})h(\mathbf{U}), \qquad n = 2, \qquad (34.4)$$

$$f(\mathbf{U}, \mathbf{V}) \ = \ \mathbf{O}, \qquad n > 2, \qquad (34.5)$$

265

where $g, h : \mathcal{V} \mapsto \mathcal{V}$ are arbitrary functions.

Proof. If we put $\mathbf{Z}_k = \mathbf{U}$ $(1 \leq k \leq 2n)$, the equation (34.1) takes the form $f(\mathbf{U}, \mathbf{U}) \equiv \mathbf{O}$.

Now, we will distinguish two possibilities:

1^0. Let $n = 2$. In this case the equation (34.1) becomes

$$f(\mathbf{Z}_1, \mathbf{Z}_2)f(\mathbf{Z}_3, \mathbf{Z}_4) \tag{34.6}$$
$$+ \quad f(\mathbf{Z}_1, \mathbf{Z}_3)f(\mathbf{Z}_4, \mathbf{Z}_2) + f(\mathbf{Z}_1, \mathbf{Z}_4)f(\mathbf{Z}_2, \mathbf{Z}_3) = \mathbf{O}.$$

This equation is the same as equation (25.8) (see also Eq. (31.9)) and according to Theorem 25.8 its general solution is given by the formula Eq. (34.4).

2^0. Let $n > 2$. If we put $\mathbf{Z}_k = \mathbf{U}$ (k odd) and $\mathbf{Z}_k = \mathbf{V}$ (k even), and if we take into consideration the property $f(\mathbf{U}, \mathbf{U}) \equiv \mathbf{O}$, then from Eq. (34.1) it follows that

$$\sum_{i=0}^{n-1} f^{n-i}(\mathbf{U}, \mathbf{V})f^i(\mathbf{V}, \mathbf{U}) = \mathbf{O}. \tag{34.7}$$

By the substitutions

$$\mathbf{Z}_1 = \mathbf{Z}_4 = \mathbf{U}, \qquad \mathbf{Z}_{2k-1} = \mathbf{U},$$
$$\mathbf{Z}_2 = \mathbf{Z}_3 = \mathbf{V}, \qquad \mathbf{Z}_{2k} = \mathbf{V}$$
$$(3 \leq k \leq n)$$

the functional equation (34.1) reduces to

$$f^{n-1}(\mathbf{U}, \mathbf{V})f(\mathbf{V}, \mathbf{U}) = \mathbf{O}. \tag{34.8}$$

Also, the following equality holds

$$f^{n-1}(\mathbf{V}, \mathbf{U})f(\mathbf{U}, \mathbf{V}) = \mathbf{O}. \tag{34.9}$$

Now, if we put

$$\mathbf{Z}_1 = \mathbf{Z}_4 = \mathbf{Z}_6 = \cdots = \mathbf{Z}_{2r} = \mathbf{Z}_{2r+2} = \mathbf{U}, \qquad \mathbf{Z}_{2k-1} = \mathbf{U},$$
$$\mathbf{Z}_2 = \mathbf{Z}_3 = \mathbf{Z}_5 = \cdots = \mathbf{Z}_{2r-1} = \mathbf{Z}_{2r+1} = \mathbf{V}, \qquad \mathbf{Z}_{2k} = \mathbf{V}$$
$$(r + 2 \leq k \leq n; \qquad 1 \leq r < n)$$

the equation (34.1) becomes

$$f^{n-r}(\mathbf{U}, \mathbf{V})f^r(\mathbf{V}, \mathbf{U}) = \mathbf{O} \qquad (1 \leq r < n). \tag{34.10}$$

According to equality (34.10), the equation (34.7) yields

$$f(\mathbf{U}, \mathbf{V}) = \mathbf{O} \qquad (n > 2).$$

This completes the proof of Theorem 34.1. $\qquad\qquad\qquad\square$

35 Higher Order Functional Equation with Complex Parameters

In this section we will generalize the results given in the previous one.

Theorem 35.1 *The general solution of the functional equation*

$$\sum_{i=1}^{2n-1} a_i F(\mathbf{Z}_1, \mathbf{Z}_{i+1}, \mathbf{Z}_{i+2}, \cdots, \mathbf{Z}_{2n+i-1}) = \mathbf{O} \tag{35.1}$$

$$(\mathbf{Z}_{2n+i-1} \equiv \mathbf{Z}_i;\ 2 \le i \le 2n - 1),$$

where a_i $(1 \le i \le 2n - 1)$ are complex parameters not all of which are equal to zero,

$$F(\mathbf{Z}_1, \mathbf{Z}_2, \cdots, \mathbf{Z}_{2n}) \tag{35.2}$$

$$= \prod_{k=1}^{n} f(\mathbf{Z}_{2k-1}, \mathbf{Z}_{2k}) \qquad (f:\ \mathcal{V}^2 \mapsto \mathcal{V};\ n > 1),$$

is given by

$$f(\mathbf{U}, \mathbf{V}) \;=\; g(\mathbf{U})g(\mathbf{V}) \qquad (n \ge 2) \quad if \quad \sum_{i=1}^{2n-1} a_i = 0, \tag{35.3}$$

$$f(\mathbf{U}, \mathbf{V}) \;=\; g(\mathbf{U})h(\mathbf{V}) - g(\mathbf{V})h(\mathbf{U}) \quad (n = 2) \tag{35.4}$$

$$if \quad a_1 = a_2 = a_3 \ (\ne 0),$$

$$f(\mathbf{U}, \mathbf{V}) \;=\; \mathbf{O} \quad in\ all\ other\ cases, \tag{35.5}$$

where $g, h:\ \mathcal{V} \mapsto \mathcal{V}$ are arbitrary functions.

Proof. First, we will prove the theorem for the case $n = 2$. Then the functional equation (35.1) becomes

$$a_1 f(\mathbf{Z}_1, \mathbf{Z}_2) f(\mathbf{Z}_3, \mathbf{Z}_4) \tag{35.6}$$

$$+ \quad a_2 f(\mathbf{Z}_1, \mathbf{Z}_3) f(\mathbf{Z}_4, \mathbf{Z}_2) + a_3 f(\mathbf{Z}_1, \mathbf{Z}_4) f(\mathbf{Z}_2, \mathbf{Z}_3) = \mathbf{O}.$$

By a cyclic permutation of the vectors $\mathbf{Z}_2, \mathbf{Z}_3, \mathbf{Z}_4$ in equation (35.6) we obtain

$$a_1 f(\mathbf{Z}_1, \mathbf{Z}_3) f(\mathbf{Z}_4, \mathbf{Z}_2) \tag{35.7}$$
$$+ \quad a_2 f(\mathbf{Z}_1, \mathbf{Z}_4) f(\mathbf{Z}_2, \mathbf{Z}_3) + a_3 f(\mathbf{Z}_1, \mathbf{Z}_2) f(\mathbf{Z}_3, \mathbf{Z}_4) = \mathbf{O},$$

$$a_1 f(\mathbf{Z}_1, \mathbf{Z}_4) f(\mathbf{Z}_2, \mathbf{Z}_3) \tag{35.8}$$
$$+ \quad a_2 f(\mathbf{Z}_1, \mathbf{Z}_2) f(\mathbf{Z}_3, \mathbf{Z}_4) + a_3 f(\mathbf{Z}_1, \mathbf{Z}_3) f(\mathbf{Z}_4, \mathbf{Z}_2) = \mathbf{O}.$$

The system of equations (35.6), (35.7) and (35.8) has a nontrivial solution if and only if the following condition is satisfied

$$\begin{vmatrix} a_1 & a_2 & a_3 \\ a_3 & a_1 & a_2 \\ a_2 & a_3 & a_1 \end{vmatrix} = 0. \tag{35.9}$$

In all other cases the general solution of the functional equation (35.6) is

$$f(\mathbf{U}, \mathbf{V}) \equiv \mathbf{O}.$$

From Eq. (35.9) it follows that

$$(a_1 + a_2 + a_3)[(a_1 - a_2)^2 + (a_2 - a_3)^2 + (a_3 - a_1)^2] = 0. \tag{35.10}$$

We will investigate the following cases:

1^0. Let $a_1 + a_2 + a_3 = 0$ and $a_1 = a_2 \ (\neq 0)$. Then the condition Eq. (35.9) is satisfied. The equation (35.6) has the form

$$f(\mathbf{Z}_1, \mathbf{Z}_2) f(\mathbf{Z}_3, \mathbf{Z}_4) + f(\mathbf{Z}_1, \mathbf{Z}_3) f(\mathbf{Z}_4, \mathbf{Z}_2) = 2 f(\mathbf{Z}_1, \mathbf{Z}_4) f(\mathbf{Z}_2, \mathbf{Z}_3). \tag{35.11}$$

By a cyclic permutation of the vectors $\mathbf{Z}_2, \mathbf{Z}_3, \mathbf{Z}_4$ from this equation we find

$$f(\mathbf{Z}_1, \mathbf{Z}_3) f(\mathbf{Z}_4, \mathbf{Z}_2) + f(\mathbf{Z}_1, \mathbf{Z}_4) f(\mathbf{Z}_2, \mathbf{Z}_3) = 2 f(\mathbf{Z}_1, \mathbf{Z}_2) f(\mathbf{Z}_3, \mathbf{Z}_4). \tag{35.12}$$

If we eliminate the term $f(\mathbf{Z}_1, \mathbf{Z}_2) f(\mathbf{Z}_3, \mathbf{Z}_4)$ from Eqs. (35.11) and (35.12), we obtain the equation

$$f(\mathbf{Z}_1, \mathbf{Z}_3) f(\mathbf{Z}_4, \mathbf{Z}_2) = f(\mathbf{Z}_1, \mathbf{Z}_4) f(\mathbf{Z}_2, \mathbf{Z}_3). \tag{35.13}$$

We denote any nonzero component of a nontrivial solution of the equation (35.11) again by f, but now f will be a scalar function of two vector arguments. For such a component there exists at least one pair of constant complex vectors $(\mathbf{A}, \mathbf{B}) \ (\mathbf{A}, \mathbf{B} \in \mathcal{V})$ such that $f(\mathbf{A}, \mathbf{B}) \neq 0$.

By putting $\mathbf{Z}_1 = \mathbf{A}$, $\mathbf{Z}_2 = \mathbf{U}$, $\mathbf{Z}_3 = \mathbf{V}$ and $\mathbf{Z}_4 = \mathbf{B}$, the equation (35.13) becomes

$$f(\mathbf{A}, \mathbf{B})f(\mathbf{U}, \mathbf{V}) = f(\mathbf{A}, \mathbf{V})f(\mathbf{B}, \mathbf{U}). \tag{35.14}$$

If we put $\mathbf{U} = \mathbf{B}$ into the last equation, we have

$$f(\mathbf{B}, \mathbf{V}) = \frac{f(\mathbf{B}, \mathbf{B})}{f(\mathbf{A}, \mathbf{B})} f(\mathbf{A}, \mathbf{V}).$$

On the basis of this equality, the equation (35.14) becomes

$$f(\mathbf{U}, \mathbf{V}) = g(\mathbf{U})g(\mathbf{V}), \tag{35.15}$$

where we put

$$\frac{\sqrt{f(\mathbf{B}, \mathbf{B})}}{f(\mathbf{A}, \mathbf{B})} f(\mathbf{A}, \mathbf{U}) = g(\mathbf{U}).$$

Really, the function given by Eq. (35.15) is a solution of the equation (35.11).

2^0. Let $a_1 + a_2 + a_3 = 0$ and $a_1 \neq a_2$. Then the condition Eq. (35.9) is satisfied.

For $a_1 + a_2 + a_3 = 0$, the equation (35.6) becomes

$$a_1 f(\mathbf{Z}_1, \mathbf{Z}_2)f(\mathbf{Z}_3, \mathbf{Z}_4) + a_2 f(\mathbf{Z}_1, \mathbf{Z}_3)f(\mathbf{Z}_4, \mathbf{Z}_2) \tag{35.16}$$
$$= (a_1 + a_2)f(\mathbf{Z}_1, \mathbf{Z}_4)f(\mathbf{Z}_2, \mathbf{Z}_3).$$

If we suppose that $a_1 = 0$, this allows us to divide by $a_2 \neq 0$ and then equation (35.16) reduces to the equation (35.13), whose general solution is the function Eq. (35.15).

Now, we will assume that $a_1 \neq 0$.

Let $f(\mathbf{U}, \mathbf{U}) \equiv \mathbf{O}$. In this case, for any nonzero component of a nontrivial solution of the equation (35.16) (denoted again by f) there exists at least one pair of constant complex vectors $(\mathbf{A}, \mathbf{B})$ such that $f(\mathbf{A}, \mathbf{B}) \neq 0$. By putting $\mathbf{Z}_1 = \mathbf{Z}_3 = \mathbf{A}$, $\mathbf{Z}_2 = \mathbf{Z}_4 = \mathbf{B}$, from the equation (35.16) we obtain

$$(a_1 + a_2)f(\mathbf{B}, \mathbf{A}) = a_1 f(\mathbf{A}, \mathbf{B}). \tag{35.17}$$

For $\mathbf{Z}_1 = \mathbf{U}$, $\mathbf{Z}_2 = \mathbf{B}$, $\mathbf{Z}_3 = \mathbf{Z}_4 = \mathbf{A}$, by virtue of the above relation Eq. (35.17), the equation (35.16) reduces to the equation

$$(a_1 - a_2)f(\mathbf{A}, \mathbf{B})f(\mathbf{U}, \mathbf{A}) = 0,$$

from which it follows that

$$f(\mathbf{U}, \mathbf{A}) = 0. \tag{35.18}$$

By the substitutions $\mathbf{Z}_1 = \mathbf{U}$, $\mathbf{Z}_2 = \mathbf{V}$, $\mathbf{Z}_3 = \mathbf{A}$, $\mathbf{Z}_4 = \mathbf{B}$, the equation (35.16), on the basis of the equality (35.18), yields

$$f(\mathbf{U}, \mathbf{V}) \equiv 0. \tag{35.19}$$

Let $f(\mathbf{U}, \mathbf{U}) \not\equiv \mathbf{O}$. In this case, for any component of the solution of the equation (35.16) (denoted again by f) for which $f(\mathbf{U}, \mathbf{U}) \not\equiv 0$ there exists at least one constant complex vector C such that $f(C, C) \neq 0$. If we put $\mathbf{Z}_1 = \mathbf{Z}_2 = \mathbf{Z}_4 = C$, $\mathbf{Z}_3 = \mathbf{U}$, from Eq. (35.16) we obtain

$$f(C, \mathbf{U}) = f(\mathbf{U}, C). \tag{35.20}$$

For $\mathbf{Z}_1 = \mathbf{Z}_2 = C$, $\mathbf{Z}_3 = \mathbf{U}$, $\mathbf{Z}_4 = \mathbf{V}$, in view of Eq. (35.20) we can write the equation (35.16) in the following form

$$f(\mathbf{U}, \mathbf{V}) = \frac{f(C, \mathbf{U}) f(C, \mathbf{V})}{f(C, C)},$$

or

$$f(\mathbf{U}, \mathbf{V}) = g(\mathbf{U}) g(\mathbf{V}), \tag{35.21}$$

with the notation $f(C, \mathbf{U}) = \sqrt{f(C, C)} g(\mathbf{U})$.

It is not hard to check that the function Eq. (35.21) is really a solution of the equation (35.16).

Since Eq. (35.21) includes the trivial solution Eq. (35.19) as well as solutions with zero components, the general solution of the equation (35.16) is given by the formula Eq. (35.21).

3^0. Let $a_1 + a_2 + a_3 \neq 0$. Adding together the equations Eqs. (35.6), (35.7) and (35.8), according to the condition $a_1 + a_2 + a_3 \neq 0$ we obtain the equation (34.6). According to Eq. (35.4) the solution of this equation has the following properties

$$f(\mathbf{U}, \mathbf{V}) = -f(\mathbf{V}, \mathbf{U}), \qquad f(\mathbf{U}, \mathbf{U}) \equiv \mathbf{O}. \tag{35.22}$$

If we put $\mathbf{Z}_1 = \mathbf{Z}_4 = \mathbf{U}$, $\mathbf{Z}_2 = \mathbf{Z}_3 = \mathbf{V}$, from Eqs. (35.6) we find

$$(a_1 - a_2) f^2(\mathbf{U}, \mathbf{V}) = \mathbf{O}. \tag{35.23}$$

We will distinguish two cases: $a_1 \neq a_2$ and $a_1 = a_2$.

In the case $a_1 \neq a_2$, starting from Eq. (35.23), we obtain Eq. (35.5). In the case $a_1 = a_2$, from the condition Eq. (35.10) we find

$$a_1 = a_2 = a_3,$$

and the equation (35.6) is transformed into equation (34.6). As stated above, the general solution of this equation is the function Eq. (35.4). Thus the theorem is proved for $n = 2$.

Now we pass to the proof of the theorem for $n > 2$.
First we will investigate the case

$$\sum_{i=1}^{2n-1} a_i \neq 0.$$

By putting $\mathbf{Z}_i = \mathbf{U}$ $(1 \leq i \leq 2n)$, from the equation (35.1) we obtain the identity

$$f(\mathbf{U}, \mathbf{U}) \equiv \mathbf{O}. \tag{35.24}$$

Next, we assume $a_1 \neq 0$. We may assume this without loss of generality since if $a_1 = 0$, then there must be at least one $a_i \neq 0$ $(2 \leq i \leq 2n - 2)$, and by a cyclic permutation of the vectors $\mathbf{Z}_2, \mathbf{Z}_3, \cdots, \mathbf{Z}_{2n}$ we may achieve that the coefficient at the term $f(\mathbf{Z}_1, \mathbf{Z}_2) f(\mathbf{Z}_3, \mathbf{Z}_4) \cdots f(\mathbf{Z}_{2n-1}, \mathbf{Z}_{2n})$ be different from zero.

By introducing the substitutions

$$\mathbf{Z}_1 = \mathbf{Z}_3 = \cdots = \mathbf{Z}_{2n-1} = \mathbf{U} \qquad \text{and} \qquad \mathbf{Z}_2 = \mathbf{Z}_4 = \cdots = \mathbf{Z}_{2n} = \mathbf{V}$$

and taking into consideration the identity Eq. (35.24), the equation (35.1) takes the form

$$\sum_{i=0}^{n-1} a_{2i+1} f^{n-i}(\mathbf{U}, \mathbf{V}) f^i(\mathbf{V}, \mathbf{U}) = \mathbf{O}. \tag{35.25}$$

For

$$\mathbf{Z}_3 = \mathbf{Z}_5 = \cdots = \mathbf{Z}_{2r+1} = \mathbf{V}, \qquad \mathbf{Z}_4 = \mathbf{Z}_6 = \cdots = \mathbf{Z}_{2r+2} = \mathbf{U},$$
$$\mathbf{Z}_1 = \mathbf{Z}_{2i-1} = \mathbf{U}, \qquad \mathbf{Z}_2 = \mathbf{Z}_{2i} = \mathbf{V}$$
$$(r + 2 \leq i \leq n; \qquad 1 \leq r < n)$$

the equation (35.1) yields

$$a_1 f^{n-r}(\mathbf{U}, \mathbf{V}) f^r(\mathbf{V}, \mathbf{U}) = \mathbf{O} \qquad (1 \leq r < n)$$

so that from Eq. (35.25) we deduce $f(\mathbf{U}, \mathbf{V}) \equiv \mathbf{O}$.

Now, we pass to the investigation of the case

$$\sum_{i=1}^{2n-1} a_i = 0.$$

If we assume $f(\mathbf{U}, \mathbf{U}) \equiv \mathbf{O}$, then the general solution of the equation (35.1) is Eq. (35.5), which may be proved as in the case $\sum_{i=1}^{2n-1} a_i \neq 0$.

If we assume that $f(\mathbf{U}, \mathbf{U}) \not\equiv \mathbf{O}$, then for each component of f (denoted again by f) such that $f(\mathbf{U}, \mathbf{U}) \not\equiv 0$ there exists at least one complex constant vector C such that $f(C, C) \neq 0$.

There is at least one index $r \in \{1, 2, \cdots, 2n - 1\}$ such that

$$\sum_{i=0}^{n-2} a_{2i+r} \neq 0 \qquad \text{where} \qquad a_j = a_{j-2n+1} \quad (j > 2n - 1).$$

In fact, if such an index did not exist, then we would have the following system of $2n - 1$ linear homogeneous equations

$$\sum_{i=0}^{n-2} a_{2i+r} = 0 \qquad (1 \leq r \leq 2n - 1),$$

which has only the trivial solution

$$a_1 = a_2 = \cdots = a_{2n-1} = 0,$$

but this contradicts the assumption that at least one of these parameters is distinct from zero.

Now we assume that

$$\sum_{i=0}^{n-2} a_{2i+1} \neq 0. \tag{35.26}$$

We may assume this because the case when

$$\sum_{i=0}^{n-2} a_{2i+1} = 0 \qquad \text{and} \qquad \sum_{i=0}^{n-2} a_{2i+r} \neq 0 \quad (r \in \{2, 3, \cdots, 2n - 1\})$$

can be reduced to the case Eq. (35.26) by a cyclic permutation of the variables $\mathbf{Z}_2, \mathbf{Z}_3, \cdots, \mathbf{Z}_{2n}$ and a simple renumeration of the variables.

By putting

$$\mathbf{Z}_{2n-1} = \mathbf{U}, \ \mathbf{Z}_{2n} = \mathbf{V}, \ \mathbf{Z}_i = C \qquad (1 \le i \le 2n - 2)$$

into Eq. (35.1), we obtain

$$\sum_{i=0}^{n-2} a_{2i+1} f^{n-1}(C, C) f(\mathbf{U}, \mathbf{V}) \tag{35.27}$$

$$+ \ \sum_{i=0}^{n-2} a_{2i+2} f^{n-2}(C, C) f(C, \mathbf{U}) f(\mathbf{V}, C)$$

$$+ \ a_{2n-1} f^{n-2}(C, C) f(C, \mathbf{U}) f(C, \mathbf{V}) = 0.$$

Since

$$a_{2n-1} = -\sum_{i=0}^{n-2} a_{2i+1} - \sum_{i=0}^{n-2} a_{2i+2},$$

according to the assumption Eq. (35.26), for $\mathbf{V} = C$, from the equation (35.27) there follows $f(C, \mathbf{U}) = f(\mathbf{U}, C)$.

By introducing the notation $f(C, \mathbf{U}) = g(\mathbf{U}) \sqrt{f(C, C)}$, from Eq. (35.27) we get

$$f(\mathbf{U}, \mathbf{V}) = g(\mathbf{U}) g(\mathbf{V}).$$

Since in the case $\sum_{i=1}^{2n-1} a_i = 0$ this function is really a solution of the equation (35.1), this means that Theorem 35.1 is completely proved. $\square$

36 Nonlinear Operator Functional Equation

In this section a nonlinear operator functional equation of k-th order will be solved.

Definition 36.1. Let Ψ_{ij} be the operator which transposes (changes the places of) the i-th and j-th argument of the function F, i.e.,

$$\Psi_{ij} F(\mathbf{Z}_1, \cdots, \mathbf{Z}_{i-1}, \mathbf{Z}_i, \mathbf{Z}_{i+1}, \cdots, \mathbf{Z}_{j-1}, \mathbf{Z}_j, \mathbf{Z}_{j+1}, \cdots, \mathbf{Z}_n) \tag{36.1}$$

$$= \ F(\mathbf{Z}_1, \cdots, \mathbf{Z}_{i-1}, \mathbf{Z}_j, \mathbf{Z}_{i+1}, \cdots, \mathbf{Z}_{j-1}, \mathbf{Z}_i, \mathbf{Z}_{j+1}, \cdots, \mathbf{Z}_n).$$

Theorem 36.1   *The general solution of the functional equation*

$$aF(\mathbf{Z}_1, \mathbf{Z}_2, \cdots, \mathbf{Z}_{kn}) \tag{36.2}$$

$$= \sum_{r=n+1}^{kn} \Psi_{nr} F(\mathbf{Z}_1, \mathbf{Z}_2, \cdots, \mathbf{Z}_{n-1}, \mathbf{Z}_n, \mathbf{Z}_{n+1}, \cdots, \mathbf{Z}_{kn}),$$

where a is a complex parameter,

$$F(\mathbf{Z}_1, \mathbf{Z}_2, \cdots, \mathbf{Z}_{kn}) = \prod_{i=0}^{k-1} f(\mathbf{Z}_{ni+1}, \mathbf{Z}_{ni+2}, \cdots, \mathbf{Z}_{ni+n}) \tag{36.3}$$

($f : \mathcal{V}^n \mapsto \mathcal{V}$; $n \geq 2$, $k \geq 2$), has components given by

$$f(\mathbf{U}_1, \mathbf{U}_2, \cdots, \mathbf{U}_n) \tag{36.4}$$

$$= \begin{vmatrix} H_1(\mathbf{U}_1) & H_1(\mathbf{U}_2) & \cdots & H_1(\mathbf{U}_n) \\ H_2(\mathbf{U}_1) & H_2(\mathbf{U}_2) & \cdots & H_2(\mathbf{U}_n) \\ \vdots & & & \\ H_n(\mathbf{U}_1) & H_n(\mathbf{U}_2) & \cdots & H_n(\mathbf{U}_n) \end{vmatrix} \quad if \quad a = k-1,$$

$$f(\mathbf{U}_1, \mathbf{U}_2, \cdots, \mathbf{U}_n) = \begin{cases} \displaystyle\prod_{i=1}^{n} K(\mathbf{U}_i) & \\ or \quad 0 \end{cases} \quad if \quad a = n(k-1), \tag{36.5}$$

$$f(\mathbf{U}_1, \mathbf{U}_2, \cdots, \mathbf{U}_n) \equiv 0 \quad if \quad a \neq r(k-1) \quad (1 \leq r \leq n), \tag{36.6}$$

where H_i ($1 \leq i \leq n$), K are arbitrary functions in $\mathcal{V}$.

Proof.   For the proof of this theorem in the case $a = k-1$, $n > 2$ we shall need the following result.

Lemma 36.2   *If $\mathbf{A}_i$ ($1 \leq i \leq n$) are constant complex vectors such that*

$$f(\mathbf{A}_1, \mathbf{A}_2, \cdots, \mathbf{A}_n) \neq 0, \tag{36.7}$$

then the following equality holds

$$f(\mathbf{A}_n, \mathbf{U}_1, \mathbf{U}_2, \cdots, \mathbf{U}_{n-2}, \mathbf{A}_n) \equiv 0. \tag{36.8}$$

Proof of Lemma 36.2. We will suppose that this is not true, *i.e.*,

$$f(\mathbf{A}_n, \mathbf{U}_1, \mathbf{U}_2, \cdots, \mathbf{U}_{n-2}, \mathbf{A}_n) \neq 0. \tag{36.9}$$

By putting

$$\mathbf{Z}_i = \mathbf{A}_i \qquad (1 \leq i \leq n),$$
$$\mathbf{Z}_{rn+1} = \mathbf{Z}_{(r+1)n} = \mathbf{A}_n \qquad (1 \leq r \leq k-1),$$
$$\mathbf{Z}_{rn+j+1} = \mathbf{U}_j \qquad (1 \leq r \leq k-1; \quad 1 \leq j \leq n-2),$$

the equation (36.2) becomes

$$(k-1)[f(\mathbf{A}_n, \mathbf{U}_1, \cdots, \mathbf{U}_{n-2}, \mathbf{A}_n)]^{k-2} \times \qquad (36.10)$$
$$\times \quad [f(\mathbf{A}_1, \mathbf{A}_2, \cdots, \mathbf{A}_{n-1}, \mathbf{A}_n)f(\mathbf{A}_n, \mathbf{U}_1, \mathbf{U}_2, \cdots, \mathbf{U}_{n-2}, \mathbf{A}_n)$$
$$+ \quad f(\mathbf{A}_1, \mathbf{A}_2, \cdots, \mathbf{A}_{n-1}, \mathbf{U}_1)f(\mathbf{A}_n, \mathbf{A}_n, \mathbf{U}_2, \cdots, \mathbf{U}_{n-2}, \mathbf{A}_n)$$
$$+ \quad f(\mathbf{A}_1, \mathbf{A}_2, \cdots, \mathbf{A}_{n-1}, \mathbf{U}_2)f(\mathbf{A}_n, \mathbf{U}_1, \mathbf{A}_n, \mathbf{U}_3, \cdots, \mathbf{U}_{n-2}, \mathbf{A}_n) + \cdots$$
$$+ \quad f(\mathbf{A}_1, \mathbf{A}_2, \cdots, \mathbf{A}_{n-1}, \mathbf{U}_{n-2})f(\mathbf{A}_n, \mathbf{U}_1, \cdots, \mathbf{U}_{n-3}, \mathbf{A}_n, \mathbf{A}_n)] = 0.$$

According to the hypothesis Eq. (36.9), from Eq. (3.10) it follows that

$$f(\mathbf{A}_1, \mathbf{A}_2, \cdots, \mathbf{A}_{n-1}, \mathbf{A}_n)f(\mathbf{A}_n, \mathbf{U}_1, \mathbf{U}_2, \cdots, \mathbf{U}_{n-2}, \mathbf{A}_n) \qquad (36.11)$$
$$+ \quad f(\mathbf{A}_1, \mathbf{A}_2, \cdots, \mathbf{A}_{n-1}, \mathbf{U}_1)f(\mathbf{A}_n, \mathbf{A}_n, \mathbf{U}_2, \cdots, \mathbf{U}_{n-2}, \mathbf{A}_n)$$
$$+ \quad f(\mathbf{A}_1, \mathbf{A}_2, \cdots, \mathbf{A}_{n-1}, \mathbf{U}_2)f(\mathbf{A}_n, \mathbf{U}_1, \mathbf{A}_n, \mathbf{U}_3, \cdots, \mathbf{U}_{n-2}, \mathbf{A}_n) + \cdots$$
$$+ \quad f(\mathbf{A}_1, \mathbf{A}_2, \cdots, \mathbf{A}_{n-1}, \mathbf{U}_{n-2})f(\mathbf{A}_n, \mathbf{U}_1, \cdots, \mathbf{U}_{n-3}, \mathbf{A}_n, \mathbf{A}_n) = 0.$$

Let $E_{n-2} = \{1, 2, 3, \cdots, n-2\}$, and let S_r $(0 < r \leq n-2)$ be a subset of the set E_{n-2} which contains r elements. For $r = n-2$ we have $S_{n-2} = E_{n-2}$. Putting into Eq. (36.2) $\mathbf{Z}_i = \mathbf{A}_n$ $(1 \leq i \leq kn)$, we obtain

$$f(\mathbf{A}_n, \mathbf{A}_n, \cdots, \mathbf{A}_n) = 0. \qquad (36.12)$$

Now, we suppose that

$$f(\mathbf{A}_n, \mathbf{V}_1, \mathbf{V}_2, \cdots, \mathbf{V}_{n-2}, \mathbf{A}_n) = 0 \qquad (36.13)$$

holds, where

$$\mathbf{V}_i = \begin{cases} \mathbf{A}_n, & i \in S_r, \\ \mathbf{Y}_i, & i \in E_{n-2} \setminus S_r. \end{cases} \qquad (36.14)$$

Under this assumption we will show that

$$f(\mathbf{A}_n, \mathbf{W}_1, \mathbf{W}_2, \cdots, \mathbf{W}_{n-2}, \mathbf{A}_n) = 0, \qquad (36.15)$$

where

$$\mathbf{W}_i = \begin{cases} \mathbf{A}_n, & i \in S_{r-1}, \\ \mathbf{Y}_i, & i \in E_{n-2} \setminus S_{r-1}. \end{cases} \tag{36.16}$$

Putting $\mathbf{U}_i = \mathbf{W}_i$ $(1 \le i \le n-2)$ into Eq. (36.11), on the basis of the hypothesis Eq. (36.13) we obtain

$$r f(\mathbf{A}_1, \mathbf{A}_2, \cdots, \mathbf{A}_{n-1}, \mathbf{A}_n) f(\mathbf{A}_n, \mathbf{W}_1, \mathbf{W}_2, \cdots, \mathbf{W}_{n-2}, \mathbf{A}_n) = 0.$$

From this relation $(r \ge 1)$ we obtain

$$f(\mathbf{A}_n, \mathbf{W}_1, \mathbf{W}_2, \cdots, \mathbf{W}_{n-2}, \mathbf{A}_n) = 0.$$

Consequently, by induction we proved that

$$f(\mathbf{A}_n, \mathbf{U}_1, \mathbf{U}_2, \cdots, \mathbf{U}_{n-2}, \mathbf{A}_n) = 0$$

if exactly r $(0 \le r \le n-2)$ elements among $\mathbf{U}_1, \mathbf{U}_2, \cdots, \mathbf{U}_{n-2}$ are equal to $\mathbf{A}_n$.

For $r = 0$ we obtain

$$f(\mathbf{A}_n, \mathbf{U}_1, \mathbf{U}_2, \cdots, \mathbf{U}_{n-2}, \mathbf{A}_n) \equiv 0,$$

which contradicts the hypothesis Eq. (36.9).

Note that if $k = 2$, we do not use the hypothesis Eq. (36.9). In this case Eqs. (36.10) and (36.11) are identical. The above proof yields directly Eq. (36.8).

This completes the proof of the lemma. $\qquad\square$

We will prove Theorem 36.1 by induction.

For $n = 2$, the functional equation (36.2) takes the form

$$\begin{aligned}
(k-1) & f(\mathbf{Z}_1, \mathbf{Z}_2) f(\mathbf{Z}_3, \mathbf{Z}_4) \cdots f(\mathbf{Z}_{2k-1}, \mathbf{Z}_{2k}) \\
= \ & f(\mathbf{Z}_1, \mathbf{Z}_3) f(\mathbf{Z}_2, \mathbf{Z}_4) \cdots f(\mathbf{Z}_{2k-1}, \mathbf{Z}_{2k}) \\
+ \ & f(\mathbf{Z}_1, \mathbf{Z}_4) f(\mathbf{Z}_3, \mathbf{Z}_2) \cdots f(\mathbf{Z}_{2k-1}, \mathbf{Z}_{2k}) \\
+ \ & \cdots \\
& \ \ \vdots \\
+ \ & f(\mathbf{Z}_1, \mathbf{Z}_{2k-1}) f(\mathbf{Z}_3, \mathbf{Z}_4) \cdots f(\mathbf{Z}_2, \mathbf{Z}_{2k}) \\
+ \ & f(\mathbf{Z}_1, \mathbf{Z}_{2k}) f(\mathbf{Z}_3, \mathbf{Z}_4) \cdots f(\mathbf{Z}_{2k-1}, \mathbf{Z}_2).
\end{aligned} \tag{36.17}$$

If we substitute $\mathbf{Z}_i = \mathbf{U}$ $(1 \leq i \leq 2k)$, the above equation becomes

$$f(\mathbf{U}, \mathbf{U}) \equiv \mathbf{O}. \tag{36.18}$$

For any nonzero component of a nontrivial solution of Eq. (36.17) (denoted again by f) there exists at least one pair of constant complex vectors $(\mathbf{A}, \mathbf{B})$ such that $f(\mathbf{A}, \mathbf{B}) \neq 0$.

Putting into the functional equation (36.17)

$$\mathbf{Z}_{2i-1} = \mathbf{A} \quad (1 \leq i \leq k), \qquad \mathbf{Z}_{2j} = \mathbf{B} \quad (2 \leq j \leq k), \qquad \mathbf{Z}_2 = \mathbf{U},$$

it takes the form

$$f^{k-1}(\mathbf{A}, \mathbf{B})[f(\mathbf{U}, \mathbf{B}) + f(\mathbf{B}, \mathbf{U})] = 0,$$

from which it follows that

$$f(\mathbf{U}, \mathbf{B}) = -f(\mathbf{B}, \mathbf{U}). \tag{36.19}$$

For $\mathbf{Z}_{2i-1} = \mathbf{A}$, $\mathbf{Z}_{2i} = \mathbf{B}$ $(1 \leq i \leq k - 1)$, $\mathbf{Z}_{2k-1} = \mathbf{U}$, $\mathbf{Z}_{2k} = \mathbf{V}$, the functional equation (36.17) by virtue of Eq. (36.19) yields

$$f^{k-1}(\mathbf{A}, \mathbf{B})f(\mathbf{U}, \mathbf{V}) = f^{k-2}(\mathbf{A}, \mathbf{B})[f(\mathbf{A}, \mathbf{U})f(\mathbf{B}, \mathbf{V}) - f(\mathbf{A}, \mathbf{V})f(\mathbf{B}, \mathbf{U})].$$

If we introduce the notations

$$\frac{f(\mathbf{A}, \mathbf{U})}{f(\mathbf{A}, \mathbf{B})} = H_1(\mathbf{U}), \qquad f(\mathbf{B}, \mathbf{U}) = H_2(\mathbf{U}),$$

we obtain that the function

$$f(\mathbf{U}, \mathbf{V}) = \begin{vmatrix} H_1(\mathbf{U}) & H_1(\mathbf{V}) \\ H_2(\mathbf{U}) & H_2(\mathbf{V}) \end{vmatrix}$$

is the general solution of the functional equation (36.2) for $n = 2$ because it includes the trivial solution $f(\mathbf{X}, \mathbf{Y}) = 0$.

Now, we suppose that Theorem 36.1 holds for $n - 1$, *i.e.*, the general solution of the functional equation

$$(k - 1)F(\mathbf{Z}_1, \mathbf{Z}_2, \cdots, \mathbf{Z}_{k(n-1)}) \tag{36.20}$$

$$= \sum_{r=n}^{k(n-1)} \Psi_{n-1,r} F(\mathbf{Z}_1, \mathbf{Z}_2, \cdots, \mathbf{Z}_{n-1}, \mathbf{Z}_n, \cdots, \mathbf{Z}_{k(n-1)}),$$

where

$$F(\mathbf{Z}_1, \mathbf{Z}_2, \cdots, \mathbf{Z}_{k(n-1)}) \tag{36.21}$$

$$= \prod_{i=0}^{k-1} f(\mathbf{Z}_{(n-1)i+1}, \mathbf{Z}_{(n-1)i+2}, \cdots, \mathbf{Z}_{(n-1)i+n-1}),$$

is given by

$$f(\mathbf{U}_1, \mathbf{U}_2, \cdots, \mathbf{U}_{n-1}) \tag{36.22}$$

$$= \begin{vmatrix} H_1(\mathbf{U}_1) & H_1(\mathbf{U}_2) & \cdots & H_1(\mathbf{U}_{n-1}) \\ H_2(\mathbf{U}_1) & H_2(\mathbf{U}_2) & \cdots & H_2(\mathbf{U}_{n-1}) \\ \vdots & & & \\ H_{n-1}(\mathbf{U}_1) & H_{n-1}(\mathbf{U}_2) & \cdots & H_{n-1}(\mathbf{U}_{n-1}) \end{vmatrix}.$$

Let $f(\mathbf{A}_1, \mathbf{A}_2, \cdots, \mathbf{A}_n) \neq 0$ (here, as usual, f denotes a nonzero component of a nontrivial solution). If we put

$$\mathbf{Z}_{ni+1} = \mathbf{A}_n \qquad (0 \le i \le k-1),$$

$$f(\mathbf{A}_n, \mathbf{Z}_2, \mathbf{Z}_3, \cdots, \mathbf{Z}_n) = g(\mathbf{Z}_2, \mathbf{Z}_3, \cdots, \mathbf{Z}_n)$$

$$(f : \mathcal{V}^n \mapsto \mathbf{C}, \quad g : \mathcal{V}^{n-1} \mapsto \mathbf{C}),$$

according to Lemma 36.2 from Eq. (36.2) we obtain

$$(k-1)g(\mathbf{Z}_2, \cdots, \mathbf{Z}_{n-1}, \mathbf{Z}_n)g(\mathbf{Z}_{n+2}, \mathbf{Z}_{n+3}, \cdots, \mathbf{Z}_{2n}) \cdots g(\mathbf{Z}_{(k-1)n+2}, \cdots, \mathbf{Z}_{kn})$$

$$= g(\mathbf{Z}_2, \cdots, \mathbf{Z}_{n-1}, \mathbf{Z}_{n+2})g(\mathbf{Z}_n, \mathbf{Z}_{n+3}, \cdots, \mathbf{Z}_{2n}) \cdots g(\mathbf{Z}_{(k-1)n+2}, \cdots, \mathbf{Z}_{kn})$$

$$+ g(\mathbf{Z}_2, \cdots, \mathbf{Z}_{n-1}, \mathbf{Z}_{n+3})g(\mathbf{Z}_{n+2}, \mathbf{Z}_n, \mathbf{Z}_{n+4}, \cdots, \mathbf{Z}_{2n}) \cdots g(\mathbf{Z}_{(k-1)n+2}, \cdots, \mathbf{Z}_{kn})$$

$$+ \quad \cdots \tag{36.23}$$

$$\vdots$$

$$+ g(\mathbf{Z}_2, \cdots, \mathbf{Z}_{n-1}, \mathbf{Z}_{kn})g(\mathbf{Z}_{n+2}, \cdots, \mathbf{Z}_{2n}) \cdots g(\mathbf{Z}_{(k-1)n+2}, \cdots, \mathbf{Z}_{kn-1}, \mathbf{Z}_n).$$

According to the inductive hypothesis, we obtain that the general solution of the equation (36.23) is given by

$$g(\mathbf{U}_1, \mathbf{U}_2, \cdots, \mathbf{U}_{n-1}) = f(\mathbf{A}_n, \mathbf{U}_1, \mathbf{U}_2, \cdots, \mathbf{U}_{n-1})$$

$$= \begin{vmatrix} H_1(\mathbf{U}_1) & H_1(\mathbf{U}_2) & \cdots & H_1(\mathbf{U}_{n-1}) \\ H_2(\mathbf{U}_1) & H_2(\mathbf{U}_2) & \cdots & H_2(\mathbf{U}_{n-1}) \\ \vdots & & & \\ H_{n-1}(\mathbf{U}_1) & H_{n-1}(\mathbf{U}_2) & \cdots & H_{n-1}(\mathbf{U}_{n-1}) \end{vmatrix}, \qquad (36.24)$$

where $H_i : \mathcal{V} \mapsto \mathbf{C}$ $(1 \leq i \leq n-1)$ are arbitrary functions.

If we put into equation (36.2)

$$\mathbf{Z}_i = \mathbf{A}_i \quad (1 \leq i \leq n-1), \qquad \mathbf{Z}_n = \mathbf{U}_1,$$

$$\mathbf{Z}_{nj+m} = \mathbf{A}_m \quad (1 \leq j \leq k-1; \; 1 \leq m \leq n),$$

$$\mathbf{Z}_{(k-1)n+1} = \mathbf{A}_n, \qquad \mathbf{Z}_{(k-1)n+r} = \mathbf{U}_r \quad (2 \leq r \leq n),$$

we obtain

$$f(\mathbf{A}_1, \mathbf{A}_2, \cdots, \mathbf{A}_{n-1}, \mathbf{A}_n) f(\mathbf{U}_1, \mathbf{U}_2, \cdots, \mathbf{U}_n) \qquad (36.25)$$

$$= \quad f(\mathbf{A}_1, \cdots, \mathbf{A}_{n-1}, \mathbf{U}_1) f(\mathbf{A}_n, \mathbf{U}_2, \cdots, \mathbf{U}_n)$$

$$- \quad f(\mathbf{A}_1, \cdots, \mathbf{A}_{n-1}, \mathbf{U}_2) f(\mathbf{A}_n, \mathbf{U}_1, \mathbf{U}_3, \cdots, \mathbf{U}_n)$$

$$- \quad \cdots$$

$$\vdots$$

$$- \quad f(\mathbf{A}_1, \cdots, \mathbf{A}_{n-1}, \mathbf{U}_n) f(\mathbf{A}_n, \mathbf{U}_2, \cdots, \mathbf{U}_{n-1}, \mathbf{U}_1).$$

On the basis of the equality (36.24) we obtain

$$f(\mathbf{A}_n, \cdots, \mathbf{U}_{i-1}, \mathbf{U}_i, \mathbf{U}_{i+1}, \cdots, \mathbf{U}_{j-1}, \mathbf{U}_j, \mathbf{U}_{j+1}, \cdots, \mathbf{U}_{n-1}) \quad (36.26)$$

$$= \quad -f(\mathbf{A}_n, \cdots, \mathbf{U}_{i-1}, \mathbf{U}_j, \mathbf{U}_{i+1}, \cdots, \mathbf{U}_{j-1}, \mathbf{U}_i, \mathbf{U}_{j+1}, \cdots, \mathbf{U}_{n-1})$$

$$(1 \leq i < j \leq n-1).$$

By using the equalities (36.24), (36.25) and (36.26) along with the notation

$$\frac{f(\mathbf{A}_1, \mathbf{A}_2, \cdots, \mathbf{A}_{n-1}, \mathbf{U})}{f(\mathbf{A}_1, \cdots, \mathbf{A}_n)} = H_n(\mathbf{U}),$$

we obtain that $f(\mathbf{U}_1, \mathbf{U}_2, \cdots, \mathbf{U}_n)$ has the form Eq. (36.4).

It remains still to show that every function of the form Eq. (36.4) is really a solution of the equation (36.2). For this purpose, we will consider the following identity

$$D(j) = \begin{vmatrix} \mathcal{H}(\mathbf{Z}) & \mathcal{O}_{n-1,n} \\ \tilde{\mathcal{H}}(\mathbf{Z}) & \tilde{\mathcal{H}}(\mathbf{Z}) \end{vmatrix} = 0, \tag{36.27}$$

where

$$\mathcal{H}(\mathbf{Z}) = \begin{pmatrix} H_1(\mathbf{Z}_1) & H_2(\mathbf{Z}_1) & \cdots & H_n(\mathbf{Z}_1) \\ H_1(\mathbf{Z}_2) & H_2(\mathbf{Z}_2) & & H_n(\mathbf{Z}_2) \\ \vdots & & & \\ H_1(\mathbf{Z}_{n-1}) & H_2(\mathbf{Z}_{n-1}) & & H_n(\mathbf{Z}_{n-1}) \end{pmatrix},$$

$$\tilde{\mathcal{H}}(\mathbf{Z}) = \begin{pmatrix} H_1(\mathbf{Z}_n) & H_2(\mathbf{Z}_n) & \cdots & H_n(\mathbf{Z}_n) \\ H_1(\mathbf{Z}_{nj+1}) & H_2(\mathbf{Z}_{nj+1}) & & H_n(\mathbf{Z}_{nj+1}) \\ \vdots & & & \\ H_1(\mathbf{Z}_{nj+n}) & H_2(\mathbf{Z}_{nj+n}) & & H_n(\mathbf{Z}_{nj+n}) \end{pmatrix}$$

and $\mathcal{O}_{n-1,n}$ is the zero $(n-1) \times n$ matrix.

According to Eq. (36.27), we conclude that the following identity holds

$$\sum_{j=1}^{k-1} D(j) \prod_{\substack{i=1 \\ i \neq j}}^{k-1} \begin{vmatrix} H_1(\mathbf{Z}_{ni+1}) & H_1(\mathbf{Z}_{ni+2}) & \cdots & H_1(\mathbf{Z}_{ni+n}) \\ H_2(\mathbf{Z}_{ni+1}) & H_2(\mathbf{Z}_{ni+2}) & \cdots & H_2(\mathbf{Z}_{ni+n}) \\ \vdots & & & \\ H_n(\mathbf{Z}_{ni+1}) & H_n(\mathbf{Z}_{ni+2}) & \cdots & H_n(\mathbf{Z}_{ni+n}) \end{vmatrix} = 0. \tag{36.28}$$

By evaluating the determinant $D(j)$ according to the Laplace rule, we show that the function given by Eq. (36.4) is really a solution of the equation (36.2). This completes the proof of Theorem 36.1 for $a = k - 1$.

Now we pass to the proof of Theorem 36.2 for $a = n(k-1)$.

First, we suppose that $f(\mathbf{U}, \mathbf{U}, \cdots, \mathbf{U}) \not\equiv \mathbf{O}$. Henceforth we denote by f any component of the function $f : \mathcal{V}^n \mapsto \mathcal{V}$ for which $f(\mathbf{U}, \mathbf{U}, \cdots, \mathbf{U}) \not\equiv 0$. For such a component there exist at least one constant complex vector C for which $f(C, C, \cdots, C) \neq 0$.

If we put $\mathbf{Z}_{n+i} = \mathbf{U}$ and substitute the rest of the variables by C, from Eq. (36.2) we obtain

$$f(C, \cdots, C, \mathbf{U}, C, \cdots, C) = f(C, C, \cdots, C, \mathbf{U}). \tag{36.29}$$

By putting $\mathbf{Z}_{n+i} = \mathbf{U}$, $\mathbf{Z}_{n+j} = \mathbf{V}$ $(j > i)$ and substituting the rest of the variables by C, from the equation (36.2) by virtue of Eq. (36.29) we obtain

$$f(C,\cdots,C,\mathbf{U},\cdots,\mathbf{V},\cdots,C) \tag{36.30}$$
$$= \frac{f(C,C,\cdots,C,\mathbf{U})f(C,C,\cdots,C,\mathbf{V})}{f(C,C,\cdots,C)}.$$

We assume that

$$f(C,\cdots,\mathbf{U}_1,\cdots,C,\mathbf{U}_2,\cdots,C,\cdots,\mathbf{U}_\nu,C,\cdots,C)$$
$$= \frac{\prod_{i=1}^{\nu} f(C,C,\cdots,C,\mathbf{U}_i)}{f^{\nu-1}(C,C,\cdots,C)}. \tag{36.31}$$

If we put $\mathbf{Z}_{n+i_1} = \mathbf{U}_1$, $\mathbf{Z}_{n+i_2} = \mathbf{U}_2,\cdots,\mathbf{Z}_{n+i_\nu} = \mathbf{U}_\nu$, $\mathbf{Z}_{n+i_{\nu+1}} = \mathbf{U}_{\nu+1}$ and substitute the rest of the variables by C, then the equation (36.2) becomes

$$\nu f(C,C,\cdots,C)f(C,\cdots,\mathbf{U}_1,\cdots,C,\mathbf{U}_2,\cdots,\mathbf{U}_\nu,C,\cdots,\mathbf{U}_{\nu+1},\cdots,C)$$
$$= f(C,C,\cdots,C,\mathbf{U}_1)f(C,\cdots,C,\cdots,\mathbf{U}_2,\cdots,\mathbf{U}_\nu,\cdots,\mathbf{U}_{\nu+1},\cdots,C)$$
$$+ f(C,C,\cdots,C,\mathbf{U}_2)f(C,\cdots,\mathbf{U}_1,\cdots,\mathbf{U}_\nu,\cdots,\mathbf{U}_{\nu+1},\cdots,C)$$
$$+ \cdots \tag{36.32}$$
$$\vdots$$
$$+ f(C,C,\cdots,C,\mathbf{U}_{\nu+1})f(C,\cdots,\mathbf{U}_1,\cdots,\mathbf{U}_2,\cdots,\mathbf{U}_\nu,\cdots,C,\cdots,C).$$

On the basis of the inductive hypothesis Eq. (36.31), from Eq. (36.32) it follows that

$$f(C,\cdots,\mathbf{U}_1,\cdots,\mathbf{U}_2,\cdots,\mathbf{U}_\nu,\cdots,\mathbf{U}_{\nu+1},\cdots,C)$$
$$= \frac{\prod_{i=1}^{\nu+1} f(C,C,\cdots,C,\mathbf{U}_i)}{f^\nu(C,C,\cdots,C)}. \tag{36.33}$$

Therefore, we proved by mathematical induction that Eq. (36.33) holds for every $\nu < n$.

By putting $\mathbf{Z}_i = C \; (1 \le i \le n)$, $\mathbf{Z}_{n+j} = \mathbf{U}_j \; (1 \le j \le n)$, $\mathbf{Z}_{2n+s} = C \; (1 \le s \le n(k-2))$, from the equation (36.2) we obtain

$$
\begin{aligned}
& nf(C, C, \cdots, C)f(\mathbf{U}_1, \mathbf{U}_2, \cdots, \mathbf{U}_n) \qquad\qquad (36.34) \\
=\ & f(C, C, \cdots, C, \mathbf{U}_1)f(C, \mathbf{U}_2, \mathbf{U}_3, \cdots, \mathbf{U}_n) \\
+\ & f(C, C, \cdots, C, \mathbf{U}_2)f(\mathbf{U}_1, C, \mathbf{U}_3, \mathbf{U}_4, \cdots, \mathbf{U}_n) \\
+\ & \cdots \\
& \vdots \\
+\ & f(C, C, \cdots, C, \mathbf{U}_n)f(\mathbf{U}_1, \mathbf{U}_2, \cdots, \mathbf{U}_{n-1}, C).
\end{aligned}
$$

On the basis of the equality (36.33), the last equality (36.34) becomes

$$
f(\mathbf{U}_1, \mathbf{U}_2, \cdots, \mathbf{U}_n) = \frac{\prod_{i=1}^{n} f(C, C, \cdots, C, \mathbf{U}_i)}{f^{n-1}(C, C, \cdots, C)}. \qquad (36.35)
$$

By introducing the notation

$$
K(\mathbf{U}) = f(C, \cdots, C, \mathbf{U}) / \sqrt[n]{f^{n-1}(C, C, \cdots, C)},
$$

we obtain that the function f in the case $a = n(k-1)$ really has the form Eq. (36.5).

Now, we suppose that $f(\mathbf{U}, \mathbf{U}, \cdots, \mathbf{U}) \equiv 0$. Next, we will need the following result.

Lemma 36.3   *If at least one of the variables* $\mathbf{U}_1, \mathbf{U}_2, \cdots, \mathbf{U}_{n-1}$ *is equal to* $\mathbf{U}_n$, *then*

$$
f(\mathbf{U}_1, \mathbf{U}_2, \cdots, \mathbf{U}_n) \equiv 0.
$$

Proof of Lemma 36.3. Let $E_{n-1} = \{1, 2, 3, \cdots, n-1\}$, and let $S_m \; (1 \le m \le n-1)$ be a subset of the set E_{n-1} which contains m elements. We have $f(\mathbf{U}_n, \mathbf{U}_n, \cdots, \mathbf{U}_n) \equiv 0$.

Now, we suppose that

$$
f(\mathbf{V}_1, \mathbf{V}_2, \cdots, \mathbf{V}_{n-1}, \mathbf{U}_n) = 0 \qquad\qquad (36.36)
$$

holds, where

$$
\mathbf{V}_i = \begin{cases} \mathbf{U}_n, & i \in S_m, \\ \mathbf{Y}_i, & i \in E_{n-1} \setminus S_m. \end{cases}
$$

We will prove that

$$f(\mathbf{W}_1, \mathbf{W}_2, \cdots, \mathbf{W}_{n-1}, \mathbf{U}_n) = 0, \tag{36.37}$$

where

$$\mathbf{W}_i = \begin{cases} \mathbf{U}_n, & i \in S_{m-1}, \\ \mathbf{Y}_i, & i \in E_{n-1} \setminus S_{m-1}. \end{cases}$$

By substituting $\mathbf{Z}_{nm+i} = \mathbf{Z}_i$ $(0 \le m \le k - 1)$ into the equation (36.2) and by putting $\mathbf{Z}_i = \mathbf{W}_i$, on the basis of the assumption (36.36) we obtain

$$(k - 1)(n - m)f^k(\mathbf{W}_1, \mathbf{W}_2, \cdots, \mathbf{W}_{n-1}, \mathbf{U}_n) = 0,$$

from which we deduce Eq. (36.37) because $k \ge 2$ and $m < n$.

Therefore, Lemma 36.3 is proved by induction. $\qquad\square$

By putting $\mathbf{Z}_{nm+i} = \mathbf{U}_i$ $(0 \le m \le k-1)$ and according to Lemma 36.3, from the equation (36.2) we obtain

$$(k - 1)(n - 1)f^k(\mathbf{U}_1, \mathbf{U}_2, \cdots, \mathbf{U}_n) = 0.$$

Since $k > 1$ and $n > 1$, from this we obtain that the function f has the form Eq. (36.5).

We may easily check that the functions given by Eq. (36.5) satisfy the functional equation (36.2). We can do this by a direct substitution of Eq. (36.5) into Eq. (36.2). This completes the proof of Theorem 36.1 for the case $a = n(k - 1)$.

Now, we will pass to the proof of Theorem 36.1 for the case $a \ne r(k - 1)$ $(1 \le r \le n)$.

In this case, Lemma 36.3 also holds.

By putting $\mathbf{Z}_{nm+i} = \mathbf{Z}_i$ $(0 \le m \le k - 1)$ and according to Lemma 36.3, from the equation (36.2) we obtain

$$[a - (k - 1)]f^k(\mathbf{Z}_1, \mathbf{Z}_2, \cdots, \mathbf{Z}_n) = \mathbf{O}.$$

Since $a \ne k - 1$, from the above equality we immediately deduce the statement of Theorem 36.1 for the case $a \ne r(k - 1)$ $(1 \le r \le n)$.

Therefore, Theorem 36.1 has been completely proved. $\qquad\square$

We have not been able to solve equation (36.2) for $a = r(k - 1)$ $(2 \le r \le n - 1)$.

Remark 36.1.   The function

$$f(\mathbf{U}_1, \mathbf{U}_2, \cdots, \mathbf{U}_n) \tag{36.38}$$

$$= \begin{vmatrix} H_1(\mathbf{U}_1) & H_1(\mathbf{U}_2) & \cdots & H_1(\mathbf{U}_s) \\ H_2(\mathbf{U}_1) & H_2(\mathbf{U}_2) & \cdots & H_2(\mathbf{U}_s) \\ \vdots & & & \\ H_s(\mathbf{U}_1) & H_s(\mathbf{U}_2) & \cdots & H_s(\mathbf{U}_s) \end{vmatrix} \prod_{i=s+1}^{n} H_1(\mathbf{U}_i),$$

where $s = n - r + 1$, H_i $(1 \leq i \leq n - r + 1)$ are arbitrary functions, is a solution of the equation (36.2) for $a = r(k - 1)$ but the question of generality of this solution remains open.

Also, for $r = n$ the function given by Eq. (36.38) becomes Eq. (36.5). If we assume that

$$\prod_{i=n+1}^{n} H_1(\mathbf{Z}_i) = \mathbf{I},$$

then the function Eq. (36.38) for $r = 1$ becomes Eq. (36.4).

All this suggests us to put forth the following hypothesis.

Hypothesis 36.1   The general solution of the functional equation (36.2) in the case $a = r(k - 1)$ $(1 \leq r \leq n)$ is given by the formula Eq. (36.38).

Chapter 10

Systems of Nonlinear Functional Equations

In this chapter complex vector systems of nonlinear partial functional equations are considered. They may be sorted in two types. The first three systems solved are considered as complex vector systems of partial quadratic functional equations with real parameters. The last three systems solved represent complex vector systems of partial functional equations of higher order without parameters. In this chapter the unknown functions depend on arguments in an arbitrary finite dimensional complex vector space $\mathcal{V}$ and take values in the field of complex numbers $\mathbf{C}$ so that their products make sense. As in Sec. 10, denote by $\mathcal{V}^0$ the space of all real vectors in $\mathcal{V}$ (thus $\mathcal{V} = \mathcal{V}^0 + i\mathcal{V}^0$) and let $\mathcal{L}(\mathcal{V}^0, \mathbf{C})$ be the space of all linear mappings $\mathcal{V}^0 \mapsto \mathbf{C}$. The results presented in this chapter were obtained in [I. B. Risteski *et al.* (submitted)].

37   Systems of Quadratic Functional Equations

Consider the system of functional equations

$$f(\mathbf{Z}_1, \mathbf{Z}_2, \mathbf{Z}_3, \cdots, \mathbf{Z}_{2n-1}, \mathbf{Z}_{2n}) + f(\mathbf{Z}_1, \mathbf{Z}_3, \mathbf{Z}_4, \cdots, \mathbf{Z}_{2n}, \mathbf{Z}_2) + \cdots \qquad (37.1)$$

$$+f(\mathbf{Z}_1, \mathbf{Z}_{2n}, \mathbf{Z}_2, \cdots, \mathbf{Z}_{2n-2}, \mathbf{Z}_{2n-1}) + \operatorname{sgn} a \left[g(\mathbf{Z}_1, \mathbf{Z}_2, \mathbf{Z}_3, \cdots, \mathbf{Z}_{2n-1}, \mathbf{Z}_{2n}) + \right.$$

$$\left. g(\mathbf{Z}_1, \mathbf{Z}_3, \mathbf{Z}_4, \cdots, \mathbf{Z}_{2n}, \mathbf{Z}_2) + \cdots + g(\mathbf{Z}_1, \mathbf{Z}_{2n}, \mathbf{Z}_2, \cdots, \mathbf{Z}_{2n-2}, \mathbf{Z}_{2n-1}) \right] = 0,$$

$$h(\mathbf{Z}_1, \mathbf{Z}_2, \mathbf{Z}_3, \cdots, \mathbf{Z}_{2n-1}, \mathbf{Z}_{2n}) + h(\mathbf{Z}_1, \mathbf{Z}_3, \mathbf{Z}_4, \cdots, \mathbf{Z}_{2n}, \mathbf{Z}_2) + \cdots$$

$$+h(\mathbf{Z}_1,\mathbf{Z}_{2n},\mathbf{Z}_2,\cdots,\mathbf{Z}_{2n-2},\mathbf{Z}_{2n-1}) + k(\mathbf{Z}_1,\mathbf{Z}_2,\mathbf{Z}_3,\cdots,\mathbf{Z}_{2n-1},\mathbf{Z}_{2n})+$$

$$k(\mathbf{Z}_1,\mathbf{Z}_3,\mathbf{Z}_4,\cdots,\mathbf{Z}_{2n},\mathbf{Z}_2) + \cdots + k(\mathbf{Z}_1,\mathbf{Z}_{2n},\mathbf{Z}_2,\cdots,\mathbf{Z}_{2n-2},\mathbf{Z}_{2n-1}) = 0,$$

where

$$f(\mathbf{Z}_1,\mathbf{Z}_2,\mathbf{Z}_3,\cdots,\mathbf{Z}_{2n-1},\mathbf{Z}_{2n}) \tag{37.2}$$
$$= [F(\mathbf{Z}_1,\mathbf{Z}_2) + F(\mathbf{Z}_3,\mathbf{Z}_4) + \cdots + F(\mathbf{Z}_{2i-1},\mathbf{Z}_{2i})] \times$$
$$\times [F(\mathbf{Z}_{2i+1},\mathbf{Z}_{2n}) + F(\mathbf{Z}_{2i+2},\mathbf{Z}_{2n-1}) + \cdots + F(\mathbf{Z}_{i+n},\mathbf{Z}_{i+n+1})],$$

$$g(\mathbf{Z}_1,\mathbf{Z}_2,\mathbf{Z}_3,\cdots,\mathbf{Z}_{2n-1},\mathbf{Z}_{2n})$$
$$= [G(\mathbf{Z}_1,\mathbf{Z}_2) + G(\mathbf{Z}_3,\mathbf{Z}_4) + \cdots + G(\mathbf{Z}_{2i-1},\mathbf{Z}_{2i})] \times$$
$$\times [G(\mathbf{Z}_{2i+1},\mathbf{Z}_{2n}) + G(\mathbf{Z}_{2i+2},\mathbf{Z}_{2n-1}) + \cdots + G(\mathbf{Z}_{i+n},\mathbf{Z}_{i+n+1})],$$

$$h(\mathbf{Z}_1,\mathbf{Z}_2,\mathbf{Z}_3,\cdots,\mathbf{Z}_{2n-1},\mathbf{Z}_{2n})$$
$$= [F(\mathbf{Z}_1,\mathbf{Z}_2) + F(\mathbf{Z}_3,\mathbf{Z}_4) + \cdots + F(\mathbf{Z}_{2i-1},\mathbf{Z}_{2i})] \times$$
$$\times [G(\mathbf{Z}_{2i+1},\mathbf{Z}_{2n}) + G(\mathbf{Z}_{2i+2},\mathbf{Z}_{2n-1}) + \cdots + G(\mathbf{Z}_{i+n},\mathbf{Z}_{i+n+1})],$$

$$k(\mathbf{Z}_1,\mathbf{Z}_2,\mathbf{Z}_3,\cdots,\mathbf{Z}_{2n-1},\mathbf{Z}_{2n})$$
$$= [G(\mathbf{Z}_1,\mathbf{Z}_2) + G(\mathbf{Z}_3,\mathbf{Z}_4) + \cdots + G(\mathbf{Z}_{2i-1},\mathbf{Z}_{2i})] \times$$
$$\times [F(\mathbf{Z}_{2i+1},\mathbf{Z}_{2n}) + F(\mathbf{Z}_{2i+2},\mathbf{Z}_{2n-1}) + \cdots + F(\mathbf{Z}_{i+n},\mathbf{Z}_{i+n+1})],$$

$F, G : \mathcal{V}^2 \mapsto \mathbf{C}$ are arbitrary complex functions, a is a real constant and $n > 2$, $1 \leq i \leq n - 1$.

Definition 37.1. A solution $(F(\mathbf{U},\mathbf{V}), G(\mathbf{U},\mathbf{V}))$ of system (37.1) is called *isotropic* if $(F(\mathbf{U},\mathbf{V}), G(\mathbf{U},\mathbf{V})) \not\equiv (0,0)$ and

$$F^2(\mathbf{U},\mathbf{V}) + G^2(\mathbf{U},\mathbf{V}) \equiv 0. \tag{37.3}$$

The *general* solution of system (37.1) includes all solutions of this system with the possible exception of the isotropic ones.

Theorem 37.1 *The general solution of the system (37.1) is given by the formulae*

$$F(\mathbf{U},\mathbf{V}) = P(\mathbf{U}) - P(\mathbf{V}), \tag{37.4}$$
$$G(\mathbf{U},\mathbf{V}) = Q(\mathbf{U}) - Q(\mathbf{V}), \tag{37.5}$$

where P and Q are arbitrary complex functions defined in $\mathcal{V}$.

Each isotropic **continuous** *solution of Eq.* (37.1) *has the form*

$$F(\mathbf{U}, \mathbf{V}) = P(\mathbf{U}) - P(\mathbf{V}), \quad G(\mathbf{U}, \mathbf{V}) = \pm i(P(\mathbf{U}) - P(\mathbf{V}))$$

$(i^2 = -1;$ *we take the same sign for all vectors* $\mathbf{U}, \mathbf{V} \in \mathcal{V})$.

Proof. Let $F(\mathbf{U}, \mathbf{V}), G(\mathbf{U}, \mathbf{V})$ represent an isotropic continuous solution of system (37.1). This means that for any pair of vectors $(\mathbf{U}, \mathbf{V}) \in \mathcal{V}^2$ we have

$$G(\mathbf{U}, \mathbf{V}) = \pm i F(\mathbf{U}, \mathbf{V}) \tag{37.6}$$

(both signs are possible). It is clear that the real dimension of $\mathcal{V}^2$ is 4 dim $\mathcal{V}$, the real codimension of the set $\{(\mathbf{U}, \mathbf{V}) \in \mathcal{V}^2 : F(\mathbf{U}, \mathbf{V}) = 0\}$ is 2. Hence its complement is a connected set in which the function $G(\mathbf{U}, \mathbf{V})/F(\mathbf{U}, \mathbf{V})$ is continuous, thus it can take just one of the values i or $-i$. Henceforth, when we write $\pm i$ (as in Eq. (37.6)), we mean only one of these two signs.

By virtue of Eq. (37.6) system (37.1) implies

$$f(\mathbf{Z}_1, \mathbf{Z}_2, \cdots, \mathbf{Z}_{2n-1}, \mathbf{Z}_{2n}) + f(\mathbf{Z}_1, \mathbf{Z}_3, \cdots, \mathbf{Z}_{2n}, \mathbf{Z}_2) \tag{37.7}$$
$$+ \cdots + f(\mathbf{Z}_1, \mathbf{Z}_{2n}, \mathbf{Z}_2, \cdots, \mathbf{Z}_{2n-1}) = 0.$$

The general solution of this equation for $n > 2$ according to [D. S. Mitrinović and S. B. Prešić (1962A)] is Eq. (37.4):

$$F(\mathbf{U}, \mathbf{V}) = P(\mathbf{U}) - P(\mathbf{V}),$$

where $P(\mathbf{U}) = F(\mathbf{U}, \mathbf{A})$ for some fixed vector $\mathbf{A}$. Then

$$G(\mathbf{U}, \mathbf{V}) = \pm i(P(\mathbf{U}) - P(\mathbf{V})).$$

Next we are looking for solutions of system (37.1) not satisfying Eq. (37.3) (*anisotropic* solutions). We will distinguish the following cases:

1°. Let $a < 0$. In this case the system of complex vector functional equations (37.1) will be

$$f(\mathbf{Z}_1, \mathbf{Z}_2, \cdots, \mathbf{Z}_{2n}) + f(\mathbf{Z}_1, \mathbf{Z}_3, \cdots, \mathbf{Z}_{2n}, \mathbf{Z}_2) \tag{37.8}$$
$$+ \cdots + f(\mathbf{Z}_1, \mathbf{Z}_{2n}, \mathbf{Z}_2, \cdots, \mathbf{Z}_{2n-1})$$
$$- \quad g(\mathbf{Z}_1, \mathbf{Z}_2, \cdots, \mathbf{Z}_{2n}) - g(\mathbf{Z}_1, \mathbf{Z}_3, \cdots, \mathbf{Z}_{2n}, \mathbf{Z}_2)$$
$$- \cdots - g(\mathbf{Z}_1, \mathbf{Z}_{2n}, \mathbf{Z}_2, \cdots, \mathbf{Z}_{2n-1}) = 0,$$

$$h(\mathbf{Z}_1, \mathbf{Z}_2, \cdots, \mathbf{Z}_{2n}) + h(\mathbf{Z}_1, \mathbf{Z}_3, \cdots, \mathbf{Z}_{2n}, \mathbf{Z}_2)$$
$$+ \cdots + h(\mathbf{Z}_1, \mathbf{Z}_{2n}, \mathbf{Z}_2, \cdots, \mathbf{Z}_{2n-1})$$
$$+ \; k(\mathbf{Z}_1, \mathbf{Z}_2, \cdots, \mathbf{Z}_{2n}) + k(\mathbf{Z}_1, \mathbf{Z}_3, \cdots, \mathbf{Z}_{2n}, \mathbf{Z}_2)$$
$$+ \cdots + k(\mathbf{Z}_1, \mathbf{Z}_{2n}, \mathbf{Z}_2, \cdots + \mathbf{Z}_{2n-1}) = 0.$$

If we substitute all the variables $\mathbf{Z}_i$ $(1 \leq i \leq 2n)$ in the system (37.8) by $\mathbf{U}$, then we obtain

$$F^2(\mathbf{U}, \mathbf{U}) - G^2(\mathbf{U}, \mathbf{U}) = 0, \qquad 2F(\mathbf{U}, \mathbf{U})G(\mathbf{U}, \mathbf{U}) = 0,$$

so that

$$F(\mathbf{U}, \mathbf{U}) = 0, \quad G(\mathbf{U}, \mathbf{U}) = 0. \tag{37.9}$$

For any nontrivial anisotropic solution of the system (37.8) there exists at least one pair of vectors $(\mathbf{A}, \mathbf{B}) \in \mathcal{V}^2$ such that

$$F^2(\mathbf{A}, \mathbf{B}) + G^2(\mathbf{A}, \mathbf{B}) \neq 0. \tag{37.10}$$

a) Let $i = 1$. If we execute the substitutions $\mathbf{Z}_5 = \mathbf{Z}_6 = \cdots = \mathbf{Z}_{2n} = \mathbf{Z}_1 = \mathbf{Z}_2 = \mathbf{A}$, $\mathbf{Z}_3 = \mathbf{B}$, $\mathbf{Z}_4 = \mathbf{U}$, on the basis of the equalities (37.9) the system (37.8) becomes

$$F(\mathbf{A}, \mathbf{B})[F(\mathbf{U}, \mathbf{A}) + F(\mathbf{A}, \mathbf{U})] - G(\mathbf{A}, \mathbf{B})[G(\mathbf{U}, \mathbf{A}) + G(\mathbf{A}, \mathbf{U})] = 0,$$
$$\tag{37.11}$$
$$G(\mathbf{A}, \mathbf{B})[F(\mathbf{U}, \mathbf{A}) + F(\mathbf{A}, \mathbf{U})] + F(\mathbf{A}, \mathbf{B})[G(\mathbf{U}, \mathbf{A}) + G(\mathbf{A}, \mathbf{U})] = 0.$$

In view of Eq. (37.10), from Eq. (37.11) it follows that

$$F(\mathbf{A}, \mathbf{U}) = -F(\mathbf{U}, \mathbf{A}), \qquad G(\mathbf{A}, \mathbf{U}) = -G(\mathbf{U}, \mathbf{A}). \tag{37.12}$$

By putting $\mathbf{Z}_5 = \mathbf{Z}_6 = \cdots = \mathbf{Z}_{2n} = \mathbf{Z}_1 = \mathbf{A}$, $\mathbf{Z}_2 = \mathbf{V}$, $\mathbf{Z}_3 = \mathbf{B}$, $\mathbf{Z}_4 = \mathbf{U}$ into Eq. (37.8), on the basis of the expressions Eqs. (37.9) and (37.12) we get

$$F(\mathbf{A}, \mathbf{B})[F(\mathbf{U}, \mathbf{V}) - F(\mathbf{U}, \mathbf{A}) + F(\mathbf{V}, \mathbf{A})] \tag{37.13}$$
$$- \; G(\mathbf{A}, \mathbf{B})[G(\mathbf{U}, \mathbf{V}) - G(\mathbf{U}, \mathbf{A}) + G(\mathbf{V}, \mathbf{A})] = 0,$$
$$G(\mathbf{A}, \mathbf{B})[F(\mathbf{U}, \mathbf{V}) - F(\mathbf{U}, \mathbf{A}) + F(\mathbf{V}, \mathbf{A})]$$
$$+ \; F(\mathbf{A}, \mathbf{B})[G(\mathbf{U}, \mathbf{V}) - G(\mathbf{U}, \mathbf{A}) + G(\mathbf{V}, \mathbf{A})] = 0,$$

whence

$$F(\mathbf{U}, \mathbf{V}) - F(\mathbf{U}, \mathbf{A}) + F(\mathbf{V}, \mathbf{A}) = 0, \quad G(\mathbf{U}, \mathbf{V}) - G(\mathbf{U}, \mathbf{A}) + G(\mathbf{V}, \mathbf{A}) = 0,$$

which implies Eqs. (37.4), (37.5):

$$F(\mathbf{U}, \mathbf{V}) = P(\mathbf{U}) - P(\mathbf{V}), \qquad G(\mathbf{U}, \mathbf{V}) = Q(\mathbf{U}) - Q(\mathbf{V}),$$

where we have introduced the notations

$$P(\mathbf{U}) = F(\mathbf{U}, \mathbf{A}), \qquad Q(\mathbf{U}) = G(\mathbf{U}, \mathbf{A}).$$

b) Let $1 < i < n - 1$. For $\mathbf{Z}_4 = \mathbf{Z}_5 = \cdots = \mathbf{Z}_{2n} = \mathbf{Z}_1 = \mathbf{A}$, $\mathbf{Z}_2 = \mathbf{B}$, $\mathbf{Z}_3 = \mathbf{U}$, the system (37.8) becomes Eq. (37.11) and again we have Eq. (37.12).

By putting $\mathbf{Z}_5 = \mathbf{Z}_6 = \cdots = \mathbf{Z}_{2n} = \mathbf{Z}_1 = \mathbf{A}$, $\mathbf{Z}_2 = \mathbf{U}$, $\mathbf{Z}_3 = \mathbf{V}$, $\mathbf{Z}_4 = \mathbf{B}$, on the basis of the equalities (37.9) and (37.12) the system (37.8) takes on the form Eq. (37.13). In view of Eq. (37.10) this again leads to Eqs. (37.4), (37.5), where $P(\mathbf{U}) = F(\mathbf{U}, \mathbf{A})$ and $Q(\mathbf{U}) = G(\mathbf{U}, \mathbf{A})$.

c) Let $i = n - 1$. If we substitute all variables $\mathbf{Z}_1$, $\mathbf{Z}_3$, $\cdots$, $\mathbf{Z}_{2n-1}$ (except for $\mathbf{Z}_2$ and $\mathbf{Z}_{2n}$) by $\mathbf{A}$, and if we put $\mathbf{Z}_2 = \mathbf{U}$, $\mathbf{Z}_{2n} = \mathbf{B}$, then the system (37.8) becomes Eq. (37.11), which implies Eq. (37.12).

If we substitute in the system (37.8)

$$\mathbf{Z}_3 = \mathbf{Z}_4 = \cdots = \mathbf{Z}_{2n-2} = \mathbf{Z}_1 = \mathbf{A}, \quad \mathbf{Z}_2 = \mathbf{V}, \quad \mathbf{Z}_{2n-1} = \mathbf{B}, \quad \mathbf{Z}_{2n} = \mathbf{U},$$

on the basis of equality (37.12) we obtain Eq. (37.13), and we find Eqs. (37.4), (37.5).

Consequently, if $a < 0$, the theorem is proved.

2°. Let $a = 0$. In this case the system (37.1) becomes

$$\begin{aligned}
& f(\mathbf{Z}_1, \mathbf{Z}_2, \mathbf{Z}_3, \cdots, \mathbf{Z}_{2n-1}, \mathbf{Z}_{2n}) && (37.14) \\
+ \ & f(\mathbf{Z}_1, \mathbf{Z}_3, \mathbf{Z}_4, \cdots, \mathbf{Z}_{2n}, \mathbf{Z}_2) + \cdots \\
+ \ & f(\mathbf{Z}_1, \mathbf{Z}_{2n}, \mathbf{Z}_2, \cdots, \mathbf{Z}_{2n-1}) = 0, \\
& h(\mathbf{Z}_1, \mathbf{Z}_2, \cdots, \mathbf{Z}_{2n}) + h(\mathbf{Z}_1, \mathbf{Z}_3, \cdots, \mathbf{Z}_{2n}, \mathbf{Z}_2) \\
& \qquad + \cdots + h(\mathbf{Z}_1, \mathbf{Z}_{2n}, \mathbf{Z}_2, \cdots, \mathbf{Z}_{2n-1}) \\
+ \ & k(\mathbf{Z}_1, \mathbf{Z}_2, \cdots, \mathbf{Z}_{2n}) + k(\mathbf{Z}_1, \mathbf{Z}_3, \cdots, \mathbf{Z}_{2n}, \mathbf{Z}_2) \\
& \qquad + \cdots + k(\mathbf{Z}_1, \mathbf{Z}_{2n}, \mathbf{Z}_2, \cdots, \mathbf{Z}_{2n-1}) = 0.
\end{aligned}$$

     Systems of Nonlinear Functional Equations

The first equation of the system (37.14) is Eq. (37.7), hence its general solution for $n > 2$ according to [D. S. Mitrinović and S. B. Prešić (1962A)] is Eq. (37.4):

$$F(\mathbf{U}, \mathbf{V}) = P(\mathbf{U}) - P(\mathbf{V}),$$

where $P(\mathbf{U}) = F(\mathbf{U}, \mathbf{A})$.

In the second equation of the system (37.14), if we put $\mathbf{Z}_1 = \mathbf{B}$ and substitute all the other variables $\mathbf{Z}_i$ ($2 \leq i \leq 2n$) by $\mathbf{A}$ so that $F(\mathbf{A}, \mathbf{B}) \neq 0$, we obtain

$$F(\mathbf{A}, \mathbf{B})G(\mathbf{A}, \mathbf{A}) = 0,$$

whence

$$G(\mathbf{A}, \mathbf{A}) = 0. \tag{37.15}$$

Now, we will distinguish three cases.

a) If $i = 1$, by putting $\mathbf{Z}_5 = \mathbf{Z}_6 = \cdots = \mathbf{Z}_{2n} = \mathbf{Z}_1 = \mathbf{A}$, $\mathbf{Z}_2 = \mathbf{V}$, $\mathbf{Z}_3 = \mathbf{B}$, $\mathbf{Z}_4 = \mathbf{U}$, the second equation of the system (37.14) in view of the equalities (37.4) and (37.15) becomes

$$
\begin{aligned}
F(\mathbf{A}, \mathbf{B})&G(\mathbf{U}, \mathbf{V}) + F(\mathbf{A}, \mathbf{B})G(\mathbf{A}, \mathbf{U}) \\
+\ &F(\mathbf{B}, \mathbf{A})G(\mathbf{A}, \mathbf{V}) + G(\mathbf{A}, \mathbf{B})F(\mathbf{U}, \mathbf{V}) \\
+\ &G(\mathbf{A}, \mathbf{B})F(\mathbf{A}, \mathbf{U}) + G(\mathbf{B}, \mathbf{A})F(\mathbf{A}, \mathbf{V}) \\
+\ &F(\mathbf{A}, \mathbf{V})G(\mathbf{A}, \mathbf{U}) + F(\mathbf{A}, \mathbf{V})G(\mathbf{U}, \mathbf{A}) = 0.
\end{aligned}
\tag{37.16}
$$

If we put $\mathbf{V} = \mathbf{A}$ into (37.16), on the basis of the expression Eq. (37.15) we obtain $G(\mathbf{A}, \mathbf{U}) = -G(\mathbf{U}, \mathbf{A})$, and then Eq. (37.16) becomes

$$G(\mathbf{U}, \mathbf{V}) = Q(\mathbf{U}) - Q(\mathbf{V}),$$

where $Q(\mathbf{U}) = G(\mathbf{U}, \mathbf{A})$.

b) In the case $1 < i < n - 1$, if we substitute all variables $\mathbf{Z}_1, \mathbf{Z}_4, \mathbf{Z}_5, \cdots$, $\mathbf{Z}_{2n-1}$, $\mathbf{Z}_{2n}$ (except for $\mathbf{Z}_2$ and $\mathbf{Z}_3$) by $\mathbf{A}$, and if we put $\mathbf{Z}_2 = \mathbf{B}$, $\mathbf{Z}_3 = \mathbf{U}$, the second equation of the system (37.14) becomes

$$F(\mathbf{A}, \mathbf{B})[G(\mathbf{A}, \mathbf{U}) + G(\mathbf{U}, \mathbf{A})] = 0,$$

whence

$$G(\mathbf{A}, \mathbf{U}) = -G(\mathbf{U}, \mathbf{A}). \tag{37.17}$$

For $\mathbf{Z}_5 = \mathbf{Z}_6 = \cdots = \mathbf{Z}_{2n} = \mathbf{Z}_1 = \mathbf{A}$, $\mathbf{Z}_2 = \mathbf{U}$, $\mathbf{Z}_3 = \mathbf{V}$, $\mathbf{Z}_4 = \mathbf{B}$, on the basis of the equality (37.17) and Eq. (37.4), the second equation of the system (37.14) takes on the following form

$$G(\mathbf{U}, \mathbf{V}) = Q(\mathbf{U}) - Q(\mathbf{V}),$$

where $Q(\mathbf{U}) = G(\mathbf{U}, \mathbf{A})$.

c) If $i = n - 1$, for $\mathbf{Z}_3 = \mathbf{Z}_4 = \cdots = \mathbf{Z}_{2n-1} = \mathbf{Z}_1 = \mathbf{A}$, $\mathbf{Z}_2 = \mathbf{U}$, $\mathbf{Z}_{2n} = \mathbf{B}$, by virtue of the equality (37.4), from the second equation of the system (37.14) we obtain Eq. (37.17).

By putting $\mathbf{Z}_3 = \mathbf{Z}_4 = \cdots = \mathbf{Z}_{2n-2} = \mathbf{Z}_1 = \mathbf{A}$, $\mathbf{Z}_2 = \mathbf{V}$, $\mathbf{Z}_{2n-1} = \mathbf{B}$, $\mathbf{Z}_{2n} = \mathbf{U}$, on the basis of the equality (37.4) and Eq. (37.17) from the second equation of the system (37.14) we find

$$G(\mathbf{U}, \mathbf{V}) = Q(\mathbf{U}) - Q(\mathbf{V}),$$

where $Q(\mathbf{U}) = G(\mathbf{U}, \mathbf{A})$, *i.e.*, Theorem 37.1 is proved if $a = 0$.

3°. If $a > 0$, then the system (37.1) becomes

$$
\begin{aligned}
& f(\mathbf{Z}_1, \mathbf{Z}_2, \mathbf{Z}_3, \cdots, \mathbf{Z}_{2n}) + f(\mathbf{Z}_1, \mathbf{Z}_3, \mathbf{Z}_4, \cdots, \mathbf{Z}_{2n}, \mathbf{Z}_2) + \cdots \\
+\ & f(\mathbf{Z}_1, \mathbf{Z}_{2n}, \mathbf{Z}_2, \cdots, \mathbf{Z}_{2n-1}) + g(\mathbf{Z}_1, \mathbf{Z}_2, \mathbf{Z}_3, \cdots, \mathbf{Z}_{2n}) \\
+\ & g(\mathbf{Z}_1, \mathbf{Z}_3, \mathbf{Z}_4, \cdots, \mathbf{Z}_{2n}, \mathbf{Z}_2) + \cdots + g(\mathbf{Z}_1, \mathbf{Z}_{2n}, \mathbf{Z}_2, \cdots, \mathbf{Z}_{2n-1}) = 0, \\
& h(\mathbf{Z}_1, \mathbf{Z}_2, \mathbf{Z}_3, \cdots, \mathbf{Z}_{2n}) + h(\mathbf{Z}_1, \mathbf{Z}_3, \mathbf{Z}_4, \cdots, \mathbf{Z}_{2n}, \mathbf{Z}_2) + \cdots \\
+\ & h(\mathbf{Z}_1, \mathbf{Z}_{2n}, \mathbf{Z}_2, \cdots, \mathbf{Z}_{2n-1}) + k(\mathbf{Z}_1, \mathbf{Z}_2, \mathbf{Z}_3, \cdots, \mathbf{Z}_{2n}) \\
+\ & k(\mathbf{Z}_1, \mathbf{Z}_3, \mathbf{Z}_4, \cdots, \mathbf{Z}_{2n}, \mathbf{Z}_2) + \cdots + k(\mathbf{Z}_1, \mathbf{Z}_{2n}, \mathbf{Z}_2, \cdots, \mathbf{Z}_{2n-1}) = 0.
\end{aligned}
\tag{37.18}
$$

If we introduce new functions S and T by the formulae

$$
\begin{aligned}
S(\mathbf{U}, \mathbf{V}) &= F(\mathbf{U}, \mathbf{V}) + G(\mathbf{U}, \mathbf{V}), \\
T(\mathbf{U}, \mathbf{V}) &= F(\mathbf{U}, \mathbf{V}) - G(\mathbf{U}, \mathbf{V}),
\end{aligned}
\tag{37.19}
$$

the system (37.18) becomes

$$s(\mathbf{Z}_1, \mathbf{Z}_2, \mathbf{Z}_3, \cdots, \mathbf{Z}_{2n}) + s(\mathbf{Z}_1, \mathbf{Z}_3, \mathbf{Z}_4, \cdots, \mathbf{Z}_{2n}, \mathbf{Z}_2) + \cdots$$

$$+ \quad s(\mathbf{Z}_1, \mathbf{Z}_{2n}, \mathbf{Z}_2, \cdots, \mathbf{Z}_{2n-1}) + t(\mathbf{Z}_1, \mathbf{Z}_2, \mathbf{Z}_3, \cdots, \mathbf{Z}_{2n}) \qquad (37.20)$$

$$+ \quad t(\mathbf{Z}_1, \mathbf{Z}_3, \mathbf{Z}_4, \cdots, \mathbf{Z}_{2n}, \mathbf{Z}_2) + \cdots + t(\mathbf{Z}_1, \mathbf{Z}_{2n}, \mathbf{Z}_2, \cdots, \mathbf{Z}_{2n-1}) = 0,$$

$$s(\mathbf{Z}_1, \mathbf{Z}_2, \mathbf{Z}_3, \cdots, \mathbf{Z}_{2n}) + s(\mathbf{Z}_1, \mathbf{Z}_3, \mathbf{Z}_4, \cdots, \mathbf{Z}_{2n}, \mathbf{Z}_2) + \cdots$$

$$+ \quad s(\mathbf{Z}_1, \mathbf{Z}_{2n}, \mathbf{Z}_2, \cdots, \mathbf{Z}_{2n-1}) - t(\mathbf{Z}_1, \mathbf{Z}_2, \mathbf{Z}_3, \cdots, \mathbf{Z}_{2n})$$

$$- \quad t(\mathbf{Z}_1, \mathbf{Z}_3, \mathbf{Z}_4, \cdots, \mathbf{Z}_{2n}, \mathbf{Z}_2) - \cdots - t(\mathbf{Z}_1, \mathbf{Z}_{2n}, \mathbf{Z}_2, \cdots, \mathbf{Z}_{2n-1}) = 0.$$

where

$$s(\mathbf{Z}_1, \mathbf{Z}_2, \cdots, \mathbf{Z}_{2n-1}, \mathbf{Z}_{2n})$$

$$= \quad [S(\mathbf{Z}_1, \mathbf{Z}_2) + S(\mathbf{Z}_3, \mathbf{Z}_4) + \cdots + S(\mathbf{Z}_{2i-1}, \mathbf{Z}_{2i})] \times$$

$$\times \quad [S(\mathbf{Z}_{2i+1}, \mathbf{Z}_{2n}) + S(\mathbf{Z}_{2i+2}, \mathbf{Z}_{2n-1}) + \cdots + S(\mathbf{Z}_{i+n}, \mathbf{Z}_{i+n+1})],$$

$$t(\mathbf{Z}_1, \mathbf{Z}_2, \cdots, \mathbf{Z}_{2n-1}, \mathbf{Z}_{2n})$$

$$= \quad [T(\mathbf{Z}_1, \mathbf{Z}_2) + T(\mathbf{Z}_3, \mathbf{Z}_4) + \cdots + T(\mathbf{Z}_{2i-1}, \mathbf{Z}_{2i})] \times$$

$$\times \quad [T(\mathbf{Z}_{2i+1}, \mathbf{Z}_{2n}) + T(\mathbf{Z}_{2i+2}, \mathbf{Z}_{2n-1}) + \cdots + T(\mathbf{Z}_{i+n}, \mathbf{Z}_{i+n+1})].$$

Comparing the equations of the system (37.20), we find

$$s(\mathbf{Z}_1, \mathbf{Z}_2, \mathbf{Z}_3, \cdots, \mathbf{Z}_{2n-1}, \mathbf{Z}_{2n}) \qquad (37.21)$$

$$+ \quad s(\mathbf{Z}_1, \mathbf{Z}_3, \mathbf{Z}_4, \cdots, \mathbf{Z}_{2n}, \mathbf{Z}_2) + \cdots$$

$$+ \quad s(\mathbf{Z}_1, \mathbf{Z}_{2n}, \mathbf{Z}_2, \cdots, \mathbf{Z}_{2n-2}, \mathbf{Z}_{2n-1}) = 0,$$

$$t(\mathbf{Z}_1, \mathbf{Z}_2, \mathbf{Z}_3, \cdots, \mathbf{Z}_{2n-1}, \mathbf{Z}_{2n})$$

$$+ \quad t(\mathbf{Z}_1, \mathbf{Z}_3, \mathbf{Z}_4, \cdots, \mathbf{Z}_{2n}, \mathbf{Z}_2) + \cdots$$

$$+ \quad t(\mathbf{Z}_1, \mathbf{Z}_{2n}, \mathbf{Z}_2, \cdots, \mathbf{Z}_{2n-2}, \mathbf{Z}_{2n-1}) = 0.$$

The general solution of the system (37.21) is

$$S(\mathbf{U}, \mathbf{V}) = M(\mathbf{U}) - M(\mathbf{V}), \qquad T(\mathbf{U}, \mathbf{V}) = H(\mathbf{U}) - H(\mathbf{V}) \qquad (37.22)$$

where M and H are arbitrary complex functions.

On the basis of the equalities (37.19) and (37.22), we obtain

$$F(\mathbf{U}, \mathbf{V}) = P(\mathbf{U}) - P(\mathbf{V}), \qquad G(\mathbf{U}, \mathbf{V}) = Q(\mathbf{U}) - Q(\mathbf{V}),$$

where

$$P(\mathbf{U}) = \frac{1}{2}[M(\mathbf{U}) + H(\mathbf{U})], \qquad Q(\mathbf{U}) = \frac{1}{2}[M(\mathbf{U}) - H(\mathbf{U})].$$

$\square$

Next we consider the system of functional equations

$$
\begin{aligned}
& f(\mathbf{Z}_1,\mathbf{Z}_2,\mathbf{Z}_3)f(\mathbf{Z}_4,\mathbf{Z}_5,\mathbf{Z}_6) + f(\mathbf{Z}_2,\mathbf{Z}_1,\mathbf{Z}_4)f(\mathbf{Z}_3,\mathbf{Z}_5,\mathbf{Z}_6) \qquad (37.23)\\
+\ & f(\mathbf{Z}_1,\mathbf{Z}_2,\mathbf{Z}_5)f(\mathbf{Z}_3,\mathbf{Z}_4,\mathbf{Z}_6) + f(\mathbf{Z}_2,\mathbf{Z}_1,\mathbf{Z}_6)f(\mathbf{Z}_3,\mathbf{Z}_4,\mathbf{Z}_5) \\
+\ & a[g(\mathbf{Z}_1,\mathbf{Z}_2,\mathbf{Z}_3)g(\mathbf{Z}_4,\mathbf{Z}_5,\mathbf{Z}_6) + g(\mathbf{Z}_2,\mathbf{Z}_1,\mathbf{Z}_4)g(\mathbf{Z}_3,\mathbf{Z}_5,\mathbf{Z}_6) \\
+\ & g(\mathbf{Z}_1,\mathbf{Z}_2,\mathbf{Z}_5)g(\mathbf{Z}_3,\mathbf{Z}_4,\mathbf{Z}_6) + g(\mathbf{Z}_2,\mathbf{Z}_1,\mathbf{Z}_6)g(\mathbf{Z}_3,\mathbf{Z}_4,\mathbf{Z}_5)] = 0, \\[1em]
& f(\mathbf{Z}_1,\mathbf{Z}_2,\mathbf{Z}_3)g(\mathbf{Z}_4,\mathbf{Z}_5,\mathbf{Z}_6) + f(\mathbf{Z}_2,\mathbf{Z}_1,\mathbf{Z}_4)g(\mathbf{Z}_3,\mathbf{Z}_5,\mathbf{Z}_6) \\
+\ & f(\mathbf{Z}_1,\mathbf{Z}_2,\mathbf{Z}_5)g(\mathbf{Z}_3,\mathbf{Z}_4,\mathbf{Z}_6) + f(\mathbf{Z}_2,\mathbf{Z}_1,\mathbf{Z}_6)g(\mathbf{Z}_3,\mathbf{Z}_4,\mathbf{Z}_5) \\
+\ & g(\mathbf{Z}_1,\mathbf{Z}_2,\mathbf{Z}_3)f(\mathbf{Z}_4,\mathbf{Z}_5,\mathbf{Z}_6) + g(\mathbf{Z}_2,\mathbf{Z}_1,\mathbf{Z}_4)f(\mathbf{Z}_3,\mathbf{Z}_5,\mathbf{Z}_6) \\
+\ & g(\mathbf{Z}_1,\mathbf{Z}_2,\mathbf{Z}_5)f(\mathbf{Z}_3,\mathbf{Z}_4,\mathbf{Z}_6) + g(\mathbf{Z}_2,\mathbf{Z}_1,\mathbf{Z}_6)f(\mathbf{Z}_3,\mathbf{Z}_4,\mathbf{Z}_5) \\
+\ & b[g(\mathbf{Z}_1,\mathbf{Z}_2,\mathbf{Z}_3)g(\mathbf{Z}_4,\mathbf{Z}_5,\mathbf{Z}_6) + g(\mathbf{Z}_2,\mathbf{Z}_1,\mathbf{Z}_4)g(\mathbf{Z}_3,\mathbf{Z}_5,\mathbf{Z}_6) \\
+\ & g(\mathbf{Z}_1,\mathbf{Z}_2,\mathbf{Z}_5)g(\mathbf{Z}_3,\mathbf{Z}_4,\mathbf{Z}_6) + g(\mathbf{Z}_2,\mathbf{Z}_1,\mathbf{Z}_6)g(\mathbf{Z}_3,\mathbf{Z}_4,\mathbf{Z}_5)] = 0,
\end{aligned}
$$

where $f, g : \mathcal{V}^3 \mapsto \mathbf{C}$ and a and b are real constants.

Definition 37.2. Let $b^2/4 + a < 0$. A solution $(f(\mathbf{U},\mathbf{V},\mathbf{W}), g(\mathbf{U},\mathbf{V},\mathbf{W}))$ of system (37.23) is called *isotropic* if

$$
\begin{aligned}
& (f(\mathbf{U},\mathbf{V},\mathbf{W}), g(\mathbf{U},\mathbf{V},\mathbf{W})) \not\equiv (0,0), && (37.24) \\
& f^2(\mathbf{U},\mathbf{V},\mathbf{W}) + bf(\mathbf{U},\mathbf{V},\mathbf{W})g(\mathbf{U},\mathbf{V},\mathbf{W}) - ag^2(\mathbf{U},\mathbf{V},\mathbf{W}) \equiv 0.
\end{aligned}
$$

The *general* solution of system (37.1) includes all solutions of this system with the possible exception of the isotropic ones.

Theorem 37.2 *The general solution of the system* (37.23) *is given by the following formulae*

1° *if* $b^2/4 + a = -p^2$ $(p > 0)$:

$$
\begin{aligned}
& 2pf(\mathbf{U},\mathbf{V},\mathbf{W}) && (37.25) \\
=\ & \Delta[H_1(\mathbf{U}), H_2(\mathbf{V}), (2p+b)H_3(\mathbf{W})] \\
-\ & \Delta[K_1(\mathbf{U}), K_2(\mathbf{V}), 2pH_3(\mathbf{W}) + bK_3(\mathbf{W})] \\
+\ & \Delta[H_1(\mathbf{U}), K_2(\mathbf{V}), 2pK_3(\mathbf{W}) - bH_3(\mathbf{W})] \\
+\ & \Delta[K_1(\mathbf{U}), H_2(\mathbf{V}), 2pK_3(\mathbf{W}) - bH_3(\mathbf{W})],
\end{aligned}
$$

$$pg(\mathbf{U}, \mathbf{V}, \mathbf{W}) \tag{37.26}$$
$$= \ \Delta[K_1(\mathbf{U}), K_2(\mathbf{V}), K_3(\mathbf{W})]$$
$$- \ \Delta[H_1(\mathbf{U}), H_2(\mathbf{V}), H_3(\mathbf{W})]$$
$$+ \ \Delta[K_1(\mathbf{U}), H_2(\mathbf{V}), H_3(\mathbf{W})]$$
$$+ \ \Delta[H_1(\mathbf{U}), K_2(\mathbf{V}), H_3(\mathbf{W})]$$

(for the continuous isotropic solutions we have

$$2pf(\mathbf{U}, \mathbf{V}, \mathbf{W}) \ = \ (2p \mp bi)\Delta[H_1(\mathbf{U}), H_2(\mathbf{V}), H_3(\mathbf{W})], \tag{37.27}$$
$$pg(\mathbf{U}, \mathbf{V}, \mathbf{W}) \ = \ \pm i\Delta[H_1(\mathbf{U}), H_2(\mathbf{V}), H_3(\mathbf{W})]);$$

2° *if* $b^2/4 + a = 0$:

$$g(\mathbf{U}, \mathbf{V}, \mathbf{W}) \tag{37.28}$$
$$= \ \Delta[H_1(\mathbf{U}), H_2(\mathbf{V}), H_3(\mathbf{W})]$$
$$+ \ \Delta[H_1(\mathbf{U}), K_2(\mathbf{V}), H_3(\mathbf{W})]$$
$$+ \ \Delta[K_1(\mathbf{U}), H_2(\mathbf{V}), H_3(\mathbf{W})],$$

$$f(\mathbf{U}, \mathbf{V}, \mathbf{W}) \tag{37.29}$$
$$= \ \Delta\left[H_1(\mathbf{U}), H_2(\mathbf{V}), H_3(\mathbf{W}) - \frac{b}{2}K_3(\mathbf{W})\right]$$
$$- \ \frac{b}{2}\Delta[H_1(\mathbf{U}), K_2(\mathbf{V}), H_3(\mathbf{W})]$$
$$- \ \frac{b}{2}\Delta[K_1(\mathbf{U}), H_2(\mathbf{V}), H_3(\mathbf{W})];$$

3° *if* $b^2/4 + a = p^2$ $(p > 0)$:

$$4pf(\mathbf{U}, \mathbf{V}, \mathbf{W}) \tag{37.30}$$
$$= \ (2p - b)\Delta[H_1(\mathbf{U}), H_2(\mathbf{V}), H_3(\mathbf{W})]$$
$$+ \ (2p + b)\Delta[K_1(\mathbf{U}), K_2(\mathbf{V}), K_3(\mathbf{W})],$$
$$2pg(\mathbf{U}, \mathbf{V}, \mathbf{W}) \tag{37.31}$$
$$= \ \Delta[H_1(\mathbf{U}), H_2(\mathbf{V}), H_3(\mathbf{W})]$$
$$- \ \Delta[K_1(\mathbf{U}), K_2(\mathbf{V}), K_3(\mathbf{W})],$$

where H_i, K_i $(i, j = 1, 2, 3)$ are arbitrary complex functions defined in $\mathcal{V}$, and

$$\Delta[H_1(\mathbf{U}), H_2(\mathbf{V}), H_3(\mathbf{W})] = \begin{vmatrix} H_1(\mathbf{U}) & H_2(\mathbf{U}) & H_3(\mathbf{U}) \\ H_1(\mathbf{V}) & H_2(\mathbf{V}) & H_3(\mathbf{V}) \\ H_1(\mathbf{W}) & H_2(\mathbf{W}) & H_3(\mathbf{W}) \end{vmatrix}.$$

Proof. If we introduce the function h by

$$h(\mathbf{U}, \mathbf{V}, \mathbf{W}) = f(\mathbf{U}, \mathbf{V}, \mathbf{W}) + \frac{b}{2} g(\mathbf{U}, \mathbf{V}, \mathbf{W}), \tag{37.32}$$

the system of functional equations (37.23) becomes

$$\begin{aligned}
& h(\mathbf{Z}_1, \mathbf{Z}_2, \mathbf{Z}_3) h(\mathbf{Z}_4, \mathbf{Z}_5, \mathbf{Z}_6) + h(\mathbf{Z}_2, \mathbf{Z}_1, \mathbf{Z}_4) h(\mathbf{Z}_3, \mathbf{Z}_5, \mathbf{Z}_6) \qquad (37.33) \\
+ \; & h(\mathbf{Z}_1, \mathbf{Z}_2, \mathbf{Z}_5) h(\mathbf{Z}_3, \mathbf{Z}_4, \mathbf{Z}_6) + h(\mathbf{Z}_2, \mathbf{Z}_1, \mathbf{Z}_6) h(\mathbf{Z}_3, \mathbf{Z}_4, \mathbf{Z}_5) \\
+ \; & (b^2/4 + a)[g(\mathbf{Z}_1, \mathbf{Z}_2, \mathbf{Z}_3) g(\mathbf{Z}_4, \mathbf{Z}_5, \mathbf{Z}_6) + g(\mathbf{Z}_2, \mathbf{Z}_1, \mathbf{Z}_4) g(\mathbf{Z}_3, \mathbf{Z}_5, \mathbf{Z}_6) \\
+ \; & g(\mathbf{Z}_1, \mathbf{Z}_2, \mathbf{Z}_5) g(\mathbf{Z}_3, \mathbf{Z}_4, \mathbf{Z}_6) + g(\mathbf{Z}_2, \mathbf{Z}_1, \mathbf{Z}_6) g(\mathbf{Z}_3, \mathbf{Z}_4, \mathbf{Z}_5)] = 0, \\[6pt]
& h(\mathbf{Z}_1, \mathbf{Z}_2, \mathbf{Z}_3) g(\mathbf{Z}_4, \mathbf{Z}_5, \mathbf{Z}_6) + h(\mathbf{Z}_2, \mathbf{Z}_1, \mathbf{Z}_4) g(\mathbf{Z}_3, \mathbf{Z}_5, \mathbf{Z}_6) \\
+ \; & h(\mathbf{Z}_1, \mathbf{Z}_2, \mathbf{Z}_5) g(\mathbf{Z}_3, \mathbf{Z}_4, \mathbf{Z}_6) + h(\mathbf{Z}_2, \mathbf{Z}_1, \mathbf{Z}_6) g(\mathbf{Z}_3, \mathbf{Z}_4, \mathbf{Z}_5) \\
+ \; & g(\mathbf{Z}_1, \mathbf{Z}_2, \mathbf{Z}_3) h(\mathbf{Z}_4, \mathbf{Z}_5, \mathbf{Z}_6) + g(\mathbf{Z}_2, \mathbf{Z}_1, \mathbf{Z}_4) h(\mathbf{Z}_3, \mathbf{Z}_5, \mathbf{Z}_6) \\
+ \; & g(\mathbf{Z}_1, \mathbf{Z}_2, \mathbf{Z}_5) h(\mathbf{Z}_3, \mathbf{Z}_4, \mathbf{Z}_6) + g(\mathbf{Z}_2, \mathbf{Z}_1, \mathbf{Z}_6) h(\mathbf{Z}_3, \mathbf{Z}_4, \mathbf{Z}_5) = 0.
\end{aligned}$$

According to [J. Aczél (1958)], we will distinguish three cases:

1°. Let $b^2/4 + a = -p^2$ $(p > 0)$. In this case the system 37.33) will be

$$\begin{aligned}
& h(\mathbf{Z}_1, \mathbf{Z}_2, \mathbf{Z}_3) h(\mathbf{Z}_4, \mathbf{Z}_5, \mathbf{Z}_6) + h(\mathbf{Z}_2, \mathbf{Z}_1, \mathbf{Z}_4) h(\mathbf{Z}_3, \mathbf{Z}_5, \mathbf{Z}_6) \quad (37.34) \\
+ \; & h(\mathbf{Z}_1, \mathbf{Z}_2, \mathbf{Z}_5) h(\mathbf{Z}_3, \mathbf{Z}_4, \mathbf{Z}_6) + h(\mathbf{Z}_2, \mathbf{Z}_1, \mathbf{Z}_6) h(\mathbf{Z}_3, \mathbf{Z}_4, \mathbf{Z}_5) \\
- \; & [k(\mathbf{Z}_1, \mathbf{Z}_2, \mathbf{Z}_3) k(\mathbf{Z}_4, \mathbf{Z}_5, \mathbf{Z}_6) + k(\mathbf{Z}_2, \mathbf{Z}_1, \mathbf{Z}_4) k(\mathbf{Z}_3, \mathbf{Z}_5, \mathbf{Z}_6) \\
+ \; & k(\mathbf{Z}_1, \mathbf{Z}_2, \mathbf{Z}_5) k(\mathbf{Z}_3, \mathbf{Z}_4, \mathbf{Z}_6) + k(\mathbf{Z}_2, \mathbf{Z}_1, \mathbf{Z}_6) k(\mathbf{Z}_3, \mathbf{Z}_4, \mathbf{Z}_5)] = 0, \\[6pt]
& h(\mathbf{Z}_1, \mathbf{Z}_2, \mathbf{Z}_3) k(\mathbf{Z}_4, \mathbf{Z}_5, \mathbf{Z}_6) + h(\mathbf{Z}_2, \mathbf{Z}_1, \mathbf{Z}_4) k(\mathbf{Z}_3, \mathbf{Z}_5, \mathbf{Z}_6) \\
+ \; & h(\mathbf{Z}_1, \mathbf{Z}_2, \mathbf{Z}_5) k(\mathbf{Z}_3, \mathbf{Z}_4, \mathbf{Z}_6) + h(\mathbf{Z}_2, \mathbf{Z}_1, \mathbf{Z}_6) k(\mathbf{Z}_3, \mathbf{Z}_4, \mathbf{Z}_5) \\
+ \; & k(\mathbf{Z}_1, \mathbf{Z}_2, \mathbf{Z}_3) h(\mathbf{Z}_4, \mathbf{Z}_5, \mathbf{Z}_6) + k(\mathbf{Z}_2, \mathbf{Z}_1, \mathbf{Z}_4) h(\mathbf{Z}_3, \mathbf{Z}_5, \mathbf{Z}_6) \\
+ \; & k(\mathbf{Z}_1, \mathbf{Z}_2, \mathbf{Z}_5) h(\mathbf{Z}_3, \mathbf{Z}_4, \mathbf{Z}_6) + k(\mathbf{Z}_2, \mathbf{Z}_1, \mathbf{Z}_6) h(\mathbf{Z}_3, \mathbf{Z}_4, \mathbf{Z}_5) = 0,
\end{aligned}$$

where

$$k(\mathbf{U}, \mathbf{V}, \mathbf{W}) = p g(\mathbf{U}, \mathbf{V}, \mathbf{W}). \tag{37.35}$$

Suppose that $(f(\mathbf{U},\mathbf{V},\mathbf{W}), g(\mathbf{U},\mathbf{V},\mathbf{W}))$ is a continuous isotropic solution of system (37.23). The relation Eq. (37.24) by virtue of the changes Eqs. (37.32) and (37.35) becomes

$$h^2(\mathbf{U},\mathbf{V},\mathbf{W}) + k^2(\mathbf{U},\mathbf{V},\mathbf{W}) \equiv 0$$

(which justifies Definition 37.2), *i.e.*,

$$k(\mathbf{U},\mathbf{V},\mathbf{W}) = \pm ih(\mathbf{U},\mathbf{V},\mathbf{W}). \tag{37.36}$$

As in Theorem 37.1 we must take the same sign for all triples of vectors $(\mathbf{U},\mathbf{V},\mathbf{W}) \in \mathcal{V}^3$.

In view of Eq. (37.36) system (37.34) becomes

$$\begin{aligned}
h(\mathbf{Z}_1,\mathbf{Z}_2,\mathbf{Z}_3)h(\mathbf{Z}_4,\mathbf{Z}_5,\mathbf{Z}_6) &+ h(\mathbf{Z}_2,\mathbf{Z}_1,\mathbf{Z}_4)h(\mathbf{Z}_3,\mathbf{Z}_5,\mathbf{Z}_6) \\
+h(\mathbf{Z}_1,\mathbf{Z}_2,\mathbf{Z}_5)h(\mathbf{Z}_3,\mathbf{Z}_4,\mathbf{Z}_6) &+ h(\mathbf{Z}_2,\mathbf{Z}_1,\mathbf{Z}_6)h(\mathbf{Z}_3,\mathbf{Z}_4,\mathbf{Z}_5) = 0.
\end{aligned} \tag{37.37}$$

The general solution of the equation (37.37) according to [L. Carlitz (1963)] is given by

$$h(\mathbf{U},\mathbf{V},\mathbf{W}) = \Delta[H_1(\mathbf{U}), H_2(\mathbf{V}), H_3(\mathbf{W})] \tag{37.38}$$
$$(H_i : \mathcal{V} \mapsto \mathbf{C},\ i = 1,2,3),$$

where

$$H_1(\mathbf{U}) = \frac{h(A,B,\mathbf{U})}{h(A,B,C)}, \quad H_2(\mathbf{U}) = \frac{h(A,\mathbf{U},C)}{h(A,B,C)}, \quad H_3(\mathbf{U}) = h(B,\mathbf{U},C)$$

are arbitrary complex functions and A, B, C are vectors from $\mathcal{V}$ such that $h(A,B,C) \neq 0$.

From Eqs. (37.32), (37.35), (37.36) we find

$$2pf(\mathbf{U},\mathbf{V},\mathbf{W}) = (2p \mp bi)h(\mathbf{U},\mathbf{V},\mathbf{W}), \quad pg(\mathbf{U},\mathbf{V},\mathbf{W}) = \pm ih(\mathbf{U},\mathbf{V},\mathbf{W}),$$

and Eq. (37.38) leads to Eq. (37.27).

Next we are looking for solutions of Eq. (37.34) not satisfying Eq. (37.36) (anisotropic solutions).

By putting $\mathbf{Z}_1 = \mathbf{Z}_2 = \cdots = \mathbf{Z}_6 = \mathbf{U}$, the system (37.34) becomes

$$h^2(\mathbf{U},\mathbf{U},\mathbf{U}) - k^2(\mathbf{U},\mathbf{U},\mathbf{U}) = 0, \quad h(\mathbf{U},\mathbf{U},\mathbf{U})k(\mathbf{U},\mathbf{U},\mathbf{U}) = 0,$$

whence

$$h(\mathbf{U},\mathbf{U},\mathbf{U}) = 0, \qquad k(\mathbf{U},\mathbf{U},\mathbf{U}) = 0. \tag{37.39}$$

For $\mathbf{Z}_1 = \mathbf{Z}_2 = \mathbf{Z}_3 = \mathbf{Z}_4 = \mathbf{U}$, $\mathbf{Z}_5 = \mathbf{Z}_6 = \mathbf{V}$, the system (37.34) yields

$$h^2(\mathbf{U}, \mathbf{U}, \mathbf{V}) - k^2(\mathbf{U}, \mathbf{U}, \mathbf{V}) = 0, \quad h(\mathbf{U}, \mathbf{U}, \mathbf{V})k(\mathbf{U}, \mathbf{U}, \mathbf{V}) = 0,$$

so that we obtain

$$h(\mathbf{U}, \mathbf{U}, \mathbf{V}) = 0, \qquad k(\mathbf{U}, \mathbf{U}, \mathbf{V}) = 0. \tag{37.40}$$

If we put $\mathbf{Z}_1 = \mathbf{Z}_4 = \mathbf{Z}_5 = \mathbf{Z}_6 = \mathbf{U}$, $\mathbf{Z}_2 = \mathbf{Z}_3 = \mathbf{V}$, from Eq. (37.34) we obtain

$$\begin{aligned}
& 2h^2(\mathbf{V}, \mathbf{U}, \mathbf{U}) + h(\mathbf{U}, \mathbf{V}, \mathbf{U})h(\mathbf{V}, \mathbf{U}, \mathbf{U}) \tag{37.41}\\
- \ & 2k^2(\mathbf{V}, \mathbf{U}, \mathbf{U}) - k(\mathbf{U}, \mathbf{V}, \mathbf{U})k(\mathbf{V}, \mathbf{U}, \mathbf{U}) = 0,\\
& 4h(\mathbf{V}, \mathbf{U}, \mathbf{U})k(\mathbf{V}, \mathbf{U}, \mathbf{U}) + h(\mathbf{U}, \mathbf{V}, \mathbf{U})k(\mathbf{V}, \mathbf{U}, \mathbf{U})\\
+ \ & h(\mathbf{V}, \mathbf{U}, \mathbf{U})k(\mathbf{U}, \mathbf{V}, \mathbf{U}) = 0.
\end{aligned}$$

By putting $\mathbf{Z}_1 = \mathbf{Z}_3 = \mathbf{V}$, $\mathbf{Z}_2 = \mathbf{Z}_4 = \mathbf{Z}_5 = \mathbf{Z}_6 = \mathbf{U}$ into Eq. (37.34), we obtain

$$\begin{aligned}
& h^2(\mathbf{V}, \mathbf{U}, \mathbf{U}) + 2h(\mathbf{V}, \mathbf{U}, \mathbf{U})h(\mathbf{U}, \mathbf{V}, \mathbf{U}) \tag{37.42}\\
- \ & k^2(\mathbf{V}, \mathbf{U}, \mathbf{U}) - 2k(\mathbf{V}, \mathbf{U}, \mathbf{U})k(\mathbf{U}, \mathbf{V}, \mathbf{U}) = 0,\\
& h(\mathbf{U}, \mathbf{V}, \mathbf{U})k(\mathbf{V}, \mathbf{U}, \mathbf{U}) + h(\mathbf{V}, \mathbf{U}, \mathbf{U})k(\mathbf{V}, \mathbf{U}, \mathbf{U})\\
+ \ & h(\mathbf{V}, \mathbf{U}, \mathbf{U})k(\mathbf{U}, \mathbf{V}, \mathbf{U}) = 0.
\end{aligned}$$

Comparing Eqs. (37.41) and (37.42), we find

$$h^2(\mathbf{V}, \mathbf{U}, \mathbf{U}) - k^2(\mathbf{V}, \mathbf{U}, \mathbf{U}) = 0, \quad h(\mathbf{V}, \mathbf{U}, \mathbf{U})k(\mathbf{V}, \mathbf{U}, \mathbf{U}) = 0,$$

i.e., we obtain

$$h(\mathbf{V}, \mathbf{U}, \mathbf{U}) = 0, \qquad k(\mathbf{V}, \mathbf{U}, \mathbf{U}) = 0. \tag{37.43}$$

By the substitutions $\mathbf{Z}_1 = \mathbf{Z}_4 = \mathbf{V}$, $\mathbf{Z}_2 = \mathbf{Z}_3 = \mathbf{Z}_5 = \mathbf{Z}_6 = \mathbf{U}$, the system (37.34) becomes

$$h^2(\mathbf{U}, \mathbf{V}, \mathbf{U}) - k^2(\mathbf{U}, \mathbf{V}, \mathbf{U}) = 0, \qquad h(\mathbf{U}, \mathbf{V}, \mathbf{U})k(\mathbf{U}, \mathbf{V}, \mathbf{U}) = 0,$$

i.e.,

$$h(\mathbf{U}, \mathbf{V}, \mathbf{U}) = 0, \qquad k(\mathbf{U}, \mathbf{V}, \mathbf{U}) = 0. \tag{37.44}$$

For any anisotropic solution of the system (37.34) there exist at least three vectors A, B, $C \in \mathcal{V}$ such that $h^2(A, B, C) + k^2(A, B, C) \neq 0$.

If we put $\mathbf{Z}_1 = \mathbf{U}$, $\mathbf{Z}_2 = \mathbf{V}$, $\mathbf{Z}_3 = A$, $\mathbf{Z}_4 = B$, $\mathbf{Z}_5 = \mathbf{Z}_6 = C$ into Eq. (37.34), on the basis of the equalities (37.39), (37.40), (37.43) and (37.44) we obtain

$$h(\mathbf{U}, \mathbf{V}, C) = -h(\mathbf{V}, \mathbf{U}, C), \qquad k(\mathbf{U}, \mathbf{V}, C) = -k(\mathbf{V}, \mathbf{U}, C). \qquad (37.45)$$

Using Eq. (37.45), the system (37.34) for $\mathbf{Z}_1 = A$, $\mathbf{Z}_2 = B$, $\mathbf{Z}_3 = \mathbf{Z}_6 = C$, $\mathbf{Z}_4 = \mathbf{U}$, $\mathbf{Z}_5 = \mathbf{V}$ yields

$$h(\mathbf{U}, \mathbf{V}, C) = h(C, \mathbf{U}, \mathbf{V}), \qquad k(\mathbf{U}, \mathbf{V}, C) = k(C, \mathbf{U}, \mathbf{V}). \qquad (37.46)$$

If we put $\mathbf{Z}_1 = A$, $\mathbf{Z}_2 = B$, $\mathbf{Z}_3 = \mathbf{Z}_5 = C$, $\mathbf{Z}_4 = \mathbf{U}$, $\mathbf{Z}_6 = \mathbf{V}$, the system (37.34) becomes

$$h(C, \mathbf{U}, \mathbf{V}) = -h(\mathbf{U}, C, \mathbf{V}), \qquad k(C, \mathbf{U}, \mathbf{V}) = -k(\mathbf{U}, C, \mathbf{V}). \qquad (37.47)$$

On the basis of the equalities (37.44), (37.45) and (37.46), if we put $\mathbf{Z}_1 = A$, $\mathbf{Z}_2 = B$, $\mathbf{Z}_3 = C$, $\mathbf{Z}_4 = \mathbf{U}$, $\mathbf{Z}_5 = \mathbf{V}$, $\mathbf{Z}_6 = \mathbf{W}$, the system (37.34) yields

$$
\begin{aligned}
h(\mathbf{U}, \mathbf{V}, \mathbf{W}) & \qquad\qquad\qquad\qquad\qquad\qquad (37.48)\\
= \;& q[H_1(\mathbf{U})F(\mathbf{V}, \mathbf{W}) - H_1(\mathbf{V})F(\mathbf{U}, \mathbf{W}) + H_1(\mathbf{W})F(\mathbf{U}, \mathbf{V})\\
- \;& K_1(\mathbf{U})G(\mathbf{V}, \mathbf{W}) + K_1(\mathbf{V})G(\mathbf{U}, \mathbf{W}) - K_1(\mathbf{W})G(\mathbf{U}, \mathbf{V})]\\
+ \;& r[H_1(\mathbf{U})G(\mathbf{V}, \mathbf{W}) - H_1(\mathbf{V})G(\mathbf{U}, \mathbf{W}) + H_1(\mathbf{W})G(\mathbf{U}, \mathbf{V})\\
+ \;& K_1(\mathbf{U})F(\mathbf{V}, \mathbf{W}) - K_1(\mathbf{V})F(\mathbf{U}, \mathbf{W}) + K_1(\mathbf{W})F(\mathbf{U}, \mathbf{V})],\\[4pt]
k(\mathbf{U}, \mathbf{V}, \mathbf{W}) & \qquad\qquad\qquad\qquad\qquad\qquad (37.49)\\
= \;& q[H_1(\mathbf{U})G(\mathbf{V}, \mathbf{W}) - H_1(\mathbf{V})G(\mathbf{U}, \mathbf{W}) + H_1(\mathbf{W})G(\mathbf{U}, \mathbf{V})\\
+ \;& K_1(\mathbf{U})F(\mathbf{V}, \mathbf{W}) - K_1(\mathbf{V})F(\mathbf{U}, \mathbf{W}) + K_1(\mathbf{W})F(\mathbf{U}, \mathbf{V})]\\
+ \;& r[H_1(\mathbf{U})F(\mathbf{V}, \mathbf{W}) - H_1(\mathbf{V})F(\mathbf{U}, \mathbf{W}) + H_1(\mathbf{W})F(\mathbf{U}, \mathbf{V})\\
- \;& K_1(\mathbf{U})G(\mathbf{V}, \mathbf{W}) + K_1(\mathbf{V})G(\mathbf{U}, \mathbf{W}) - K_1(\mathbf{W})G(\mathbf{U}, \mathbf{V})],
\end{aligned}
$$

where we have introduced the notations

$$
\begin{aligned}
H_1(\mathbf{U}) \;&=\; h(A, B, \mathbf{U}), \qquad K_1(\mathbf{U}) = k(A, B, \mathbf{U}),\\
F(\mathbf{U}, \mathbf{V}) \;&=\; h(\mathbf{U}, \mathbf{V}, C), \qquad G(\mathbf{U}, \mathbf{V}) = k(\mathbf{U}, \mathbf{V}, C), \qquad (37.50)\\
q \;&=\; \frac{h(A, B, C)}{h^2(A, B, C) + k^2(A, B, C)}, \quad r = \frac{k(A, B, C)}{h^2(A, B, C) + k^2(A, B, C)}.
\end{aligned}
$$

For $\mathbf{Z}_1 = \mathbf{Z}_4 = C$, $\mathbf{Z}_2 = \mathbf{S}$, $\mathbf{Z}_3 = \mathbf{T}$, $\mathbf{Z}_5 = \mathbf{U}$, $\mathbf{Z}_6 = \mathbf{V}$ in view of the notations Eq. (37.50), the system (37.34) becomes

$$F(\mathbf{S},\mathbf{T})F(\mathbf{U},\mathbf{V}) + F(\mathbf{S},\mathbf{U})F(\mathbf{V},\mathbf{T}) + F(\mathbf{S},\mathbf{V})F(\mathbf{T},\mathbf{U}) \quad (37.51)$$
$$- \quad G(\mathbf{S},\mathbf{T})G(\mathbf{U},\mathbf{V}) - G(\mathbf{S},\mathbf{U})G(\mathbf{V},\mathbf{T}) - G(\mathbf{S},\mathbf{V})G(\mathbf{T},\mathbf{U}) = 0,$$
$$F(\mathbf{S},\mathbf{T})G(\mathbf{U},\mathbf{V}) + F(\mathbf{S},\mathbf{U})G(\mathbf{V},\mathbf{T}) + F(\mathbf{S},\mathbf{V})G(\mathbf{T},\mathbf{U})$$
$$+ \quad G(\mathbf{S},\mathbf{T})F(\mathbf{U},\mathbf{V}) + G(\mathbf{S},\mathbf{U})F(\mathbf{V},\mathbf{T}) + G(\mathbf{S},\mathbf{V})F(\mathbf{T},\mathbf{U}) = 0.$$

The general solution of the system (37.51) according to [P. M. Vasić (1963)] is given by the formulae

$$\begin{aligned}
F(\mathbf{U},\mathbf{V}) &= \Delta[H_1(\mathbf{U}), qH_3'(\mathbf{V}) + rK_3'(\mathbf{V})] \quad (37.52)\\
&+ \Delta[K_2(\mathbf{U}), rH_3'(\mathbf{V}) - qK_3'(\mathbf{V})],\\
G(\mathbf{U},\mathbf{V}) &= \Delta[K_2(\mathbf{U}), qH_3'(\mathbf{V}) + rK_3'(\mathbf{V})]\\
&+ \Delta[H_2(\mathbf{U}), qK_3'(\mathbf{V}) - rH_3'(\mathbf{V})],
\end{aligned}$$

On the basis of the equalities (37.32), (37.35), (37.48), (37.49) and (37.52) we obtain Eqs. (37.25) and (37.26), with the following notations

$$H_3(\mathbf{U}) = (q^2 - r^2)H_3'(\mathbf{U}) + 2qrK_3'(\mathbf{U}), \quad K_3(\mathbf{U}) = (r^2 - q^2)K_3'(\mathbf{U}) + 2qrH_3'(\mathbf{U}).$$

The functions f and g given by the formulae Eqs. (37.25) and (37.26) are really solutions of the system (37.23). Consequently, in the case $b^2/4 + a = -p^2$ $(p > 0)$ all functions f and g which are solutions of the system (37.23) and do not satisfy Eq. (37.24) are determined by the formulae Eqs. (37.25) and (37.26). The functions H_i, K_i $(i,j = 1,2,3)$ which appear in these solutions are arbitrary complex functions defined in $\mathcal{V}$.

2°. Let $b^2/4 + a = 0$. In this case the first equation of the system (37.33) takes on the form Eq. (37.37):

$$h(\mathbf{Z}_1, \mathbf{Z}_2, \mathbf{Z}_3)h(\mathbf{Z}_4, \mathbf{Z}_5, \mathbf{Z}_6) \quad + \quad h(\mathbf{Z}_2, \mathbf{Z}_1, \mathbf{Z}_4)h(\mathbf{Z}_3, \mathbf{Z}_5, \mathbf{Z}_6)$$
$$+ h(\mathbf{Z}_1, \mathbf{Z}_2, \mathbf{Z}_5)h(\mathbf{Z}_3, \mathbf{Z}_4, \mathbf{Z}_6) \quad + \quad h(\mathbf{Z}_2, \mathbf{Z}_1, \mathbf{Z}_6)h(\mathbf{Z}_3, \mathbf{Z}_4, \mathbf{Z}_5) = 0,$$

and its general solution according to [L. Carlitz (1963)] is given by the equation (37.38):

$$h(\mathbf{U}, \mathbf{V}, \mathbf{W}) = \Delta[H_1(\mathbf{U}), H_2(\mathbf{V}), H_3(\mathbf{W})] \quad (H_i : \mathcal{V} \mapsto \mathbf{C},\ i = 1,2,3),$$

where

$$H_1(\mathbf{U}) = \frac{h(A,B,\mathbf{U})}{h(A,B,C)}, \quad H_2(\mathbf{U}) = \frac{h(A,\mathbf{U},C)}{h(A,B,C)}, \quad H_3(\mathbf{U}) = h(B,\mathbf{U},C)$$

are arbitrary complex functions and A, B, C are vectors from $\mathcal{V}$ such that $h(A,B,C) \neq 0$.

If we put $\mathbf{Z}_1 = \mathbf{Z}_2 = \mathbf{Z}_4 = \mathbf{Z}_5 = C$, $\mathbf{Z}_3 = \mathbf{U}$, $\mathbf{Z}_6 = \mathbf{V}$, the second equation of the system (37.33) yields

$$h(\mathbf{U},C,\mathbf{V})g(C,C,C) = 0. \tag{37.53}$$

Using the equality $h(\mathbf{U},C,\mathbf{V}) = -h(\mathbf{U},\mathbf{V},C)$, which follows from the equation (37.38), and by putting $\mathbf{U} = A$, $\mathbf{V} = B$, from Eq. (37.53) we obtain $g(C,C,C) = 0$.

For $\mathbf{Z}_1 = A$, $\mathbf{Z}_2 = B$, $\mathbf{Z}_3 = \mathbf{Z}_4 = \mathbf{Z}_5 = C$, $\mathbf{Z}_6 = \mathbf{U}$ the second equation of the system (37.33) becomes $g(C,C,\mathbf{U}) = 0$. If we put $\mathbf{Z}_1 = A$, $\mathbf{Z}_2 = B$, $\mathbf{Z}_3 = \mathbf{Z}_5 = \mathbf{Z}_6 = C$, $\mathbf{Z}_4 = \mathbf{U}$, the same equation yields $g(\mathbf{U},C,C) = 0$.

Finally, the substitution $\mathbf{Z}_1 = \mathbf{Z}_4 = \mathbf{Z}_5 = C$, $\mathbf{Z}_2 = \mathbf{U}$, $\mathbf{Z}_3 = A$, $\mathbf{Z}_6 = B$ reduces the second equation of the system (37.33) to $g(C,\mathbf{U},C) = 0$.

By putting $\mathbf{Z}_1 = \mathbf{U}$, $\mathbf{Z}_2 = \mathbf{V}$, $\mathbf{Z}_3 = A$, $\mathbf{Z}_4 = B$, $\mathbf{Z}_5 = \mathbf{Z}_6 = C$, on the basis of the above result we obtain

$$g(\mathbf{U},\mathbf{V},C) = -g(\mathbf{V},\mathbf{U},C). \tag{37.54}$$

For $\mathbf{Z}_1 = A$, $\mathbf{Z}_2 = B$, $\mathbf{Z}_3 = \mathbf{Z}_6 = C$, $\mathbf{Z}_4 = \mathbf{U}$, $\mathbf{Z}_5 = \mathbf{V}$, the same equation yields

$$g(\mathbf{U},\mathbf{V},C) = g(C,\mathbf{U},\mathbf{V}). \tag{37.55}$$

Also, based on the previous results, for $\mathbf{Z}_1 = A$, $\mathbf{Z}_2 = B$, $\mathbf{Z}_3 = \mathbf{Z}_5 = C$, $\mathbf{Z}_4 = \mathbf{U}$, $\mathbf{Z}_6 = \mathbf{V}$, the second equation of the system (37.33) is reduced to

$$g(C,\mathbf{U},\mathbf{V}) = -g(\mathbf{U},C,\mathbf{V}). \tag{37.56}$$

By putting $\mathbf{Z}_1 = A$, $\mathbf{Z}_2 = B$, $\mathbf{Z}_3 = C$, $\mathbf{Z}_4 = \mathbf{U}$, $\mathbf{Z}_5 = \mathbf{V}$, $\mathbf{Z}_6 = \mathbf{W}$, on the basis of Eqs. (37.54), (37.55) and (37.56) the second equation of the system (37.33) becomes

$$h(A,B,C)[g(\mathbf{U},\mathbf{V},\mathbf{W}) - H_1(\mathbf{U})g(\mathbf{V},\mathbf{W},C) \tag{37.57}$$
$$+ \quad H_1(\mathbf{V})g(\mathbf{U},\mathbf{W},C) - H_1(\mathbf{W})g(\mathbf{U},\mathbf{V},C)] + g(A,B,C)h(\mathbf{U},\mathbf{V},\mathbf{W})$$
$$- \quad g(A,B,\mathbf{U})h(\mathbf{V},\mathbf{W},C) + g(A,B,\mathbf{V})h(\mathbf{U},\mathbf{W},C)$$
$$- \quad g(A,B,\mathbf{W})h(\mathbf{U},\mathbf{V},C) = 0.$$

For $\mathbf{Z}_1 = \mathbf{Z}_4 = C$, $\mathbf{Z}_2 = \mathbf{S}$, $\mathbf{Z}_3 = \mathbf{T}$, $\mathbf{Z}_5 = \mathbf{U}$, $\mathbf{Z}_6 = \mathbf{V}$ and with the notations Eq. (37.50) the system (37.33) is reduced to

$$F(\mathbf{S},\mathbf{T})F(\mathbf{U},\mathbf{V}) + F(\mathbf{S},\mathbf{U})F(\mathbf{V},\mathbf{T}) + F(\mathbf{S},\mathbf{V})F(\mathbf{T},\mathbf{U}) \quad = \quad 0,$$
$$F(\mathbf{S},\mathbf{T})G(\mathbf{U},\mathbf{V}) + F(\mathbf{S},\mathbf{U})G(\mathbf{V},\mathbf{T}) + F(\mathbf{S},\mathbf{V})G(\mathbf{T},\mathbf{U})$$
$$+G(\mathbf{S},\mathbf{T})F(\mathbf{U},\mathbf{V}) + G(\mathbf{S},\mathbf{U})F(\mathbf{V},\mathbf{T}) + G(\mathbf{S},\mathbf{V})F(\mathbf{T},\mathbf{U}) \quad = \quad 0.$$

The solution of this system according to [P. M. Vasić (1963)] is given by the formulae

$$F(\mathbf{U},\mathbf{V}) \quad = \quad \Delta[H_2(\mathbf{U}), H_3(\mathbf{V})], \tag{37.58}$$
$$G(\mathbf{U},\mathbf{V}) \quad = \quad \Delta[K_2(\mathbf{U}), H_3(\mathbf{V})] + \Delta[H_2(\mathbf{U}), K_3'(\mathbf{V}) - mH_3'(\mathbf{V})],$$

where

$$K_2(\mathbf{U}) = \frac{g(A,\mathbf{U},C)}{h(A,B,C)}, \quad K_3'(\mathbf{U}) = g(B,\mathbf{U},C), \quad m = \frac{g(A,B,C)}{h(A,B,C)}.$$

On the basis of the equalities (37.57) and (37.58) we obtain Eq. (37.28), where

$$K_3(\mathbf{U}) = K_3'(\mathbf{U}) - 2mH_3(\mathbf{U}), \quad K_1(\mathbf{U}) = \frac{g(A,B,\mathbf{U})}{h(A,B,C)}.$$

On the basis of the expressions Eqs. (37.32), (37.52) and (37.28) we get Eq. (37.29).

The functions given by Eqs. (37.28) and (37.29) are really solutions of the system (37.23). Consequently, the general solution of the system (37.23) in the case $b^2 + 4a = 0$ is given by the formulae Eqs. (37.28) and (37.29), where H_i, K_j $(i, j = 1, 2, 3)$ are arbitrary complex functions defined in $\mathcal{V}$ and m is an arbitrary constant.

3°. Let $b^2/4 + a = p^2$ $(p > 0)$. Introducing the new functions M and N by the formulae

$$M(\mathbf{U}, \mathbf{V}, \mathbf{W}) = h(\mathbf{U}, \mathbf{V}, \mathbf{W}) + pg(\mathbf{U}, \mathbf{V}, \mathbf{W}),$$
$$N(\mathbf{U}, \mathbf{V}, \mathbf{W}) = h(\mathbf{U}, \mathbf{V}, \mathbf{W}) - pg(\mathbf{U}, \mathbf{V}, \mathbf{W}), \tag{37.59}$$

the system (37.33) becomes

$$\begin{aligned}
&M(\mathbf{Z}_1, \mathbf{Z}_2, \mathbf{Z}_3)M(\mathbf{Z}_4, \mathbf{Z}_5, \mathbf{Z}_6) + M(\mathbf{Z}_2, \mathbf{Z}_1, \mathbf{Z}_4)M(\mathbf{Z}_3, \mathbf{Z}_5, \mathbf{Z}_6) \\
+&M(\mathbf{Z}_2, \mathbf{Z}_1, \mathbf{Z}_5)M(\mathbf{Z}_3, \mathbf{Z}_4, \mathbf{Z}_6) + M(\mathbf{Z}_2, \mathbf{Z}_1, \mathbf{Z}_6)M(\mathbf{Z}_3, \mathbf{Z}_4, \mathbf{Z}_5) \\
+&N(\mathbf{Z}_1, \mathbf{Z}_2, \mathbf{Z}_3)N(\mathbf{Z}_4, \mathbf{Z}_5, \mathbf{Z}_6) + N(\mathbf{Z}_2, \mathbf{Z}_1, \mathbf{Z}_4)N(\mathbf{Z}_3, \mathbf{Z}_5, \mathbf{Z}_6) \\
+&N(\mathbf{Z}_1, \mathbf{Z}_2, \mathbf{Z}_5)N(\mathbf{Z}_3, \mathbf{Z}_4, \mathbf{Z}_6) + N(\mathbf{Z}_2, \mathbf{Z}_1, \mathbf{Z}_6)N(\mathbf{Z}_3, \mathbf{Z}_4, \mathbf{Z}_5) = 0,
\end{aligned} \tag{37.60}$$

$$\begin{aligned}
&M(\mathbf{Z}_1, \mathbf{Z}_2, \mathbf{Z}_3)M(\mathbf{Z}_4, \mathbf{Z}_5, \mathbf{Z}_6) + M(\mathbf{Z}_2, \mathbf{Z}_1, \mathbf{Z}_4)M(\mathbf{Z}_3, \mathbf{Z}_5, \mathbf{Z}_6) \\
+&M(\mathbf{Z}_1, \mathbf{Z}_2, \mathbf{Z}_5)M(\mathbf{Z}_3, \mathbf{Z}_4, \mathbf{Z}_6) + M(\mathbf{Z}_2, \mathbf{Z}_1, \mathbf{Z}_6)M(\mathbf{Z}_3, \mathbf{Z}_4, \mathbf{Z}_5) \\
-&N(\mathbf{Z}_1, \mathbf{Z}_2, \mathbf{Z}_3)N(\mathbf{Z}_4, \mathbf{Z}_5, \mathbf{Z}_6) - N(\mathbf{Z}_2, \mathbf{Z}_1, \mathbf{Z}_4)N(\mathbf{Z}_3, \mathbf{Z}_5, \mathbf{Z}_6) \\
-&N(\mathbf{Z}_1, \mathbf{Z}_2, \mathbf{Z}_5)N(\mathbf{Z}_3, \mathbf{Z}_4, \mathbf{Z}_6) - N(\mathbf{Z}_2, \mathbf{Z}_1, \mathbf{Z}_6)N(\mathbf{Z}_3, \mathbf{Z}_4, \mathbf{Z}_5) = 0,
\end{aligned}$$

Comparing the equations of the system (37.60), we find

$$\begin{aligned}
&M(\mathbf{Z}_1, \mathbf{Z}_2, \mathbf{Z}_3)M(\mathbf{Z}_4, \mathbf{Z}_5, \mathbf{Z}_6) + M(\mathbf{Z}_2, \mathbf{Z}_1, \mathbf{Z}_4)M(\mathbf{Z}_3, \mathbf{Z}_5, \mathbf{Z}_6) \\
+&M(\mathbf{Z}_1, \mathbf{Z}_2, \mathbf{Z}_5)M(\mathbf{Z}_3, \mathbf{Z}_4, \mathbf{Z}_6) + M(\mathbf{Z}_2, \mathbf{Z}_1, \mathbf{Z}_6)M(\mathbf{Z}_3, \mathbf{Z}_4, \mathbf{Z}_5) = 0,
\end{aligned} \tag{37.61}$$

$$\begin{aligned}
&N(\mathbf{Z}_1, \mathbf{Z}_2, \mathbf{Z}_3)N(\mathbf{Z}_4, \mathbf{Z}_5, \mathbf{Z}_6) + N(\mathbf{Z}_2, \mathbf{Z}_1, \mathbf{Z}_4)N(\mathbf{Z}_3, \mathbf{Z}_5, \mathbf{Z}_6) \\
+&N(\mathbf{Z}_1, \mathbf{Z}_2, \mathbf{Z}_5)N(\mathbf{Z}_3, \mathbf{Z}_4, \mathbf{Z}_6) + N(\mathbf{Z}_2, \mathbf{Z}_1, \mathbf{Z}_6)N(\mathbf{Z}_3, \mathbf{Z}_4, \mathbf{Z}_5) = 0.
\end{aligned}$$

The general solution of this system according to [L. Carlitz (1963)] will be

$$M(\mathbf{U}, \mathbf{V}, \mathbf{W}) = \Delta[H_1(\mathbf{U}), H_2(\mathbf{V}), H_3(\mathbf{W})],$$
$$N(\mathbf{U}, \mathbf{V}, \mathbf{W}) = \Delta[K_1(\mathbf{U}), K_2(\mathbf{V}), K_3(\mathbf{W})], \tag{37.62}$$

where H_i, K_j $(i, j = 1, 2, 3)$ are arbitrary complex functions defined in $\mathcal{V}$.

On the basis of the equalities (37.32), (37.59) and (37.62), we find Eqs. (37.30) and (37.31).

The functions given by Eqs. (37.30) and (37.31) are solutions of the system (37.23).

Consequently, the general solution of the system (37.23) in the case $b^2/4 + a = p^2$ $(p > 0)$ is given by Eqs. (37.30) and (37.31) where H_i, K_j $(i, j = 1, 2, 3)$ are arbitrary complex functions defined in $\mathcal{V}$. $\square$

Now we will prove the following general result.

Theorem 37.3   *The general solution of the system of functional equations*

$$
\begin{aligned}
& f(\mathbf{Z}_1, \mathbf{Z}_2, \cdots, \mathbf{Z}_{n-1}, \mathbf{Z}_n) f(\mathbf{Z}_{n+1}, \mathbf{Z}_{n+2}, \cdots, \mathbf{Z}_{2n}) && (37.63) \\
-\ & f(\mathbf{Z}_1, \mathbf{Z}_2, \cdots, \mathbf{Z}_{n-1}, \mathbf{Z}_{n+1}) f(\mathbf{Z}_n, \mathbf{Z}_{n+2}, \mathbf{Z}_{n+3}, \cdots, \mathbf{Z}_{2n}) \\
-\ & f(\mathbf{Z}_1, \mathbf{Z}_2, \cdots, \mathbf{Z}_{n-1}, \mathbf{Z}_{n+2}) f(\mathbf{Z}_{n+1}, \mathbf{Z}_n, \mathbf{Z}_{n+3}, \cdots, \mathbf{Z}_{2n}) - \cdots \\
-\ & f(\mathbf{Z}_1, \mathbf{Z}_2, \cdots, \mathbf{Z}_{n-1}, \mathbf{Z}_{2n}) f(\mathbf{Z}_{n+1}, \mathbf{Z}_{n+2}, \cdots, \mathbf{Z}_{2n-1}, \mathbf{Z}_n) \\
+\ & \operatorname{sgn} a\, [g(\mathbf{Z}_1, \mathbf{Z}_2, \cdots, \mathbf{Z}_{n-1}, \mathbf{Z}_n) g(\mathbf{Z}_{n+1}, \mathbf{Z}_{n+2}, \cdots, \mathbf{Z}_{2n}) \\
-\ & g(\mathbf{Z}_1, \mathbf{Z}_2, \cdots, \mathbf{Z}_{n-1}, \mathbf{Z}_{n+1}) g(\mathbf{Z}_n, \mathbf{Z}_{n+2}, \mathbf{Z}_{n+3}, \cdots, \mathbf{Z}_{2n}) \\
-\ & g(\mathbf{Z}_1, \mathbf{Z}_2, \cdots, \mathbf{Z}_{n-1}, \mathbf{Z}_{n+2}) g(\mathbf{Z}_{n+1}, \mathbf{Z}_n, \mathbf{Z}_{n+3}, \cdots, \mathbf{Z}_{2n}) - \cdots \\
-\ & g(\mathbf{Z}_1, \mathbf{Z}_2, \cdots, \mathbf{Z}_{n-1}, \mathbf{Z}_{2n}) g(\mathbf{Z}_{n+1}, \mathbf{Z}_{n+2}, \cdots, \mathbf{Z}_{2n-1}, \mathbf{Z}_n)] = 0, \\[4pt]
& f(\mathbf{Z}_1, \mathbf{Z}_2, \cdots, \mathbf{Z}_{n-1}, \mathbf{Z}_n) g(\mathbf{Z}_{n+1}, \mathbf{Z}_{n+2}, \cdots, \mathbf{Z}_{2n}) \\
-\ & f(\mathbf{Z}_1, \mathbf{Z}_2, \cdots, \mathbf{Z}_{n-1}, \mathbf{Z}_{n+1}) g(\mathbf{Z}_n, \mathbf{Z}_{n+2}, \mathbf{Z}_{n+3}, \cdots, \mathbf{Z}_{2n}) \\
-\ & f(\mathbf{Z}_1, \mathbf{Z}_2, \cdots, \mathbf{Z}_{n-1}, \mathbf{Z}_{n+2}) g(\mathbf{Z}_{n+1}, \mathbf{Z}_n, \mathbf{Z}_{n+3}, \cdots, \mathbf{Z}_{2n}) - \cdots \\
-\ & f(\mathbf{Z}_1, \mathbf{Z}_2, \cdots, \mathbf{Z}_{n-1}, \mathbf{Z}_{2n}) g(\mathbf{Z}_{n+1}, \mathbf{Z}_{n+2}, \cdots, \mathbf{Z}_{2n-1}, \mathbf{Z}_n) \\
+\ & g(\mathbf{Z}_1, \mathbf{Z}_2, \cdots, \mathbf{Z}_{n-1}, \mathbf{Z}_n) f(\mathbf{Z}_{n+1}, \mathbf{Z}_{n+2}, \cdots, \mathbf{Z}_{2n}) \\
-\ & g(\mathbf{Z}_1, \mathbf{Z}_2, \cdots, \mathbf{Z}_{n-1}, \mathbf{Z}_{n+1}) f(\mathbf{Z}_n, \mathbf{Z}_{n+2}, \mathbf{Z}_{n+3}, \cdots, \mathbf{Z}_{2n}) \\
-\ & g(\mathbf{Z}_1, \mathbf{Z}_2, \cdots, \mathbf{Z}_{n-1}, \mathbf{Z}_{n+2}) f(\mathbf{Z}_{n+1}, \mathbf{Z}_n, \mathbf{Z}_{n+3}, \cdots, \mathbf{Z}_{2n}) - \cdots \\
-\ & g(\mathbf{Z}_1, \mathbf{Z}_2, \cdots, \mathbf{Z}_{n-1}, \mathbf{Z}_{2n}) f(\mathbf{Z}_{n+1}, \mathbf{Z}_{n+2}, \cdots, \mathbf{Z}_{2n-1}, \mathbf{Z}_n) = 0,
\end{aligned}
$$

where $f, g : \mathcal{V}^n \mapsto \mathbf{C}$ and a is a real constant, is given by the following formulae:

$1^{o}.$ *if* $a < 0$,

$$
\begin{aligned}
f(\mathbf{U}_1, \mathbf{U}_2, \cdots, \mathbf{U}_n) &= \Delta[H_1(\mathbf{U}_1), H_2(\mathbf{U}_2), \cdots, H_n(\mathbf{U}_n)] && (37.64) \\
&\quad + \Delta[K_1(\mathbf{U}_1), K_2(\mathbf{U}_2), \cdots, K_n(\mathbf{U}_n)], \\
g(\mathbf{U}_1, \mathbf{U}_2, \cdots, \mathbf{U}_n) &= i\{\Delta[K_1(\mathbf{U}_1), K_2(\mathbf{U}_2), \cdots, K_n(\mathbf{U}_n)] && (37.65) \\
&\quad - \Delta[H_1(\mathbf{U}_1), H_2(\mathbf{U}_2), \cdots, H_n(\mathbf{U}_n)]\};
\end{aligned}
$$

$2^o.$ *if* $a > 0,$

$$f(\mathbf{U}_1, \mathbf{U}_2, \cdots, \mathbf{U}_n) = \Delta[H_1(\mathbf{U}_1), H_2(\mathbf{U}_2), \cdots, H_n(\mathbf{U}_n)] \quad (37.66)$$
$$+ \Delta[K_1(\mathbf{U}_1), K_2(\mathbf{U}_2), \cdots, K_n(\mathbf{U}_n)],$$

$$g(\mathbf{U}_1, \mathbf{U}_2, \cdots, \mathbf{U}_n) = \Delta[H_1(\mathbf{U}_1), H_2(\mathbf{U}_2), \cdots, H_n(\mathbf{U}_n)] \quad (37.67)$$
$$- \Delta[K_1(\mathbf{U}_1), K_2(\mathbf{U}_2), \cdots, K_n(\mathbf{U}_n)];$$

3° *if* $a = 0,$

$$f(\mathbf{U}_1, \mathbf{U}_2, \cdots, \mathbf{U}_n) = \Delta[H_1(\mathbf{U}_1), H_2(\mathbf{U}_2), \cdots, H_n(\mathbf{U}_n)] \quad (37.68)$$

$$= \begin{vmatrix} H_1(\mathbf{U}_1) & H_1(\mathbf{U}_2) & \cdots & H_1(\mathbf{U}_n) \\ H_2(\mathbf{U}_1) & H_2(\mathbf{U}_2) & \cdots & H_2(\mathbf{U}_n) \\ \vdots & & & \\ H_n(\mathbf{U}_1) & H_n(\mathbf{U}_2) & \cdots & H_n(\mathbf{U}_n) \end{vmatrix},$$

$$g(\mathbf{U}_1, \mathbf{U}_2, \cdots, \mathbf{U}_n) \quad (37.69)$$
$$= \Delta[K_1(\mathbf{U}_1), H_2(\mathbf{U}_2), H_3(\mathbf{U}_3), \cdots, H_n(\mathbf{U}_n)]$$
$$+ \Delta[H_1(\mathbf{U}_1), K_2(\mathbf{U}_2), H_3(\mathbf{U}_3), \cdots, H_n(\mathbf{U}_n)] + \cdots$$
$$+ \Delta[H_1(\mathbf{U}_1), H_2(\mathbf{U}_2), \cdots, H_{n-1}(\mathbf{U}_{n-1}), K_n(\mathbf{U}_n)],$$

where H_i, K_i $(1 \leq i, j \leq n)$ *are arbitrary complex functions defined in* $\mathcal{V}$.

Proof. $1^o.$ Let $a < 0.$ Introducing new functions M and N by the formulae

$$M(\mathbf{U}_1, \mathbf{U}_2, \cdots, \mathbf{U}_n) \quad (37.70)$$
$$= f(\mathbf{U}_1, \mathbf{U}_2, \cdots, \mathbf{U}_n) + ig(\mathbf{U}_1, \mathbf{U}_2, \cdots, \mathbf{U}_n),$$
$$N(\mathbf{U}_1, \mathbf{U}_2, \cdots, \mathbf{U}_n)$$
$$= f(\mathbf{U}_1, \mathbf{U}_2, \cdots, \mathbf{U}_n) - ig(\mathbf{U}_1, \mathbf{U}_2, \cdots, \mathbf{U}_n),$$

$(i^2 = -1)$ the system (37.63) becomes

$$M(\mathbf{Z}_1, \mathbf{Z}_2, \cdots, \mathbf{Z}_{n-1}, \mathbf{Z}_n)M(\mathbf{Z}_{n+1}, \mathbf{Z}_{n+2}, \mathbf{Z}_{n+3}, \cdots, \mathbf{Z}_{2n}) \quad (37.71)$$
$$= M(\mathbf{Z}_1, \mathbf{Z}_2, \cdots, \mathbf{Z}_{n-1}, \mathbf{Z}_{n+1})M(\mathbf{Z}_n, \mathbf{Z}_{n+2}, \mathbf{Z}_{n+3}, \cdots, \mathbf{Z}_{2n})$$
$$+ M(\mathbf{Z}_1, \mathbf{Z}_2, \cdots, \mathbf{Z}_{n-1}, \mathbf{Z}_{n+2})M(\mathbf{Z}_{n+1}, \mathbf{Z}_n, \mathbf{Z}_{n+3}, \cdots, \mathbf{Z}_{2n}) + \cdots$$
$$+ M(\mathbf{Z}_1, \mathbf{Z}_2, \cdots, \mathbf{Z}_{n-1}, \mathbf{Z}_{2n})M(\mathbf{Z}_{n+1}, \mathbf{Z}_{n+2}, \cdots, \mathbf{Z}_{2n-1}, \mathbf{Z}_n),$$

$$N(\mathbf{Z}_1, \mathbf{Z}_2, \cdots, \mathbf{Z}_{n-1}, \mathbf{Z}_n)N(\mathbf{Z}_{n+1}, \mathbf{Z}_{n+2}, \mathbf{Z}_{n+3}, \cdots, \mathbf{Z}_{2n})$$
$$= \; N(\mathbf{Z}_1, \mathbf{Z}_2, \cdots, \mathbf{Z}_{n-1}, \mathbf{Z}_{n+1})N(\mathbf{Z}_n, \mathbf{Z}_{n+2}, \mathbf{Z}_{n+3}, \cdots, \mathbf{Z}_{2n})$$
$$+ \; N(\mathbf{Z}_1, \mathbf{Z}_2, \cdots, \mathbf{Z}_{n-1}, \mathbf{Z}_{n+2})N(\mathbf{Z}_{n+1}, \mathbf{Z}_n, \mathbf{Z}_{n+3}, \cdots, \mathbf{Z}_{2n}) + \cdots$$
$$+ \; N(\mathbf{Z}_1, \mathbf{Z}_2, \cdots, \mathbf{Z}_{n-1}, \mathbf{Z}_{2n})N(\mathbf{Z}_{n+1}, \mathbf{Z}_{n+2}, \cdots, \mathbf{Z}_{2n-1}, \mathbf{Z}_n).$$

On the basis of the equalities (37.71) and (37.70), we find Eqs. (37.64) and (37.65), where H_i, $K_j : \mathcal{V} \mapsto \mathbf{C}$ are arbitrary complex functions.

$2°$. Let $a > 0$. Introducing the functions M and N by

$$M(\mathbf{U}_1, \mathbf{U}_2, \cdots, \mathbf{U}_n) \;=\; f(\mathbf{U}_1, \mathbf{U}_2, \cdots, \mathbf{U}_n) + g(\mathbf{U}_1, \mathbf{U}_2, \cdots, \mathbf{U}_n),$$
$$\tag{37.72}$$
$$N(\mathbf{U}_1, \mathbf{U}_2, \cdots, \mathbf{U}_n) \;=\; f(\mathbf{U}_1, \mathbf{U}_2, \cdots, \mathbf{U}_n) - g(\mathbf{U}_1, \mathbf{U}_2, \cdots, \mathbf{U}_n),$$

the system (37.63) is reduced to Eq. (37.71). On the basis of the equalities (37.71) and (37.72) there follow Eqs. (37.66) and (37.67).

$3°$. Let $a = 0$. In this case, the first equation of the system (37.63) is reduced to

$$f(\mathbf{Z}_1, \mathbf{Z}_2, \cdots, \mathbf{Z}_{n-1}, \mathbf{Z}_n)f(\mathbf{Z}_{n+1}, \mathbf{Z}_{n+2}, \cdots, \mathbf{Z}_{2n}) \tag{37.73}$$
$$= \; f(\mathbf{Z}_1, \mathbf{Z}_2, \cdots, \mathbf{Z}_{n-1}, \mathbf{Z}_{n+1})f(\mathbf{Z}_n, \mathbf{Z}_{n+2}, \mathbf{Z}_{n+3}, \cdots, \mathbf{Z}_{2n})$$
$$+ \; f(\mathbf{Z}_1, \mathbf{Z}_2, \cdots, \mathbf{Z}_{n-1}, \mathbf{Z}_{n+2})f(\mathbf{Z}_{n+1}, \mathbf{Z}_n, \mathbf{Z}_{n+3}, \cdots, \mathbf{Z}_{2n}) + \cdots$$
$$+ \; f(\mathbf{Z}_1, \mathbf{Z}_2, \cdots, \mathbf{Z}_{n-1}, \mathbf{Z}_{2n})f(\mathbf{Z}_{n+1}, \mathbf{Z}_{n+2}, \cdots, \mathbf{Z}_{2n-1}, \mathbf{Z}_n).$$

According to [P. M. Vasić (1963)], the general solution of the equation (37.73) is given by Eq. (37.68). Further in the proof of this theorem, we will use the following

Lemma 37.4 *If* $\mathbf{A}_i \in \mathcal{V}$ $(1 \leq i \leq n)$ *are such that*

$$f(\mathbf{A}_1, \mathbf{A}_2, \cdots, \mathbf{A}_n) \neq 0,$$

then the following equality holds

$$f(\mathbf{A}_n, \mathbf{U}_1, \mathbf{U}_2, \cdots, \mathbf{U}_{n-2}, \mathbf{A}_n) = 0. \tag{37.74}$$

Proof of Lemma 37.4. By putting

$$\mathbf{Z}_i = \mathbf{A}_i \qquad (1 \leq i \leq n),$$

$$\mathbf{Z}_{n+1} = \mathbf{Z}_{2n} = \mathbf{A}_n, \quad \mathbf{Z}_j = \mathbf{U}_{j-n-1} \quad (j = n+2, n+3, \cdots, 2n-1),$$

the equality (37.73) becomes

$$
\begin{aligned}
&f(\mathbf{A}_1, \mathbf{A}_2, \cdots, \mathbf{A}_{n-1}, \mathbf{A}_n) f(\mathbf{A}_n, \mathbf{U}_1, \mathbf{U}_2, \cdots, \mathbf{U}_{n-2}, \mathbf{A}_n) \qquad (37.75)\\
+\ &f(\mathbf{A}_1, \mathbf{A}_2, \cdots, \mathbf{A}_{n-1}, \mathbf{U}_1) f(\mathbf{A}_n, \mathbf{A}_n, \mathbf{U}_2, \cdots, \mathbf{U}_{n-2}, \mathbf{A}_n)\\
+\ &f(\mathbf{A}_1, \mathbf{A}_2, \cdots, \mathbf{A}_{n-1}, \mathbf{U}_2) f(\mathbf{A}_n, \mathbf{U}_1, \mathbf{A}_n, \mathbf{U}_3, \cdots, \mathbf{U}_{n-2}, \mathbf{A}_n) + \cdots\\
+\ &f(\mathbf{A}_1, \mathbf{A}_2, \cdots, \mathbf{A}_{n-1}, \mathbf{U}_{n-2}) f(\mathbf{A}_n, \mathbf{U}_1, \cdots, \mathbf{U}_{n-3}, \mathbf{A}_n, \mathbf{A}_n) = 0.
\end{aligned}
$$

Since the formulation of the lemma is exactly the same as of Lemma 36.2, and equation (37.75) is identical with Eq. (36.11), the proof of the lemma is completed in the same way as of Lemma 36.2. $\qquad\square$

We will prove that Eq. (37.69) holds by induction.

If $n = 2$, then

$$
f(\mathbf{U}_1, \mathbf{U}_2) = \Delta[H_1(\mathbf{U}_1), H_2(\mathbf{U}_2)], \qquad (37.76)
$$

where

$$
H_1(\mathbf{U}_1) = \frac{f(\mathbf{A}, \mathbf{U}_1)}{f(\mathbf{A}, \mathbf{B})}, \qquad H_2(\mathbf{U}_2) = f(\mathbf{B}, \mathbf{U}_2)
$$

and

$$
f(\mathbf{Z}_1, \mathbf{Z}_2)g(\mathbf{Z}_3, \mathbf{Z}_4) - f(\mathbf{Z}_1, \mathbf{Z}_3)g(\mathbf{Z}_2, \mathbf{Z}_4) - f(\mathbf{Z}_1, \mathbf{Z}_4)g(\mathbf{Z}_3, \mathbf{Z}_2) \qquad (37.77)
$$

$$
+ g(\mathbf{Z}_1, \mathbf{Z}_2)f(\mathbf{Z}_3, \mathbf{Z}_4) - g(\mathbf{Z}_1, \mathbf{Z}_3)f(\mathbf{Z}_2, \mathbf{Z}_4) - g(\mathbf{Z}_1, \mathbf{Z}_4)f(\mathbf{Z}_3, \mathbf{Z}_2) = 0.
$$

For any nontrivial solution of the equation (37.77) there exists at least one pair of complex vectors $(\mathbf{A}, \mathbf{B})$ such that $f(\mathbf{A}, \mathbf{B}) \neq 0$.

If we put $\mathbf{Z}_1 = \mathbf{A}$, $\mathbf{Z}_3 = \mathbf{B}$, $\mathbf{Z}_2 = \mathbf{U}_1$ and $\mathbf{Z}_4 = \mathbf{U}_2$, then the equation (37.77) takes the following form

$$
g(\mathbf{U}_1, \mathbf{U}_2) = \frac{f(\mathbf{A}, \mathbf{U}_1)}{f(\mathbf{A}, \mathbf{B})} g(\mathbf{B}, \mathbf{U}_2) - \frac{f(\mathbf{A}, \mathbf{U}_2)}{f(\mathbf{A}, \mathbf{B})} g(\mathbf{B}, \mathbf{U}_1) \qquad (37.78)
$$

$$
+ \frac{g(\mathbf{A}, \mathbf{U}_1)}{f(\mathbf{A}, \mathbf{B})} f(\mathbf{B}, \mathbf{U}_2) - \frac{g(\mathbf{A}, \mathbf{U}_2)}{f(\mathbf{A}, \mathbf{B})} f(\mathbf{B}, \mathbf{U}_1) - \frac{g(\mathbf{A}, \mathbf{B})}{f(\mathbf{A}, \mathbf{B})} f(\mathbf{U}_1, \mathbf{U}_2).
$$

Using the notations

$$
\frac{g(\mathbf{A}, \mathbf{U}_1)}{f(\mathbf{A}, \mathbf{B})} = K_1(\mathbf{U}_1), \quad g(\mathbf{B}, \mathbf{U}_1) = K_2'(\mathbf{U}_1), \quad \frac{g(\mathbf{A}, \mathbf{B})}{f(\mathbf{A}, \mathbf{B})} = -m,
$$

on the basis of the relations Eqs. (37.78) and (37.76) we find

$$\begin{aligned}
g(\mathbf{U}_1, \mathbf{U}_2) &= \Delta[H_1(\mathbf{U}_1), K_2'(\mathbf{U}_2)] + \Delta[K_1(\mathbf{U}_1), H_2(\mathbf{U}_2)] \\
&+ m\Delta[H_1(\mathbf{U}_1), H_2(\mathbf{U}_2)],
\end{aligned}$$

i.e.,

$$g(\mathbf{U}_1, \mathbf{U}_2) = \Delta[K_1(\mathbf{U}_1), H_2(\mathbf{U}_2)] + \Delta[H_1(\mathbf{U}_1), K_2(\mathbf{U}_2)],$$

where

$$K_2(\mathbf{U}) = K_2'(\mathbf{U}) + mH_2(\mathbf{U}).$$

Assume that this theorem holds for $n - 1$, so that the general solution of the functional equation

$$\begin{aligned}
& f(\mathbf{Z}_1, \cdots, \mathbf{Z}_{n-2}, \mathbf{Z}_{n-1})g(\mathbf{Z}_n, \mathbf{Z}_{n+1}, \cdots, \mathbf{Z}_{2n-2}) && (37.79) \\
-\ & f(\mathbf{Z}_1, \cdots, \mathbf{Z}_{n-2}, \mathbf{Z}_n)g(\mathbf{Z}_{n-1}, \mathbf{Z}_{n+1}, \cdots, \mathbf{Z}_{2n-2}) \\
-\ & f(\mathbf{Z}_1, \cdots, \mathbf{Z}_{n-2}, \mathbf{Z}_{n+1})g(\mathbf{Z}_n, \mathbf{Z}_{n-1}, \mathbf{Z}_{n+2}, \cdots, \mathbf{Z}_{2n-2}) - \cdots \\
-\ & f(\mathbf{Z}_1, \cdots, \mathbf{Z}_{n-2}, \mathbf{Z}_{2n-2})g(\mathbf{Z}_n, \mathbf{Z}_{n+1}, \cdots, \mathbf{Z}_{2n-3}, \mathbf{Z}_{n-1}) \\
+\ & g(\mathbf{Z}_1, \cdots, \mathbf{Z}_{n-2}, \mathbf{Z}_{n-1})f(\mathbf{Z}_n, \mathbf{Z}_{n+1}, \cdots, \mathbf{Z}_{2n-2}) \\
-\ & g(\mathbf{Z}_1, \cdots, \mathbf{Z}_{n-2}, \mathbf{Z}_n)f(\mathbf{Z}_{n-1}, \mathbf{Z}_{n+1}, \cdots, \mathbf{Z}_{2n-2}) \\
-\ & g(\mathbf{Z}_1, \cdots, \mathbf{Z}_{n-2}, \mathbf{Z}_{n+1})f(\mathbf{Z}_n, \mathbf{Z}_{n-1}, \mathbf{Z}_{n+2}, \cdots, \mathbf{Z}_{2n-2}) - \cdots \\
-\ & g(\mathbf{Z}_1, \cdots, \mathbf{Z}_{n-2}, \mathbf{Z}_{2n-2})f(\mathbf{Z}_n, \mathbf{Z}_{n+1}, \cdots, \mathbf{Z}_{2n-3}, \mathbf{Z}_{n-1}) = 0
\end{aligned}$$

is introduced by

$$\begin{aligned}
& g(\mathbf{U}_1, \mathbf{U}_2, \ldots, \mathbf{U}_{n-1}) && (37.80) \\
=\ & \Delta[K_1(\mathbf{U}_1), H_2(\mathbf{U}_2), \cdots, H_{n-1}(\mathbf{U}_{n-1})] \\
+\ & \Delta[H_1(\mathbf{U}_1), K_2(\mathbf{U}_2), H_3(\mathbf{U}_3), \cdots, H_{n-1}(\mathbf{U}_{n-1})] + \cdots \\
+\ & \Delta[H_1(\mathbf{U}_1), H_2(\mathbf{U}_2), \cdots, H_{n-2}(\mathbf{U}_{n-2}), K_{n-1}(\mathbf{U}_{n-1})].
\end{aligned}$$

Let $f(\mathbf{A}_1, \mathbf{A}_2, \cdots, \mathbf{A}_n) \neq 0$. If we put $\mathbf{Z}_1 = \mathbf{A}_n$, $\mathbf{Z}_{n+1} = \mathbf{A}_n$,

$$\begin{aligned}
f(\mathbf{A}_n, \mathbf{U}_2, \mathbf{U}_3, \cdots, \mathbf{U}_n) &= F(\mathbf{U}_2, \mathbf{U}_3, \cdots, \mathbf{U}_n), \\
g(\mathbf{A}_n, \mathbf{U}_2, \mathbf{U}_3, \cdots, \mathbf{U}_n) &= G(\mathbf{U}_2, \mathbf{U}_3, \cdots, \mathbf{U}_n),
\end{aligned}$$

the second equation of the system (37.63), according to the lemma, becomes

$$F(\mathbf{Z}_2, \mathbf{Z}_3, \cdots, \mathbf{Z}_{n-1}, \mathbf{Z}_n)G(\mathbf{Z}_{n+2}, \mathbf{Z}_{n+3}, \cdots, \mathbf{Z}_{2n}) \qquad (37.81)$$

$$
\begin{aligned}
&- \ F(\mathbf{Z}_2, \mathbf{Z}_3, \cdots, \mathbf{Z}_{n-1}, \mathbf{Z}_{n+2}) G(\mathbf{Z}_n, \mathbf{Z}_{n+3}, \cdots, \mathbf{Z}_{2n}) \\
&- \ F(\mathbf{Z}_2, \mathbf{Z}_3, \cdots, \mathbf{Z}_{n-1}, \mathbf{Z}_{n+3}) G(\mathbf{Z}_{n+2}, \mathbf{Z}_n, \mathbf{Z}_{n+4}, \cdots, \mathbf{Z}_{2n}) - \cdots \\
&- \ F(\mathbf{Z}_2, \mathbf{Z}_3, \cdots, \mathbf{Z}_{n-1}, \mathbf{Z}_{2n}) G(\mathbf{Z}_{n+2}, \cdots, \mathbf{Z}_{2n-1}, \mathbf{Z}_{2n}) \\
&+ \ G(\mathbf{Z}_2, \mathbf{Z}_3, \cdots, \mathbf{Z}_{n-1}, \mathbf{Z}_n) F(\mathbf{Z}_{n+2}, \mathbf{Z}_{n+3}, \cdots, \mathbf{Z}_{2n}) \\
&- \ G(\mathbf{Z}_2, \mathbf{Z}_3, \cdots, \mathbf{Z}_{n-1}, \mathbf{Z}_{n+2}) F(\mathbf{Z}_n, \mathbf{Z}_{n+3}, \cdots, \mathbf{Z}_{2n}) \\
&- \ G(\mathbf{Z}_2, \mathbf{Z}_3, \cdots, \mathbf{Z}_{n-1}, \mathbf{Z}_{n+3}) F(\mathbf{Z}_{n+2}, \mathbf{Z}_n, \mathbf{Z}_{n+4}, \cdots, \mathbf{Z}_{2n}) - \cdots \\
&- \ G(\mathbf{Z}_2, \mathbf{Z}_3, \cdots, \mathbf{Z}_{n-1}, \mathbf{Z}_{2n}) F(\mathbf{Z}_{n+2}, \cdots, \mathbf{Z}_{2n-1}, \mathbf{Z}_n) = 0.
\end{aligned}
$$

On the basis of the equalities (37.79) and (37.80), we find the general solution of the equation (37.81) which is

$$
\begin{aligned}
&G(\mathbf{U}_1, \mathbf{U}_2, \cdots, \mathbf{U}_{n-1}) \qquad\qquad\qquad\qquad (37.82) \\
={}& g(\mathbf{A}_n, \mathbf{U}_1, \mathbf{U}_2, \cdots, \mathbf{U}_{n-1}) \\
={}& \Delta[(K_1(\mathbf{U}_1), H_2(\mathbf{U}_2), \cdots, H_{n-1}(\mathbf{U}_{n-1})] \\
+{}& \Delta[H_1(\mathbf{U}_1), K_2(\mathbf{U}_2), H_3(\mathbf{U}_3), \cdots, H_{n-1}(\mathbf{U}_{n-1})] + \cdots \\
+{}& \Delta[H_1(\mathbf{U}_1), H_2(\mathbf{U}_2), \cdots, H_{n-2}(\mathbf{U}_{n-2}), K_{n-1}(\mathbf{U}_{n-1})],
\end{aligned}
$$

where $H_i(\mathbf{U}), K_j(\mathbf{U})$ $(1 \le i, j \le n - 1)$ are arbitrary complex functions defined in $\mathcal{V}$.

If we put

$$
\mathbf{Z}_i = \mathbf{A}_i \quad (1 \le i \le n - 1),
$$

$$
\mathbf{Z}_n = \mathbf{U}_1, \quad \mathbf{Z}_{n+1} = \mathbf{A}_n, \quad \mathbf{Z}_{n+k} = \mathbf{U}_k \quad (2 \le k \le n)
$$

into the second equation of the system (37.63), we obtain

$$
\begin{aligned}
&f(\mathbf{A}_1, \mathbf{A}_2, \cdots, \mathbf{A}_n) g(\mathbf{U}_1, \mathbf{U}_2, \cdots, \mathbf{U}_n) \qquad\qquad (37.83) \\
={}& f(\mathbf{A}_1, \mathbf{A}_2, \cdots, \mathbf{A}_{n-1}, \mathbf{U}_1) g(\mathbf{A}_n, \mathbf{U}_2, \cdots, \mathbf{U}_n) \\
-{}& f(\mathbf{A}_1, \mathbf{A}_2, \cdots, \mathbf{A}_{n-1}, \mathbf{U}_2) g(\mathbf{A}_n, \mathbf{U}_1, \mathbf{U}_3, \cdots, \mathbf{U}_n) - \cdots \\
-{}& f(\mathbf{A}_1, \mathbf{A}_2, \cdots, \mathbf{A}_{n-1}, \mathbf{U}_n) g(\mathbf{A}_n, \mathbf{U}_2, \cdots, \mathbf{U}_{n-1}, \mathbf{U}_1) \\
+{}& g(\mathbf{A}_1, \mathbf{A}_2, \cdots, \mathbf{A}_{n-1}, \mathbf{U}_1) f(\mathbf{A}_n, \mathbf{U}_2, \cdots, \mathbf{U}_n) \\
-{}& g(\mathbf{A}_1, \mathbf{A}_2, \cdots, \mathbf{A}_{n-1}, \mathbf{U}_2) f(\mathbf{A}_n, \mathbf{U}_1, \mathbf{U}_3, \cdots, \mathbf{U}_n) - \cdots \\
-{}& g(\mathbf{A}_1, \mathbf{A}_2, \cdots, \mathbf{A}_{n-1}, \mathbf{U}_n) f(\mathbf{A}_n, \mathbf{U}_2, \cdots, \mathbf{U}_{n-1}, \mathbf{U}_1) \\
-{}& g(\mathbf{A}_1, \mathbf{A}_2, \cdots, \mathbf{A}_n) f(\mathbf{U}_1, \mathbf{U}_2, \cdots, \mathbf{U}_n).
\end{aligned}
$$

On the basis of the properties Eqs. (37.68) and (37.82), we obtain

$$f(\mathbf{A}_n, \cdots, \mathbf{U}_{i-1}, \mathbf{U}_i, \mathbf{U}_{i+1}, \cdots, \mathbf{U}_{j-1}, \mathbf{U}_j, \mathbf{U}_{j+1}, \cdots, \mathbf{U}_{n-1})$$
$$= - f(\mathbf{A}_n, \cdots, \mathbf{U}_{i-1}, \mathbf{U}_j, \mathbf{U}_{i+1}, \cdots, \mathbf{U}_{j-1}, \mathbf{U}_i, \mathbf{U}_{j+1}, \cdots, \mathbf{U}_{n-1}),$$

$$(37.84)$$

$$g(\mathbf{A}_n, \cdots, \mathbf{U}_{i-1}, \mathbf{U}_i, \mathbf{U}_{i+1}, \cdots, \mathbf{U}_{j-1}, \mathbf{U}_j, \mathbf{U}_{j+1}, \cdots, \mathbf{U}_{n-1})$$
$$= - g(\mathbf{A}_n, \cdots, \mathbf{U}_{i-1}, \mathbf{U}_j, \mathbf{U}_{i+1}, \cdots, \mathbf{U}_{j-1}, \mathbf{U}_i, \mathbf{U}_{j+1}, \cdots, \mathbf{U}_{n-1})$$
$$(1 \le i < j \le n-1).$$

By introducing the notations

$$\frac{f(\mathbf{A}_1, \cdots, \mathbf{A}_{n-1}, \mathbf{U})}{f(\mathbf{A}_1, \cdots, \mathbf{A}_n)} = H_1(\mathbf{U}), \qquad \frac{g(\mathbf{A}_1, \cdots, \mathbf{A}_{n-1}, \mathbf{U})}{f(\mathbf{A}_1, \cdots, \mathbf{A}_n)} = K_1(\mathbf{U}),$$

$$\frac{g(\mathbf{A}_1, \cdots, \mathbf{A}_n)}{f(\mathbf{A}_1, \cdots, \mathbf{A}_n)} = -m,$$

on the basis of the equalities (37.84), (37.83), (37.82) and (37.68) we find

$$
\begin{aligned}
g(\mathbf{U}_1, \mathbf{U}_2, \cdots, \mathbf{U}_n) \;=\;& \Delta[K_1(\mathbf{U}_1), H_2(\mathbf{U}_2), \cdots, H_n(\mathbf{U}_n)] \\
+\;& \Delta[H_1(\mathbf{U}_1), K_2(\mathbf{U}_2), H_3(\mathbf{U}_3), \cdots, H_n(\mathbf{U}_n)] + \cdots \\
+\;& \Delta[H_1(\mathbf{U}_1), \cdots, H_{n-1}(\mathbf{U}_{n-1}), K_n(\mathbf{U}_n)] \\
+\;& m\Delta[H_1(\mathbf{U}_1), H_2(\mathbf{U}_2), \cdots, H_n(\mathbf{U}_n)],
\end{aligned}
$$

i.e., $g(U_1, \mathbf{U}_2, \cdots, \mathbf{U}_n)$ is given by Eq. (37.69), where $K_n(\mathbf{U}_n) + mH_n(\mathbf{U}_n)$ is replaced by $K_n(\mathbf{U}_n)$.

Now we will prove that the functions given by Eq. (37.64) and (37.65), (37.66) and (37.67), (37.68) and (37.69) are solutions of the system (37.63). We introduce the following notation similar to that of Sec. 36:

$$D(F_1, F_2, \cdots, F_n, G_1, G_2, \cdots, G_n) = \begin{vmatrix} \mathcal{F}(\mathbf{U}) & \mathcal{O}_{n-1,n} \\ \tilde{\mathcal{F}}(\mathbf{U}) & \tilde{\mathcal{G}}(\mathbf{U}) \end{vmatrix},$$

where

$$\mathcal{F}(\mathbf{U}) = \begin{pmatrix} F_1(\mathbf{U}_1) & F_2(\mathbf{U}_1) & \cdots & F_n(\mathbf{U}_1) \\ F_1(\mathbf{U}_2) & F_2(\mathbf{U}_2) & \cdots & F_n(\mathbf{U}_2) \\ \vdots & & & \\ F_1(\mathbf{U}_{n-1}) & F_2(\mathbf{U}_{n-1}) & \cdots & F_n(\mathbf{U}_{n-1}) \end{pmatrix},$$

$$\tilde{\mathcal{F}}(\mathbf{U}) = \begin{pmatrix} F_1(\mathbf{U}_n) & F_2(\mathbf{U}_n) & \cdots & F_n(\mathbf{U}_n) \\ F_1(\mathbf{U}_{n+1}) & F_2(\mathbf{U}_{n+1}) & \cdots & F_n(\mathbf{U}_{n+1}) \\ \vdots & & & \\ F_1(\mathbf{U}_{2n}) & F_2(\mathbf{U}_{2n}) & \cdots & F_n(\mathbf{U}_{2n}) \end{pmatrix},$$

$\tilde{\mathcal{G}}(\mathbf{U})$ is given by a formula of the same form with F_j replaced by G_j ($1 \leq j \leq n$) and $\mathcal{O}_{n-1,n}$ is the zero $(n-1) \times n$ matrix.

Let

$$D(H_1, H_2, \cdots, H_n, H_1, H_2, \cdots, H_n) = 0,$$

then it will be

$$\begin{aligned}
& D(H_1, H_2, \cdots, H_n, H_1, H_2, \cdots, H_n) \\
+ \ & D(H_1, H_2, \cdots, H_n, K_1, K_2, \cdots, K_n) \\
+ \ & D(K_1, K_2, \cdots, K_n, H_1, H_2, \cdots, H_n) \\
+ \ & D(K_1, K_2, \cdots, K_n, K_1, K_2, \cdots, K_n) \\
+ \ & D(H_1, H_2, \cdots, H_n, H_1, H_2, \cdots, H_n) \\
- \ & D(H_1, H_2, \cdots, H_n, K_1, K_2, \cdots, K_n) \\
- \ & D(K_1, K_2, \cdots, K_n, H_1, H_2, \cdots, H_n) \\
+ \ & D(K_1, K_2, \cdots, K_n, K_1, K_2, \cdots, K_n) = 0, \\[1em]
& D(H_1, H_2, \cdots, H_n, H_1, H_2, \cdots, H_n) \\
+ \ & D(K_1, K_2, \cdots, K_n, H_1, H_2, \cdots, H_n) \\
- \ & D(H_1, H_2, \cdots, H_n, K_1, K_2, \cdots, K_n) \\
- \ & D(K_1, K_2, \cdots, K_n, K_1, K_2, \cdots, K_n) \\
+ \ & D(H_1, H_2, \cdots, H_n, H_1, H_2, \cdots, H_n) \\
- \ & D(K_1, K_2, \cdots, K_n, H_1, H_2, \cdots, H_n) \\
+ \ & D(H_1, H_2, \cdots, H_n, K_1, K_2, \cdots, K_n) \\
- \ & D(K_1, K_2, \cdots, K_n, K_1, K_2, \cdots, K_n) = 0.
\end{aligned}$$

By the Laplace rule, if we evaluate the determinants which appear in the previous identities, we obtain that the functions given by Eqs. (37.66) and (37.67) in the case $a > 0$ are a solution of the system (37.63).

In the case $a < 0$, we can analogously show that the functions Eq. (37.64) and (37.65) represent a solution of the system (37.63).

If $a = 0$, we will show that the functions Eqs. (37.68) and (37.69) are a solution of the system (37.63).

The function Eq. (37.68) satisfies the first equation of the system (37.63).

Now we will prove that the functions Eqs. (37.68) and (37.69) satisfy the second equation of the system (37.63).

Let us consider in this case the following identity

$$D(H_1 + K_1, H_2, \cdots, H_n, H_1 + K_1, H_2, \cdots, H_n) \qquad (37.85)$$
$$+ \quad D(H_1, H_2 + K_2, H_3, \cdots, H_n, H_1, H_2 + K_2, H_3, \cdots, H_n) + \cdots$$
$$+ \quad D(H_1, H_2, \cdots, H_n + K_n, H_1, H_2, \cdots, H_n + K_n) = 0.$$

By the evaluation of the determinants which are found in the identity Eq. (37.85), we obtain that Eqs. (37.68) and (37.69) satisfy the second equation of the system (37.63). Consequently, the functions Eqs. (37.68) and (37.69) represent a solution of the system (37.63) if $a = 0$. $\square$

38 Systems of Higher Order Functional Equations

Theorem 38.1 *The general solution of the system of functional equations*

$$f(Z_1, Z_2)f(Z_3, Z_4) + f(Z_1, Z_3)f(Z_4, Z_2) + f(Z_1, Z_4)f(Z_2, Z_3) = 0,$$

$$g(Z_1, Z_2)g(Z_3, Z_4) + g(Z_1, Z_3)g(Z_4, Z_2) + g(Z_1, Z_4)g(Z_2, Z_3)$$
$$+ \quad g(Z_1, Z_2)f^{3+2i}(Z_3, Z_4) + g(Z_1, Z_3)f^{3+2i}(Z_4, Z_2)$$
$$+ \quad g(Z_1, Z_4)f^{3+2i}(Z_2, Z_3) + g(Z_3, Z_4)f^{3+2i}(Z_1, Z_2)$$
$$+ \quad g(Z_4, Z_2)f^{3+2i}(Z_1, Z_3) + g(Z_2, Z_3)f^{3+2i}(Z_1, Z_4)$$
$$+ \quad (3 + 2i)f(Z_1, Z_2)f(Z_1, Z_3)f(Z_1, Z_4)f(Z_3, Z_4)f(Z_4, Z_2)f(Z_2, Z_3) \times$$
$$\times \quad [f^2(Z_1, Z_2)f^2(Z_3, Z_4) + f(Z_1, Z_2)f(Z_1, Z_3)f(Z_3, Z_4)f(Z_4, Z_2)$$
$$+ \quad f^2(Z_1, Z_3)f^2(Z_4, Z_2)]^i = 0, \qquad (38.1)$$

$$h(Z_1, Z_2)h(Z_3, Z_4) + h(Z_1, Z_3)h(Z_4, Z_2) + h(Z_1, Z_4)h(Z_2, Z_3)$$
$$+ \quad h(Z_1, Z_2)[g(Z_3, Z_4) + f^{3+2i}(Z_3, Z_4)]^{3+2j}$$
$$+ \quad h(Z_1, Z_3)[g(Z_4, Z_2) + f^{3+2i}(Z_4, Z_2)]^{3+2j}$$
$$+ \quad h(Z_1, Z_4)[g(Z_2, Z_3) + f^{3+2i}(Z_2, Z_3)]^{3+2j}$$

$$
\begin{aligned}
+ \quad &(3+2j)[g(\mathbf{Z}_1,\mathbf{Z}_2)+f^{3+2i}(\mathbf{Z}_1,\mathbf{Z}_2)][g(\mathbf{Z}_1,\mathbf{Z}_3)+f^{3+2i}(\mathbf{Z}_1,\mathbf{Z}_3)] \times \\
\times \quad &[g(\mathbf{Z}_1,\mathbf{Z}_4)+f^{3+2i}(\mathbf{Z}_1,\mathbf{Z}_4)][g(\mathbf{Z}_3,\mathbf{Z}_4)+f^{3+2i}(\mathbf{Z}_3,\mathbf{Z}_4)] \times \\
\times \quad &[g(\mathbf{Z}_4,\mathbf{Z}_2)+f^{3+2i}(\mathbf{Z}_4,\mathbf{Z}_2)][g(\mathbf{Z}_2,\mathbf{Z}_3)+f^{3+2i}(\mathbf{Z}_2,\mathbf{Z}_3)] \times \\
\times \quad &\{[g(\mathbf{Z}_1,\mathbf{Z}_4)+f^{3+2i}(\mathbf{Z}_1,\mathbf{Z}_4)]^2[g(\mathbf{Z}_2,\mathbf{Z}_3)+f^{3+2i}(\mathbf{Z}_2,\mathbf{Z}_3)]^2 \\
- \quad &[g(\mathbf{Z}_1,\mathbf{Z}_2)+f^{3+2i}(\mathbf{Z}_1,\mathbf{Z}_2)][g(\mathbf{Z}_1,\mathbf{Z}_3)+f^{3+2i}(\mathbf{Z}_1,\mathbf{Z}_3)] \times \\
\times \quad &[g(\mathbf{Z}_3,\mathbf{Z}_4)+f^{3+2i}(\mathbf{Z}_3,\mathbf{Z}_4)][g(\mathbf{Z}_4,\mathbf{Z}_2)+f^{3+2i}(\mathbf{Z}_4,\mathbf{Z}_2)]\}^j = 0,
\end{aligned}
$$

where f, g, $h : \mathcal{V}^2 \mapsto \mathbf{C}$ are arbitrary complex functions and $i, j \in \{0,1,2\}$, is given by the formulae

$$
f(\mathbf{U},\mathbf{V}) = \begin{vmatrix} F_1(\mathbf{U}) & F_2(\mathbf{U}) \\ F_1(\mathbf{V}) & F_2(\mathbf{V}) \end{vmatrix}, \tag{38.2}
$$

$$
g(\mathbf{U},\mathbf{V}) = \begin{vmatrix} G_1(\mathbf{U}) & G_2(\mathbf{U}) \\ G_1(\mathbf{V}) & G_2(\mathbf{V}) \end{vmatrix} - \begin{vmatrix} F_1(\mathbf{U}) & F_2(\mathbf{U}) \\ F_1(\mathbf{V}) & F_2(\mathbf{V}) \end{vmatrix}^{3+2i}, \tag{38.3}
$$
$$
i \in \{0,1,2\}
$$

$$
h(\mathbf{U},\mathbf{V}) = \begin{vmatrix} H_1(\mathbf{U}) & H_2(\mathbf{U}) \\ H_1(\mathbf{V}) & H_2(\mathbf{V}) \end{vmatrix} - \begin{vmatrix} G_1(\mathbf{U}) & G_2(\mathbf{U}) \\ G_1(\mathbf{V}) & G_2(\mathbf{V}) \end{vmatrix}^{3+2j}, \tag{38.4}
$$
$$
j \in \{0,1,2\}
$$

where F_k, G_k, H_k $(k=1,2)$ are arbitrary complex functions defined in $\mathcal{V}$.

Proof. The first functional equation of the system (38.1) according to [D. S. Mitrinović and S. B. Prešić (1962B); D. S. Mitrinović *et al.* (1963)] has a general solution given by Eq. (38.2).

The complex function $f(\mathbf{U},\mathbf{V})$ satisfied the conditions

$$
f(\mathbf{U},\mathbf{U}) \equiv 0, \quad f(\mathbf{U},\mathbf{V}) + f(\mathbf{V},\mathbf{U}) = 0. \tag{38.5}
$$

The functions $g(\mathbf{U},\mathbf{V})$ and $h(\mathbf{U},\mathbf{V})$ from the second, respectively third equation of the system (38.1) also satisfy the same conditions for $\mathbf{Z}_1 = \mathbf{Z}_2 = \mathbf{Z}_3 = \mathbf{Z}_4 = \mathbf{U} \neq \mathbf{O}$ or $\mathbf{Z}_1 = \mathbf{Z}_4 = \mathbf{U}$, $\mathbf{Z}_2 = \mathbf{Z}_3 = \mathbf{V}$.

Since $f^{3+2i}(\mathbf{Z}_r,\mathbf{Z}_k)$ have distinct powers in the second equation of the system (38.1) for $i \in \{0,1,2\}$, we obtain that the first equation of the system (38.1) implies

$$
\begin{aligned}
&f^3(\mathbf{Z}_1,\mathbf{Z}_2)f^3(\mathbf{Z}_3,\mathbf{Z}_4) + f^3(\mathbf{Z}_1,\mathbf{Z}_3)f^3(\mathbf{Z}_4,\mathbf{Z}_2) + f^3(\mathbf{Z}_1,\mathbf{Z}_4)f^3(\mathbf{Z}_2,\mathbf{Z}_3) \\
= \quad &3f(\mathbf{Z}_1,\mathbf{Z}_2)f(\mathbf{Z}_1,\mathbf{Z}_3)f(\mathbf{Z}_1,\mathbf{Z}_4)f(\mathbf{Z}_3,\mathbf{Z}_4)f(\mathbf{Z}_4,\mathbf{Z}_2)f(\mathbf{Z}_2,\mathbf{Z}_3),
\end{aligned} \tag{38.6}
$$

$$f^5(\mathbf{Z}_1,\mathbf{Z}_2)f^5(\mathbf{Z}_3,\mathbf{Z}_4) + f^5(\mathbf{Z}_1,\mathbf{Z}_3)f^5(\mathbf{Z}_4,\mathbf{Z}_2) + f^5(\mathbf{Z}_1,\mathbf{Z}_4)f^5(\mathbf{Z}_2,\mathbf{Z}_3)$$
$$= \frac{5}{2}f(\mathbf{Z}_1,\mathbf{Z}_2)f(\mathbf{Z}_1,\mathbf{Z}_3)f(\mathbf{Z}_1,\mathbf{Z}_4)f(\mathbf{Z}_3,\mathbf{Z}_4)f(\mathbf{Z}_4,\mathbf{Z}_2)f(\mathbf{Z}_2,\mathbf{Z}_3)$$
$$\times \; [f^2(\mathbf{Z}_1,\mathbf{Z}_2)f^2(\mathbf{Z}_3,\mathbf{Z}_4)+f^2(\mathbf{Z}_1,\mathbf{Z}_3)f^2(\mathbf{Z}_4,\mathbf{Z}_2)+f^2(\mathbf{Z}_1,\mathbf{Z}_4)f^2(\mathbf{Z}_2,\mathbf{Z}_3)],$$

$$f^7(\mathbf{Z}_1,\mathbf{Z}_2)f^7(\mathbf{Z}_3,\mathbf{Z}_4) + f^7(\mathbf{Z}_1,\mathbf{Z}_3)f^7(\mathbf{Z}_4,\mathbf{Z}_2) + f^7(\mathbf{Z}_1,\mathbf{Z}_4)f^7(\mathbf{Z}_2,\mathbf{Z}_3)$$
$$= \frac{7}{4}f(\mathbf{Z}_1,\mathbf{Z}_2)f(\mathbf{Z}_1,\mathbf{Z}_3)f(\mathbf{Z}_1,\mathbf{Z}_4)f(\mathbf{Z}_3,\mathbf{Z}_4)f(\mathbf{Z}_4,\mathbf{Z}_2)f(\mathbf{Z}_2,\mathbf{Z}_3) \times$$
$$\times \; [f^2(\mathbf{Z}_1,\mathbf{Z}_2)f^2(\mathbf{Z}_3,\mathbf{Z}_4)+f^2(\mathbf{Z}_1,\mathbf{Z}_3)f^2(\mathbf{Z}_4,\mathbf{Z}_2)+f^2(\mathbf{Z}_1,\mathbf{Z}_4)f^2(\mathbf{Z}_2,\mathbf{Z}_3)]^2.$$

If we apply the formulae Eq. (38.6) to the second functional equation of the system (38.1), then we obtain the following equation

$$\begin{aligned}
& [y(\mathbf{Z}_1,\mathbf{Z}_2) + f^{3+2i}(\mathbf{Z}_1,\mathbf{Z}_2)][y(\mathbf{Z}_3,\mathbf{Z}_4) + f^{3+2i}(\mathbf{Z}_3,\mathbf{Z}_4)] \quad (38.7)\\
+ \; & [g(\mathbf{Z}_1,\mathbf{Z}_3) + f^{3+2i}(\mathbf{Z}_1,\mathbf{Z}_3)][g(\mathbf{Z}_4,\mathbf{Z}_2) + f^{3+2i}(\mathbf{Z}_4,\mathbf{Z}_2)] \\
+ \; & [g(\mathbf{Z}_1,\mathbf{Z}_4) + f^{3+2i}(\mathbf{Z}_1,\mathbf{Z}_4)][g(\mathbf{Z}_2,\mathbf{Z}_3) + f^{3+2i}(\mathbf{Z}_2,\mathbf{Z}_3)] = 0.
\end{aligned}$$

By repeating the previous procedure to the last equation of the system (38.1), we obtain

$$\begin{aligned}
& \{h(\mathbf{Z}_1,\mathbf{Z}_2) + [g(\mathbf{Z}_1,\mathbf{Z}_2) + f^{3+2i}(\mathbf{Z}_1,\mathbf{Z}_2)]^{3+2j}\} \times \quad (38.8)\\
\times \; & \{h(\mathbf{Z}_3,\mathbf{Z}_4) + [g(\mathbf{Z}_3,\mathbf{Z}_4) + f^{3+2i}(\mathbf{Z}_3,\mathbf{Z}_4)]^{3+2j}\} \\
+ \; & \{h(\mathbf{Z}_1,\mathbf{Z}_3) + [g(\mathbf{Z}_1,\mathbf{Z}_3) + f^{3+2i}(\mathbf{Z}_1,\mathbf{Z}_3)]^{3+2j}\} \times \\
\times \; & \{h(\mathbf{Z}_4,\mathbf{Z}_2) + [g(\mathbf{Z}_4,\mathbf{Z}_2) + f^{3+2i}(\mathbf{Z}_4,\mathbf{Z}_2)]^{3+2j}\} \\
+ \; & \{h(\mathbf{Z}_1,\mathbf{Z}_4) + [g(\mathbf{Z}_1,\mathbf{Z}_4) + f^{3+2i}(\mathbf{Z}_1,\mathbf{Z}_4)]^{3+2j}\} \times \\
\times \; & \{h(\mathbf{Z}_2,\mathbf{Z}_3) + [g(\mathbf{Z}_2,\mathbf{Z}_3) + f^{3+2i}(\mathbf{Z}_2,\mathbf{Z}_3)]^{3+2j}\} = 0.
\end{aligned}$$

The system of vector functional equations (38.1) with the unknown functions f, g, $h : \mathcal{V}^2 \mapsto \mathbf{C}$ and $i, j \in \{0, 1, 2\}$ is equivalent to the system formed by the first equation of the system (38.1) and the equations (38.7) and (38.8), whose general solution is given by the equations (38.2), (38.3) and (38.4). $\qquad\square$

Theorem 38.2 *The general solution of the following system of functional equations*

$$f(\mathbf{Z}_1, \mathbf{Z}_2, \mathbf{Z}_3)f(\mathbf{Z}_4, \mathbf{Z}_5, \mathbf{Z}_6) + f(\mathbf{Z}_2, \mathbf{Z}_1, \mathbf{Z}_4)f(\mathbf{Z}_3, \mathbf{Z}_5, \mathbf{Z}_6)$$
$$+ \quad f(\mathbf{Z}_1, \mathbf{Z}_2, \mathbf{Z}_5)f(\mathbf{Z}_3, \mathbf{Z}_4, \mathbf{Z}_6) + f(\mathbf{Z}_2, \mathbf{Z}_1, \mathbf{Z}_6)f(\mathbf{Z}_3, \mathbf{Z}_4, \mathbf{Z}_5) = 0,$$

$$g(\mathbf{Z}_1, \mathbf{Z}_2, \mathbf{Z}_3)g(\mathbf{Z}_4, \mathbf{Z}_5, \mathbf{Z}_6) + g(\mathbf{Z}_2, \mathbf{Z}_1, \mathbf{Z}_4)g(\mathbf{Z}_3, \mathbf{Z}_5, \mathbf{Z}_6) \qquad (38.9)$$
$$+ \quad g(\mathbf{Z}_1, \mathbf{Z}_2, \mathbf{Z}_5)g(\mathbf{Z}_3, \mathbf{Z}_4, \mathbf{Z}_6) + g(\mathbf{Z}_2, \mathbf{Z}_1, \mathbf{Z}_6)g(\mathbf{Z}_3, \mathbf{Z}_4, \mathbf{Z}_5)$$
$$+ \quad g(\mathbf{Z}_1, \mathbf{Z}_2, \mathbf{Z}_3)f^3(\mathbf{Z}_4, \mathbf{Z}_5, \mathbf{Z}_6) + g(\mathbf{Z}_4, \mathbf{Z}_5, \mathbf{Z}_6)f^3(\mathbf{Z}_1, \mathbf{Z}_2, \mathbf{Z}_3)$$
$$+ \quad g(\mathbf{Z}_2, \mathbf{Z}_1, \mathbf{Z}_4)f^3(\mathbf{Z}_3, \mathbf{Z}_5, \mathbf{Z}_6) + g(\mathbf{Z}_3, \mathbf{Z}_5, \mathbf{Z}_6)f^3(\mathbf{Z}_2, \mathbf{Z}_1, \mathbf{Z}_4)$$
$$+ \quad g(\mathbf{Z}_1, \mathbf{Z}_2, \mathbf{Z}_5)f^3(\mathbf{Z}_3, \mathbf{Z}_4, \mathbf{Z}_6) + g(\mathbf{Z}_3, \mathbf{Z}_4, \mathbf{Z}_6)f^3(\mathbf{Z}_1, \mathbf{Z}_2, \mathbf{Z}_5)$$
$$+ \quad g(\mathbf{Z}_2, \mathbf{Z}_1, \mathbf{Z}_6)f^3(\mathbf{Z}_3, \mathbf{Z}_4, \mathbf{Z}_5) + g(\mathbf{Z}_3, \mathbf{Z}_4, \mathbf{Z}_5)f^3(\mathbf{Z}_2, \mathbf{Z}_1, \mathbf{Z}_6)$$
$$+ \quad 3f(\mathbf{Z}_1, \mathbf{Z}_2, \mathbf{Z}_3)f(\mathbf{Z}_2, \mathbf{Z}_1, \mathbf{Z}_4)f(\mathbf{Z}_4, \mathbf{Z}_5, \mathbf{Z}_6)f(\mathbf{Z}_3, \mathbf{Z}_5, \mathbf{Z}_6) \times$$
$$\times \quad [f(\mathbf{Z}_1, \mathbf{Z}_2, \mathbf{Z}_5)f(\mathbf{Z}_3, \mathbf{Z}_4, \mathbf{Z}_6) + f(\mathbf{Z}_2, \mathbf{Z}_1, \mathbf{Z}_6)f(\mathbf{Z}_3, \mathbf{Z}_4, \mathbf{Z}_5)]$$
$$+ \quad 3f(\mathbf{Z}_1, \mathbf{Z}_2, \mathbf{Z}_5)f(\mathbf{Z}_2, \mathbf{Z}_1, \mathbf{Z}_6)f(\mathbf{Z}_3, \mathbf{Z}_4, \mathbf{Z}_6)f(\mathbf{Z}_3, \mathbf{Z}_4, \mathbf{Z}_5) \times$$
$$\times \quad [f(\mathbf{Z}_1, \mathbf{Z}_2, \mathbf{Z}_3)f(\mathbf{Z}_4, \mathbf{Z}_5, \mathbf{Z}_6) + f(\mathbf{Z}_2, \mathbf{Z}_1, \mathbf{Z}_4)f(\mathbf{Z}_3, \mathbf{Z}_5, \mathbf{Z}_6)] = 0,$$

where $f,\ g : \mathcal{V}^3 \mapsto \mathbf{C}$ *are arbitrary complex functions, is given by the following formulae*

$$f(\mathbf{U}, \mathbf{V}, \mathbf{W}) = \begin{vmatrix} F_1(\mathbf{U}) & F_2(\mathbf{U}) & F_3(\mathbf{U}) \\ F_1(\mathbf{V}) & F_2(\mathbf{V}) & F_3(\mathbf{V}) \\ F_1(\mathbf{W}) & F_2(\mathbf{W}) & F_3(\mathbf{W}) \end{vmatrix}, \qquad (38.10)$$

$$g(\mathbf{U}, \mathbf{V}, \mathbf{W}) \qquad\qquad\qquad\qquad (38.11)$$
$$= \begin{vmatrix} G_1(\mathbf{U}) & G_2(\mathbf{U}) & G_3(\mathbf{U}) \\ G_1(\mathbf{V}) & G_2(\mathbf{V}) & G_3(\mathbf{V}) \\ G_1(\mathbf{W}) & G_2(\mathbf{W}) & G_3(\mathbf{W}) \end{vmatrix} - \begin{vmatrix} F_1(\mathbf{U}) & F_2(\mathbf{U}) & F_3(\mathbf{U}) \\ F_1(\mathbf{V}) & F_2(\mathbf{V}) & F_3(\mathbf{V}) \\ F_1(\mathbf{W}) & F_2(\mathbf{W}) & F_3(\mathbf{W}) \end{vmatrix}^3,$$

where F_i, G_i $(i = 1, 2, 3)$ *are arbitrary complex functions defined in* $\mathcal{V}$.

Proof.   For the first equation of the system (38.9), the general solution is given by Eq. (38.10).

We can rearrange the terms in the second equation of the system (38.9) as follows:

$$[g(\mathbf{Z}_1, \mathbf{Z}_2, \mathbf{Z}_3) + f^3(\mathbf{Z}_1, \mathbf{Z}_2, \mathbf{Z}_3)][g(\mathbf{Z}_4, \mathbf{Z}_5, \mathbf{Z}_6) + f^3(\mathbf{Z}_4, \mathbf{Z}_5, \mathbf{Z}_6)] \qquad (38.12)$$

$$+ \quad [g(\mathbf{Z}_2, \mathbf{Z}_1, \mathbf{Z}_4) + f^3(\mathbf{Z}_2, \mathbf{Z}_1, \mathbf{Z}_4)][g(\mathbf{Z}_3, \mathbf{Z}_5, \mathbf{Z}_6) + f^3(\mathbf{Z}_3, \mathbf{Z}_5, \mathbf{Z}_6)]$$
$$+ \quad [g(\mathbf{Z}_1, \mathbf{Z}_2, \mathbf{Z}_5) + f^3(\mathbf{Z}_1, \mathbf{Z}_2, \mathbf{Z}_5)][g(\mathbf{Z}_3, \mathbf{Z}_4, \mathbf{Z}_6) + f^3(\mathbf{Z}_3, \mathbf{Z}_4, \mathbf{Z}_6)]$$
$$+ \quad [g(\mathbf{Z}_2, \mathbf{Z}_1, \mathbf{Z}_6) + f^3(\mathbf{Z}_2, \mathbf{Z}_1, \mathbf{Z}_6)][g(\mathbf{Z}_3, \mathbf{Z}_4, \mathbf{Z}_5) + f^3(\mathbf{Z}_3, \mathbf{Z}_4, \mathbf{Z}_5)] = 0$$

by using the identity

$$A^3 + B^3 + C^3 + D^3 = 3AB(C+D) + 3CD(A+B) \quad \text{if} \quad A+B+C+D = 0.$$

Therefore, the system (38.9) with two unknown functions f, $g : \mathcal{V}^3 \mapsto \mathbf{C}$ is equivalent to the system formed by the first equation of the system (38.9) and the equation (38.12), whose general solution is given by Eqs. (38.10) and (38.11). $\qquad\square$

Theorem 38.3 *The general continuous solution of the following system of functional equations*

$$f_{11}(\mathbf{Z}_1, \mathbf{Z}_2 + \mathbf{Z}_3) + f_{21}(\mathbf{Z}_2, \mathbf{Z}_3 + \mathbf{Z}_1) = f_{31}(\mathbf{Z}_3, \mathbf{Z}_1 + \mathbf{Z}_2),$$

$$f_{12}(\mathbf{Z}_1, \mathbf{Z}_2 + \mathbf{Z}_3) + 2f_{11}(\mathbf{Z}_1, \mathbf{Z}_2 + \mathbf{Z}_3)f_{21}(\mathbf{Z}_2, \mathbf{Z}_3 + \mathbf{Z}_1) \quad (38.13)$$
$$+ \quad f_{22}(\mathbf{Z}_2, \mathbf{Z}_3 + \mathbf{Z}_1) = f_{32}(\mathbf{Z}_3, \mathbf{Z}_1 + \mathbf{Z}_2),$$

$$f_{13}(\mathbf{Z}_1, \mathbf{Z}_2 + \mathbf{Z}_3) + 3f_{12}(\mathbf{Z}_1, \mathbf{Z}_2 + \mathbf{Z}_3)f_{21}(\mathbf{Z}_2, \mathbf{Z}_3 + \mathbf{Z}_1)$$
$$+ \quad 3f_{11}(\mathbf{Z}_1, \mathbf{Z}_2 + \mathbf{Z}_3)f_{22}(\mathbf{Z}_2, \mathbf{Z}_3 + \mathbf{Z}_1)$$
$$+ \quad f_{23}(\mathbf{Z}_2, \mathbf{Z}_3 + \mathbf{Z}_1) = f_{33}(\mathbf{Z}_3, \mathbf{Z}_1 + \mathbf{Z}_2),$$

$$f_{14}(\mathbf{Z}_1, \mathbf{Z}_2 + \mathbf{Z}_3) + 4f_{13}(\mathbf{Z}_1, \mathbf{Z}_2 + \mathbf{Z}_3)f_{21}(\mathbf{Z}_2, \mathbf{Z}_3 + \mathbf{Z}_1)$$
$$+ \quad 6f_{12}(\mathbf{Z}_1, \mathbf{Z}_2 + \mathbf{Z}_3)f_{22}(\mathbf{Z}_2, \mathbf{Z}_3 + \mathbf{Z}_1)$$
$$+ \quad 4f_{11}(\mathbf{Z}_1, \mathbf{Z}_2 + \mathbf{Z}_3)f_{23}(\mathbf{Z}_2, \mathbf{Z}_3 + \mathbf{Z}_1)$$
$$+ \quad f_{24}(\mathbf{Z}_2, \mathbf{Z}_3 + \mathbf{Z}_1) = f_{34}(\mathbf{Z}_3, \mathbf{Z}_1 + \mathbf{Z}_2),$$

$$f_{15}(\mathbf{Z}_1, \mathbf{Z}_2 + \mathbf{Z}_3) + 5f_{14}(\mathbf{Z}_1, \mathbf{Z}_2 + \mathbf{Z}_3)f_{21}(\mathbf{Z}_2, \mathbf{Z}_3 + \mathbf{Z}_1)$$
$$+ \quad 10f_{13}(\mathbf{Z}_1, \mathbf{Z}_2 + \mathbf{Z}_3)f_{22}(\mathbf{Z}_2, \mathbf{Z}_3 + \mathbf{Z}_1)$$
$$+ \quad 10f_{12}(\mathbf{Z}_1, \mathbf{Z}_2 + \mathbf{Z}_3)f_{23}(\mathbf{Z}_2, \mathbf{Z}_3 + \mathbf{Z}_1)$$
$$+ \quad 5f_{11}(\mathbf{Z}_1, \mathbf{Z}_2 + \mathbf{Z}_3)f_{24}(\mathbf{Z}_2, \mathbf{Z}_3 + \mathbf{Z}_1)$$
$$+ \quad f_{25}(\mathbf{Z}_2, \mathbf{Z}_3 + \mathbf{Z}_1) = f_{35}(\mathbf{Z}_3, \mathbf{Z}_1 + \mathbf{Z}_2),$$

where f_{1i}, f_{2i}, $f_{3i} : \mathcal{V}^2 \mapsto \mathbf{C}$ $(1 \leq i \leq 5)$ are arbitrary complex functions, is given by the formulae

$$
\begin{aligned}
f_{i1}(\mathbf{U}, \mathbf{V}) \;=\;& F_1(\mathbf{U}+\mathbf{V})\,\mathrm{Re}\,(2\mathbf{U}-\mathbf{V}) + F_2(\mathbf{U}+\mathbf{V})\,\mathrm{Im}\,(2\mathbf{U}-\mathbf{V}) \\
+\;& G_i(\mathbf{U}+\mathbf{V}) \qquad (i=1,2), \hspace{3cm} (38.14)
\end{aligned}
$$

$$
\begin{aligned}
f_{31}(\mathbf{U}, \mathbf{V}) \;=\;& F_1(\mathbf{U}+\mathbf{V})\,\mathrm{Re}\,(\mathbf{V}-2\mathbf{U}) + F_2(\mathbf{U}+\mathbf{V})\,\mathrm{Im}\,(\mathbf{V}-2\mathbf{U}) \\
+\;& G_1(\mathbf{U}+\mathbf{V}) + G_2(\mathbf{U}+\mathbf{V}),
\end{aligned}
$$

$$
\begin{aligned}
f_{i2}(\mathbf{U}, \mathbf{V}) \;=\;& L_1(\mathbf{U}+\mathbf{V})\,\mathrm{Re}\,(2\mathbf{U}-\mathbf{V}) + L_2(\mathbf{U}+\mathbf{V})\,\mathrm{Im}\,(2\mathbf{U}-\mathbf{V}) \\
+\;& H_i(\mathbf{U}+\mathbf{V}) + f_{i1}^2(\mathbf{U}, \mathbf{V}) \qquad (i=1,2), \hspace{1.5cm} (38.15)
\end{aligned}
$$

$$
\begin{aligned}
f_{32}(\mathbf{U}, \mathbf{V}) \;=\;& L_1(\mathbf{U}+\mathbf{V})\,\mathrm{Re}\,(\mathbf{V}-2\mathbf{U}) + L_2(\mathbf{U}+\mathbf{V})\,\mathrm{Im}\,(\mathbf{V}-2\mathbf{U}) \\
+\;& H_1(\mathbf{U}+\mathbf{V}) + H_2(\mathbf{U}+\mathbf{V}) + f_{31}^2(\mathbf{U}, \mathbf{V}),
\end{aligned}
$$

$$
\begin{aligned}
f_{i3}(\mathbf{U}, \mathbf{V}) \;=\;& M_1(\mathbf{U}+\mathbf{V})\,\mathrm{Re}\,(2\mathbf{U}-\mathbf{V}) + M_2(\mathbf{U}+\mathbf{V})\,\mathrm{Im}\,(2\mathbf{U}-\mathbf{V}) \\
+\;& K_i(\mathbf{U}+\mathbf{V}) + 3f_{i2}^2(\mathbf{U}, \mathbf{V})f_{i1}(\mathbf{U}, \mathbf{V}) \\
-\;& 2f_{i1}^3(\mathbf{U}, \mathbf{V}) \qquad (i=1,2), \hspace{3cm} (38.16)
\end{aligned}
$$

$$
\begin{aligned}
f_{33}(\mathbf{U}, \mathbf{V}) \;=\;& M_1(\mathbf{U}+\mathbf{V})\,\mathrm{Re}\,(\mathbf{V}-2\mathbf{U}) + M_2(\mathbf{U}+\mathbf{V})\,\mathrm{Im}\,(\mathbf{V}-2\mathbf{U}) \\
+\;& K_1(\mathbf{U}+\mathbf{V}) + K_2(\mathbf{U}+\mathbf{V}) \\
+\;& 3f_{32}(\mathbf{U}, \mathbf{V})f_{31}(\mathbf{U}, \mathbf{V}) - 2f_{31}^3(\mathbf{U}, \mathbf{V}),
\end{aligned}
$$

$$
\begin{aligned}
f_{i4}(\mathbf{U}, \mathbf{V}) \;=\;& P_1(\mathbf{U}+\mathbf{V})\,\mathrm{Re}\,(2\mathbf{U}-\mathbf{V}) + P_2(\mathbf{U}+\mathbf{V})\,\mathrm{Im}\,(2\mathbf{U}-\mathbf{V}) \\
+\;& Q_i(\mathbf{U}+\mathbf{V}) + 4f_{i3}(\mathbf{U}, \mathbf{V})f_{i1}(\mathbf{U}, \mathbf{V}) - 12f_{i2}(\mathbf{U}, \mathbf{V})f_{i1}^2(\mathbf{U}, \mathbf{V}) \\
+\;& 3f_{i2}^2(\mathbf{U}, \mathbf{V}) + 6f_{i1}^4(\mathbf{U}, \mathbf{V}) \quad (i=1,2), \hspace{1.5cm} (38.17)
\end{aligned}
$$

$$
\begin{aligned}
f_{34}(\mathbf{U}, \mathbf{V}) \;=\;& P_1(\mathbf{U}+\mathbf{V})\,\mathrm{Re}\,(\mathbf{V}-2\mathbf{U}) + P_2(\mathbf{U}+\mathbf{V})\,\mathrm{Im}\,(\mathbf{V}-2\mathbf{U}) \\
+\;& Q_1(\mathbf{U}+\mathbf{V}) + Q_2(\mathbf{U}+\mathbf{V}) + 4f_{33}(\mathbf{U}, \mathbf{V})f_{31}(\mathbf{U}, \mathbf{V}) \\
-\;& 12f_{32}(\mathbf{U}, \mathbf{V})f_{31}^2(\mathbf{U}, \mathbf{V}) + 3f_{32}^2(\mathbf{U}, \mathbf{V}) + 6f_{31}^4(\mathbf{U}, \mathbf{V}),
\end{aligned}
$$

$$
\begin{aligned}
f_{i5}(\mathbf{U}, \mathbf{V}) \;=\;& R_1(\mathbf{U}+\mathbf{V})\,\mathrm{Re}\,(2\mathbf{U}-\mathbf{V}) + R_2(\mathbf{U}+\mathbf{V})\,\mathrm{Im}\,(2\mathbf{U}-\mathbf{V}) \\
+\;& S_i(\mathbf{U}+\mathbf{V}) + 5f_{i4}(\mathbf{U}, \mathbf{V})f_{i1}(\mathbf{U}, \mathbf{V}) \\
-\;& 20f_{i3}(\mathbf{U}, \mathbf{V})f_{i1}^2(\mathbf{U}, \mathbf{V}) + 60f_{i2}(\mathbf{U}, \mathbf{V})f_{i1}^3(\mathbf{U}, \mathbf{V}) \\
-\;& 30f_{i2}^2(\mathbf{U}, \mathbf{V})f_{i1}(\mathbf{U}, \mathbf{V}) + 10f_{i3}(\mathbf{U}, \mathbf{V})f_{i2}(\mathbf{U}, \mathbf{V}) \\
-\;& 24f_{i1}^5(\mathbf{U}, \mathbf{V}) \qquad (i=1,2), \hspace{3cm} (38.18)
\end{aligned}
$$

$$
\begin{aligned}
f_{35}(\mathbf{U}, \mathbf{V}) \;=\;& R_1(\mathbf{U}+\mathbf{V})\,\mathrm{Re}\,(\mathbf{V}-2\mathbf{U}) + R_2(\mathbf{U}+\mathbf{V})\,\mathrm{Im}\,(\mathbf{V}-2\mathbf{U}) \\
+\;& S_1(\mathbf{U}+\mathbf{V}) + S_2(\mathbf{U}+\mathbf{V}) \\
+\;& 5f_{34}(\mathbf{U}, \mathbf{V})f_{31}(\mathbf{U}, \mathbf{V}) - 20f_{33}(\mathbf{U}, \mathbf{V})f_{31}^2(\mathbf{U}, \mathbf{V})
\end{aligned}
$$

$$+ \quad 60f_{32}(\mathbf{U},\mathbf{V})f_{31}^3(\mathbf{U},\mathbf{V}) - 30f_{32}^2(\mathbf{U},\mathbf{V})f_{31}(\mathbf{U},\mathbf{V})$$
$$+ \quad 10f_{33}(\mathbf{U},\mathbf{V})f_{32}(\mathbf{U},\mathbf{V}) - 24f_{31}^5(\mathbf{U},\mathbf{V}),$$

$F_i,\ L_i,\ M_i,\ P_i,\ R_i : \mathcal{V} \mapsto \mathcal{L}(\mathcal{V}^0, \mathbf{C}),\ G_i,\ H_i,\ K_i,\ Q_i,\ S_i :\ \mathcal{V} \mapsto \mathbf{C}\ (i = 1, 2)$ *are arbitrary complex functions.*

Proof. The general continuous solution of the first equation of the system (38.13) according to [S. B. Prešić and D. Ž. Djoković (1961)] is given by Eqs. (38.14).

Now, if we use the first equation of the system (38.13), then the second equation takes on the following form

$$[f_{12}(\mathbf{Z}_1, \mathbf{Z}_2 + \mathbf{Z}_3) - f_{11}^2(\mathbf{Z}_1, \mathbf{Z}_2 + \mathbf{Z}_3)] \qquad (38.19)$$
$$+ \quad [f_{22}(\mathbf{Z}_2, \mathbf{Z}_3 + \mathbf{Z}_1) - f_{21}^2(\mathbf{Z}_2, \mathbf{Z}_3 + \mathbf{Z}_1)]$$
$$= \quad [f_{32}(\mathbf{Z}_3, \mathbf{Z}_1 + \mathbf{Z}_2) - f_{31}^2(\mathbf{Z}_3, \mathbf{Z}_1 + \mathbf{Z}_2)].$$

This equation has a form like the first equation of the system (38.13).

If we take into account the previous two equations, then the third equation of the system (38.13) becomes

$$[f_{13}(\mathbf{Z}_1, \mathbf{Z}_2 + \mathbf{Z}_3) - 3f_{12}(\mathbf{Z}_1, \mathbf{Z}_2 + \mathbf{Z}_3)f_{11}(\mathbf{Z}_1, \mathbf{Z}_2 + \mathbf{Z}_3) \qquad (38.20)$$
$$+ \quad 2f_{11}^3(\mathbf{Z}_1, \mathbf{Z}_2 + \mathbf{Z}_3)] + [f_{23}(\mathbf{Z}_2, \mathbf{Z}_3 + \mathbf{Z}_1)$$
$$- \quad 3f_{22}(\mathbf{Z}_2, \mathbf{Z}_3 + \mathbf{Z}_1)f_{21}(\mathbf{Z}_2, \mathbf{Z}_3 + \mathbf{Z}_1) + 2f_{21}^3(\mathbf{Z}_2, \mathbf{Z}_3 + \mathbf{Z}_1)]$$
$$= \quad [f_{33}(\mathbf{Z}_3, \mathbf{Z}_1 + \mathbf{Z}_2) - 3f_{32}(\mathbf{Z}_3, \mathbf{Z}_1 + \mathbf{Z}_2)f_{31}(\mathbf{Z}_3, \mathbf{Z}_1 + \mathbf{Z}_2)$$
$$+ \quad 2f_{31}^3(\mathbf{Z}_3, \mathbf{Z}_1 + \mathbf{Z}_2)].$$

The equation (38.20) has also a form like the first equation of the system (38.13).

If we apply the above two equations, then the fourth equation of the system (38.13) takes on the form

$$[f_{14}(\mathbf{Z}_1, \mathbf{Z}_2 + \mathbf{Z}_3) - 4f_{13}(\mathbf{Z}_1, \mathbf{Z}_2 + \mathbf{Z}_3)f_{11}(\mathbf{Z}_1, \mathbf{Z}_2 + \mathbf{Z}_3) \qquad (38.21)$$
$$+ \quad 12f_{12}(\mathbf{Z}_1, \mathbf{Z}_2 + \mathbf{Z}_3)f_{11}^3(\mathbf{Z}_1, \mathbf{Z}_2 + \mathbf{Z}_3)$$
$$- \quad 3f_{12}^2(\mathbf{Z}_1, \mathbf{Z}_2 + \mathbf{Z}_3) - 6f_{11}^4(\mathbf{Z}_1, \mathbf{Z}_2 + \mathbf{Z}_3)]$$
$$+ \quad [f_{24}(\mathbf{Z}_2, \mathbf{Z}_3 + \mathbf{Z}_1) - 4f_{23}(\mathbf{Z}_2, \mathbf{Z}_3 + \mathbf{Z}_1)f_{21}(\mathbf{Z}_2, \mathbf{Z}_3 + \mathbf{Z}_1)$$
$$+ \quad 12f_{22}(\mathbf{Z}_2, \mathbf{Z}_3 + \mathbf{Z}_1)f_{21}^2(\mathbf{Z}_2, \mathbf{Z}_3 + \mathbf{Z}_1)$$
$$- \quad 3f_{22}^2(\mathbf{Z}_2, \mathbf{Z}_3 + \mathbf{Z}_1) - 6f_{21}^4(\mathbf{Z}_2, \mathbf{Z}_3 + \mathbf{Z}_1)]$$

$$
\begin{aligned}
=\ & [f_{34}(\mathbf{Z}_3,\mathbf{Z}_1+\mathbf{Z}_2)-4f_{33}(\mathbf{Z}_3,\mathbf{Z}_1+\mathbf{Z}_2)f_{31}(\mathbf{Z}_3,\mathbf{Z}_1+\mathbf{Z}_2)\\
+\ & 12f_{32}(\mathbf{Z}_3,\mathbf{Z}_1+\mathbf{Z}_2)f_{31}^{2}(\mathbf{Z}_3,\mathbf{Z}_1+\mathbf{Z}_2)\\
-\ & 3f_{32}^{2}(\mathbf{Z}_3,\mathbf{Z}_1+\mathbf{Z}_2)-6f_{31}^{4}(\mathbf{Z}_3,\mathbf{Z}_1+\mathbf{Z}_2)].
\end{aligned}
$$

Thus also the equation (38.21) has a form like the first equation of the system (38.13).

Similarly, the last equation of the system (38.13) is reduced to the following equation

$$
\begin{aligned}
& [f_{15}(\mathbf{Z}_1,\mathbf{Z}_2+\mathbf{Z}_3)-5f_{14}(\mathbf{Z}_1,\mathbf{Z}_2+\mathbf{Z}_3)f_{11}(\mathbf{Z}_1,\mathbf{Z}_2+\mathbf{Z}_3) \quad (38.22)\\
+\ & 20f_{13}(\mathbf{Z}_1,\mathbf{Z}_2+\mathbf{Z}_3)f_{11}^{2}(\mathbf{Z}_1,\mathbf{Z}_2+\mathbf{Z}_3)\\
-\ & 60f_{12}(\mathbf{Z}_1,\mathbf{Z}_2+\mathbf{Z}_3)f_{11}^{3}(\mathbf{Z}_1,\mathbf{Z}_2+\mathbf{Z}_3)\\
+\ & 30f_{12}^{2}(\mathbf{Z}_1,\mathbf{Z}_2+\mathbf{Z}_3)f_{11}(\mathbf{Z}_1,\mathbf{Z}_2+\mathbf{Z}_3)\\
-\ & 10f_{13}(\mathbf{Z}_1,\mathbf{Z}_2+\mathbf{Z}_3)f_{12}(\mathbf{Z}_1,\mathbf{Z}_2+\mathbf{Z}_3)+24f_{11}^{5}(\mathbf{Z}_1,\mathbf{Z}_2+\mathbf{Z}_3)]\\
+\ & [f_{25}(\mathbf{Z}_2,\mathbf{Z}_3+\mathbf{Z}_1)-5f_{24}(\mathbf{Z}_2,\mathbf{Z}_3+\mathbf{Z}_1)f_{21}(\mathbf{Z}_2,\mathbf{Z}_3+\mathbf{Z}_1)\\
+\ & 20f_{23}(\mathbf{Z}_2,\mathbf{Z}_3+\mathbf{Z}_1)f_{21}^{2}(\mathbf{Z}_2,\mathbf{Z}_3+\mathbf{Z}_1)\\
-\ & 60f_{22}(\mathbf{Z}_2,\mathbf{Z}_3+\mathbf{Z}_1)f_{21}^{2}(\mathbf{Z}_2,\mathbf{Z}_3+\mathbf{Z}_1)\\
+\ & 30f_{22}^{2}(\mathbf{Z}_2,\mathbf{Z}_3+\mathbf{Z}_1)f_{21}(\mathbf{Z}_2,\mathbf{Z}_3+\mathbf{Z}_1)\\
-\ & 10f_{23}(\mathbf{Z}_2,\mathbf{Z}_3+\mathbf{Z}_1)f_{22}(\mathbf{Z}_2,\mathbf{Z}_3+\mathbf{Z}_1)+24f_{21}^{5}(\mathbf{Z}_2,\mathbf{Z}_3+\mathbf{Z}_1)\\
=\ & [f_{35}(\mathbf{Z}_3,\mathbf{Z}_1+\mathbf{Z}_2)-5f_{34}(\mathbf{Z}_3,\mathbf{Z}_1+\mathbf{Z}_2)f_{31}(\mathbf{Z}_3,\mathbf{Z}_1+\mathbf{Z}_2)\\
+\ & 20f_{33}(\mathbf{Z}_3,\mathbf{Z}_1+\mathbf{Z}_2)f_{31}^{2}(\mathbf{Z}_3,\mathbf{Z}_1+\mathbf{Z}_2)\\
-\ & 60f_{32}(\mathbf{Z}_3,\mathbf{Z}_1+\mathbf{Z}_2)f_{31}^{3}(\mathbf{Z}_3,\mathbf{Z}_1+\mathbf{Z}_2)\\
+\ & 30f_{32}^{2}(\mathbf{Z}_3,\mathbf{Z}_1+\mathbf{Z}_2)f_{31}(\mathbf{Z}_3,\mathbf{Z}_1+\mathbf{Z}_2)\\
-\ & 10f_{33}(\mathbf{Z}_3,\mathbf{Z}_1+\mathbf{Z}_2)f_{32}(\mathbf{Z}_3,\mathbf{Z}_1+\mathbf{Z}_2)+f_{31}^{5}(\mathbf{Z}_3,\mathbf{Z}_1+\mathbf{Z}_2)],
\end{aligned}
$$

which has a form like the first equation of the system (38.13).

Thus system (38.13) with five functional equations and fifteen unknown functions $\quad f_{ij}:\mathcal{V}^2\mapsto\mathbf{C}\quad(i=1,2,3;\ j=1,2,3,4,5)$ is equivalent to the system formed by the first equation of the system (38.13) and the equations (38.19), (38.20), (38.21) and (38.22), whose general continuous solution is given by Eqs. (38.14), (38.15), (38.16), (38.17) and (38.18). $\square$

Such systems have not been taken into account in the references [J. Aczél (1966); Ya. Brodskiy and A. K. Slipenko (1987); L. I. Davidov (1977); M. Ghermǎnescu (1960); D. S. Mitrinović and J. E. Pečarić (1991)].

Bibliography

Aczél, J. (1958) "Vorlesungen über Funktionalgleichungen I", *Math. Nachr.* **19**, 87.

Aczél, J. (1966) "Lectures on Functional Equations and Their Applications", Academic Press, New York–London.

Aczél, J., Ghermănescu, M. and Hosszú, M. (1960) "On cyclic equations", *Publ. Math. Inst. Hung. Acad. Sci.* **5A**, 215.

Brodskiy, Ya. and Slipenko, A. K. (1987) "Functional Equations", Kiev (in Russian).

Carlitz, L. (1963) "A special functional equation", *Univ. Beograd. Publ. Elektrotehn. Fak. Ser. Mat. Fiz.* **97–99**, 1.

Davidov, L. I. (1977) "Functional Equations", Sofia (in Bulgarian).

Djoković, D. Ž. (1961) "Solution d'une équation fonctionnelle cyclique", *Bull. Soc. Math. Phys. R. P. Serbie* **13**, 185.

Djoković, D. Ž. (1962) "Sur l'équation fonctionnelle $f(x_3, x_1 + x_2) - f(x_1, x_2 + x_3) + f(x_2, x_1) - f(x_2, x_3) = 0$", *Univ. Beograd. Publ. Elektrotehn. Fak. Ser. Mat. Fiz.* **70–76**, 7.

Djoković, D. Ž. (1964) "Generalization of a result of Aczél, Ghermănescu and Hosszú", *Publ. Math. Inst. Hung. Acad. Sci.* **9A**, 51.

Djoković, D. Ž. (1965) "General solution of a functional equation", *Univ. Beograd. Publ. Elektrotehn. Fak. Ser. Mat. Fiz.* **132–142**, 55.

Djoković, D. Ž. (1965) "A special cyclic functional equation", *Univ. Beograd. Publ. Elektrotehn. Fak. Ser. Mat. Fiz.* **143–155**, 45.

Djoković, D. Ž., Djordjević, R. Ž. and Vasić, P. M. (1966) "On a class of functional equations", *Publ. Inst. Math. Beograd* **6**, (20), 65.

Djordjević, R. Ž. (1965) "Solution d'une équation fonctionnelle à plusieurs fonctions inconnues", *Univ. Beograd. Publ. Elektrotehn. Fak. Ser. Mat. Fiz.* **132–142**, 51.

Djordjević, R. Ž. (1965) "Solution d'une système d'équations fonctionnelles linéaires", *Univ. Beograd. Publ. Elektrotehn. Fak. Ser. Mat. Fiz.* **143–**

155, 61.

Djordjević, R. Ž. and Vasić, P. M. (1967) "O jednoj klasi funkcionalnih jednačina", *Mat. Vesnik* **4**, (19), 33.

Eichhorn, W. (1963) "Lösung einer Klasse von Funktionalgleichungssystemen", *Arch. Math.* **14**, 266.

Fréchet, M. (1909) "Une définition fonctionnelles des polynômes", *Nouv. Ann. Math.* **9**, (4), 145.

Gheorghiu, O. E. (1963) "Sur une équation fonctionelle matricielle", *C. R. Acad. Sci. Paris* **256**, 3562.

Ghermănescu, M. (1940) "Sur quelques équations fonctionnelles linéaires", *Bull. Soc. Math. France* **68**, 109.

Ghermănescu, M. (1953) "Un sistem de equaţii funcţionale", *Acad. R. P. Române. Bul. Şti. Secţ. Şti. Mat. Fiz.* **5**, 575.

Ghermănescu, M. (1955) "Sisteme de Ecuaţii Funcţionale Liniare de Primul Ordin", Bucureşti.

Ghermănescu, M. (1960) "Ecuaţii Funcţionale", Editura Academiei Republicii Populare Române, Bucureşti.

Hosszú, M. (1961) "A lineáris függvényegyenletek egy osztályáról", *Magyar Tud. Akad. Mat. Fiz. Oszt. Közl.* **11**, 249.

Hosszú, M. (1963) "On a class of functional equations", *Publ. Inst. Math. Beograd* **3**, (17), 53.

Kuczma, M. (1968) "Functional Equations in a Single Variable", *Monografie Matematyczne* **46**, Państwowe Wydawnictwo Naukowe, Warsaw.

Kurepa, S. (1956) "On some functional equations", *Glasnik Mat. Fiz. Astr.* **11**, 3.

Kuwagaki, A. (1962) "Sur l'équation fonctionelle de Cauchy pour les matrices", *J. Math. Soc. Japan* **14**, 359.

MacLane, S. (1963) "Homology", Die Grundlehren der mathematischen Wissenschaften **114**, Academic Press, Inc., Publishers, New York, Springer-Verlag, Berlin–Göttingen–Heidelberg.

Mitrinović, D. S. (1963) "Équation fonctionelle à fonctions inconnues dont toutes ne dépendent pas du même nombre d'arguments", *Univ. Beograd. Publ. Elektrotehn. Fak. Ser. Mat. Fiz.* **115–121**, 29.

Mitrinović, D. S. (1963) "Équations fonctionelles cycliques généralisées", *C. R. Acad. Sci. Paris* **257**, 2951.

Mitrinović, D. S. (1963) "Équations fonctionelles linéaires paracycliques de première espèce", *Publ. Inst. Math. Beograd* **3**, (17), 115.

Mitrinović, D. S. (with the collaboration of Vasić, P. M.) (1986) "Diferencijalne Jednačine. Zbornik Zadataka i Problema", Naučna Knjiga, Beograd.

Mitrinović, D. S. and Djoković, D. Ž. (1961) "Sur quelques équations fonctionnelles", *Publ. Inst. Math. Beograd* **1**, (15), 67.

Mitrinović, D. S. and Djoković, D. Ž. (1963) "Neki nerešeni problemi u teoriji funkcionalnih jednačina", *Mat. Bibliot.* **25**, 153.

Mitrinović, D. S. and Pečarić, J. E. (1991) "Ciklične Nejednakosti i Ciklične Funkcionalne Jednačine", Naučna Knjiga, Beograd.

Mitrinović, D. S. and Prešić, S. B. (1962) "Sur Une Équation Fonctionnelle Cyclique Non Linéaire", *C. R. Acad. Sci. Paris* **254**, 611.

Mitrinović, D. S. and Prešić, S. B. (1962) "Sur Une Équation Fonctionnelle Cyclique d'Ordre Supérieur", *Univ. Beograd. Publ. Elektrotehn. Fak. Ser. Mat. Fiz.* **70–76**, 1.

Mitrinović, D. S., Prešić, S. B. and Vasić, P. M. (1963) "Sur Deux Équations Fonctionnelles Cycliques Non Linéaires", *Bull. Soc. Math. Phys. R. P. Serbie* **15**, 3.

Prešić, S. B. and Djoković, D. Ž. (1961) "Sur une equation fonctionnelle", *Bull. Soc. Math. Phys. R. P. Serbie* **13**, 149.

Risteski, I. B. (to appear) "Solution of a class of complex vector linear functional equations", *Missouri J. Math. Sci.*

Risteski, I. B. (to appear) "Matrix method for solving of linear complex vector functional equations", *Internat. J. of Math. and Math. Sci.*

Risteski, I. B. (submitted) "Some higher order complex vector functional equations".

Risteski, I. B. and Covachev, V. C. (2000) "On some general classes of partial linear complex vector functional equations", *Sci. Univ. Tokyo J. Math.* **36**, 105.

Risteski, I. B. and Covachev, V. C. (submitted) "On some quadratic vector functional equations".

Risteski, I. B. and Covachev, V. C. (submitted) "Some modified quadratic complex vector functional equations".

Risteski, I. B. and Covachev, V. C. (submitted) "Some expanded quadratic complex vector functional equations".

Risteski, I. B., Trenčevski, K. G. and Covachev, V. C. (1999) "A simple functional operator", *New York J. Math.* **5**, 139.

Risteski, I. B., Trenčevski, K. G. and Covachev, V. C. (2000) "Cyclic complex vector functional equations", *Appl. Sci.* **2**, 13.

Risteski, I. B., Trenčevski, K. G. and Covachev, V. C. (2000) "On a linear vector functional equation", in: *Some Problems of Applied Mathematics*, Eds. A. Ashyralyev, H. A. Yurtsever, Fatih University Publications, Istanbul, 174.

Risteski, I. B., Trenčevski, K. G. and Covachev, V. C. (2001) "An expanded class of parametric linear complex vector functional equations", *Balkan J. Geom. Appl.* **6**.

Risteski, I. B., Trenčevski, K. G. and Covachev, V. C. (2001) "Systems of Functional Equations", in: *Applications of Mathematics in Engineering and Economics'26*, Eds. B. I. Cheshankov, M. D. Todorov, Heron Press, Sofia, 113.

Risteski, I. B., Trenčevski, K. G. and Covachev, V. C. (to appear) "On some simple complex vector functional equations", *J. of Univ. of Braşov.*

Risteski, I. B., Trenčevski, K. G. and Covachev, V. C. (to appear) "On a class of parametric partial linear complex vector functional equations", *An. Univ. Craiova. Ser. Mat. Informat.*

Risteski, I. B., Trenčevski, K. G. and Covachev, V. C. (submitted) "Complex vector systems of partial functional equations".

Trenčevski, K. G., Risteski, I. B. and Covachev, V. C. (1999) "On two complex functional equations of Fréchet's type", in: *Biomathematics, Bioinformatics & Applications of Functional, Differential, Difference Equations*, Eds. Haydar Akça, Valéry Covachev & Elena Litsyn, Akdeniz Univ., 180.

Valiron, G. (1945) "Equations Fonctionelles—Applications", Paris.

Vasić, P. M. (1963) "Sur un système d'équations fonctionnelles", *Glasnik Mat. Fiz. Astr.* **18**, 229.

Vasić, P. M. (1963) "Équation fonctionnelle d'un certain type de déterminants", *C. R. Acad. Sci. Paris* **256**, 1898.

Vasić, P. M. (1965) "Sur une classe d'équations fonctionelles linéaires", *C. R. Acad. Sci. Paris* **260**, 5666.

Vasić, P. M. (1967) "Résolution d'une classe d'équations fonctionelles linéaires", *Univ. Beograd. Publ. Elektrotehn. Fak. Ser. Mat. Fiz.* **181–196**, 53.

Vasić, P. M. and Djordjević, R. Ž. (1965) "Sur l'équation fonctionelle cyclique généralisée", *Univ. Beograd. Publ. Elektrotehn. Fak. Ser. Mat. Fiz.* **132–142**, 33.

Vasić, P. M., Janić, R. R. and Djordjević, R. Ž. (1965) "Sur les équations fonctionelles linéaires paracycliques de première espèce", *Univ. Beograd. Publ. Elektrotehn. Fak. Ser. Mat. Fiz.* **132–142**, 39.

Index

cyclic matrix (matrices), 176, 178,
  187, 190, 191, 192, 198
cyclic operator, 21, 25
cyclic permutation, 108, 177, 178,
  182, 183, 191, 198, 200, 268, 271,
  273

equation
  binomial, 200, 201
  characteristic, 201
  matrix, x, 173
  reduced, 201
  scalar, 225, 233, 234, 236, 238, 248
  transformed, 221, 224, 226

FE = (complex vector) functional
  equation
Cauchy (matrix) FE, 78
cyclic FEs, vii, ix, 3
  basic cyclic FEs, vii, ix, 3, 5, 20,
      25, 175, 190
  condensed cyclic FEs, vii, ix, 3, 72
  derived cyclic FEs, vii, ix, 3, 6, 8,
      10, 19, 21
  paracyclic FEs, viii, ix, x, 3, 29,
      197
  semicyclic FEs, vii, ix, 3, 65, 66

special cyclic FEs, vii, ix, 3, 69
expanded parametric FE, viii, ix, 121,
  142
FEs with constant parameters, viii,
  ix, 121
FEs with distinct functions, viii
FEs with operations between
  arguments, vii, ix, 75
FEs with several unknown functions,
  ix, 99
FEs with two operations, viii, ix, 111
Fréchet's FEs, viii, ix, 106, 109, 110
general expanded parametric FE, viii,
  ix, 121, 161
general parametric FE, viii, ix, 121
generalized equation of operator type,
  vii
generalized FE, ix, 81
higher order FEs, viii, x, 265
  with complex parameters, x, 265,
      267
  without parameters, x, 265
linear FEs, vii, ix, 1
(linear) FEs with constant coefficients,
  viii, x, 173
  homogeneous, viii, x, 173, 175, 192,
      195

nonhomogeneous, viii, x, 173, 190, 192, 196
nonlinear FE(s), vii, x, 217
nonlinear operator FE, x, 273
operator FE, vii, ix, 75, 78, 80
quadratic FEs, viii, x, 219
  basic quadratic FEs, x, 231
  expanded quadratic FEs, viii, x, 247
    with alternating sign between the functions, x, 247, 258, 259
    with functional arguments, x, 247
    with the same sign between the functions, x, 247, 251
    generalized (expanded) quadratic FE, x, 247, 262
  modified quadratic FEs, viii, x, 231, 238, 239, 241, 244
  permuted quadratic FE, x, 231, 234
  special (real) quadratic FE, x, 219, 227, 229, 230
special parametric FE, viii, ix, 121, 135
systems of (linear) FEs, viii, x, 205, 208, 210, 211, 213, 214, 215, 216
  systems of (nonlinear) FEs, viii, xi, 285, 293, 303, 311, 313, 314, 315, 318
  quadratic, viii, xi, 285
  higher order, viii, xi, 285, 311

homogeneous system of equations, 175

Laplace rule, 280, 310
linear homogeneous equations, 272

matrix approach (method), viii, 173

Pexider form, 92

solution
  anisotropic, 287, 288, 296, 297
  isotropic, 286, 287, 293, 294, 296
  nontrivial, 176, 178, 225, 227, 228, 229, 230, 232, 235, 236, 244, 248, 250, 252, 255, 268, 269, 277, 278, 288, 306
  reproductive, 180, 181, 186, 187, 188, 189
  trivial, 225, 227, 230, 232, 235, 248, 250, 252, 255, 270, 272, 277